IMPORTANT NOTICE

Chemistry in the Community (ChemCom) is intended for use by high school students in the classroom laboratory under the direct supervision of a qualified chemistry teacher. The experiments described in this book involve substances that may be harmful if they are misused or if the procedures described are not followed. Read cautions carefully and follow all directions. Do not use or combine any substances or materials not specifically called for in carrying out experiments. Other substances are mentioned for educational purposes only and should not be used by students unless the instructions specifically so indicate.

The materials, safety information, and procedures contained in this book are believed to be reliable. This information and these procedures should serve only as a starting point for good laboratory practices, and they do not purport to specify minimal legal standards or to represent the policy of the American Chemical Society. No warranty, guarantee, or representation is made by the American Chemical Society as to the accuracy or specificity of the information contained herein, and the American Chemical Society assumes no responsibility in connection therewith. The added safety information is intended to provide basic guidelines for safe practices. It cannot be assumed that all necessary warnings and precautionary measures are contained in the document and that other additional information and measures may not be required.

PREFACE

Since 1988 when the first edition of *Chemistry in the Community* (*ChemCom*) was published, *ChemCom* has been successfully used by more than 1,000,000 students in many types of high schools with a wide range of students. A major goal of *Chemistry in the Community* is to increase the scientific literacy of high school chemistry students by emphasizing chemistry's impact on society. Developed by the American Chemical Society (ACS) with financial support by the National Science Foundation and several ACS funding sources, the writing of *ChemCom* third edition was assisted by an Editorial Advisory Board of high school chemistry teachers, university chemistry professors, and chemists from industry and federal agencies.

Briefly, *Chemistry in the Community* is designed to help students:

- recognize and understand the importance of chemistry to their lives;

- develop problem-solving techniques and critical thinking skills to apply chemical principles in order to make decisions about scientific and technological issues; and

- acquire an awareness of the potential as well as limitations of science and technology.

This third edition of *Chemistry in the Community* follows the templates provided by the two preceding editions. Chemical principles are still presented on a "need-to-know" basis. As in previous editions, each of the eight units begins with a significant socio-technological issue that then permeates the unit as a framework around which the appropriate chemistry is introduced. Chemical principles are woven within the tapestry of chemistry applied to a particular socio-technological issue. Each unit deals in some way with a community and its chemistry—a community that may be the school, town, region, or the world, all of us as passengers on Spaceship Earth. And, as before, each unit finishes with a consolidating Putting It All Together activity. Laboratory activities continue as integral parts of each unit, not stand-alone exercises. This third edition, however, has some changes that are more significant than those from the first to the second editions. Major changes are:

- The Putting It All Together activities are all new except that for the Water unit.

- The Nuclear unit has been refocused to emphasize everyday applications of nuclear technologies rather than nuclear armaments.

- The previous Health unit has been redone. The new title—Personal Chemistry and Choices—expresses the new focus on personal choices and risk analysis involving personal chemistry, including a new section on sunscreens and a UV index.

- Laboratory activities are now predominantly microscale—for safety and for cost reduction of materials and their disposal. Larger scale versions of the laboratory activities are available in the Teacher's Guide.

Four new laboratory activities are included: Separation by Distillation, Food Energy in a Peanut, Food Coloring Analysis, and Sunscreens.

• Indoor air quality, a new topic, serves as the Putting It All Together for the Air unit.

• The Chemical Industry unit now has two industries—a nitrogen-products firm and an aluminum-based company—each vying to come into Riverwood. This unit (and the textbook) concludes with a Putting It All Together Riverwood town council meeting, like that which completes the opening unit (Supplying Our Water Needs).

• The Career Vignettes of individuals who use chemistry in widely diverse careers are mostly new and are now accompanied by related questions.

As in its predecessor editions, the third edition presents the major concepts, vocabulary, laboratory techniques, and thinking skills expected in a high school introductory chemistry course. But *ChemCom* provides a greater variety and number of student-oriented activities than is customary, each unit containing three types of decision-making activities and problem-solving exercises. This edition also includes updated information on many chemical topics so as to provide contemporary data.

It is the hope and expectation of those who have created this third edition of *Chemistry in the Community* that users of this textbook will complete the course with a positive, enduring understanding and appreciation of the fascinating range of chemistry and the daily impact that it has on their community and the world.

CREDITS

ChemCom is the product of teamwork involving individuals from all over the United States over the past decade. The American Chemical Society is pleased to recognize all who contributed to *ChemCom.*

The team responsible for the third edition of *ChemCom* is listed on the copyright page. Individuals who contributed to the initial development of *ChemCom,* to the first edition in 1988, and to the second edition in 1993 are listed below.

Principal Investigator:
W. T. Lippincott

Project Manager:
Sylvia Ware

Chief Editor:
Henry Heikkinen

Contributing Editor:
Mary Castellion

Assistant to Contributing Editor:
Arnold Diamond

Editor of Teacher's Guide:
Thomas O'Brien

Revision Team:
Diane Bunce, Gregory Crosby, David Holzman, Thomas O'Brien, Joan Senyk, Thomas Wysocki

Editorial Advisory Board:
Glenn Crosby, James DeRose, Dwaine Eubanks, W. T. Lippincott (*ex officio*), Lucy McCorkle, Jeanne Vaughn, Sylvia Ware (*ex officio*)

Writing Team:
Rosa Balaco, James Banks, Joan Beardsley, William Bleam, Kenneth Brody, Ronald Brown, Diane Bunce, Becky Chambers, Alan DeGennaro, Patricia Eckfeldt, Dwaine Eubanks (dir.), Henry Heikkinen (dir.), Bruce Jarvis (dir.), Dan Kallus, Jerry Kent, Grace McGuffie, David Newton (dir.), Thomas O'Brien, Andrew Pogan, David Robson, Amado Sandoval, Joseph Schmuckler (dir.), Richard Shelly, Patricia Smith, Tamar Susskind, Joseph Tarello, Thomas Warren, Robert Wistort, Thomas Wysocki

Steering Committee:
Alan Cairncross, William Cook, Derek Davenport, James DeRose, Anna Harrison (ch.), W. T. Lippincott (*ex officio*), Lucy McCorkle, Donald McCurdy, William Mooney, Moses Passer, Martha Sager, Glenn Seaborg, John Truxall, Jeanne Vaughn

Consultants:
Alan Cairncross, Michael Doyle, Donald Fenton, Conrad Fernelius, Victor Fratalli, Peter Girardot, Glen Gordon, Dudley Herron, John Hill, Chester Holmlund, John Holman, Kenneth Kolb, E. N. Kresge, David Lavallee, Charles Lewis, Wayne Marchant, Joseph Moore, Richard Millis, Kenneth Mossman, Herschel Porter, Glenn Seaborg, Victor Viola, William West, John Whitaker

Synthesis Committee:
Diane Bunce, Dwaine Eubanks, Anna Harrison, Henry Heikkinen, John Hill, Stanley Kirschner, W. T. Lippincott (*ex officio*), Lucy McCorkle, Thomas O'Brien, Ronald Perkins, Sylvia Ware (*ex officio*), Thomas Wysocki

Evaluation Team:
Ronald Anderson, Matthew Bruce, Frank Sutman (dir.)

Field Test Coordinator:
Sylvia Ware

Field Test Workshops:
Dwaine Eubanks

ACS also offers thanks to the National Science Foundation for its support of the initial development of *ChemCom,* and to NSF project officers Mary Ann Ryan and John Thorpe for their comments, suggestions, and unfailing support.

Second Edition Revision Team

Project Manager:
Keith Michael Shea

Chief Editor:
Henry Heikkinen

Assistant to Chief Editor:
Wilbur Bergquist

Editor of Teacher's Guide:
Jon Malmin

Second Edition Editorial Advisory Board:
Diane Bunce, Henry Heikkinen (*ex officio*), S. Allen Heininger, Donald Jones (chair), Jon Malmin, Paul Mazzocchi, Bradley Moore, Carolyn Morse, Keith Michael Shea (*ex officio*), Sylvia Ware (*ex officio*)

Teacher Reviewers of First Edition:
Vincent Bono, New Dorp High School, New York; Charles Butterfield, Brattle Union High School, Vermont; Regis Goode, Spring Valley High School, South Carolina; George Gross, Union High School, New Jersey; C. Leonard Himes, Edgewater High School, Florida; Gary Hurst, Standley Lake High School, Colorado; Jon Malmin, Peninsula High School, Washington; Maureen Murphy, Essex Junction Educational Center, Vermont; Keith Michael Shea, Hinsdale Central High School, Illinois; Betsy Ross Uhing, Grand Island Senior High School, Nebraska; Jane Voth-Palisi, Concord High School, New Hampshire; Terri Wahlberg, Golden High School, Colorado

Safety Consultant:
Stanley Pine

Editorial:
The Stone Cottage

Design:
Bonnie Baumann

Art:
Additional art for this edition by Seda Sookias Maurer

A SPECIAL NOTE
TO STUDENTS

This is likely your first chemistry course. But without realizing it, you have been associated with chemistry for quite some time. Chemical reactions are all around you in your daily living, involved in foods, fuels, fabrics, and medicines. Chemistry is the study of substances in our world and is responsible for the materials that make up where you live, how you are transported, what you eat and wear, even you yourself.

As society depends more on technology, that is, the application of science, the decisions that individuals, communities, and nations make will rely more heavily on citizens—not just those who are scientists—to understand the scientific phenomena and principles required by such decisions. As a future voter, you can bring your chemical knowledge to decisions that require it. This textbook, *Chemistry in the Community (ChemCom),* was developed to present ways for you to apply chemical concepts to help you to understand the chemistry behind some important socio-technological issues.

Each of the eight *ChemCom* units has issues that concern your life, your community, and the relation of chemistry to them. You will be involved in laboratory activities and other problem-solving exercises that ask you to apply your chemical knowledge to a particular problem. You will seek solutions and evaluate the consequences of those you propose, because chemistry applied to communities is not without consequences.

Through *ChemCom,* we hope that you will better understand and appreciate your world. You may decide to study more chemistry. But whether you do or not, we hope that through your *ChemCom* course, you will experience the excitement and fascination that chemistry brings to those of us in chemistry and chemical education. You will begin your study of chemistry as a student immersed in a water problem in the fictional town of Riverwood. We begin with a newspaper article about a water-related emergency in Riverwood, an issue that is the theme for the entire first unit. Can chemistry help to solve Riverwood's problem? Welcome to *ChemCom* and to Riverwood!

Conrad Stanitski
Editor, Third Edition

SAFETY IN THE LABORATORY

In *ChemCom* you will frequently perform laboratory activities. While no human activity is completely risk free, if you use common sense as well as chemical sense, and follow the rules of laboratory safety, you should encounter no problems. Chemical sense is an extension of common sense. Sensible laboratory conduct won't happen by memorizing a list of rules, any more than a perfect score on a written driver's test ensures an excellent driving record. The true "driver's test" of chemical sense is your actual conduct in the laboratory.

The following safety pointers apply to all laboratory activities. For your personal safety and that of your classmates, make following these guidelines second nature in the laboratory. Your teacher will point out any special safety guidelines that apply to each activity.

If you understand the reasons behind them, these safety rules will be easy to remember and to follow. So, for each listed safety guideline:

- Identify a similar rule or precaution that applies in everyday life—for example in cooking, repairing or driving a car, or playing a sport.

- Briefly describe possible harmful results if the rule is not followed.

RULES OF LABORATORY CONDUCT

1. Perform laboratory work only when your teacher is present. Unauthorized or unsupervised laboratory experimenting is not allowed.

2. Your concern for safety should begin even before the first activity. Always read and think about each laboratory assignment before starting the laboratory activity.

3. Know the location and use of all safety equipment in your laboratory. These should include the safety shower, eye wash, first-aid kit, fire extinguisher, blanket, exits, and evacuation routes.

4. Wear a laboratory coat or apron and impact/splashproof goggles for all laboratory work. Wear shoes (rather than sandals or open-toed shoes) and tie back loose hair. Shorts or short skirts must not be worn.

5. Clear your benchtop of all unnecessary materials such as books and clothing before starting your work.

6. Check chemical labels twice to make sure you have the correct substance and the correct concentration of a solution. Some chemical formulas and names may differ by only a letter or a number.

7. You may be asked to transfer some laboratory chemicals from a common bottle or jar to your own test tube or beaker. Do not return any excess material to its original container unless authorized by your teacher.

8. Avoid unnecessary movement and talk in the laboratory.

9. Never taste laboratory materials. Do not bring gum, food, or drinks into the laboratory. Do not put fingers, pens, or pencils in your mouth while in the laboratory.

10. If you are instructed to smell something, do so by fanning some of the vapor toward your nose. Do not place your nose near the opening of the container. Your teacher will show the correct technique.

11. Never look directly down into a test tube; view the contents from the side. Never point the open end of a test tube toward yourself or your neighbor. Never heat a test tube *directly* in a Bunsen burner flame.

12. Any laboratory accident, however small, should be reported immediately to your teacher.

13. In case of a chemical spill on your skin or clothing, rinse the affected area with plenty of water. If the eyes are affected, water-washing must begin immediately and continue for at least 10 to 15 minutes. Professional assistance must be obtained.

14. Minor skin burns should be placed under cold, running water.

15. When discarding or disposing of used materials, carefully follow the instructions provided.

16. Return equipment, chemicals, aprons, and protective goggles to their designated locations.

17. Before leaving the laboratory, make sure that gas lines and water faucets are shut off.

18. Wash your hands before leaving the laboratory.

19. If in doubt, ask!

Supplying Our Water Needs

- ► Can we continue to obtain enough water to supply our needs?

- ► Can we get sufficiently pure water?

- ► How do everyday decisions affect the quality and quantity of our water supplies?

- ► How can chemistry help explain water's personal and societal importance?

Fish Kill Causes Water Emergency in Riverwood
Severe Water Rationing in Effect

By Lori Katz
Staff writer of *Riverwood News*

Citing possible health hazards, Mayor Edward Cisko announced today that the city will stop withdrawal of water from the Snake River and will temporarily shut down the Riverwood water treatment plant. He also announced cancellation of the community's "Fall Fish-In" that was to begin Friday. River water will not be pumped into the water treatment plant for at least three days, starting at 6 P.M. tonight. No plans were announced for rescheduling the annual fishing tournament.

During the pumping station shutdown, water engineers and chemists from the County Sanitation Commission and Environmental Protection Agency (EPA) will search for the cause of a fish kill discovered yesterday at the base of Snake River Dam, two miles upstream from Riverwood's water pumping station.

"There's no cause for alarm, because preliminary tests show no present danger to townspeople. However, consensus at last night's emergency town council meeting was to start a thorough investiga-

tion of the situation immediately," said Mayor Cisko.

The alarm was sounded when Jane Abelson, 15, and Chad Wong, 16—both students at Riverwood High School—found many dead fish floating in a favorite fishing spot. "We thought maybe someone had poured poison in the reservoir or dam," explained Wong.

Mary Steiner, biology teacher at Riverwood High School, accompanied the students back to the river. "We hiked downstream for almost a mile. Dead fish of all kinds were

washed up on the banks and caught in the rocks as far as we could see," Abelson reported.

Mrs. Steiner contacted the County Sanitation Commission, which made preliminary tests for substances that might have killed the fish. Chief Engineer Hal Cooper reported at last night's emergency meeting that the water samples looked completely clear and that no toxic substances were found. However, he indicated some concern. "We can't say for

see Fish Kill page 3

Fish Kill   from page 2

sure that our present water supply is safe until the reason for the fish kill is known. It's far better that we take no chances until we know the water is safe," Cooper advised.

Arrangements are being made for drinking water to be trucked from Mapleton, with the first shipments due to arrive in Riverwood at 10 A.M. tomorrow. Distribution points are listed on page 8, along with guidelines on saving and using water during the emergency.

Mayor Cisko gave assurances that essential municipal services would not be affected by the water crisis. Specifically, he stated that the fire department would have access to adequate supplies of water to meet any firefighting needs.

Riverwood schools will be closed Monday, Tuesday, and Wednesday. No other closings or cancellations are known at this time. Local TV and radio stations will announce any as they occur.

The town council reached agreement to stop drawing water from the Snake River after five hours of heated debate. Councilman Henry McLatchen (who is also a member of the Chamber of Commerce) described the decision as a highly emotional and unnecessary reaction. He cited the great financial loss that Riverwood's motel and restaurant owners will suffer from the fish-in cancellation, as well as the potential loss of future tourism dollars as the result of adverse publicity. He and others sharing his view were outvoted by the majority, who expressed concern

that the fish kill might have public health implications.

A public meeting will be held at 8 P.M. tonight at Town Hall. Dr. Margaret Brooke of State University, an expert on water systems, will answer questions concerning water safety and use. Dr. Brooke has agreed to aid the County Sanitation Commission in explaining the situation to the town.

Asked how long the water emergency would last, Dr. Brooke refused to speculate before talking with chemists conducting the investigation. EPA scientists, in addition to collecting and analyzing water samples, will examine dead fish to determine if a disease caused their deaths.

Townspeople React to Fish Kill and Water Crisis

By Juan Hernandez
Staff contributor to *Riverwood News*

In a series of person-on-the-street interviews, Riverwood citizens expressed their feelings about the crisis. "It doesn't bother me," said nine-year-old Jimmy Hendricks. "I'm just going to drink milk and canned fruit juice."

"I knew it was just a matter of time before they killed the fish," complained Harmon Lewis, a lifelong resident of the Fieldstone Acres area. Lewis, who traces his ancestry back to original county settlers, still gets his water from a well and will be unaffected by the water crisis. Lewis drew his well water by hand until 1967. He installed his present system after area development brought electricity to his property. "I wouldn't have done even that except for the arthritis in my shoulders," said Lewis.

He plans to pump enough well water to supply the children's ward at Community Hospital if the emergency continues more than a few days.

Bob and Ruth Hardy of Hardy's Ice Cream Parlor expressed annoyance at the inconvenience, but felt relieved about council actions. They were anxious to know the reason for the fish kill and its possible effects on future water supplies.

The Hardys' daughter Toni, who loves to fish, was worried that the late summer fishing season would be ruined. Toni and her father won first prize last year in the Chamber of Commerce's angling competition.

Don Harris, owner of the Uptown Motel, expressed concern regarding the health of town residents, as well as the loss of business due to the tournament cancellation. "I always earn a reasonable amount of money from this event and will most likely have to get a loan to pay my bills in the spring."

The unexpected school vacation was "great" according to twelve-year-old David Price. Asked why he thought schools needed to be closed because of a water shortage, Price said all he could think of was that "the drinking fountains won't work."

Elmo Turner, whose yard and flower beds have won Garden Club recognition for the last five years, felt reassured on one point. Because this past summer was so wet, grass-watering is unnecessary, and lawns are not in danger of drying out due to water rationing.

Riverwood will be without adequate water for three days. As these newspaper articles indicate, the water emergency has created understandable concern among Riverwood citizens, town officials, and business owners. What caused the fish kill? Does the fish kill mean that the community's water supply poses hazards to humans? We will follow the town's progress in answering these questions as we learn about water's properties.

Even though Riverwood is imaginary, its problems have already been faced by residents of real communities. In fact, two water-related challenges confront all of us. Can we continue to obtain *enough* water to supply our needs? Can we get sufficiently *pure* water? These two questions are the major themes of this unit. Such challenges require us to understand water's chemistry, uses, and importance.

Water has many important properties (characteristics). Some of its properties, such as color and taste, can be observed by the unaided senses. Others must be observed indirectly, by measurement and calculation.

Measurements and calculations are useful only when everyone uses the same units. So, to be consistent, scientists in all countries report their measurements and calculations in units of the metric system.

A.1 MEASUREMENT AND THE METRIC SYSTEM

Metric units were first introduced in France more than 100 years ago. A modernized form of the metric system was internationally adopted in 1960. It is called "SI," which is an abbreviation of the French name, *Le Système International d'Unités.* SI units are used by scientists in all nations, including the United States. You are already familiar with some SI units, such as the gram, degree Celsius, and second. Other SI units you will encounter in your study of chemistry (such as the pascal, joule, and mole) may be new to you. We will explain each unit when it is first used.

In the following laboratory activity, you will make measurements of length and volume in SI units. The SI unit of length is the **meter** (symbolized by m). Most doorways are about two meters high.

Many lengths we may wish to measure are either much larger or smaller than a meter. SI prefixes have been defined to indicate different fractions and multiples of all SI units, including the meter. Important prefixes for our present use are **deci-,** meaning one-tenth (1/10); **centi-,** meaning one one-hundredth (1/100) (recall that a cent is one one-hundredth of a dollar); and **milli-,** meaning one one-thousandth (1/1,000).

An audiocassette cartridge has a width of one decimeter (dm); its thickness is about one centimeter (cm). A millimeter (mm) is approximately the thickness of a compact disk (CD).

The derived SI unit for volume is the cubic meter (m^3). You can visualize one cubic meter as the space occupied by a box one meter on each edge. (The volume of a cube is calculated as length × width × height.) This 1-m^3 box would be big enough to hold a very large dog comfortably. This is too large a volume unit for convenient use in chemistry!

Consider a smaller cube, one decimeter (dm) on each edge. The volume of this cube is one cubic decimeter (dm^3). Although a cubic decimeter may not be familiar to you, you probably know it by another name—a **liter** (L). For example, the volume of a large bottle of soda can be given as 2 L or 2 dm^3. One cubic decimeter (or one liter) of volume is exactly equal to 1,000 cm^3 (cubic centimeters). A full-scale photo of one cubic centimeter is shown in Figure 1 (page 6). You may know the cubic centimeter by another name—the **milliliter** (mL). Because the liter and the milliliter are more familiar, we will use these units for volume.

Let's summarize the metric (SI) units for volume by considering a can of soft drink labeled 12 fluid ounces. If we "think metric," this is 355 cm^3 or 355 mL of beverage. Using larger units, we have 0.355 dm^3

▶ *1 dm = 0.1 m*
1 cm = 0.01 m
1 mm = 0.001 m

1 mL = 1 cm^3

1,000 cm^3 = 1 dm^3
1,000 mL = 1 L

Can you think of a use for each measuring device?

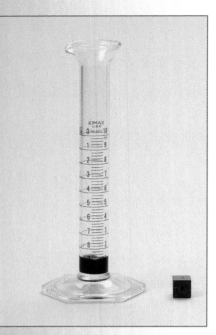

Figure 1 One cubic centimeter (cm³) equals one milliliter (mL).

▶ *In such problems, include units; then multiply and divide the units just like numbers. This approach is helpful in solving many kinds of calculations.*

(0.355 L) of the drink. Notice that this change in metric units involved just a shift in decimal point location. This illustrates one advantage of metric units over customary U.S. units.

The following exercises will help you become more familiar with common metric length and volume units.

YOUR TURN

Meters and Liters

Examine a ruler graduated in millimeters. Note the 10 markings between each centimeter (cm) mark. These smaller markings represent millimeter (mm) divisions, where 10 mm = 1 cm, or 1 mm = 0.1 cm.

A small paper clip is 8 mm wide. What is this width in centimeters? Because 10 mm equals 1 cm, there's a "10 times" difference (one decimal place) between units of millimeters and centimeters. Thus the answer must be either 80 cm or 0.8 cm, depending on which direction the decimal point moves. Because it would take *10* mm to equal 1 cm, the *8*-mm paper clip must be slightly *less* than 1 cm. The answer must be 0.8 cm. Thus to convert the answer from millimeters to centimeters we just move the decimal point one place to the left.

The conversion of 8 mm to centimeters can also be written like this:

$$8 \text{ mm} \times \frac{1 \text{ cm}}{10 \text{ mm}} = 0.8 \text{ cm}$$

The same paper clip is 3.2 cm long. What is this length in millimeters? This answer can be reasoned out just as we did for the first question, or written out:

$$3.2 \text{ cm} \times \frac{10 \text{ mm}}{1 \text{ cm}} = 32 \text{ mm}$$

1. Measure the diameter of a penny, a nickel, a dime, and a quarter. Report each diameter
   a. in millimeters.
   b. in centimeters.

2. Sketch a square, 10 cm on each side. Now imagine a three-dimensional box or cube that is made up of six of these squares. The volume of a cube can be found by multiplying its length × width × height. What would be the total volume of this cube expressed
   a. in cubic centimeters (cm³)?
   b. in milliliters (mL)?
   c. in liters (L)?
   d. Grocery-store sugar cubes each have a volume of about 1 cm³. How many of these cubes could you pack into the cube with 10-cm sides?

3. Read the labels on some common beverage containers, such as a carton of milk, fruit juice can, or a soft drink, to see how volume is indicated.

   a. Can you find any beverage containers that list *only* U.S. customary volume units such as the quart, pint, or fluid ounce? If so, describe them.

   b. Describe at least three beverage containers that specify their volumes in *both* customary U.S. units and SI units. Include each container's volume expressed in both types of units.

4. a. State at least one advantage of SI units over customary U.S. units. (*Hint:* How are units—such as length—subdivided in each system? How are units of volume related to units of length?)

   b. Can you think of any disadvantages to using SI units? Explain your answer.

Even such an apparently simple idea as "water use" can present some fascinating puzzles, particularly when data on the volumes of water involved are included. For example, consider the following *ChemQuandary*.

CHEMQUANDARY

WATER, WATER EVERYWHERE

It requires 120 L of water to produce a 1.3-L can of juice and it takes about 450 L of water to place a single egg on your breakfast plate. What explanation can you give for these two statements?

A.2 LABORATORY ACTIVITY: FOUL WATER

GETTING READY

If you haven't already done so, carefully read *Safety in the Laboratory*, found on page xviii.

The purpose of this activity is to purify a sample of "foul" water, producing as much "clean" water as possible. *Do not test the purity of the water by drinking it.*

Three water purification procedures will be used: (1) oil–water separation, (2) sand filtration, and (3) charcoal adsorption/filtration. (**Filtration** is a general term for separating solid particles from a liquid by passing the mixture through a material that retains the solid particles. The liquid collected after filtration is called the **filtrate.**)

Prepare a table similar to the one on page 8 in your laboratory notebook (provide more space for your entries).

Data Table						
	Color	Clarity	Odor	Presence of Oil	Presence of Solids	Volume, mL
Before treatment						
After oil–water separation						
After sand filtration						
After charcoal adsorption and filtration						

Figure 2 Funnel in clay triangle.

PROCEDURE

1. Obtain approximately 100 mL of foul water, provided by your teacher. Measure its volume precisely with a graduated cylinder; record the value (with units) in your data table.

2. Examine the properties of your sample: color, odor, clarity, presence of solids or oily regions. Record your observations in the "Before treatment" section of your data table.

Oil–Water Separation

Oil and water do not noticeably dissolve in each other. If oil and water are mixed and left undisturbed, two layers form—the oil floats on top of the water.

1. Place a funnel in a clay triangle supported by a ring clamp and ring stand. Attach a rubber hose to the funnel tip as shown in Figure 2.

2. Close the rubber tube by nipping it with your fingers (or by using a pinch clamp). Shake or stir the foul-water sample. Then pour about half the sample into the funnel and let it stand for a few seconds until the liquid layers separate. (Gentle tapping may encourage oil droplets to float free.)

3. Carefully open the tube to release the lower layer into a 150-mL beaker. When the lower layer has drained out, quickly close the rubber tube.

4. Drain the remaining layer into a second 150-mL beaker.

5. Repeat Steps 2–4 using the remainder of your sample, adding each liquid to the correct beaker.

6. Dispose of the top, oily layer as instructed by your teacher. Observe the properties of the remaining layer and measure its volume. Record your observations and data. Save the water sample for the next procedure.

7. Wash the funnel with soap and water.

Sand Filtration

A sand filter traps solid impurities that are too large to fit between sand grains.

1. Using a straightened paper clip, poke small holes in the bottom of a paper cup (Figure 3).

2. Add pre-moistened gravel and sand layers to the cup as shown in Figure 4. (The bottom gravel prevents the sand from washing through the holes. The top gravel keeps the sand from churning up when the sample is poured in.)

3. *Gently* pour the sample to be filtered into the cup. Catch the filtrate (filtered water) in a beaker as it drains through.

4. Dispose of the used sand and gravel according to your teacher's instructions. Do *not* pour sand or gravel into the sink!

5. Observe the properties and measure the volume of the water. Record your results. Save the water sample for the next procedure.

Charcoal Adsorption/Filtration

Charcoal **adsorbs** (attracts and holds on its surface) many substances that could give water a bad taste, odor, or cloudy appearance. Fish tanks include charcoal filters for the same purpose.

1. Fold a piece of filter paper as shown in Figure 5.

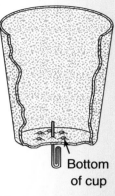

Figure 3 Preparing paper cup.

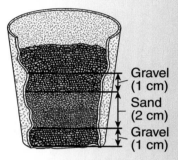

Figure 4 Sand filtration.

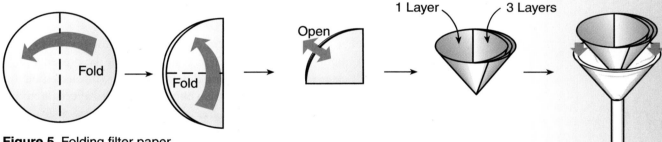

Figure 5 Folding filter paper.

2. Place the folded filter paper in a funnel. Wet the paper slightly so it adheres to the funnel cone.

3. Place the funnel in a clay triangle supported by a ring clamp (see Figure 2, page 8). Lower the ring clamp so the funnel stem extends 2–3 cm inside a 150-mL beaker.

4. Place one teaspoon of charcoal in a 125- or 250-mL Erlenmeyer flask.

5. Pour the water sample into the flask. Shake vigorously. Then gently pour the liquid through the filter paper. Keep the liquid level below the top of the filter paper—no liquid should flow between the filter paper and the funnel.

6. If the filtrate is darkened by small charcoal particles, refilter the liquid. Use a clean piece of moistened filter paper.

7. When you are satisfied with the appearance and odor of your purified water sample, pour it into a graduated cylinder. Observe and record the properties and the final volume of the sample.

8. Wash your hands thoroughly before leaving the laboratory.

CALCULATIONS

Complete the following calculations. Record your work and answers in your notebook.

1. What percent of the original foul-water sample did you recover as "pure" water?

$$\text{Percent of water purified} = \frac{\text{Vol. of water purified}}{\text{Vol. of foul-water sample}} \times 100\%$$

2. What volume of liquid did you lose during purification?

3. What percent of your original foul-water sample was lost during purification?

POST-LAB ACTIVITIES

1. Your teacher will demonstrate another water purification technique, called *distillation*.

   a. Write a complete description of the steps in distillation.

   b. Why did your teacher discard the first portion of recovered liquid?

   c. Why was some liquid left in the distilling flask at the end of the demonstration?

2. Your teacher will test the *electrical conductivity* of the purified water samples obtained by your class. This test checks for the presence of dissolved, electrically charged particles in the water (discussed on pages 32–33). Your teacher will also check the electrical conductivity of samples of distilled water and tap water. What do these tests suggest about the purity of your water sample?

QUESTIONS

1. Compare your water purification experiences and data with those of other lab teams. How should the success of various teams be judged?

2. Distillation is not used by municipal water treatment plants. Why?

A.3 YOU DECIDE: INFORMATION GATHERING

Keep a diary of water use in your home for three days. Using a chart similar to the one on page 11, record the number of times various water-use activities occur. Ask each family member to help you.

Check the activities listed on the chart. If family members use water in other ways during the three days, add these to your diary. Estimate the quantities of water used by those activities.

The data you gather will help you see how much water you and your family use every day, and for what purposes. You may be surprised at the amount you use, which will lead you to think about the next important question, Why is water such an important substance?

Data Table			
	Day 1	Day 2	Day 3
Number of persons in family			
Number of baths			
Number of showers Length in minutes			

Number of washing machine loads
    Low setting
    High setting			

Dish washing
    Number of times by hand
    Number of times by dishwasher			

Number of toilet flushes			
Watering lawn Time in hours			
Number of car washes			
Cooking and drinking Number of cups			
Running water in sink Number of minutes			
Other uses and number of each			

A.4 WATER AND HEALTH

Living organisms require a continual supply of water. A human must drink about two liters (roughly two quarts) of water-containing liquids daily to replace water losses through bodily excretions and evaporation from skin and lungs. You can live 50 to 60 days without food but only 5 to 10 days without water.

Early humans simply drank water from the nearest river or stream with few harmful effects. However, as cities were built, this practice became risky. Wastes were dumped, or washed by rain, into the same streams from which people drank.

As the number of people increased, there was less time for natural processes to purify dirty water before someone drank it. Eventually, mysterious outbreaks of fatal diseases occurred, such as in London in the 1850s, when the Thames River became contaminated with cholera bacteria from sewage.

Today, water quality has become everyone's concern. We can no longer rely on obtaining pure water from streams and lakes; most often we must purify the water ourselves. Because we use large quantities, we must be sure not only that we have enough water, but also that it is pure enough for our needs. How can this be accomplished? We will begin to answer that question by examining our nation's overall use of water and some typical ways our communities and industries use this resource.

A.5 WATER USES

Do we use so much water that we are in danger of running out? The answer is both no and yes. The total water available to us is far more than enough. Each day, some 15 trillion liters (4 trillion gallons) of rain or snow fall in the United States. Only 10% is used by humans. The rest flows back into large bodies of water, evaporates, and falls again as part of a perpetual *water cycle.* However, the distribution of rain and snowfall does not necessarily correspond to the locations of high water use.

There are also regional differences in the way we affect the water cycle. In the eastern half of the nation, 88% of used water is returned to natural waterways. However, in the western half of the country only 48% is returned to waterways, with the rest evaporating directly to the air. Why should there be such a great difference?

In eastern states, rainwater soaking into the ground provides much of the moisture needed by crops. However, in many other parts of the country much less rain falls, and so irrigation water must be obtained from streams or wells, many of which are far from the croplands. Most irrigation water evaporates from the leaves of growing plants. The rest evaporates directly from spray irrigation or from the moist soil, before plants can use it. This evaporated moisture is blown by prevailing winds to condense and fall, days later, on eastern crops.

YOUR TURN

U.S. Water Use

Refer to Figure 6 (page 13) in answering these questions.

1. In the United States, what is the greatest single water use in
   a. the East?   c. the Midwest?   e. Alaska?
   b. the South?   d. the West?   f. Hawaii?

2. Suggest reasons for differences in water use among these six U.S. regions.

3. Explain the differences in how water is used in the East and the West. Think about where most people live and where most of the nation's factories and farms are. What other regional differences help explain patterns of water use?

According to authorities, an average U.S. family of four uses about 1360 liters (360 gallons) of water *daily.* This value represents direct, measurable use, but beyond that are many hidden or indirect uses you probably have never considered. Each time you eat bread, a hamburger, or an

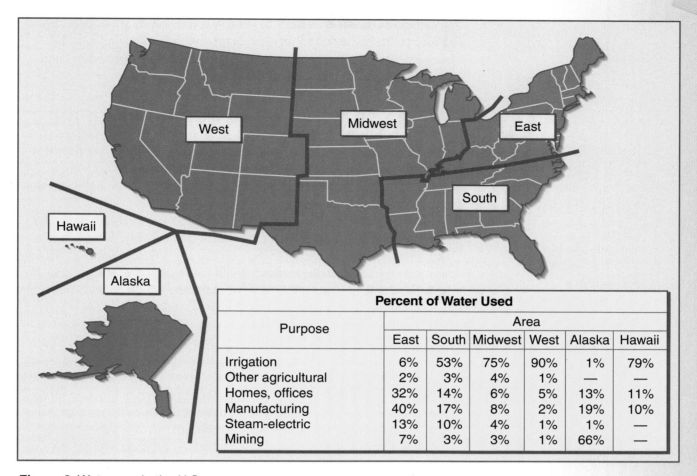

Percent of Water Used

Purpose	East	South	Midwest	West	Alaska	Hawaii
Irrigation	6%	53%	75%	90%	1%	79%
Other agricultural	2%	3%	4%	1%	—	—
Homes, offices	32%	14%	6%	5%	13%	11%
Manufacturing	40%	17%	8%	2%	19%	10%
Steam-electric	13%	10%	4%	1%	1%	—
Mining	7%	3%	3%	1%	66%	—

Figure 6 Water use in the U.S.

egg, you are "using" water! Why? Because water was needed to grow and process the food.

Let's reconsider *ChemQuandary: Water, Water Everywhere* (page 7). At first glance the volumes of water mentioned in this *ChemQuandary* probably seemed absurd. How could so much water be required to provide you with one egg or one can of fruit juice?

These examples illustrate two typical indirect (hidden) uses of water. The chicken that laid the egg needed drinking water. Water was used to grow the chicken's feed. Even the small quantities of water used for other purposes in processing add up when billions of eggs are consumed.

In one Riverwood newspaper article you read earlier, one youth said he'd just drink milk and canned fruit juice until the water was turned back on. However, drinking a glass of water consumes much less water than does drinking the same amount of canned fruit juice. Why should that be so? Because the liquid *in* the can is insignificant when compared to the water used for making the metal can itself. That's where the mysterious 120 L of water mentioned in the *ChemQuandary* arises!

Though we depend on large quantities of water, we are scarcely aware of it because, in normal times, when faucet taps are turned on water flows freely—in Riverwood or in your own community. How does this happen? What is the source of this plentiful supply? Let's review and extend what you already know by taking an imaginary "upstream" journey through water pipes back to the water's origin.

A.6 BACK THROUGH THE WATER PIPES

If you live in a city or town, your residence's water pipes are linked to underground water pipes. These pipes bring water downhill from a reservoir or a water tower, which is usually located at the highest point in town, to all the faucets.

Water in the water tower has been cleaned and purified at a water treatment plant. It may enter the treatment plant from a reservoir, a lake, river, stream, or well. If your community's water source is a river or stream, you use *surface water,* which flows on top of the ground. If the water comes from a well, it is *groundwater,* which collects underground and must be pumped to the surface. About 20% of the water supply in the United States is groundwater.

Immense supplies of groundwater flow far beneath your feet. The world's supply of groundwater is many times greater than all the water held in the world's rivers and lakes. Even when surface soil appears dry and dusty, millions of liters of fresh water are stored below in regions called *aquifers.* These porous rock structures act like sponges, in some cases holding water for thousands of years.

Neither groundwater nor surface water is completely pure. As water flows over the ground to join a stream, or seeps far into the soil to become groundwater, it dissolves small amounts of soil and rock.

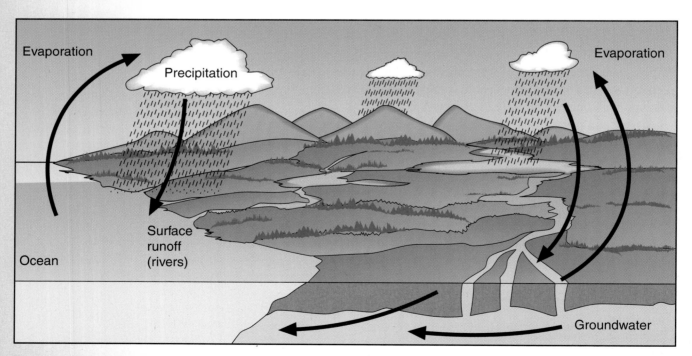

Figure 7 The hydrologic cycle.

These dissolved substances are usually not removed at the water treatment plant; they are harmless in the amounts normally found. In fact, some (such as iron, zinc, and calcium ions) are essential in small quantities to human health or improve the flavor of water.

If you live in a rural area, your home probably has its own water supply system. A well driven deep into an aquifer brings water to the surface by an electric pump. A small pressurized tank holds the water before it enters the pipes in your home.

Let's continue tracing water to its origins. Most water falls to the earth as rain (and, if it's cold enough, as snow). Most of the rain that falls on flat ground soaks directly into the soil, seeping downward to join the groundwater. When it falls on sloping terrain, much of it runs downhill before it soaks into the soil.

Water that does not become groundwater runs off into waterways. It collects into brooks, which flow into wider streams, rivers, and lakes. A substantial quantity of this water eventually finds its way to the ocean.

Where does rainwater come from? Rain falls as tiny water droplets from clouds formed when heat energy from the sun causes surface water to evaporate. Upon cooling at higher elevations, the evaporated water forms droplets that gather in large quantities to become clouds. This endless sequence of events is called the water cycle. Another name for it is the **hydrologic cycle** (Figure 7 on opposite page).

We can go back even further—to where water *originally* came from. Most scientists believe our modern world's supply of liquid water originally was formed by the condensation of water vapor. That vapor had been released from volcanoes on the young planet's surface. Within the volcanoes, the vapor had been formed by the chemical combination of hydrogen and oxygen gases. Geologists believe the total amount of water in the world has remained constant for billions of years, and has simply continued cycling through the environment.

A.7 WHERE IS THE EARTH'S WATER?

According to scientists, about 97% of Earth's total water is stored in the oceans. The next largest storage place is not as obvious. A common incorrect guess is in rivers and lakes. Actually, the second largest quantity of water is stored in Earth's ice caps and glaciers! Figure 8 shows how the world's supply of water is distributed.

Water may be found not only in different places, but also in three different **states** (forms), as it goes through its cycle. Water vapor in the air is in the **gaseous state.** In some regions of the United States, water in the gaseous state is the high humidity that contributes to human discomfort during the summer. Water is most obvious in the **liquid state,** found in lakes, rivers, oceans, clouds, and rain. Water in the **solid state** may occur as ice, snowflakes, frost, and even hail.

Water vapor will fill any size container because, as a gas, it has no fixed volume. It also has no fixed shape, but takes the shape of the vessel containing it. On the other hand, a sample of liquid water has a definite volume. But liquid water has no shape of its own; it takes the shape of the container into

If the world's total water supply were in a 200-L drum …

Rivers: 0.0001%

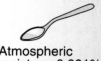

Atmospheric moisture: 0.001%

Lakes: 0.009%

Groundwater: 0.62%

Glaciers and ice caps: 2.11%

Ocean water: 97.2%

Figure 8 Distribution of the world's water supply.

CHEMISTRY AT WORK

Purifying Water Means More Than Going With the Flow

When you drink from a water fountain, do you ever wonder where the water came from? In some parts of the country, drinking water is provided by people like **Phil Noe.**

Phil is Chief Plant Operator at Island Water Association (IWA), which provides water for Florida's Sanibel and Captiva Islands. "We have a limited supply of fresh water," says Phil, "so we've built a plant that lets us use water from our aquifers."

After pumping the brackish, undrinkable water up from the Suwanee Aquifer to IWA's processing plant, Phil and his coworkers remove almost all of the salt and minerals, using a process known as *reverse osmosis.* This results in water that is more pure than in many mountain streams. (See diagram comparing osmosis with reverse osmosis).

Between 700 and 900 feet underground, permeable layers of sand and limestone called **aquifers** contain large quantities of water. Water pumped up from these aquifers contains an average of 3,000 ppm total dissolved solids (TDS).

◆ *Tapping a hidden water source.*

Feed water is pumped through hundreds of feet of tubes containing salt-filtering membranes. In the process, the purified water (called the **permeate**) separates from the salts and other dissolved solids. Eventually, the water is purified to levels as low as 100 ppm TDS.

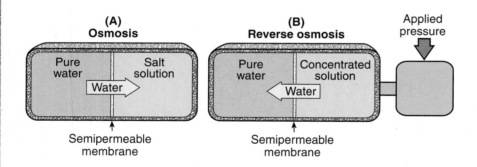

◆ *(A) Osmosis occurs naturally when water in a dilute solution passes through a semipermeable membrane into a concentrated solution. (B) Reverse osmosis: Pressure must be applied to a concentrated solution in order to force water through a semipermeable membrane into a dilute solution.*

which it is poured. In the solid state, a water (ice) sample has both a fixed volume and a shape of its own. (Though we are discussing water here, the characteristics of other gases, liquids, and solids are similar to these.)

We are very fortunate in most of the United States to have an abundant water supply at present. We turn on the tap, use what we need, and go on about our business, not thinking about how much water we use. When the water supply is shut off, as in Riverwood, it's usually for a short time.

◆ After filtration to catch the larger solids, the water passes through equipment where reverse osmosis is used to produce purified water.

◆ Saving up for a windy day.

Because Florida is subject to hurricanes and other tropical storms, Sanibel Island's water system maintains 15 million gallons of purified water in storage. As a result, during brief periods of severe weather island customers can manage without the plant being in operation.

◆ Aquatic wildlife (manatee) is unaffected by adding water back to the water cycle.

Eighty percent of IWA's feed water is converted into drinkable water. The other 20% is pumped out into the Gulf of Mexico. This water, a salt solution, is lower in salt and mineral concentrations than the gulf's water and is harmless to marine life.

- Most of the Earth's water is in the oceans, but it is not drinkable in that form. What areas of the United States might have to use reverse osmosis of ocean water to obtain their municipal water? What factors would influence that decision?

Photographs courtesy of Island Water Association (except photograph of manatee © Art Wolfe, Tony Stone Images)

But, suppose there were a drought lasting several years. Or suppose the shortage were a perpetual problem, as it is in many areas. Priority would go to "survival" uses of available water. Some nonessential water uses would probably stop entirely.

What water uses would you give up first? Refer to your completed water-use diary. How much water did *you* use in your home during the three-day period? For what purposes? Let's examine your water-use data.

A.8 YOU DECIDE: WATER USE ANALYSIS

Table 1 lists typical quantities of water used in the home. Use the table to answer the following questions.

QUESTIONS

1. Estimate the total water volume (in liters) used by your household during the three days.

2. What was the average amount of water (in liters) used by *each* family member over the *three* days? *Note:* To find the average, divide the total volume of water used in three days by the number of family members. This answer will be in units of liters per person per three days.

3. On average, how much water (in liters) did *each* family member use in *one* day?

4. Compare the average water volume used daily by each person in your household (Answer 3) with the reported average water volume used daily by each person in the United States, 340 L. What reasons can you give to explain any differences between your average and the national average?

▶ *Table 1 indicates that a regular showerhead delivers 19 L each minute. In English system units, that's 5 gallons per minute, or 25 gallons for a five-minute shower!*

Table 1	Water Required for Typical Activities	
Bathing (per bath)		130 L
Showering (per min)		
Regular showerhead		19 L
Water-efficient showerhead		9 L
Cooking and drinking		
Per 10 cups of water		2 L
Flushing toilet (per flush)		
Conventional toilet		19 L
"Water-saver" toilet		13 L
"Low-flow" toilet		6 L
Watering lawn (per hour)		1130 L
Washing clothes (per load)		
Low setting		72 L
High setting		170 L
Washing dishes (per load)		
By hand (with water running)		114 L
By hand (washing and rinsing in dishpans)		19 L
By machine (full cycle)		61 L
By machine (short cycle)		26 L
Washing car (running hose)		680 L
Running water in sink (per min)		
Conventional faucet		19 L
Water-saving faucet		9 L

You are now quite aware of the average amount of water you use daily. Suppose you had to live with much less. How would you ration your allowance for survival and comfort? This is exactly the question Riverwood residents had to answer.

A.9 YOU DECIDE: RIVERWOOD WATER USE

Riverwood authorities have severely rationed your home water supply for three days while possible causes of the fish kill are investigated. The County Sanitation Commission recommends cleaning and rinsing your bathtub, adding a tight stopper, then filling the tub with water. That water must be used by your family for everything except drinking and cooking during the three-day period. Water for drinking and cooking will be trucked in from Mapleton.

Assuming that your family has one tub of water, 375 L (100 gal), to use during these three days, consider this list of typical water uses:

- washing car, floors, windows, pets
- bathing, showering, washing hair, washing hands
- washing clothes, dishes
- cooking and preparing food
- drinking
- watering indoor plants, outdoor plants, lawn
- flushing toilet

1. a. Which water uses could you completely avoid?
    b. What would be the consequences?

2. a. For which tasks could you reduce your water use?
    b. How?

3. Impurities added by some water uses may not interfere with reusing the same water for other purposes. For example, you might decide to save hand-washing water, using it later to bathe your dog.
    a. For which activities could you use such impure water ("gray water")?
    b. From which prior uses could this water be taken?

Clearly, pure water is a valuable resource no one should take for granted. Unfortunately, water is easily contaminated. In the next part of this unit we will examine why.

БЕРЕГИТЕ ЧИСТУЮ ВОДУ!

A poster designed and drawn by a Russian high school student. The poster's caption (in Russian) is translated as "Save Clean Water."

? PART A: SUMMARY QUESTIONS

1. For each item, select the better answer:
   a. Is the thickness of a dime closer to 1 mm or to 1 cm?
   b. Is the volume of a glass of water closer to 250 dm³ or to 250 cm³?
   c. Is the diameter of a pencil closer to 7 mm or to 7 cm?
   d. Is one gallon of gasoline closer to 38 mL or to 3.8 L?

2. Complete these conversions:
   a. 735 cm³ = ? mL = ? L
   b. 10.7 mm = ? cm

3. Describe some implications if there were a lack of water
   a. in your family's home.
   b. in your town.
   c. in your region of the country.
   d. in the entire United States.

4. a. What does "indirect water use" mean?
   b. List at least two indirect uses of water needed to put a loaf of bread on your table.

5. The total supply of water in the world has been the same for billions of years. Explain why.

6. Name a serious epidemic that can be caused by drinking polluted water.

7. Consider Earth's large ocean water volume and also your laboratory experience in purifying a water sample.
   a. Why is there worldwide concern about water's availability and purity?
   b. What could be done to use more of Earth's water?
   c. How could more ocean water be used?

 EXTENDING YOUR KNOWLEDGE (OPTIONAL)

- One of water's unique characteristics is that it occurs in all three states (solid, liquid, and gas) in the range of temperatures found on our planet. How would the hydrologic cycle be different if this were not true?

- Using an encyclopedia or other reference, compare the maximum and minimum temperatures naturally found on the surfaces of Earth, the moon, and Venus. The large amount of water on Earth serves to limit the natural temperature range found on this planet. Suggest ways that water does this. As a start, find out what the terms heat of fusion, heat capacity, and heat of vaporization mean.

- Look up the normal freezing point, boiling point, heat of fusion, and heat of vaporization of ammonia (NH_3). If a planet's life forms were made up mostly of ammonia rather than of water, what special survival problems might they face? What temperature range would this planet need to support "life"?

- Write a report on the development and spread of SI metrics and its precursor, the metric system.

Meeting Raises Fish Kill Concerns

By Carol Simmons
Staff writer of *Riverwood News*

An Environmental Protection Agency (EPA) scientist, Dr. Harold Schmidt, reported last night that evidence so far indicates no danger, past or future, to Riverwood water users. Dr. Schmidt presented his report at a Riverwood Town Hall public meeting attended by over 300 concerned citizens.

Dr. Margaret Brooke, a State University water-systems specialist, helped interpret information and answer questions regarding the still-mysterious Snake River fish kill. Local physician Dr. Jason Martingdale and Riverwood High School home economics teacher Alicia Green joined the speakers during the question-and-answer session following the reports.

Dr. Brooke confirmed that preliminary water-sample analyses have shown no likely cause for the fish kill. She reported that EPA chemists will collect water samples at hourly intervals today to look for any unusual fluctuations in dissolved oxygen levels. Although fish don't breathe as we do, Dr. Brooke explained, they must take in an adequate amount of oxygen dissolved in water through their gills.

Dr. Schmidt expressed regret that the fishing tournament had to be cancelled but strongly supported the town council's action, saying that it was the safest course in the long run. He reported that "nothing has yet been found to determine the cause of the fish kill," but that "all fish found and examined since the initial discovery of the kill show unexpected and puzzling signs of biological damage. These signs include hemorrhaging and small bubbles under the skin along the lateral line." His laboratory is presently looking into the implications of these findings. In response to questions, Dr. Schmidt advised against rescheduling the fishing tournament yet.

Concerning possible causes of the fish kill, Dr. Brooke reported that "it must be something dissolved in the water, because suspended materials filtered from the water show nothing unusual." She emphasized that water's unique properties make the question of the fish kill quite complicated. Important factors to consider include the relative amounts of various substances that water can dissolve, and the effect water temperature has on solubility. She

expressed confidence, however, that further studies would shed light on the situation.

Dr. Martingdale reassured citizens that "thus far, no illness reported by either physicians or the hospital can be linked to drinking water." Ms. Green offered water-conservation tips for housekeeping and cooking that could make life easier for inconvenienced citizens. She distributed an information sheet that will also be available at the town hall office.

Mayor Edward Cisko confirmed that water supplies will again be trucked in from Mapleton today and expressed hope that the crisis will last no longer than three days.

Those attending the meeting appeared to be dealing with the emergency in good spirits. Citizens leaving Town Hall expressed a variety of opinions.

"I'll never take my tap water for granted again," said Trudy Anderson, a resident of southern Riverwood. "I thought scientists would have all the answers," puzzled Robert Morgan of Morgan Enterprises. "They don't know either! What's going on here? There's certainly more involved in all of this than I ever imagined."

B.1 PHYSICAL PROPERTIES OF WATER

Water is a familiar substance. We drink it, wash with it, swim in it, and sometimes grumble when it falls from the sky. We are so accustomed to water that most of us are unaware it is one of the rarest and most unusual substances in the universe. As space probes have explored our solar system and collected data as pictures, we have learned that water is almost totally absent on Mars, Jupiter and other planets and moons. Earth, on the other hand, is usually half-surrounded by water-laden clouds. In addition, more than 70% of the Earth's surface is covered by oceans having an average depth of more than three kilometers (two miles).

▶ *Kilo is the prefix meaning 1,000. 1 km = 1,000 m.*

To understand the issues involved in the fish kill, we need to know more about water itself. First, you will take a look at water's ***physical properties***—those distinctive characteristics shared by all water samples. (Later, you will see some of the chemical properties of water. Those properties can be studied only by chemically changing water into other substances.)

Water is the only ordinary liquid found naturally in our environment. That is partly because so many things dissolve easily in water; many other liquids are actually water solutions. Even water that seems pure never is entirely so. Surface water contains dissolved minerals as well as other substances. Even the distilled water used in steam irons and car batteries contains dissolved gases from the atmosphere. So does rainwater.

▶ *Solutions in which water is the solvent are called* aqueous *solutions.*

How many states of water are shown in this photograph?

Pure water is clear, colorless, odorless, and tasteless. Some tap water samples have a characteristic taste and even a slight odor, caused by other substances dissolved in the water. (You can confirm this by boiling, then refrigerating some distilled water. When you compare its taste with the taste of chilled tap water, you may notice that "pure" distilled water tastes flat.)

Density is an important physical property that can help identify a substance. To determine it, you need two measurements of a sample of matter—volume and mass. The SI unit of mass used most often in

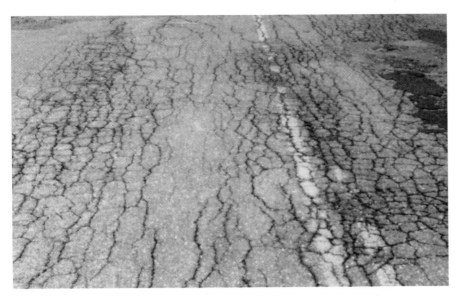

A series of freezes and thaws have broken up the pavement of this highway.

chemistry is the **gram** (g). A nickel coin has a mass of about 5 g. You have already been introduced to a commonly used unit of volume, the milliliter (mL). **Density** is the ratio of a substance's mass to its volume (mass/volume.) Thus, the *mass* of material present in one unit of *volume* of the material is its density.

At 4 °C, the mass of 1.0 mL of liquid water is 1.0 g. Thus, the density of water is about one gram per milliliter (1.0 g/mL). Correspondingly, 8.0 grams of water occupy 8.0 mL; the density is still 1.0 g/mL; (8.0 g/8.0 mL = 1.0 g/mL.)

How does the density of water compare with that of other familiar liquids? Corn oil has a density of 0.92 g/mL. That means one milliliter of corn oil has a mass of 0.92 g. Thus, a given volume of corn oil has less mass than the same volume of water. That's why corn oil floats on the surface of water.

Some other liquids are more dense than water. Typically, antifreeze for a car has a density of 1.11 g/mL. Therefore, a given volume of antifreeze has more mass than the same volume of water. Liquid mercury, at a density of 13.6 g/mL, is one of the densest common substances. Table 2 shows some densities of other materials.

State influences density: gases are much less dense than liquids, and the solid form of a substance is usually denser than its liquid form. However, ice is an important exception. As a water sample freezes, it expands to occupy a larger volume, and so one milliliter of ice has a density of 0.92 g/mL. That is slightly less than the density of liquid water (1.0 g/mL). As a result, ice floats on water. If ice had a greater density than liquid water and sank to the bottom as it froze, aquatic life in rivers and lakes could not survive.

This special property of water is also responsible for much erosion. Rainwater seeps into tiny cracks in rocks, freezes, then expands and cracks them further. After many seasons, with other factors contributing, rocks become soil. In a similar fashion, highway pavement is broken.

▶ *Sometimes a substance is said to be "lighter" than another of the same size. It is more correct to say that it is less dense, not "lighter."*

Table 2	Densities (g/mL)
Solids	
Cork	0.24
Ice	0.92
Aluminum	2.70
Iron	7.86
Liquids	
Gasoline	0.67
Water	1.00
Glycerine	1.27
Gases	
(at 25 °C, 1 atm pressure)	
Hydrogen	0.00008
Oxygen	0.0013
Carbon dioxide	0.0018

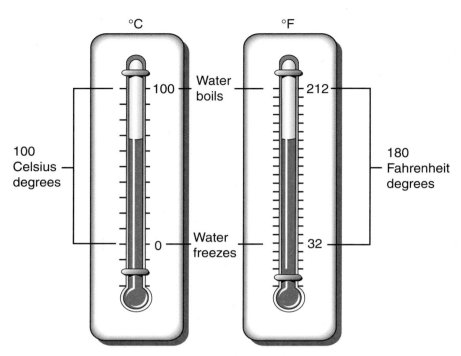

Figure 9 Celsius and Fahrenheit temperature scales. Reference temperatures for water are based on one atmosphere of pressure.

Figure 10 Water has a high surface tension, allowing a needle to float on its surface. It also causes water beads to form on a newly polished car surface.

This type of erosion is especially rapid because of the effect of traffic on the cracked pavement. You will notice that potholes are particularly bad following a winter of many freezes and thaws.

Among the other important physical properties of water are its **boiling point** and **freezing point.** The Celsius temperature scale, used in most of the world, divides the interval between the freezing point of water and its boiling point into 100 parts. Water's freezing point is defined as 0 °C, and its normal boiling point as 100 °C.

The Fahrenheit temperature scale is still used in the United States. Water boils at 212 °F and freezes at 32 °F on this scale; there are 180 divisions between these freezing and boiling temperatures (Figure 9).

Evaporation and boiling are related physical properties of liquids. In both processes, liquid changes to gas. However, evaporation occurs at all temperatures—but only from the surface of the liquid. At higher temperatures, evaporation occurs more rapidly. During boiling, by contrast, gaseous water (water vapor) forms *under the surface* of the liquid. Because the vapor is less dense than the liquid, it rises to the surface and escapes as steam. We cannot see steam, but it condenses into visible clouds of tiny droplets as it contacts cooler air.

Water has an unusually high boiling point, which is responsible for its being a liquid at normal temperatures. The life-supporting form of water is its liquid state.

Another very important and unusual physical property of water is its high **surface tension.** This property causes water to form spherical drops and to form a curved surface in a small container. Surface tension enables a water bug to dart across the surface of a quiet pool and makes it possible to float a dry needle on the surface of a dish of water (Figure 10).

B.2 MIXTURES AND SOLUTIONS

Samples of water collected from waterways for testing are often cloudy. Or, they may look clear but contain invisible substances. The samples of Riverwood's water, for example, contained things other than pure water. But were they dissolved in the water, or temporarily mixed in it? The difference could be important to Riverwood's future.

If two or more substances combine, yet retain their individual properties, a **mixture** has been produced. A mixture of salt and pepper is described as **heterogeneous** because it is not completely uniform throughout. A pepper and water mixture is also heterogeneous.

Mixing salt and water has a very different result—a **solution** is formed. In water, the salt crystals dissolve by separating into particles so small that they cannot be seen even at high magnification. These particles become uniformly mingled with the particles of water, producing a **homogeneous** mixture, one that is uniform throughout. All solutions are homogeneous mixtures. In the salt solution, the salt is the **solute** (the substance to be dissolved) and the water is the **solvent** (the dissolving agent).

Water mixtures are classified according to the size of particles dispersed in the water. **Suspensions** are mixtures containing relatively large, easily seen particles. The particles remain suspended for a while after stirring, but then settle out or form layers within the liquid. Suspensions are classified as heterogeneous mixtures because they are not uniform throughout.

Muddy water includes suspended particles of soil and other matter. A solid layer forms when muddy water is left undisturbed. Even after several days without stirring, the liquid still appears slightly cloudy. Some very small particles remain distributed throughout the water. This type of mixture is called a **colloid.** The tiny particles, just large enough to produce the cloudy appearance, are called colloidal particles.

Milk is a colloid with small butterfat particles dispersed in water. These colloidal butterfat particles are not visible to the unaided eye. Thus, milk can be classified as homogeneous, leading to the familiar term *homogenized milk.* Under high magnification, however, individual butterfat globules can be seen floating in the water. Milk no longer appears homogeneous (Figure 11).

▶ *A heterogeneous mixture's composition varies.*

▶ *Suspension (largest solute particles)*
↓
colloid (intermediate-sized solute particles)
↓
solution (smallest solute particles)

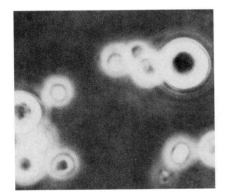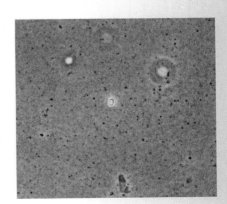

Figure 11 Fat globules can be seen under magnification, so that milk no longer looks homogeneous. *Left:* Whole milk under 100× magnification. *Center:* Whole milk under 10× magnification. *Right:* Skimmed milk under 10× magnification.

The Tyndall effect is apparent in the colloidal suspension on the left. The solution on the right is not colloidal, and so the light beam cannot be seen.

There's a simple way to decide whether the particles in a mixture are large enough to consider the liquid a colloid rather than a solution. *When a light beam shines through a colloidal liquid, the beam's path can be clearly seen in the liquid.* The particles are too small to see, but are large enough to reflect light off their surfaces. This reflection is called the **Tyndall effect,** named after the Irish physicist who first explained the phenomenon. *Solutions do not show a Tyndall effect.* The particles in a solution are so small that they do not settle out and cannot be seen even at high magnification.

The foul-water sample you purified earlier illustrates all three types of mixtures: (1) The sample certainly contained some particles large enough to form a suspension. (2) The sample's persistent cloudiness suggests that suspended colloidal particles were also present. (3) Even the final clear, purified sample had atmospheric gases and electrically charged particles dissolved in it. Thus, your "purified" water was actually a solution.

Now, you can put your knowledge of water mixtures to work. The following laboratory activity will give you some firsthand experience in distinguishing among these types of mixtures.

B.3 LABORATORY ACTIVITY: MIXTURES

GETTING READY

In this laboratory activity, you will examine four different water-containing mixtures and classify each as a suspension, a colloid, a solution, or a combination. You will filter each sample and look for the Tyndall effect in both the filtered and unfiltered samples. Particles in a suspension can be separated by filtration, while those in a colloid or a solution are too small to be retained by the filter paper. The Tyndall effect will reveal the presence of colloidal particles.

In your laboratory notebook, prepare data tables for entering your observations on the original samples and filtrates, and your conclusions about each mixture.

Data Table 1				ORIGINAL SAMPLES			
					Mixture Classification		
Well	Color	Clarity (Clear or Cloudy?)	Settle Out?	Tyndall Effect	Suspension	Colloid	Solution
A1							
A3							
A5							
C1							

Data Table 2					FILTRATES			
						Mixture Classification		
Well	Color	Clarity (Clear or Cloudy?)	Settle Out?	Tyndall Effect	Cotton Filter	Suspension	Colloid	Solution
D2								
D4								
D6								
B6								

PROCEDURE

1. Put on your goggles.

2. Obtain a 24-well wellplate and a set of Beral pipets containing the four unknown water mixtures. You will use only the wells designated in Figure 12.

3. Half-fill well A1 with mixture 1.
   Half-fill well A3 with mixture 2.
   Half-fill well A5 with mixture 3.
   Half-fill well C1 with mixture 4.

4. Carefully examine each of the mixtures. Record your observations in Data Table 1.

   a. Are any of the mixtures colored?

   b. Which mixtures are clear? Cloudy?

   c. Do particles settle out of any of the mixtures?

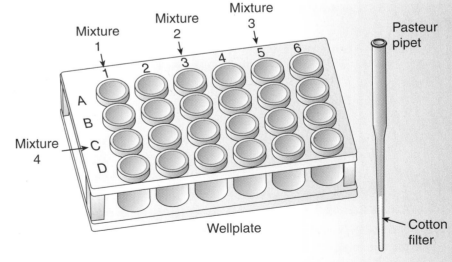

Figure 12 24-well wellplate with designated wells.

5. Perform a light-scattering Tyndall effect test on each mixture to determine whether it is a colloid. If the mixture is relatively clear, shine a light beam through the side of the wellplate and observe from the top. If the mixture is cloudy, shine the light beam across the liquid surface and observe the wellplate from the side. Can you see the light beam passing through the sample? Record your observations. Based on Steps 4 and 5, make an initial classification of each mixture as a suspension, colloid, or solution.

6. Prepare four filters, one for each of the mixtures. To prepare a filter, pack cotton (but not too tightly) into the tip of a Pasteur pipet. Use a new filter for each mixture.

7. Add 25–30 drops of mixture 1 from the Beral pipet to the filter. Collect the filtrate (the liquid that passes through the filter) into well D2 of the wellplate.

8. Separately filter mixtures 2, 3, and 4. Collect mixture 2 filtrate in well D4, mixture 3 filtrate in well D6, and mixture 4 filtrate in well B6 (see Figure 12, page 27).

9. Examine each of the cotton filters. Were any particles removed from the original mixtures by filtration? (Discoloration of the cotton does not necessarily mean that a solid was trapped.) Record your observations in Data Table 2.

10. Repeat Steps 4 and 5 with each collected filtrate. Record your observations in Data Table 2.

11. Based on these observations, make a final classification of each of the original mixtures as a suspension, colloid, or solution.

   a. Do any of the final classifications differ from the initial ones in Step 5?

   b. If so, account for the differences.

12. Dispose of the wellplate contents and return the equipment according to directions given by your teacher.

13. Wash your hands thoroughly before leaving the laboratory.

B.4 MOLECULAR VIEW OF WATER

So far in our investigation of water, we have focused on observing its properties with our unaided senses. To understand why water has such special properties, however, we must examine it at the level of its *atoms.*

All samples of matter are composed of atoms. Oxygen is considered an *element* because it is composed of only oxygen atoms. Likewise, the element hydrogen contains only hydrogen atoms. Approximately 90 elements are found in nature, each with its own unique type of atom and having unique identifying properties.

Water is not classified as an element because it has atoms of two types—oxygen and hydrogen. Nor is water classified as a mixture, because its properties are different from those of either oxygen or

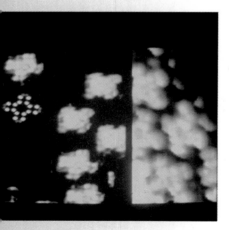

A scanning tunneling microscope image of copper-phthalocyanine. The image reveals individual atoms in the distinctive four-leaf-clover-shaped molecule.

Atoms of hydrogen and oxygen.   A molecule of water.

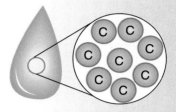

Figure 13 Enlarged view (not to scale) of a drop of food coloring. Each food coloring molecule is represented by Ⓒ.

hydrogen. Also, water cannot be separated into oxygen and hydrogen by simple physical means.

Instead, water is an example of a substance called a **compound**. To date, chemists have identified over 12 million compounds. Every compound is composed of atoms of two or more different elements linked together by **chemical bonds**. Two hydrogen atoms and one oxygen atom bond together to form a unit called a water **molecule**. A molecule is the smallest unit of a substance that retains the properties of that substance. Even one drop of water contains an unimaginably large number of water molecules.

The following *Your Turn* will give you a chance to apply an atomic and molecular view to explain a variety of common observations.

YOUR TURN

Matter at the Microlevel

If you add a few drops of food coloring to an undisturbed glass of water, you can see the color slowly spread out. If you stir this mixture, the color will become lighter and evenly distributed. How can this be explained in terms of molecules?

The food coloring contains molecules that appear colored. (See Figure 13.) After the drop of food coloring enters the water, food coloring molecules begin to spread among the colorless water molecules (Figure 14). Eventually a uniformly colored solution forms (Figure 15).

Now read the following observations and decide how each can be explained in terms of molecules. Feel free to draw sketches to clarify your explanations.

1. a. A regular balloon filled with helium decreases in volume as time passes.
    b. Metal-foil (Mylar) balloons shrink much more slowly than do plastic ones.

2. The solid form of a substance is usually denser than the liquid form, which is, in turn, much denser than the gaseous form.

3. A bottle of perfume is opened in a room. Soon the perfume's odor is apparent far away from the bottle.

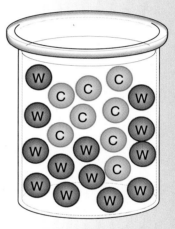

Figure 14 Food coloring molecules, Ⓒ, begin spreading among water molecules, Ⓦ.

Figure 15 Uniform solution of food coloring and water.

A Look at Water and Its Contaminants

B.5 SYMBOLS, FORMULAS, AND EQUATIONS

Table 3
Common Elements

Element Name	Symbol
Aluminum	Al
Bromine	Br
Calcium	Ca
Carbon	C
Chlorine	Cl
Cobalt	Co
Gold	Au
Hydrogen	H
Iodine	I
Iron	Fe
Lead	Pb
Magnesium	Mg
Mercury	Hg
Nickel	Ni
Nitrogen	N
Oxygen	O
Phosphorus	P
Potassium	K
Silver	Ag
Sodium	Na
Sulfur	S
Tin	Sn

To represent atoms, elements, and compounds on paper, a useful international chemical language has developed. The "letters" in this language are **chemical symbols.** Each element is assigned a chemical symbol of one or two letters. Only the first letter of the symbol is capitalized. For example, C is the symbol for the element carbon and Ca is the symbol for the element calcium. Symbols for some common elements are listed in Table 3.

The "words" or **chemical formulas** of this chemical language each represent a different chemical substance. In the chemical formula of any substance, symbols represent each element present.

Subscripts (numbers that are written below the normal line of letters) indicate how many atoms of each element are in one molecule or unit of the substance. For example, as you might know, the chemical formula for water is H_2O. The subscript 2 indicates that each water molecule contains two hydrogen atoms. Each water molecule also contains one oxygen atom. However, the subscript 1 is not included in chemical formulas.

"Sentences," in the special language of chemistry, are **chemical equations.** Each chemical equation summarizes the details of a particular chemical reaction. **Chemical reactions** involve the breaking and forming of chemical bonds, causing atoms to become rearranged into new substances. These new substances have different properties than the original material. Not only are their physical properties different, but the chemical reactions in which they participate are also different. In other words, the new substances formed in a chemical reaction have different **chemical properties** than those of the reactants. The chemical equation for the formation of water

$$2\,H_2 \quad + \quad O_2 \quad \rightarrow \quad 2\,H_2O$$

hydrogen	oxygen		water
reactants			*product*

shows that two hydrogen molecules ($2\,H_2$) and one oxygen molecule (O_2) react to produce ($\rightarrow$) two molecules of water ($2\,H_2O$). The original substances in a chemical reaction are called the **reactants;** their formulas are always written on the left. The new substance or substances formed from the rearrangement of the reactant atoms are called **products;** their formulas are always written on the right. Note that this equation, like all chemical equations, is balanced—the total number of atoms (four H atoms and two O atoms) is the same for both reactants and products.

YOUR TURN

Working with Symbols, Formulas, and Equations

Let's look at the kind of information available from a simple chemical equation and the formulas included in it:

$$N_2 \quad + \quad 3\,H_2 \quad \rightarrow \quad 2\,NH_3$$

nitrogen	hydrogen	ammonia

▶ *Household ammonia is made by dissolving gaseous ammonia in water.*

First, we can complete an "atom inventory" based on this chemical equation:

$$N_2 \quad + \quad 3\,H_2 \quad \rightarrow \quad 2\,NH_3$$

2 N atoms   6 H atoms   =   2 N atoms and 6 H atoms

Note that the total atoms of N (the element nitrogen) and H (the element hydrogen) remain unchanged during this chemical reaction.

We can also interpret the equation in terms of molecules: One molecule of N_2 reacts with three molecules of H_2 to produce two molecules of the compound NH_3, called ammonia. We also note that molecules of nitrogen (N_2) and hydrogen (H_2) are each made up of two atoms, while ammonia molecules are each composed of four atoms—one nitrogen atom and three hydrogen atoms.

1. Name the element represented by each symbol:

   a. P   c. N   e. Br   g. Na

   b. Ni   d. Co   f. K   h. Fe

2. For each formula, name the elements present and give the number of atoms of each element shown in the formula.

   a. H_2O_2 Hydrogen peroxide (antiseptic)

   b. $CaCl_2$ Calcium chloride (de-icing salt for roads)

   c. $NaHCO_3$ Sodium hydrogen carbonate (baking soda)

   d. H_2SO_4 Sulfuric acid (battery acid)

3. The following chemical equation shows the formation of the air pollutant nitric oxide, NO:

$$N_2 + O_2 \rightarrow 2\,NO$$

   a. Interpret this equation in terms of molecules.

   b. Complete an atom inventory for this equation.

We must consider another question before we return to our discussion of "foul water" and the problems in Riverwood: Why is the water molecule in Figure 16 (page 32) represented by that particular shape, rather than by some other shape? That is, does the arrangement of atoms in a molecule (called ***molecular structure***) affect the substance's observed properties?

B.6 THE ELECTRICAL NATURE OF MATTER

You have already had direct experience with matter's electrical nature, most probably without realizing it. Clothes often display "static cling" when they are taken from the dryer. Pieces of apparel stick firmly together, and can be separated only with effort. The shock you sometimes receive after walking across a rug and touching a metal doorknob is another reminder of matter's electrical nature. If two inflated balloons are rubbed against your hair, both will attract your hair, but will repel each other. (This is best seen when the humidity is low.)

The electrical properties of matter can be summarized as follows:

Like charges repel.   ←─(+) (+)─→ or ←─(−) (−)─→

Unlike charges attract.   (+)─→←─(−)

What are these positive and negative charges? How do they relate to the idea of atoms and molecules? For now, you need to know only a few key points:

- Neutral (uncharged) atoms contain equal numbers of positively charged particles (called **protons**) and negatively charged particles (called **electrons**). In addition, most atoms contain one or more neutral particles (called **neutrons**).

- Positive–negative attractions between the protons of one atom and the electrons of another atom provide the "glue" that holds different atoms together. This "glue" helps explain **chemical bonding.**

A variety of observations suggest that a hydrogen atom can bond to only one other atom at a time. By contrast, an oxygen atom can form two bonds, and can thus be joined to two hydrogen atoms. The result: H_2O!

Experiments suggest that water molecules are electrically **polar.** A polar molecule has an uneven distribution of electrical charge. This means that each molecule has a positive region on one end and a negative region on the other end. Evidence also suggests that the water molecule has a

Figure 16 Two models of hydrogen, oxygen, and water molecules. Atoms in molecules are held together by chemical bonds, shown to the right by springs and sticks.

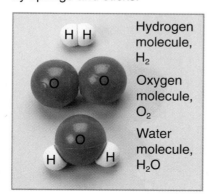

Hydrogen molecule, H_2

Oxygen molecule, O_2

Water molecule, H_2O

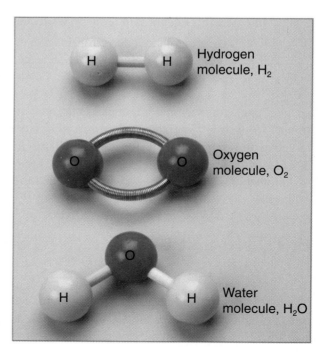

Hydrogen molecule, H_2

Oxygen molecule, O_2

Water molecule, H_2O

bent or V-shape (see Figures 16 and 17), rather than a linear, stick-like shape as in H—O—H. The "oxygen end" is an electrically negative region (having a *greater* concentration of electrons, shown as δ^-, delta minus), compared to the two "hydrogen ends" which are electrically positive (shown as δ^+). The entire water molecule is still electrically neutral, even though the electrons are not evenly distributed in its structure.

The combined effects of water's molecular shape and electrical polarity help explain many properties of water described earlier. For example, because unlike charges attract, the oppositely charged "ends" of neighboring water molecules attract each other, causing them to stick together. This gives water its high boiling point. (It takes a large quantity of thermal energy to separate the attracted liquid water molecules to form water vapor.) Water's high surface tension and reduced density when it crystallizes as ice can also be explained by its molecular shape and electrical polarity.

Polar water molecules are attracted to other polar substances and to substances composed of electrically charged particles. These attractions make it possible for water to dissolve a great variety of substances.

In certain substances, such as common table salt, the smallest particles are neither uncharged atoms nor molecules. Atoms can gain or lose electrons to form negatively or positively charged particles called **ions.** Compounds that are called **ionic compounds** or ionic substances are composed of positive and negative ions. There are always enough positively and negatively charged ions so that the *total* positive charge equals the *total* negative charge.

In solid ionic compounds, such as table salt, the ions are held together in crystals by attraction between negative and positive charges. The resulting compound has no overall electric charge. When an ionic compound dissolves in water, individual ions become separated from each other and disperse in the water. The designation (aq) following the symbol for an ion, as in Na^+(aq), means that the ion is present in water (aqueous) solution.

B.7 PURE AND IMPURE WATER

We are now ready to return to the problem of Riverwood's fish kill. Recall that various Riverwood residents had different ideas about the cause of the problem. For example, longtime resident Harmon Lewis was sure the cause was pollution of the river water. Let's examine water pollution in greater detail.

Families in most U.S. cities and towns receive clean, but not pure, water at an extremely low cost. Check the cost for your own area: If you use municipal water, use your family's water bill to find the current cost per gallon. Divide that number by 3.8 (the number of liters in one gallon) to find your cost for one liter of water.

It's useless to insist on *pure* water. The cost of processing water to make it completely pure would be prohibitively high. Even if costs were not a problem it would still be impossible to have pure water. The atmospheric gases nitrogen (N_2), oxygen (O_2), and carbon dioxide (CO_2) will always dissolve in the water to some extent.

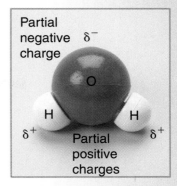

Figure 17 Polarity of a water molecule. The δ^+ and δ^- indicate partial charges. Because these charges balance, the molecule as a whole is electrically neutral.

Na	*electrically neutral sodium atom*
Na$^+$	*sodium ion*
Cl	*electrically neutral chlorine atom*
Cl$^-$	*chloride ion*
Na$^+$Cl$^-$	*sodium chloride (table salt)*

In fact, it's unnecessary and even undesirable to remove all substances from the water we use. Some dissolved gases and minerals give water a pleasing taste. Fish and other water-dwelling creatures depend on the oxygen gas dissolved in the water for survival. Some dissolved substances also promote human health. For example, adding a small amount of certain chlorine-containing compounds ensures that harmful bacteria are not living in a municipal water supply.

Unfortunately, not all substances that become suspended or dissolved in water are desirable there. High concentrations of substances containing iron (Fe) ions give water a bad taste and cause unwanted deposits in pipes and fixtures. Some compounds containing sulfur (S) give water a bad odor. Compounds containing poisonous ions such as mercury (Hg), lead (Pb), cadmium (Cd), and arsenic (As) can dissolve in water. Even at relatively low concentrations, these elements can be harmful to human health.

Manufactured substances, such as pesticides, and commercial and industrial waste products (particularly solvents) can also find their way into drinking water with harmful effects. Even sunlight can produce potentially harmful substances in chlorinated water containing certain types of contaminants.

Our real challenge is to ensure that our water supplies do not become polluted with unwanted substances. This is accomplished by preventing such substances from entering our water supplies and removing those already present.

B.8 LABORATORY ACTIVITY: WATER TESTING

GETTING READY

Chemists can detect and identify ions in water solution in several different ways. In this activity you will use some chemical tests to check for the presence of certain ions in aqueous solution. Positively charged ions have a deficiency of electrons and are called *cations;* negatively charged ions have an excess of electrons and are called *anions.* You will investigate two different cations and two different anions.

The tests you will perform are *confirming tests.* That is, if the test is positive it confirms that the ion in question is present. In each confirming test you will look for a change in solution color, or for the appearance of an insoluble material called a *precipitate.* A negative test (no color or precipitate) doesn't necessarily mean the ion in question is not present. The ion may simply be present in such a small amount that the color or precipitate cannot be seen. The tests you will do are classified as *qualitative* tests, ones that identify the presence or absence of a particular substance in a sample. This is in contrast to *quantitative* tests that determine the amount of a specific substance present in a sample.

You will test for the presence of iron(III), Fe^{3+}, and calcium, Ca^{2+}, cations, as well as chloride, Cl^-, and sulfate, SO_4^{2-}, anions. (Notice that when the positive or negative charge of an ion is more than one,

Data Table			
Solutions	**Color**	**Precipitate**	**Is Ion Present?**
Fe^{3+} reference			
Tap water			
Control			
Ca^{2+} reference			
Tap water			
Control			
Cl^- reference			
Tap water			
Control			
SO_4^{2-} reference			
Tap water			
Control			

► *There are two types of iron ions: Fe^{2+} is designated Fe(II); Fe^{3+} is Fe(III).*

the amount of the charge is shown with the plus or minus sign for the ion.) You'll perform each confirming test on three different samples:

- a *reference solution* (known to contain the ion of interest)
- *tap water* (which may or may not contain the ion)
- a *control* (distilled water, known *not* to contain the ion).

Prepare a data table similar to the one shown above in your laboratory notebook.

Here are some suggestions to guide your ion analysis:

1. If the ion is in tap water, it will probably be present in a lower amount than in the same volume of reference solution. Thus, the color or quantity of precipitate produced in the tap water sample will be less than in the reference solution.

2. When completing an ion test, mix the well's contents thoroughly, using a toothpick or small glass stirring rod. Do not use the same toothpick or stirring rod in other samples.

3. In a confirming test based on color change, so few color-producing ions may be present that you can't be sure the reaction actually occurred. Here are several ways to decide whether the expected color is actually present:

- Place a sheet of white paper behind or under the wellplate to make any color more visible.
- Compare the color of the control (distilled water) test to that of tap water. Distilled water doesn't contain any of the ions tested. Thus, even a faint color in the tap water confirms that the ion is present.

4. In a confirming test involving a precipitate, you may be uncertain whether a precipitate is present even after thoroughly mixing the solutions. Placing the wellplate on a black or dark surface makes the precipitate more visible.

PROCEDURE

1. Use a 24-well wellplate. The wells in Row A contain reference solutions; Row B wells contain tap water; Row C wells contain distilled water (the control).

2. Add 20 drops of iron(III) (Fe^{3+}) reference solution to well A1.

   Add 20 drops of calcium ion (Ca^{2+}) reference solution to well A2.

   Add 20 drops of chloride ion (Cl^-) reference solution to well A3.

   Add 20 drops of sulfate ion (SO_4^{2-}) reference solution to well A4.

3. Add 20 drops of tap water to each of wells B1, B2, B3, and B4.

4. Add 20 drops of distilled water to each of wells C1, C2, C3, and C4.

Iron(III) Ion Test (Fe^{3+})

1. Add one or two drops of potassium thiocyanate (KSCN) test solution to wells A1, B1, and C1.

2. Record your observations in the data table. The confirming test you observed for Fe^{3+} can be represented as follows:

$$Fe^{3+}(aq) \quad + \quad SCN^-(aq) \quad \rightarrow \quad [FeSCN]^{2+}(aq)$$

| iron (III) ion | thiocyanate ion | iron (III) thiocyanate ion |
| (reference solution) | (test solution) | (red color) |

▶ *Only ions that take part in the reaction are included in this type of equation.*

Calcium Ion Test (Ca^{2+})

Avoid contact of the acetic acid solution with skin.

1. Add three drops of acetic acid ($HC_2H_3O_2$) to wells A2, B2, and C2.

2. Add three drops of sodium oxalate ($Na_2C_2O_4$) test solution to wells A2, B2, and C2.

3. Record your observations in the data table.

The confirming test you observed for Ca^{2+} can be represented as follows:

$$Ca^{2+}(aq) \quad + \quad C_2O_4^{2-}(aq) \quad \rightarrow \quad CaC_2O_4(s)$$

| calcium ion | oxalate ion | calcium oxalate |
| (reference solution) | (test solution) | (precipitate) |

▶ *The (s) means a solid.*

Chloride Ion Test (Cl^-)

1. Add three drops of silver nitrate ($AgNO_3$) test solution to wells A3, B3, and C3.

2. Record your observations. If chloride ions are present, the following reaction takes place:

$$Cl^-(aq) \quad + \quad Ag^+(aq) \quad \rightarrow \quad AgCl(s)$$

| chloride ion | silver ion | silver chloride |
| (reference solution) | (test solution) | (precipitate) |

Sulfate Ion Test (SO_4^{2-})

1. Add three drops of barium chloride ($BaCl_2$) test solution to wells A4, B4, and C4.

2. Record your observations in the data table. The confirming test for SO_4^{2-} can be represented by the following equation:

$$SO_4^{2-}(aq) \quad + \quad Ba^{2+}(aq) \quad \rightarrow \quad BaSO_4(s)$$

| sulfate ion | barium ion | barium sulfate |
| (reference solution) | (test solution) | (precipitate) |

3. Discard the wellplate contents as directed by your teacher.

4. Wash your hands thoroughly before leaving the laboratory.

QUESTIONS

1. a. Why was a control used in each test?

   b. Why was distilled water chosen as the control?

2. Describe some difficulties associated with the use of qualitative tests.

3. These tests cannot absolutely confirm the absence of an ion. Why?

4. How might your observations have changed if the wellplate was not cleaned and dried before testing?

In a *ChemCom* lab.

B.9 YOU DECIDE: THE RIVERWOOD MYSTERY

Your teacher will divide the class into groups composed of four or five students. Each group will complete this decision-making activity. Afterward, the entire class will compare and discuss the answers obtained by each group.

A Look at Water and Its Contaminants

CHEMISTRY AT WORK

Environmental Cleanup: It's a Dirty Job . . . But That's the Point

Wayne Crayton spends his summers touring exotic islands in the Aleutian Islands chain off the coast of Alaska. But it's not just an adventure. It's also his job.

As an Environmental Contaminants Specialist with the U.S. Army Corps of Engineers, Alaska District, Wayne investigates areas previously used as military bases and fueling stations. Wayne and the team review and assess the damage (if any) that contaminants have done to key areas used by wildlife, and then develop plans to fix the problems.

To plan for an investigation, Wayne's team reviews information to determine what they're likely to find at a given site. For example, historical documents about the site will show whether they should be looking for petroleum residues or other contaminants; aerial photography and records from earlier investigations will show them specific areas that potentially have contaminants.

At the site, the team collects soil, sediment, and water samples at the actual location where a contaminant was originally introduced to the environment, as well as samples from the area over which it might have spread. They may also collect small mammals or fish that have been exposed to the contaminants. After they have collected the necessary samples, they return home quickly, because some samples can degrade or change characteristics soon after collection.

During the rest of the year, Wayne works in an office in Anchorage, analyzing and interpreting data and test results from the field investigations. He and his team calculate concentrations of hazardous chemicals, including organochlorines, PCBs, pesticides, petroleum residues, and trace metals. Next they determine whether any of these chemicals present a risk to humans or the surrounding ecosystems.

Using these results, Wayne and his team recommend solutions for removing and treating contaminated soil and other material. Other times, Wayne's team decides the best solution is to do nothing, because the cleanup itself could destroy wetlands, disturb endangered wildlife, or have other negative effects. The Corps of Engineers uses the team's recommendations to direct the work of a contractor who performs the actual cleanup.

Assemble Your Crew . . .

Wayne's team is *interdisciplinary,* meaning that it includes specialists from a variety of fields. If you were helping to pick the next field investigation team, you'd need to think about the mission, and then choose your team carefully. Wayne, for example, has a B.S. in biology and an M.S. in botany. What other areas of expertise might you look for?

Photographs courtesy of U.S. Army Corps of Engineers

At the beginning of this unit, you read newspaper articles describing the Riverwood fish kill and several citizens' reactions to it. Among those interviewed were Harmon Lewis, a longtime resident of Riverwood, and Dr. Margaret Brooke, a water-systems scientist. These two individuals had very different reactions to the fish kill. Harmon Lewis was angry, certain that human activity had caused the fish kill, probably by some sort of pollution. Dr. Brooke refused even to speculate about the cause of the fish kill until she had run some tests. Which of these two positions comes closer to your own reaction at this point?

Let's investigate this issue further.

1. Reread the fish-kill newspaper reports located at the beginning of Parts A and B. List all *facts* (not opinions) concerning the fish kill found in these articles. Scientists often refer to facts as **data.** Data are objective pieces of information. They do not include interpretation.

2. List at least five factual questions you would want answered before you could decide on possible causes of the fish kill. Some typical questions: Do barges or commercial boats travel on the Snake River? Were any shipping accidents on the river reported recently?

3. Look over your two lists—one of facts and the other of questions.
   a. At this point, which possible fish-kill causes can you rule out as unlikely? Why?
   b. Can you suggest any probable cause? Be as specific as possible.

Later in this unit you will have an opportunity to test the reasoning you used in answering these questions.

B.10 WHAT ARE THE POSSIBILITIES?

The activities you just completed (gathering data, seeking patterns or regularities among the data, suggesting possible reasons to account for the data) are typical of scientists' approaches in attempting to solve problems. Such scientific methods are a combination of systematic, step-by-step procedures and logic, as well as occasional hunches and guesses.

A fundamental and difficult part of scientists' work is knowing what questions to ask. You have listed some questions that might be asked concerning the cause of the fish kill. Such questions help focus a scientist's thinking. Often a large problem can be reduced to several smaller problems or questions, each of which is more easily managed and solved.

The number of possible causes for the fish kill is large. Scientists investigating this problem must find ways to eliminate some causes and zero in on more promising ones. They try to either disprove all but one cause or to produce conclusive proof in support of a specific cause.

Water-systems analyst Dr. Brooke reported that suspended materials had already been eliminated as possible causes of the fish kill. She concluded that if the actual cause were water-related, it must be due to something dissolved in the water.

In the next part of this unit we will examine several major categories of water-soluble substances and consider how they might be involved in the fish kill. The mystery of the Riverwood fish kill will be confronted at last!

? PART B: SUMMARY QUESTIONS

1. a. List three unusual properties of water. Indicate to which state(s) of water—solid, liquid, or gas—each property applies.

    b. Briefly describe the importance of each property to life on this planet.

2. When gasoline and water are mixed, they form two distinct layers. Which liquid—gasoline or water—will be found in the top layer? (*Hint:* Refer to Table 2, page 23.)

3. Decide which material in each pair has the higher density:

    a. ice or liquid water

    b. lead or aluminum

    c. silver chloride (AgCl) precipitate or liquid water (*Hint:* Check your Chloride Ion Test observations in the *Water Testing* activity.)

    d. hot air or cold air

4. Identify each of the following as either a solution, suspension, or colloid.

    a. a medicine that says "shake before using"

    b. rubbing alcohol

    c. Italian salad dressing

    d. mayonnaise

    e. a cola soft drink

    f. an oil-based paint

    g. milk

5. Using what you know about chemical notation, which of the following is an element, and which is a compound?

    a. CO   b. Co

6. a. Is 100% pure, chemical-free water possible?

    b. What do we actually want when we ask for "pure" water?

7. Name the elements in each substance. List the number of atoms of each element shown in the substance's formula.

    a. Phosphoric acid, H_3PO_4 (used in some soft drinks and to produce some types of fertilizer)

    b. Sodium hydroxide, NaOH (found in some drain cleaners)

    c. Sulfur dioxide, SO_2 (a by-product of burning some types of coal)

8. The following chemical equation represents the burning of methane, CH_4.

$$CH_4 + 2\,O_2 \rightarrow CO_2 + 2\,H_2O$$

    a. Interpret this equation in terms of molecules.

    b. Complete an atom inventory for this equation.

EXTENDING YOUR KNOWLEDGE (OPTIONAL)

- The Celsius temperature scale is used in scientific work worldwide, and is used in everyday life by most of the world's citizens. However, Fahrenheit temperatures are still used in weather reports in the United States. From the normal freezing and boiling points of water shown on these two scales, try to develop an equation that converts a Celsius temperature to its Fahrenheit equivalent. (*Hint:* Recall that there are 180 Fahrenheit degrees, but only 100 Celsius degrees, between the freezing point and boiling point of water.)

- The density of petroleum-based oil is approximately 0.9 g/mL. By contrast, the density of water is 1.0 g/mL. What implications does this density difference have for an "oil spill" in a body of water? Consider the implications for aquatic life, oil-spill cleanup operations, and the possibility of an oil fire.

- Some elements in Table 3 (page 30) have symbols that are not based on their modern names (such as K for potassium). Look up their historical names and explain the origin of their symbols.

This land in Nevada is desert except in the green irrigated areas.

C INVESTIGATING THE CAUSE OF THE FISH KILL

The immediate challenge facing investigators of the Riverwood fish kill is to decide whether some dissolved substance was responsible for the crisis.

C.1 SOLUBILITY

SOLUBILITY OF SOLIDS

How much of a substance will dissolve in a given amount of water? Imagine preparing a water solution of potassium nitrate, KNO_3. You pour water into a beaker and then add a scoopful of solid, white potassium nitrate crystals. As you stir the water, the solid crystals dissolve and disappear. The resulting solution remains colorless and clear. In this solution, water is the solvent, and potassium nitrate is the solute.

You add a second scoopful of potassium nitrate crystals to the beaker and stir. These crystals also dissolve. But as you continue adding more solid without increasing the amount of water, eventually some potassium nitrate crystals remain on the bottom of the beaker, no matter how long you stir. There is a maximum quantity of potassium nitrate that will dissolve in a given amount of water at room temperature.

The **solubility** of a substance in water refers to the maximum quantity of the substance, expressed in grams, for example, that will dissolve in a certain quantity of water (for example, 100 g) at a specified temperature. The solubility of potassium nitrate, for example, might be expressed in units of "grams per 100 g water" at a certain temperature.

The term "soluble" is actually a relative term, because everything is soluble in water to some extent. The term "insoluble," therefore, really refers to substances that are only very, very slightly soluble in water. Chalk, for example, is insoluble in water.

Both the size of the solute crystals and the vigor and duration of stirring help determine the time needed for a solute to dissolve at a given temperature. However, these factors do not affect *how much* substance will eventually dissolve.

When a solvent holds as much of a solute as it normally can at a given temperature, we say the solution is **saturated**. A solution that contains less than this amount of dissolved solute at that temperature is called an **unsaturated** solution.

Figure 18 shows the maximum mass (in grams) of three ionic compounds that will dissolve in 100 g of water at temperatures from 0 °C to 100 °C. The plotted line for each solute is called a **solubility curve**.

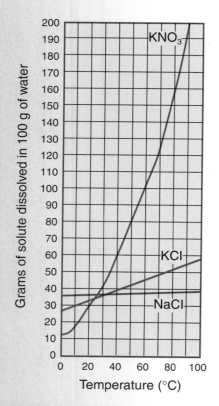

Figure 18 Relationship between solute solubility in water and temperature

Let's look first at the solubility curve for potassium nitrate (KNO_3) in Figure 18. The graph indicates that, at 80 °C, 160 g of potassium nitrate will dissolve in 100 g of water, forming a saturated solution. Thus, the *solubility* of potassium nitrate in 80 °C water is 160 g per 100 g water. By contrast, potassium nitrate's solubility in 20 °C water is about 30 g per 100 g water. (Check your ability to "read" this value on the graph.)

Note that the solubility curve for sodium chloride (NaCl) is nearly a horizontal line. This means that the solubility of sodium chloride is affected very little by changes in the solution's temperature. In contrast, the curve for potassium nitrate (KNO_3) rises steeply as temperature increases, showing that its solubility in water is greatly influenced by temperature.

Each point on a solubility curve represents a saturated solution. For example, the graph shows that about 39 g of sodium chloride dissolved in 100 g of water at 100 °C produces a saturated solution.

Any point on a graph below a solubility curve represents an unsaturated solution. For example, a solution containing 80 g of potassium nitrate and 100 g of water at 60 °C is an unsaturated solution. (Locate this point on the graph; note that it falls below the potassium nitrate solubility curve.)

If you could cool this solution to 40 °C (moving to the left on the graph) without forming any solid crystals, you would have a **supersaturated** solution of potassium nitrate at this lower temperature. (Note that this new point lies above the solubility curve for potassium nitrate.) A supersaturated solution contains more dissolved solute than a saturated solution. Agitating the solution, however, would cause 18 g of solid potassium nitrate crystals to appear and settle to the bottom of the beaker. The remaining liquid would then contain 62 g (80 g − 18 g) of solute per 100 g water—a stable, saturated solution at 40 °C. Table 4 gives the quantities of potassium nitrate in saturated, unsaturated, and supersaturated solutions at 20 °C.

Table 4	Solutions of KNO_3
Solution Description	**Quantity of Solute/Solvent**
Unsaturated KNO_3 solution (at 20 °C)	Less than 30 g/100 g water
Saturated KNO_3 solution (at 20 °C)	30 g/100 g water
Supersaturated KNO_3 solution (at 20 °C)	More than 30 g/100 g water

SOLUBILITY OF GASES

We have seen that the solubility of a solid in water often decreases when the water temperature is lowered. The solubility behavior of gases in water is quite different. As water temperature decreases, a gas becomes *more* soluble! Because of the importance of oxygen gas to aquatic life, let's see what happens to its solubility when water temperature changes.

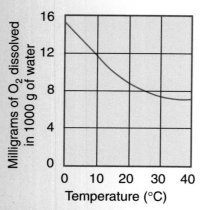

Figure 19 Solubility curve for O_2 gas in water in contact with air.

▶ *Be sure to note differences in axis labels when you compare graphs.*

Figure 19 shows the solubility curve for oxygen gas, plotted as milligrams of oxygen dissolved per 1,000 g of water. Note that the solubility of oxygen at 0 °C is about twice its solubility at 30 °C. The values for oxygen solubility are much smaller than the values shown in Figure 18 for solid solutes. For example, at 20 °C, about 37 g of sodium chloride will dissolve in 100 g of water. By contrast, only about 9 mg (0.009 g) of oxygen gas will dissolve in *ten* times as much water—1,000 g of water—at this temperature. It's clear that gases are *far* less soluble in water than many ionic solids are.

The solubility of a gas depends not only on water temperature, but also on gas pressure. Gas solubility is directly proportional to the pressure of the gas on the liquid. That is, if the pressure of the gas were doubled, the amount of gas that dissolves would also double. You see one effect of this every time you open a can or bottle of carbonated soft drink as dissolved carbon dioxide gas (CO_2) escapes from the liquid in a rush of bubbles. To increase the amount of carbon dioxide dissolved in the beverage, carbon dioxide gas was originally forced into the container at high pressure just before it was sealed. When the can or bottle is opened, gas pressure on the liquid drops back to atmospheric pressure. Dissolved carbon dioxide gas escapes from the liquid until it reaches its lower solubility at this lower pressure. When the fizzing stops, you might think that the beverage has "gone flat." Actually, the excess carbon dioxide gas has simply escaped into the air; the resulting solution is merely saturated.

The following exercise will help you become more familiar with the information found in the solubility graphs in Figures 18 (page 42) and 19. Refer to the appropriate graph to answer each question.

YOUR TURN

Solubility and Solubility Curves

What is the solubility of potassium nitrate at 80 °C? The answer is found by locating, in Figure 18, the intersection of the potassium nitrate curve with the vertical line representing 80 °C. The solubility is found by following the horizontal line to the left. Thus, the solubility of potassium nitrate in water at 80 °C is 160 g per 100 g of water.

At what temperature will the solubility of dissolved oxygen be 10 mg per 1,000 g of water? See Figure 19. Think of the space between 8 and 12 mg/1,000 g on the axis in Figure 19 as divided into two equal parts; follow an imaginary horizontal line at "10 mg/1,000 g" to its intersection with the curve. Follow a vertical line down to the *x* axis. As the line falls half-way between 10 °C and 20 °C, the desired temperature must be about 15 °C.

It's possible to calculate the solubility of a substance in various amounts of water. The solubility curve indicated that 160 g of potassium nitrate will dissolve in 100 g of water at 80 °C. How much potassium nitrate will dissolve in 200 g of water at this temperature? We can "reason" the answer as follows: The amount of solvent (water) has doubled from 100 g to 200 g. That means that twice as much solute can be dissolved. Thus: 2×160 g $KNO_3 = 320$ g KNO_3. Done! The calculation can also be written out as a simple proportion, giving the same answer:

$$\left(\frac{160 \text{ g KNO}_3}{100 \text{ g H}_2\text{O}}\right) = \left(\frac{x \text{ g KNO}_3}{200 \text{ g H}_2\text{O}}\right)$$

$$x = \left(\frac{160 \text{ g KNO}_3}{100 \text{ g H}_2\text{O}}\right) \times (200 \text{ g H}_2\text{O}) = 320 \text{ g KNO}_3$$

1. Refer to Figure 18 in answering these questions:
   a. What mass (in grams) of potassium nitrate (KNO_3) will dissolve in 100 g of water at 50 °C?
   b. What mass of potassium chloride (KCl) will dissolve in 100 g of water at this temperature?

2. Refer to Figure 18 in answering these questions:
   a. We dissolve 25 g of potassium nitrate in 100 g of water at 30 °C, thus producing an unsaturated solution. How much *more* potassium nitrate (in grams) must be added to form a saturated solution at 30 °C?
   b. What is the minimum mass (in grams) of 30 °C water needed to dissolve 25 g of potassium nitrate?

3. Refer to Figure 19 in answering these questions: What mass of oxygen gas can be dissolved in 1,000 g of water
   a. at 30 °C?   b. at 20 °C?

4. What mass of oxygen gas can be dissolved in 100 g of water at 20 °C?

C.2 SOLUTION CONCENTRATION

Some substances pose no problems when dissolved in water at very low concentrations, and may even be beneficial. However, larger dissolved quantities of these same substances may be harmful. For example, the level of chlorine in a swimming pool and the amount of fluoride in a municipal water system must be carefully measured and controlled. For these applications and countless others, we need to specify exact, numerical concentrations of solutions.

Solution concentration refers to the quantity of solute dissolved in a specific quantity of solvent or solution. You have already worked with one type of concentration expression. The water solubility graphs reported solution concentrations as a mass of a substance dissolved in a given mass of water.

Another way to express concentration is with percentages. For example, dissolving 5 g of table salt in 95 g of water produces 100 g of solution, or a 5% salt solution (by mass).

$$\left(\frac{5 \text{ g salt}}{100 \text{ g solution}}\right) \times 100\% = 5\%$$

"Percent" means parts per hundred parts. Thus, a 5% salt solution *could* also be reported as five parts per hundred of salt (5 pph salt), even though percent is much more commonly used.

To prepare 10 times as much salt solution of the *same* concentration, you would need to dissolve 10 times more salt (50 g) in 10 times more water (950 g) to make 1,000 g of solution.

$$\left(\frac{50 \text{ g salt}}{1,000 \text{ g solution}}\right) \times 100\% = 5\%$$

This solution can be described as containing 50 parts per thousand of salt (50 ppt salt), or, because it has the same concentration as the original solution, as 5% (5 pph) salt.

For solutions involving much smaller quantities of solute (such as dissolved gases in water), concentration units of parts per million, ppm, are useful. What's the concentration of the 5% salt solution expressed in ppm? Because 5% of 1 million is 50,000, a 5% salt solution is 50,000 parts per million.

The notion of concentration is part of our daily lives. Interpreting recipes, adding antifreeze to an automobile, or mixing a cleaning solution all require using solution concentrations. The following exercises will help you review solution concentration terminology and ideas. You will also gain experience with the chemist's view of this concept.

▶ $\dfrac{5}{100} = \dfrac{50}{1000} = \dfrac{50,000}{1,000,000}$

5% (5 pph) = 50 ppt = 50,000 ppm

YOUR TURN

Describing Solution Concentrations

Solution concentrations can be used in calculations, just as we used solubilities in *Your Turn: Solubility and Solubility Curves*. For example, in medicine a common intravenous (IV) saline solution contains 4.55 g NaCl dissolved in 495.45 g of sterilized distilled water. What is the concentration of this solution, expressed as grams of NaCl per 100 g of solution? We can calculate the answer as follows:

$$\frac{4.55 \text{ g NaCl}}{4.55 \text{ g NaCl} + 495.45 \text{ g water}} = \frac{4.55 \text{ g NaCl}}{500 \text{ g solution}}$$

Therefore, for 100 grams of solution, we have: 0.91 g NaCl—1/5 the amount of NaCl; 99.09 g water—1/5 the mass of water; and 100 g solution—1/5 the mass of solution.

▶ 4.55 g NaCl × $\dfrac{1}{5}$ = 0.91 g NaCl.

$$\frac{0.91 \text{ g NaCl}}{0.91 \text{ g NaCl} + 99.09 \text{ g water}} = \frac{0.91 \text{ g NaCl}}{100 \text{ g solution}} = 0.91 \text{ pph NaCl}$$

The concentration of this solution can be expressed as 0.91 g of NaCl per 100 g of solution. The concentration of this solution in mass percent is found as follows, using the total mass of the solution, which is 0.91 g NaCl + 99.09 g of water = 100 g solution:

$$\frac{0.91 \text{ g NaCl}}{100 \text{ g solution}} \times 100\% = 0.91\%$$

The solution is 0.91% NaCl by mass—or it could be expressed as 0.91 pph, 9.1 ppt, or 9,100 ppm.

Consider this second example: One teaspoon of sucrose, which has a mass of 10 g, is dissolved in 240 g of water.

 a. What is the concentration of the solution, expressed as grams of sucrose per 100 g of solution (percent sucrose by mass, or pph)?

 b. What is the concentration in ppt?

These two questions can be answered as follows:

 a. Because we have 10 g of sucrose and 240 g of water, the solution has a total mass of 250 g.

$$\left(\frac{10 \text{ g sucrose}}{250 \text{ g solution}}\right) \times 100\% = 4\% \text{ sucrose by mass}$$
$$(4 \text{ pph sucrose})$$

 b. Because 4/100 = 40/1,000, this solution has a concentration of 40 parts per thousand (ppt) sucrose. Or, reasoning another way, 4% of 1,000 (0.04 × 1,000) is 40—so the concentration is 40 ppt.

1. One teaspoon of table sugar (sucrose) is dissolved in a cup of water. Identify

 a. the solute.

 b. the solvent.

2. What is the percent concentration sucrose (or pph sucrose) in each solution?

 a. 17 g sucrose is dissolved in 183 g water.

 b. 34 g sucrose is dissolved in 366 g water.

3. What does the term "saturated solution" mean?

4. A water sample at 25 °C contains 8.4 ppm dissolved oxygen. The water is heated in an open pan to a higher temperature.

 a. Would the dissolved oxygen concentration in the warmer water be greater or less than that in the original water?

 b. Why? (*Hint:* Refer to Figure 19.)

C.3 OXYGEN SUPPLY AND DEMAND

All animals need oxygen gas (O_2) to survive. Although an oxygen atom is present in every water molecule (the O in H_2O), animals cannot remove this oxygen—the oxygen atom is strongly bonded to the hydrogen atoms. Aquatic organisms such as fish, frogs, insect larvae, and bacteria must have a continuous supply of oxygen gas *dissolved* in the water they inhabit.

Was a shortage of dissolved oxygen gas in the Snake River responsible for the Riverwood fish kill? To explore this possibility, we must consider several factors. How much oxygen gas (or other substances) will dissolve in water? How does water temperature affect the amount of dissolved oxygen? How much oxygen do various aquatic creatures actually consume?

Some oxygen used by fish and other aquatic creatures comes from oxygen gas that dissolves directly into the water from air above the calm water surface. Additional oxygen gas mixes into the water by ***aeration,***

Dissolved oxygen in water is not accessible to humans.

which occurs when water plunges over a dam, flows across boulders in a stream, or breaks as waves on a beach, forming water–air "froth."

Oxygen gas is also added to natural waters through *photosynthesis,* the process by which green plants and ocean plankton make sugars from carbon dioxide and water in the presence of sunlight. During daylight hours, aquatic green plants constantly produce the sugar glucose. Oxygen gas, another product of photosynthesis, is released from aquatic green plants to the surrounding water. The overall chemical equation for the formation of glucose ($C_6H_{12}O_6$) and oxygen gas through photosynthesis is

$$\text{Energy (from sunlight)} + 6\,CO_2 + 6\,H_2O \rightarrow C_6H_{12}O_6 + 6\,O_2$$

The organisms living in a water environment continuously compete for the available oxygen. Oxygen-consuming bacteria (*aerobic bacteria*) feed on solid wastes of larger animals and their dead remains. Aerobic bacteria also feed on certain industry-produced substances found in the water. Such *biodegradable* substances are broken down to simpler substances by these aerobic bacteria.

If the water contains large quantities of biodegradable materials, bacteria thrive and multiply. The resulting bacterial "population explosion" places greater demands on the available dissolved oxygen. Aquatic creatures needing the largest amounts of dissolved oxygen are at greatest risk if the bacterial population gets too large. Their survival is then in question.

The minimum concentration of dissolved oxygen (abbreviated *DO*) needed to support aquatic life depends on the type of animal being considered. Fish, for example, cannot live in water having a dissolved

▶ *A biodegradable material is one that can be broken down by natural organisms in the environment.*

oxygen level less than 0.004 grams per 1,000 g of solution. This concentration of 0.004 parts per thousand (ppt) is more conveniently expressed as four parts per million, 4 ppm (0.004/1,000 = 4/1,000,000).

A concentration as small as four parts per million is difficult to visualize, but we'll try. Assume that a water sample is represented by a million pennies, stacked one on top of another. This stack would rise 1.6 kilometers (one mile) high! Just four pennies in this stack would represent the minimum level of dissolved oxygen needed for fish to survive. It's a very small, but absolutely essential quantity!

If water's dissolved oxygen concentration decreases, the species of fish requiring more oxygen will migrate to other water regions or die. Unfortunately, such species include desirable sport fish, such as pike and trout. Figure 20 summarizes the relative dissolved oxygen requirements of several fish species.

When studying systems in nature, we must consider many factors. This is particularly true in exploring possible reasons for a community problem such as the fish kill in Riverwood. We've considered several factors that would influence the dissolved oxygen concentration. Can you think of any others?

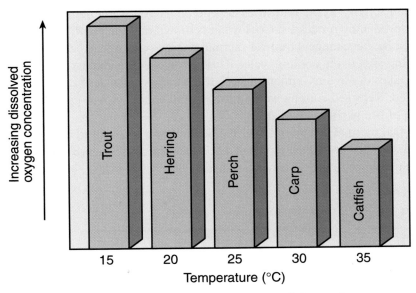

Figure 20 Dissolved oxygen required for various fish species.

C.4 TEMPERATURE AND GAS SOLUBILITY

In Section C.1 we noted that water temperature helps determine the maximum amount of oxygen gas that water can dissolve. This dissolved oxygen concentration, in turn, helps determine water's ability to support oxygen-consuming creatures. Water temperature also affects the amount of oxygen actually needed by aquatic organisms.

Fish are "cold-blooded" animals. That is, their body temperatures rise and fall with the surrounding water temperature. Body temperature affects the fish's metabolism, the complex series of interrelated chemical reactions that keep the fish alive. A 10 °C temperature rise roughly doubles the rates or speeds of many chemical reactions. Cooling a system by 10 °C slows down the rates of such reactions by a similar factor. (Thus, for example, food spoilage—involving complex chemical reactions—is significantly slowed down when foods are stored in a refrigerator rather than left at room temperature.)

If water temperature increases, fish body temperatures will also increase. Chemical reactions inside the fish then speed up; they become more active. Fish eat more and swim faster. As a result, they use up more dissolved oxygen. Bodily processes of aerobic bacteria also speed up in warmer water, thus increasing their oxygen consumption.

During warm summer months, competition among water inhabitants for dissolved oxygen can become quite severe. With rising temperatures, bacteria and fish require more oxygen. But warmer water is unable to dissolve as much oxygen gas. After a long stretch of hot summer days, some streams experience large fish kills, in which hundreds of fish suffocate. Table 5 summarizes the maximum water temperatures at which selected fish species can survive.

High lake or river water temperatures can sometimes be traced to human activity. Many industries depend on natural water bodies to cool heat-producing processes. Cool water is drawn into the plant. Devices called heat exchangers transfer thermal energy (heat energy) from the processing area to the cooling water. The heated water is then released back into lakes or streams, either immediately or after the water has partially cooled. Released cooling water must be cool enough not to upset the balance of life in rivers and lakes.

Did thermal pollution lower the Snake River's dissolved oxygen level, thus leading to the fish kill? Let's examine dissolved oxygen data obtained prior to the Riverwood fish kill.

Table 5 Survival		
	Maximum Temperature	
Fish	**°C**	**°F**
Trout	15	59
Perch and pike	24	75
Carp	32	90
Catfish	34	93

C.5 YOU DECIDE: TOO MUCH, TOO LITTLE?

Your teacher will divide the class into groups of four or five students each. All groups will read the following passage and use the information supplied to complete the exercises. After groups have finished their work, the class will compare their conclusions.

Joseph Fisker of the County Sanitation Commission has measured dissolved oxygen levels in the Snake River for 18 months. These data help the Sanitation Commission monitor the river's water quality. Mr. Fisker takes daily measurements at 9 A.M. under the bridge near Community Hospital, at a water depth of one-half meter.

In addition, Fisker records the water temperature and consults a table to find the DO that would produce a saturated water solution at that temperature.

Let's examine Snake River data for last year and also for this summer (Tables 6 and 7, page 51).

Table 6	Last Year's DOs	
Month	Average Water Temperature (°C)	Average Dissolved Oxygen (ppm)
January	2	12.7
February	3	12.5
March	7	11.0
April	8	10.6
May	9	10.4
June	11	9.8
July	19	9.2
August	20	9.2
September	19	9.2
October	11	10.6
November	7	11.0
December	7	11.0

Table 7	This Summer's DOs	
Month	Average Water Temperature (°C)	Average Dissolved Oxygen (ppm)
June	14	10.2
July	16	9.6
August	18	9.6

GRAPHING HINTS

A graph of experimental data often highlights regularities or patterns among the values. Here are some suggestions on preparing and interpreting such graphs.

- Choose your scale so the graph becomes large enough to fill most of the available graph paper space.

- Each regularly spaced division on the graph paper should equal some convenient, constant value. In general, each interval between graph paper lines should have a value easily "divided by eye," such as 1, 2, 5, or 10, rather than a value such as 6, 7, 9, or 14.

- An axis scale does *not* need to start at "zero," particularly if the plotted values cluster in a narrow range not near zero. For example, if all values to be plotted on the *x* axis are between 0.50 and 0.60, the *x*-axis scale can be drawn as shown in Figure 21.

- Label each axis with the quantity and unit being graphed. For example, a scale might be labeled "Temperature (°C)." Plot each point. Then draw a small circle around each point, like this: ⊙ If you plot more than one curve on the same graph, distinguish each set of points by using a different color or geometric shape, such as △, ▽, or ⊡. After you have plotted all the points, draw the best possible "smooth line" (a line passing through as many points as possible). Give your graph a suitable title.

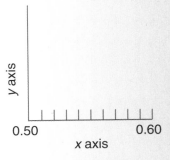

Figure 21 A possible *x*-axis scale.

Table 8	Hourly DOs	
Time	**Water Temperature °C**	**Dissolved Oxygen (ppm)**
8 A.M.	21	9.1
9	21	9.1
10	21	9.1
11	21	9.1
12	22	9.2
1 P.M.	23	9.3
2	23	9.3
3	23	9.2
4	23	9.2
5	23	9.2
6	23	9.2
7	23	9.2
8	22	9.2
9	22	9.2
10	22	9.2
11	21	9.1
12	21	9.1
1 A.M.	21	9.1
2	19	9.0
3	19	9.0
4	19	9.0
5	19	9.0
6	19	9.0
7	19	9.0

1. Prepare a graph of the Snake River's monthly dissolved oxygen levels during last year (Table 6, page 51). Label the *y* axis as dissolved oxygen level in ppm, and the *x* axis as water temperature in °C. How is the dissolved oxygen level related to the water temperature?

2. On the same graph, plot the dissolved oxygen levels and river temperatures collected during this year's summer months (Table 7, page 51). Use a different plotting symbol for this set of points.

3. a. Compare the dissolved oxygen concentrations measured in December and June. How do you explain this difference?

   b. How do you account for the similar concentrations in March and November?

4. Compare the average dissolved oxygen concentration in August of this year with that of August last year. What reasons might explain the difference?

QUESTIONS

In September, soon after the Riverwood fish kill, the County Sanitation Commission invited the Environmental Protection Agency to help with the river water analysis. The EPA sent Marilyn Crocker to measure dissolved oxygen in the Snake River hourly for one day. The goal was to detect any short-term changes in either the temperature or the dissolved oxygen concentration. Ms. Crocker decided to measure dissolved oxygen at the same location used by Mr. Fisker. Her data are listed in Table 8 and graphed in Figures 22 and 23.

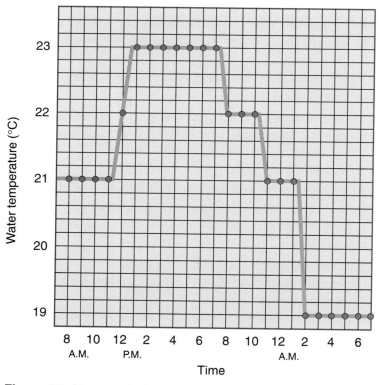

Figure 22 Changes in Snake River water temperature during one day.

1. a. Compare the two graphs. Is any pattern apparent in either graph?

   b. Can you explain any pattern that you detect?

   c. Compare the DO levels during daylight and nighttime hours. How do you account for this difference?

2. Calculate the average water temperature and the average concentration of DO for this one-day period.

3. No daily water temperatures or DO levels were reported during September this year. Thus, the only possible comparison is between the average for one day in September this year and the monthly average for all of September last year. Is this a valid comparison? Why or why not?

4. Now consider the one-day Snake River measurements. Which do you think provides more useful information—the average temperature and DO values, or the maximum and minimum values? Give reasons to support your answer.

5. The DO concentrations needed for saturated water solutions at various temperatures are provided in Table 9. Use this table to decide whether the DO is below, at, or above the saturation level for each measurement in Table 8 (page 52). You'll also need the following formula:

$$\text{Percent of saturation} = \frac{\text{ppm DO measured}}{\text{ppm DO for saturation}} \times 100\%$$

For example, at 8 A.M. during the one-day measurements, the water temperature was 21 °C and the dissolved oxygen concentration was 9.1 ppm. According to Table 9, 9.0 ppm dissolved oxygen is a saturated solution at 21 °C.

Table 9 DO Needed	
Water °C	100% Oxygen Saturation (ppm)
0	14.6
1	14.2
2	13.9
3	13.5
4	13.2
5	12.8
6	12.5
7	12.2
8	11.9
9	11.6
10	11.3
11	11.1
12	10.8
13	10.6
14	10.4
15	10.2
16	9.9
17	9.7
18	9.5
19	9.3
20	9.2
21	9.0
22	8.8
23	8.7
24	8.5
25	8.4

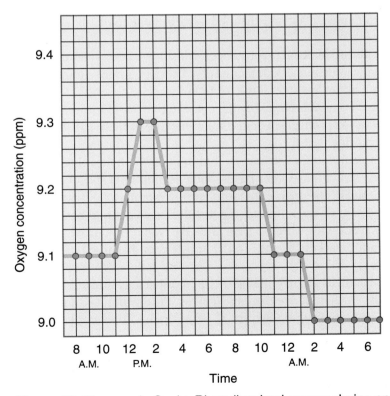

Figure 23 Changes in Snake River dissolved oxygen during one day.

$$\text{Percent of saturation} = \frac{9.1 \text{ ppm}}{9.0 \text{ ppm}} = \times 100\% = 101\%$$

So at 8 A.M., the saturation level of dissolved oxygen was 101%—slightly supersaturated.

6. Acceptable and unacceptable dissolved oxygen levels in river water for fish life are given below:

125% or more
of saturation: Too high for survival of some fish species

80–124%:   Excellent for survival of most fish species

60–79%:   Adequate for survival of most fish species

Below 60%:   Too low; most fish species die

Based on this information, do the collected data suggest that Snake River's dissolved oxygen level range is acceptable for supporting fish life?

7. Is the amount of dissolved oxygen in the Snake River a likely cause of the Riverwood fish kill? Explain your answer.

C.6 ACID CONTAMINATION

We will now turn to a classification scheme that plays a major role in water quality and the life of aquatic organisms—the acidity or basicity of a water sample.

Acids and *bases* can be identified in the laboratory by certain characteristic properties. For example, the vegetable dye litmus turns blue in a basic solution, and red in an acidic solution. Other vegetable dyes, including red cabbage juice, also have distinctively different colors in acid and base solutions.

Many acids are molecular substances. Most have one or more hydrogen atoms that can be released rather easily. These hydrogen atoms are usually written at the left side of the formula for an acid (see Table 10, page 55). Some compounds lacking the characteristic acid formula still dissolve in water to produce an acidic solution. One of these substances is ammonium chloride, NH_4Cl.

Many bases are ionic substances that include the hydroxide ion (OH^-) in their structures. Other bases, such as ammonia (NH_3) and baking soda ($NaHCO_3$, sodium bicarbonate), contain no OH^- ions but still produce basic water solutions. Such substances act like bases because they react with water to generate OH^- ions. Human blood has enough sodium bicarbonate and similar substances dissolved in it to be slightly basic.

Some substances display neither acidic nor basic characteristics. Chemists classify them as *neutral* substances. Pure water, sodium chloride ($NaCl$), and table sugar (sucrose, $C_{12}H_{22}O_{11}$) are all examples of neutral compounds.

When an acid and a base react with each other the characteristic properties of both substances are destroyed. Such a reaction is called *neutralization.*

▶ *Basic solutions are also called "alkaline" solutions.*

Table 10		Some Common Acids and Bases
Name	**Formula**	**Use**
Acids		
Acetic acid	$HC_2H_3O_2$	In vinegar (typically a 5% solution of acetic acid)
Carbonic acid	H_2CO_3	In carbonated soft drinks
Hydrochloric acid	HCl	Used in removing scale buildup from boilers and for cleaning materials
Nitric acid	HNO_3	Used in the manufacture of fertilizers, dyes, and explosives
Phosphoric acid	H_3PO_4	Added to some soft drinks to give a tart flavor; also used in the manufacture of fertilizers and detergents
Sulfuric acid	H_2SO_4	Largest volume industrial chemical; present in automobile battery fluid
Bases		
Calcium hydroxide	$Ca(OH)_2$	Present in mortar, plaster, and cement; used in paper pulping and dehairing animal hides
Magnesium hydroxide	$Mg(OH)_2$	Active ingredient in milk of magnesia
Potassium hydroxide	KOH	Used in the manufacture of some liquid soaps
Sodium hydroxide	$NaOH$	A major industrial chemical; active ingredient in some drain and oven cleaners; used to convert animal fats into soap

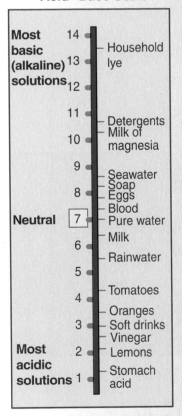

Acid–Base Scale

Figure 24 The pH of some common materials.

The acidic, basic, or neutral character of a solution can be measured and reported using the **pH scale.** Nearly all pH values are in the range of 0 to 14, although some very acidic or basic solutions may be outside this range. At 25 °C, *a pH of 7 indicates a neutral solution. Any pH values less than 7 indicate acids; the lower the pH, the more acidic the solution. Solutions with pH values greater than 7 are basic; the higher the pH, the more basic the solution.* The pH values of some common materials are shown in Figure 24. EPA standards specify that the pH of drinking water supplies should be within the pH range of 6.5 to 8.5.

Rainwater is naturally slightly acidic. This is because the atmosphere includes substances—carbon dioxide (CO_2) for one—that produce an acidic solution when dissolved in water. When acidic or basic contamination causes problems in a body of water, it is usually due to substances or wastes from human activities such as coal mining or industrial processing.

Investigating the Cause of the Fish Kill

Most fish can tolerate a pH range from 5 to 9 in lake or river water. People who fish seriously in freshwater try to catch fish in water between pH 6.5 and pH 8.2. Snake River measurements near Riverwood revealed that the stream pH during the fish kill ranged from 6.7 to 6.9, well within acceptable limits. Thus, we can dismiss an abnormal pH as the cause of the fish kill.

Our next major candidate for investigation as a Snake River contaminant will be so-called heavy metal ions. The following overview provides some useful background.

C.7 IONS AND IONIC COMPOUNDS

We noted earlier that ions are electrically charged particles, some charged positively and others negatively. Positive ions are called **cations,** negative ions **anions.** An ion can be a charged individual atom, such as Na^+ or Cl^-, or a group of bonded atoms, such as NO_3^-, which also possesses an electrical charge.

▶ *This explains why the relative numbers of chloride ions are different in sodium chloride and calcium chloride, as you will soon see.*

Solid sodium chloride, NaCl, consists of equal numbers of positive sodium ions (Na^+) and negative chloride ions (Cl^-) arranged in three-dimensional networks that form crystals (Figure 25). An ionic compound such as calcium chloride, $CaCl_2$ (road de-icing salt), presents a similar picture. However, unlike sodium ions, calcium ions (Ca^{2+}) each have a charge of 2+. Table 11 (page 57) lists the formulas and names of common cations and anions.

The name of an ionic compound is composed of two parts. The cation is named first, then the anion. As the table suggests, many cations have the same name as their original elements. Anions composed of a single atom, however, have the last few letters of the element's name changed to the suffix *-ide.* For example, the negative ion formed from fluorine (F) is fluor*ide* (F^-). Thus, KF is named potassium fluoride.

You can easily write formulas for ionic compounds, if you follow a simple rule. *The correct formula contains the fewest positive and negative ions needed to make the total electrical charge zero.* In sodium chloride the ion charges are 1+ and 1−. Because one ion of each type results in a total charge of zero, the formula for sodium chloride must be NaCl.

When the two ion charges do not add up to zero, add ions of either type until the charges fully cancel. In calcium chloride, one Ca^{2+} ion has a charge of 2+. Each chloride ion has a charge of 1−; two Cl^- ions are needed to equal 2−. Thus, *two* chloride ions (2 Cl^-) are needed for each calcium ion (Ca^{2+}). The subscript 2 written after chlorine's symbol in the formula indicates this. Thus, the formula for calcium chloride is $CaCl_2$.

Formulas for compounds containing ***polyatomic*** (many atom) ions follow this same basic rule. However, if more than one polyatomic ion is needed to bring the total charge to zero, the polyatomic ion formula is enclosed in parentheses before the needed subscript is added. Ammonium sulfate is composed of ammonium (NH_4^+) and sulfate anion (SO_4^{2-}) ions. Two ammonium ions with a total charge of 2+ are needed to match the 2− charge of sulfate. Thus, the formula for ammonium sulfate is written $(NH_4)_2SO_4$.

The following exercises will help you recognize ionic compounds and practice naming and writing formulas for them.

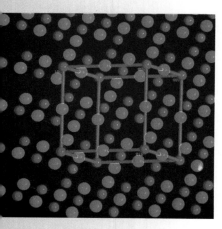

Figure 25 Sodium chloride (NaCl) crystal shown in space-filling modeling. Sodium ions are shown in pink, chloride ions in green.

Table 11 — Common Ions

Cations

+1 Charge		+2 Charge		+3 Charge	
Formula	**Name**	**Formula**	**Name**	**Formula**	**Name**
Na^+	Sodium	Mg^{2+}	Magnesium	Al^{3+}	Aluminum
K^+	Potassium	Ca^{2+}	Calcium	Fe^{3+}	Iron(III)*
NH_4^+	Ammonium	Ba^{2+}	Barium		
		Zn^{2+}	Zinc		
		Cd^{2+}	Cadmium		
		Hg^{2+}	Mercury(II)*		
		Cu^{2+}	Copper(II)*		
		Pb^{2+}	Lead(II)*		
		Fe^{2+}	Iron(II)*		

Anions

−1 Charge		−2 Charge		−3 Charge	
Formula	**Name**	**Formula**	**Name**	**Formula**	**Name**
F^-	Fluoride	O^{2-}	Oxide	PO_4^{3-}	Phosphate
Cl^-	Chloride	S^{2-}	Sulfide		
Br^-	Bromide	SO_4^{2-}	Sulfate		
I^-	Iodide	SO_3^{2-}	Sulfite		
NO_3^-	Nitrate	SeO_4^{2-}	Selenate		
NO_2^-	Nitrite	CO_3^{2-}	Carbonate		
OH^-	Hydroxide				
HCO_3^-	Hydrogen carbonate (bicarbonate)				

Micrograph of sodium chloride crystals (NaCl) at 70× magnification. Note the characteristic cubic geometry.

*Some metals form ions with one charge under certain conditions and a different charge under different conditions. To specify the charge for these metal ions, Roman numerals are used in parentheses after the metal's name.

YOUR TURN

Ionic Compounds

In writing the formula for an ionic compound, first decide how many of each type of ion are needed to make their total electric charge add to zero. Then write the formula for the cation followed by the formula for the anion. Add any needed subscripts and parentheses.

Prepare a chart describing the composition of each ionic compound described below. Your chart should have four columns, as shown in the sample. Refer to Table 11 as needed to complete this exercise.

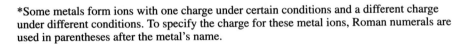

Data Table				
	Cation	**Anion**	**Formula**	**Name**
1.	K^+	Cl^-	KCl	Potassium chloride

(Complete this chart for items 2 through 10 that follow)

1. Potassium chloride is the primary ingredient in the table-salt substitutes designed for people on low-sodium diets. (Answers for this item are already provided in the sample data table; follow the same pattern for the other ionic compounds below.)

2. $CaSO_4$ is a component of plaster.

3. A substance composed of Ca^{2+} and PO_4^{3-} ions is found in some brands of phosphorus-containing fertilizer. This substance is also a major component of bones and teeth.

4. Ammonium nitrate, a rich source of nitrogen, is often used in fertilizer mixtures.

5. Iron(III) chloride is used in water purification.

6. $Al_2(SO_4)_3$ is another substance used to purify water in some municipalities.

7. Baking soda is an ionic compound composed of sodium ions and hydrogen carbonate ions.

8. Magnesium hydroxide is called milk of magnesia.

9. A compound composed of Fe^{3+} and O^{2-} is a principal component of rust.

10. Limestone and marble are two common forms of the compound calcium carbonate.

C.8 DISSOLVING IONIC COMPOUNDS

The dissolving of a substance in water, either from the Snake River or from the tap in your kitchen, is like a tug of war. A solid substance will dissolve if its particles are attracted strongly enough to water molecules to release the particles from the crystal. To dissolve, attractive forces between particles at the solid crystal surface must be overcome for the ions to separate from the crystal and move into the solvent.

Water molecules are attracted to the ions located on the surface of an ionic solid. As we have noted earlier, water molecules have both an electrically negative region (the oxygen end) and an electrically positive region (the hydrogen end). The water molecule's negative (oxygen) end is attracted to the crystal's positive ions. The positive (hydrogen) ends of other water molecules are attracted to the negative ions. When an ionic crystal dissolves in water, the ions leave the crystal and become surrounded by water molecules. That is, they dissolve in the water. Figure 26 (page 59) illustrates such a dissolving process. If positive–negative attractions among cations and anions in the crystal are sufficiently strong, an ionic compound may be only slightly soluble in water.

With this background on ions in solution, you are now prepared to consider whether certain dissolved metal ions were responsible for the Riverwood fish kill.

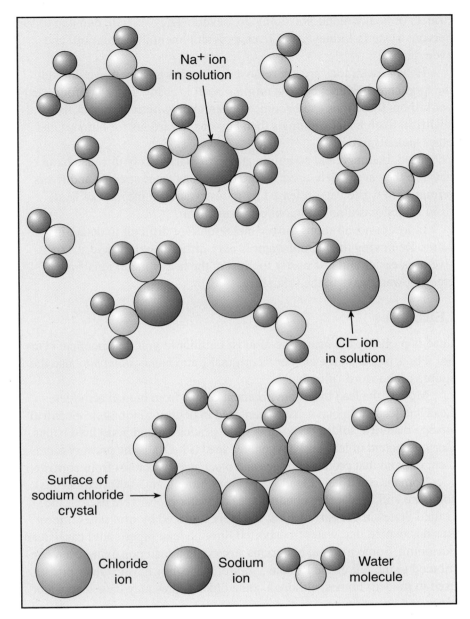

Figure 26 Sodium chloride dissolving in water.

C.9 HEAVY METAL ION CONTAMINATION

Many ions of metallic elements, including iron (Fe^{2+}), potassium (K^+), calcium (Ca^{2+}), and magnesium (Mg^{2+}), are essential to human health. As much as 10% of our requirements for these elements is obtained from minerals dissolved in drinking water.

Other metallic elements, termed ***heavy metals*** because their atoms are generally heavier than those of essential metallic elements, can also dissolve in water as ions. The heavy metal ions causing greatest concern in water are cations of lead (Pb^{2+}), mercury (Hg^{2+}), and cadmium (Cd^{2+}). These elements are toxic, even in small amounts. Their ions can attach to proteins in our bodies, prohibiting the proteins from functioning properly.

Effects of heavy metal poisoning are severe. They include damage to the nervous system, kidneys, and liver, probable mental retardation, and even death.

Heavy metal ions are not removed as wastes as they move through the food chain. Even when the original level is as low as two to four parts per billion, these ions can become concentrated within bodies of fish and shellfish. Such aquatic animals then become hazardous for humans and other animals to eat.

Lead, mercury, and cadmium are particularly likely to cause harm because they are widely used. These metal ions enter streams and rivers primarily in wastewater released by industry and (in the case of lead) from leaded-gasoline automobile exhaust in air.

In low concentrations, heavy metal ions are difficult to detect in water. Removing them from water is very difficult and thus very expensive. It is easier and less costly to prevent the heavy metal ions from entering waterways in the first place.

LEAD (Pb)

Lead is probably the heavy metal most familiar to you. Its chemical symbol, Pb, is based on the element's original Latin name *plumbum,* also the source of the word "plumber."

Most of the lead in our environment comes from human activities. Lead and lead compounds have been used in pottery, automobile electrical storage batteries, solder, cooking vessels, pesticides, and household paints. One compound of lead and oxygen, red lead (Pb_3O_4), is the primary ingredient of paint that protects bridges and other steel structures from corrosion.

Romans constructed lead water pipes more than 2,000 years ago; some are still in working condition. Lead water pipes were used in the United States in the early 1800s, but were replaced by iron pipes after it was discovered that water transported through lead pipes could cause lead poisoning. Some medical historians suspect that lead poisoning even contributed to the fall of the Roman Empire. Copper or plastic water pipes are used in modern homes, although some older homes may have lead water pipes and shower pans.

Until recently a molecular lead compound, tetraethyl lead, $(C_2H_5)_4Pb$, was added to automobile gasoline to produce a better burning fuel. Unfortunately, such lead enters the atmosphere through auto exhaust. The phaseout of most leaded gasoline since the 1970s has reduced lead emissions to date by more than 90%. Even so, some studies suggest that most urban children's primary exposure to lead is through soil and dust that were originally contaminated by use of leaded gasoline. In homes built before 1978, the flaking of leaded paint is another source.

Most drinking water in the United States contains dissolved lead ions at about half the U.S. Environmental Protection Agency (EPA) allowable limit of 0.05 ppm. However, a 1993 EPA report indicated that the water supplies of a significant number of U.S. municipalities exceeded the 0.05 ppm limit for lead.

Most of the lead we take in is excreted in the urine. However, children are particularly at risk in terms of lead poisoning. In 1991, the

Centers for Disease Control lowered the minimum level of lead in a child's blood required for medical intervention from 25 to 10 micrograms per deciliter of blood (0.01 ppm). The EPA estimates that in the United States about one out of six children under the age of six has a blood level of lead above the 10 micrograms per deciliter.

MERCURY (Hg)

Mercury is the only metallic element that is a liquid at room temperature. In fact, its symbol comes from the Latin *hydrargyrum*, meaning quick silver or liquid silver.

Mercury has several important uses, some due specifically to its liquid state. Mercury, an excellent electrical conductor, is used in noiseless light switches. It is also found in medical and weather thermometers, thermostats, mercury-vapor street lamps, fluorescent light bulbs, and in some paints. As a pure liquid, elemental mercury is not particularly dangerous. However, liquid mercury slowly evaporates and its vapor is quite hazardous to health. Thus, you should always avoid direct exposure to mercury.

► *Mercury compounds were eliminated in 1990 from all U.S. indoor latex paints.*

Some mercury compounds are useful as antiseptics, fungicides, and pesticides. You may have used mercurochrome on a wound to kill bacteria. The toxicity of mercury compounds to bacteria extends also to humans. In the eighteenth and nineteenth centuries, mercury(II) nitrate, $Hg(NO_3)_2$, was used in making felt hats. After unintentionally absorbing this compound for many years, hatmakers often suffered from mercury poisoning. Symptoms included numbness, staggered walk, tunnel vision, and brain damage, giving rise to the expression "mad as a hatter."

Organic compounds containing the methylmercury ion (CH_3Hg^+) are highly poisonous. (Organic compounds are composed of carbon atoms attached to other atoms, predominantly hydrogen.) The methylmercury is produced from metallic mercury by **anaerobic bacteria** (bacteria that do not require oxygen). A danger to humans can arise if mercury-containing industrial wastes in acidic river-bottom sediments are converted to these highly toxic methylmercury compounds.

The danger of mercury poisoning from aqueous sources is an occasional local problem where mercury-containing waste is discharged into acidic waters. For example, in the 1950s villagers near Minamata Bay, Japan suffered serious brain and nerve damage, and many died from eating fish contaminated with methylmercury discharged from a local chemical manufacturing plant. Fortunately, the concentration of methylmercury ion in U.S. water supplies is usually quite low.

CADMIUM (Cd)

Cadmium's properties are similar to those of zinc (Zn); the two are usually found together. Galvanized steel is electrically plated with zinc, which normally contains about 1% cadmium. Cadmium is used in photography and in paint manufacturing. It is also a component of nickel–cadmium batteries (called NiCad batteries) and certain special solders.

Cadmium is a very toxic element. Some cases of cadmium poisoning have been traced to cadmium-plated food utensils. In low doses it produces headaches, coughing, and vomiting. With extended exposure, cadmium ions

Collecting water samples.

Investigating the Cause of the Fish Kill

(Cd^{2+}) can accumulate in the liver and kidneys, causing irreversible damage. Cadmium ions may also replace some calcium ions (Ca^{2+}) in bones, resulting in painful bone disorders. Such cadmium poisoning struck several hundred people in northern Japan in the 1960s, when cadmium associated with zinc mine drainage caused cadmium ions to enter a local river.

In the United States, most drinking water contains very little cadmium. Concentrations of cadmium in the U.S. drinking water have been estimated to be less than or about equal to 1 ppb, considerably less than the EPA threshold limit of 10 ppb. However, cadmium ions can leach into groundwater supplies or home water systems from galvanized water pipes and solder (found in older homes), particularly if the water is slightly acidic.

Low cadmium-ion concentrations found in most drinking water supplies do not seem to cause harm, although some evidence connects such concentrations to high blood pressure. Caution is certainly justified, however, because cadmium taken into the body remains there. Thus, bodily cadmium levels increase slowly year by year.

Cadmium creates greater human health problems when it is inhaled than when it is ingested. Tobacco smoke is a major source of inhaled cadmium.

C.10 YOU DECIDE: HEAVY METAL IONS

Following the Riverwood fish kill, County Sanitation Commission members, working with the EPA and other water-quality scientists, gathered concentration data on certain ions in Snake River water. Table 12 lists ions EPA officials decided to measure in the Snake River as possible fish-kill causes. The table also includes levels of these ions measured six months ago, their present levels, and EPA limits for freshwater aquatic and human life.

Table 12	Ion Concentrations in the Snake River			
Ion	Concentration Six Months Ago (ppm)	Present Concentration (ppm)	EPA Limit for Freshwater Aquatic Life (ppm)	EPA Limit for Humans (ppm)
Arsenic	0.0002	0.0002	0.44	0.05
Cadmium	0.0001	0.001	0.0015	0.01
Lead	0.01	0.02	0.074	0.05
Mercury	0.0004	0.0001	0.0041	0.05
Selenium	0.004	0.008	0.26	0.01
Chloride	52.4	51.6	No limit	250.0
Nitrate (as N)	2.1	1.9	No limit	10.0
Sulfate	34.0	35.1	No limit	250.0

1. Which ions decreased in concentration in the Snake River over the past six months?
2. Which ions have increased in concentration over the past six months?

3. Calculate an aquatic life *risk factor* for Snake River ions that have increased in concentration. Use this equation:

$$\text{Risk factor} = \frac{\text{Present concentration of ion}}{\text{EPA concentration limit}}$$

Any ion that exceeds the EPA limit for freshwater aquatic life will have a risk factor larger than 1. Ions below the EPA limit will have risk factors smaller than 1.

4. Which ion has the largest risk factor for freshwater aquatic life?

5. Calculate the risk factor for human life for ions that have increased in concentration.

6. Do you think Riverwood residents should be concerned about possible health effects of ions that have increased in concentration? Why or why not?

Analysis of more extensive Snake River ion-concentration data convinced County Sanitation Commission and EPA representatives that the fish kill was not caused by excessively high concentrations of any hazardous ions. So the Riverwood mystery continues!

In addition to dissolved ionic substances, another class of solutes could serve as the source of Riverwood's water problem. The presence of certain molecular substances in Snake River water might provide the answer to the fish kill. These are the next focus of our investigation.

C.11 MOLECULAR SUBSTANCES IN THE RIVER

Most molecules contain atoms of nonmetallic elements. These atoms are held together by the attraction of one atom's nucleus for another atom's electrons. However, the difference in electron attraction between the two atoms is not great enough for electrons to be transferred from one atom to another. Thus, ions are not formed, as happens when electrons are transferred from metallic to nonmetallic atoms.

Unlike ionic substances, which are solid crystals at normal conditions, a molecular substance may be a solid, liquid, or a gas at room temperature. Molecular substances like oxygen (O_2) and carbon dioxide (CO_2) have little attraction among their molecules; these substances are gases at normal conditions. Molecular substances like alcohol (C_2H_5OH) and water (H_2O) have larger between-molecule attractions; these are liquids at normal conditions. Other molecular substances such as table sugar (sucrose, $C_{12}H_{22}O_{11}$) are solids at normal conditions; these substances have even greater between-molecule attractions.

The attraction of a substance's molecules for each other compared to their attraction for water molecules helps determine how soluble the substance will be in water. But what causes these attractions? The distribution of electrical charge within the molecules has a great deal to do with it.

As you know (page 33), the "oxygen end" of a water molecule is electrically negative compared to its positive "hydrogen end." That is, water molecules are polar. Such charge separation is found in many molecules whose atoms have differing attraction for electrons.

Polar molecules tend to dissolve readily in water, which is a polar solvent. For example, water is a good solvent for sugar and alcohol, which are both composed of polar molecules. Similarly, **nonpolar** substances are good solvents for other nonpolar molecules. For example, nonpolar cleaning fluids are used to "dry clean" clothes. They readily dissolve away nonpolar body oils found in the fabric. By contrast, nonpolar molecules (such as those of oil and gasoline) do not dissolve well in polar solvents (such as water and alcohol).

These general patterns of solubility behavior—polar dissolving in polar, nonpolar dissolving in nonpolar—are often summarized in the statement "like dissolves like." The rule can be applied to explain why nonpolar molecules are ineffective solvents for ionic or polar molecular substances.

Were dissolved molecular contaminants present in Snake River water when the fish died? Most likely yes—at least in some small amounts. Were they responsible for the fish kill? That depends on which molecular contaminants were present and at what concentrations. That, in turn, depends on how each molecular solute interacts with the polar solvent water.

In the following laboratory activity you will compare the solubilities of typical molecular and ionic substances.

C.12 LABORATORY ACTIVITY: SOLVENTS

GETTING READY

In this laboratory activity you will investigate the solubilities of seven solutes in two different solvents—water (H_2O), a polar solvent, and hexane (C_6H_{14}), a nonpolar solvent. You will interpret your observations in terms of "like dissolves like" (see above).

Prepare a data table, as illustrated below, in your laboratory notebook. This table provides names and chemical formulas of the seven solutes you will investigate.

Data Table		
	Solubility in	
Solute	**Water (H_2O)**	**Hexane (C_6H_{14})**
Urea [$CO(NH_2)_2$]		
Iodine (I_2)		
Ammonium chloride (NH_4Cl)		
Naphthalene ($C_{10}H_{18}$)		
Copper(II) sulfate ($CuSO_4$)		
Ethanol (C_2H_5OH)		
Sodium chloride ($NaCl$)		

▶ A nonpolar molecule has an even distribution of electric charge.

▶ The familiar saying "oil and water don't mix" has a chemical basis!

PROCEDURE

1. Using a graduated cylinder, measure and pour 5 mL of water into a test tube. Mark the height of this volume on the test tube with a marking pencil. Place the same height mark on six other test tubes. These marks will allow you to estimate 5-mL volumes without using a graduated cylinder.

2. Add 5 mL of water to each test tube.

3. Test the water solubility of each of the seven solutes by adding a different solute to each test tube. For liquid solutes, add 1 mL (about 20 drops). Transfer a matchhead-size sample of each solid solute, using a metal spatula or wood splint. You must use a wood splint for iodine; discard the splint after use.

4. Gently mix each test tube's contents by firmly "tapping" the tube.

5. Judge how well each solute dissolved in the polar solvent water. Record your observations, using this key: S = soluble; SS = slightly soluble; IN = insoluble.

6. Discard all test tube contents, following your teacher's directions.

7. Wash and thoroughly dry the test tubes.

8. Repeat Steps 2 through 7, using nonpolar hexane as the solvent, rather than water.

9. Wash your hands thoroughly before leaving the laboratory.

Marking 5-mL heights on test tubes.

QUESTIONS

1. Which solutes were more soluble in water than in hexane?

2. Which were more soluble in hexane than in water?

3. Explain the observations summarized by your answers to Questions 1 and 2.

4. Did any solutes produce unexpected results? If so, briefly describe them. Can you suggest reasons for this behavior?

Tapping a forefinger against a test tube mixes its contents.

? PART C: SUMMARY QUESTIONS

1. Explain why a bottle of warm soda produces more "fizz" when opened than a bottle of cold soda does.

2. Explain the phrase "like dissolves like."

3. Why does table salt (NaCl) dissolve in water but not in cooking oil?

4. From each pair below, select the water source more likely to contain the greater amount of dissolved oxygen. Give a reason for each choice.

   a. A river with rapids or a quiet lake

   b. A lake in spring or the same lake in summer

   c. A lake containing only catfish or a lake containing trout

5. Refer to Figure 18 (page 42) to answer the following questions.

  a. What mass of potassium nitrate, KNO_3, must be dissolved in 100 mL of water at 50 °C to make a saturated solution?

  b. We dissolve 50 g potassium nitrate in 100 mL of water at 30 °C. Is this solution saturated, unsaturated, or supersaturated? How would you describe the same solution if it were heated to 60 °C?

6. A water sample has a dissolved oxygen concentration of 9 ppm. What does "ppm" mean?

7. Seawater has a pH of about 8.6. Is seawater acidic, basic, or neutral?

8. Name these ionic compounds:

  a. $NaNO_3$ (used in meat processing)

  b. $MgSO_4$ (Epsom salt)

  c. Al_2O_3 (coating on an aluminum metal surface, protecting the underlying metal)

  d. $BaSO_4$ (used when taking X-rays of the gastrointestinal system)

9. A 35-g sample of ethanol is dissolved in 115 g of water. What is the concentration of the ethanol, expressed as grams ethanol per 100 g solution?

10. What are heavy metals? Why are they such an environmental problem? List some general effects of heavy-metal poisoning.

 # EXTENDING YOUR KNOWLEDGE (OPTIONAL)

• Organic pesticides such as DDT can become concentrated in fatty tissues of animals. Explain this effect using the "like dissolves like" rule. Why don't water-soluble substances become concentrated in fatty tissues in a similar way?

• Prepare a report on the biological magnification (or concentration) of DDT in food chains, and the banning of DDT from use in this country.

As part of its water-quality monitoring program, the U.S. Environmental Protection Agency (EPA) employs a specially modified Huey helicopter to obtain samples along ocean beaches.

D WATER PURIFICATION AND TREATMENT

By now you probably share an interest in resolving the Snake River fish-kill mystery with Riverwood residents. However, you have a bigger stake in the quality of your *own* community's water supply. We will temporarily leave Riverwood to discuss how natural and community purification systems can jointly ensure the safety of community water supplies, including those in your community.

Until the late 1800s, Americans obtained water from local ponds, wells, and rainwater holding tanks. Wastewater was discarded in dry wells or leaching cesspools (pits lined with broken stone), or was just dumped on the ground. Human wastes were usually thrown into holes or receptacles lined with brick or stone. These were either replaced or emptied periodically.

By 1880, about one-quarter of U.S. urban households had flush toilets; municipal sewer systems were soon constructed. However, as recently as 1909, nearly all sewer wastes were released without treatment into streams and lakes from which water supplies were drawn downstream. Many community leaders believed that natural waters would purify themselves indefinitely.

As you might expect, waterborne diseases increased as the concentration of intestinal bacteria in the drinking water rose. As a result, water filtering and chlorinating of water supplies soon began. However, municipal sewage—the combined waterborne wastes of a community—remained generally untreated. Today, with larger quantities of sewage being generated and with extensive recreational use of natural waters, sewage treatment has become essential.

To act intelligently about water use, whether in Riverwood or in our own community, we need to know how clean our water is, how it can be brought up to the quality we require, and how waterborne community wastes can be treated. An emergency or crisis shouldn't be necessary to focus our attention on these issues. Let's look first at ways that water is naturally purified.

D.1 NATURAL WATER PURIFICATION

We noted that many early community leaders believed natural waters would, if left alone, purify themselves. Under some conditions they do!

Fish Kill Remains a Mystery

In a brief communication to the *Riverwood News*, Dr. Harold Schmidt, an EPA scientist investigating the recent Snake River fish kill, reported more negative results. Extensive chemical tests apparently failed to reveal any unusual levels of organic compounds in the river water. Thus, pesticides, fertilizers, and industrial wastes have been ruled out as culprits in the mysterious death of the fish. More details will be reported as they become available.

Nature's water cycle, the hydrologic cycle, was briefly described on page 15. Recall that thermal energy from the sun causes water to evaporate from oceans and other water sources, leaving behind dissolved minerals and other substances carried by the water. Water vapor rises, condenses to tiny droplets in clouds, and, depending on the temperature, eventually falls as rain or snow. It then joins surface water or seeps into the ground to become groundwater. Eventually groundwater may become surface water and evaporate again, continuing the cycle.

Raindrops and snowflakes are nature's purest form of water, containing only dissolved atmospheric gases. Unfortunately, human activities release a number of gases into the air, making present-day rain less pure than it was in earlier times.

When raindrops strike soil, the rainwater quickly loses its relative purity. Organic substances deposited by living creatures become suspended or dissolved in the rainwater. A few centimeters below the soil surface, bacteria feed on these substances, converting organic materials to carbon dioxide, water, and other simple compounds. Such bacteria thus help repurify the water.

As water seeps farther into the ground, it usually passes through gravel, sand, and even rock. Your first laboratory activity in this unit (page 7) demonstrated that gravel and sand can act as a water filter. In the ground, waterborne bacteria and suspended matter are removed by such filtration.

In summary, three basic processes make up nature's water purification system:

- *Evaporation,* followed by *condensation,* removes nearly all dissolved substances.

- *Bacterial action* converts dissolved organic contaminants to a few simple compounds.

- *Filtration* through sand and gravel removes nearly all suspended matter from the water.

Given appropriate conditions, we could depend solely on nature to purify our water. "Pure" rainwater is our best natural supply of clean water. If water seeping through the ground encounters enough bacteria for a long enough time, all natural organic contaminants can be removed. Flowing through sufficient sand and gravel will remove suspended matter. However, nature's system cannot be overloaded if it is to work well.

If groundwater is slightly acidic (pH less than 7) and passes through rocks containing slightly soluble magnesium and calcium minerals, a problem arises. Chemical reactions with these minerals may *add* substances to the water rather than *remove* them. In this case the water may contain a high concentration of dissolved minerals.

Water containing an excess of calcium (Ca^{2+}), magnesium (Mg^{2+}), or iron(III) (Fe^{3+}) ions does not form a soapy lather easily and is therefore called *hard water.* Because hard water can cause a variety of problems (described later), it is important to remove these ions from solution. The process of removing Ca^{2+}, Mg^{2+}, or Fe^{3+} from water, known as *water softening,* results in water that readily forms a lather with soap.

Water cleaned by nature is not always safe to drink.

D.2 LABORATORY ACTIVITY: WATER SOFTENING

GETTING READY

In this laboratory activity you will explore several ways of softening water. You will compare the effectiveness of three water treatments for removing calcium ions from a hard-water sample: sand filtration, treatment with Calgon®, and treatment with an ion-exchange resin.

Calgon (which contains sodium hexametaphosphate, $Na_6P_6O_{18}$) and similar commercial products "remove" hard-water cations by causing them to become part of larger soluble anions, for example:

$$2\ Ca^{2+}(aq) \quad + \quad (P_6O_{18})^{6-} \quad \rightarrow \quad [Ca_2(P_6O_{18})]^{2-}$$

calcium ion   hexametaphosphate   calcium
from hard water   hexametaphosphate ion

Calgon also contains sodium carbonate, Na_2CO_3, which softens water by removing hard-water cations as precipitates such as calcium carbonate, $CaCO_3$ (see below). Calcium carbonate solid particles are washed away with the rinse water.

Another water-softening method involves **ion exchange.** Hard water is passed through an ion-exchange resin like those found in home water-softening units. The resin consists of millions of tiny, insoluble, porous beads capable of attracting positive ions (cations). Cations causing water hardness are retained on the ion-exchange resin; cations that do not cause hardness (usually Na^+) are released from the resin into the water to take their place. We will consider such water-softening procedures in greater detail following this laboratory activity.

Two laboratory tests will help you decide whether the water has been softened. The first involves the reaction between calcium cation and carbonate anion (added as sodium carbonate, Na_2CO_3, solution), forming a calcium carbonate precipitate:

$$Ca^{2+}(aq) \quad + \quad CO_3^{2-}(aq) \quad \rightarrow \quad CaCO_3(s)$$

calcium ion   carbonate ion   calcium carbonate
in hard water   from sodium carbonate   precipitate

The second laboratory test is to note the effect of adding soap to the water to form a lather.

In your laboratory notebook, prepare a data table as illustrated below:

Data Table				
	Filter Paper	**Filter Paper and Sand**	**Filter Paper and Calgon**	**Filter Paper and Ion-Exchange Resin**
Reaction with sodium carbonate (Na_2CO_3)				
Degree of cloudiness (turbidity) with Ivory® soap				
Height of suds				

1. Prepare the equipment as shown in Figure 27. Lower each funnel stem tip into a test tube supported in a test tube rack.

2. Fold four pieces of filter paper; insert one in each funnel. Number the funnels 1 to 4.

3. Funnel 1 should contain only the filter paper; it serves as the control (Hard-water ions in solution cannot be removed by filter paper.) Fill funnel 2 one-third full of sand. Fill funnel 3 one-third full of Calgon. Fill funnel 4 one-third full of ion-exchange resin.

4. Pour about 5 mL of hard water into each funnel. Do not pour any water over the top of the filter paper or between the filter paper and the funnel wall.

5. Collect the filtrates in the test tubes. *Note:* The Calgon filtrate may appear blue due to other additives in the softener. This will cause no problem. However, if the filtrate appears cloudy, some Calgon powder may have passed through the filter paper. In this case, use a new piece of filter paper and re-filter the test tube liquid.

6. Add 10 drops of sodium carbonate (Na_2CO_3) solution to each filtrate. Does a precipitate form? Record your observations. A cloudy precipitate indicates that the Ca^{2+} ion (a hard-water cation) was not removed.

7. Discard the test tube solutions. Clean the test tubes thoroughly with tap water and rinse with distilled water. Do *not* empty or clean the funnels; they're used in the next step.

8. Pour another 5-mL hard-water sample through each funnel, collecting the filtrates in clean test tubes. Each filtrate should be the same volume.

9. Add one drop of Ivory-brand liquid hand soap (not liquid detergent) to each test tube.

10. Stir each tube gently. Wipe the stirring rod before inserting it into another test tube.

11. Compare the cloudiness (*turbidity*) of the four soap solutions. Record your observations. The greater the turbidity, the greater the quantity of soap that dispersed. The quantity of dispersed soap determines the cleaning effectiveness of the solution.

12. Shake each test tube vigorously, as shown by your teacher. The more suds that form, the softer the water. Measure the height of suds in each tube and record your observations.

13. Wash your hands thoroughly before leaving the laboratory.

Figure 27 Filtration apparatus.

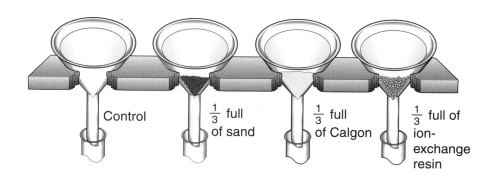

Control $\frac{1}{3}$ full of sand $\frac{1}{3}$ full of Calgon $\frac{1}{3}$ full of ion-exchange resin

QUESTIONS

1. Which was the most effective water-softening method? Suggest why this worked best.

2. What relationship can you describe between the amount of hard-water ion (Ca^{2+}) remaining in the filtrate and the solubility of Ivory-brand liquid soap?

3. What effect does this relationship have on the cleansing action of the soap?

4. Explain the advertising claim that Calgon prevents "bathtub ring." Base your answer on observations made in this laboratory activity.

D.3 HARD WATER AND WATER SOFTENING

You have now investigated several ways to soften hard water, using the calcium ion (Ca^{2+}) as a typical hard-water ion. River water usually contains low concentrations of three hard-water ions—iron(III) (Fe^{3+}), calcium (Ca^{2+}), and magnesium (Mg^{2+}). Groundwater, flowing over limestone, chalk, and other calcium-, magnesium-, and iron-containing minerals, often has much higher concentrations of these hard-water ions (see Figure 28 and Table 13).

Laboratory activity Questions 2 and 3 highlighted some household problems caused by hard water. First, hard water interferes with soap's

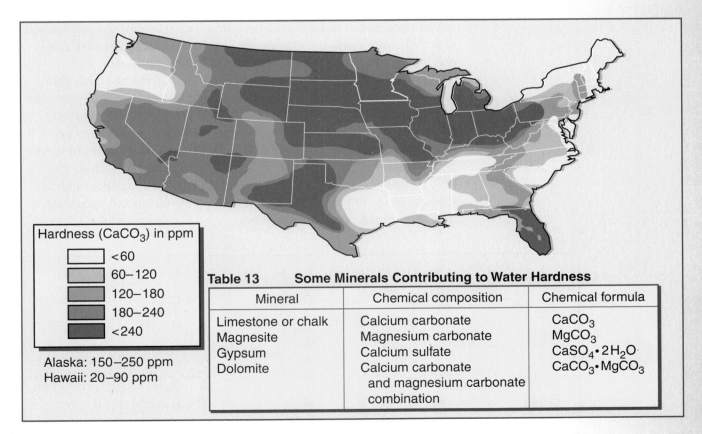

Hardness ($CaCO_3$) in ppm

	<60
	60–120
	120–180
	180–240
	<240

Alaska: 150–250 ppm
Hawaii: 20–90 ppm

Table 13   Some Minerals Contributing to Water Hardness

Mineral	Chemical composition	Chemical formula
Limestone or chalk	Calcium carbonate	$CaCO_3$
Magnesite	Magnesium carbonate	$MgCO_3$
Gypsum	Calcium sulfate	$CaSO_4 \cdot 2H_2O$
Dolomite	Calcium carbonate and magnesium carbonate combination	$CaCO_3 \cdot MgCO_3$

Figure 28 U.S. groundwater hardness.

► The "hardness" of water samples can be classified as follows:

Soft
    < 120 ppm CaCO₃

Moderately hard
    120–350 ppm CaCO₃

Very hard
    > 350 ppm CaCO₃

cleaning action. When soap mixes with soft water, it disperses to form a cloudy solution with a sudsy layer on top. In hard water, however, soap reacts with hard-water ions, forming insoluble compounds (precipitates). These insoluble compounds appear as solid flakes or a sticky scum—the source of a "bathtub ring." This precipitated soap is not available for cleaning. Worse yet, soap curd can deposit on clothes, skin, and hair. The structural formula of this objectionable substance, formed through reaction of soap with calcium ions, is shown in Figure 29.

If hydrogen carbonate (bicarbonate, HCO_3^-) ions are present in hard water, boiling the water causes solid calcium carbonate ($CaCO_3$) to form. This removes undesirable calcium ions and thus softens the water. The solid calcium carbonate, however, produces rock-like scale inside tea kettles and household hot water heaters. This scale (the same compound found in marble and limestone) acts as thermal insulation. Heat flow to the water is partially blocked. More time and energy are required to heat the water. Such deposits can also form in home water pipes. In older homes with this problem, water flow can be greatly reduced.

Fortunately, as the laboratory activity demonstrated, it's possible to soften hard water by removing some calcium, magnesium, or iron(III) ions. Adding sodium carbonate to hard water, (as in the laboratory activity) was an early method to soften water. Sodium carbonate (Na_2CO_3), known as washing soda, was added to laundry water along with the clothes and soap. Hard-water ions, precipitated as calcium carbonate ($CaCO_3$) and magnesium carbonate ($MgCO_3$), were washed away with the rinse water. Other water softeners in common use are borax, trisodium phosphate, and sodium hexametaphosphate (Calgon). As you learned in the laboratory activity, Calgon does not tie up hard-water ions as a precipitate, but rather as an ion that doesn't react with soap.

Synthetic detergents act like soap, but do not form insoluble compounds with hard-water ions. Most cleaning products sold today contain such detergents. Unfortunately, many early detergents were not easily decomposed by bacteria in the environment—that is, they were not biodegradable. At times, mountains of foamy suds were observed in natural waterways. These early detergents also contained phosphate ions (PO_4^{3-}) that encouraged extensive algae growth, choking lakes and streams. Many of today's detergents are biodegradable and phosphate-free; they do not cause these problems.

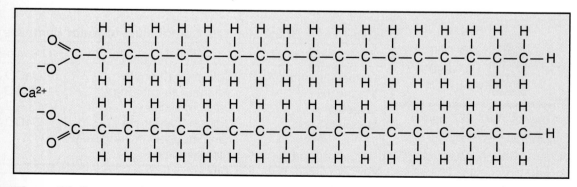

Figure 29 Structural formula of a typical soap scum. This substance is calcium stearate.

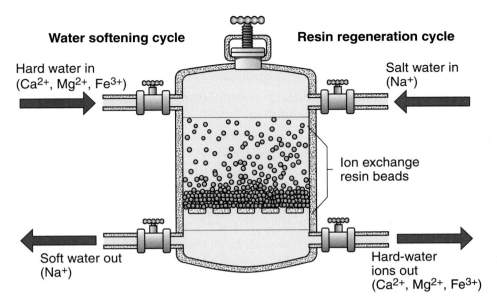

Water softening cycle

Hard water in
(Ca^{2+}, Mg^{2+}, Fe^{3+})

Resin regeneration cycle

Salt water in
(Na^+)

Ion exchange
resin beads

Soft water out
(Na^+)

Hard-water
ions out
(Ca^{2+}, Mg^{2+}, Fe^{3+})

Figure 30 Ion exchange water softener cycles.

The brown stains in this swimming pool are mineral deposits from the water.

▶ *It may not be economical to soften water unless its hardness is more than 200 ppm $CaCO_3$.*

▶ *A water-softening unit typically uses 5–6 pounds of sodium chloride (salt) for one regeneration.*

If you live in a hard-water region, your home plumbing may include a **water softener.** Hard water flows through a large tank filled with an ion-exchange resin similar to the resin you used in the water-softening activity. Initially, the resin is filled with sodium cations. Calcium and magnesium cations in the hard water are attracted to the resin and become attached to it. At the same time, sodium cations leave the resin to dissolve in the water. Thus, undesirable hard-water ions are *exchanged* for sodium ions (Figure 30), which do not react to form soap curd or water pipe scale.

Eventually, of course, the resin fills with hard-water ions and must be **regenerated.** Concentrated salt water (containing sodium ions and chloride ions) flows through the resin. This process replaces hard-water ions held on the resin with sodium ions. The released hard-water ions are washed down the drain with excess salt water. Because this process takes several hours, it is usually completed at night. After the resin is regenerated and water softener valves reset, the softener is again ready to exchange ions with incoming water.

Water softeners are most often installed in individual homes. Other water treatment occurs at a municipal level, both in Riverwood and other cities. How are community water supplies made drinkable? How is wastewater treated before it is returned to the environment? Such questions are our next concern.

D.4 MUNICIPAL WATER PURIFICATION

Today, many rivers are both a source of municipal water and a place to dump wastewater (sewage). Therefore, water must be cleaned twice— once before we use it, and again after we use it. Pre-use cleaning, called **water treatment,** takes place at a municipal filtration and treatment plant.

Figure 31 diagrams typical water treatment steps. Each step in the figure is briefly described below:

- *Screening.* Metal screens prevent fish, sticks, soda cans, and other large objects from entering the water treatment plant.

- *Pre-chlorination.* Chlorine, a powerful disinfecting agent, is added to kill disease-causing organisms.

- *Flocculation.* Crystals of alum—aluminum sulfate, $Al_2(SO_4)_3$—and slaked lime—calcium hydroxide, $Ca(OH)_2$—are added to remove suspended particles such as colloidal clay from the water. Such suspended particles give water an unpleasant, murky appearance. The added substances react to form aluminum hydroxide, $Al(OH)_3$, a sticky, jellylike material which traps the colloidal particles.

- *Settling.* The aluminum hydroxide (with trapped colloidal particles) and other solids remaining in the water are allowed to settle to the tank bottom.

- *Sand filtration.* Any remaining suspended materials that did not settle out are removed here. (This process should remind you of a procedure used to purify your foul-water sample.)

- *Post-chlorination.* The water's chlorine concentration is adjusted so residual chlorine remains in the water, protecting it from bacterial infestation.

- *Optional further treatment.* Depending on community regulations, one or more additional steps might also occur before water leaves the treatment plant.

- *Aeration.* Sometimes water is sprayed into the air to remove odors and improve its taste.

- *pH adjustment.* Well water may be acidic enough to slowly dissolve metallic water pipes. This not only shortens pipe life, but may also cause cadmium (Cd^{2+}) and other undesirable ions to enter the water

Figure 31 Diagram of a municipal water purification plant.

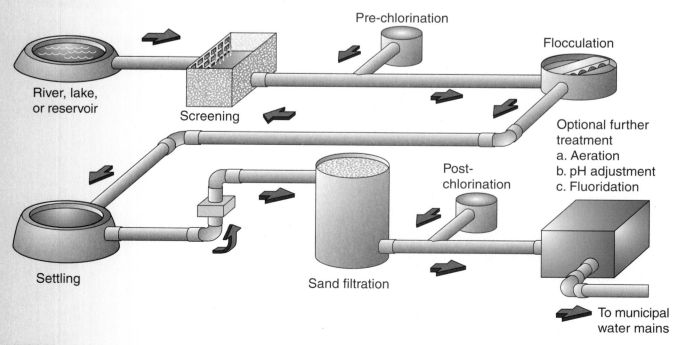

Pre-chlorination

Flocculation

River, lake, or reservoir

Screening

Optional further treatment
a. Aeration
b. pH adjustment
c. Fluoridation

Post-chlorination

Settling

Sand filtration

To municipal water mains

A water treatment plant.

supply. Calcium oxide (CaO), a basic substance, may be added to neutralize such acidic water, thus raising its pH.

- *Fluoridation.* Up to about 1 ppm of fluoride ion (F^-) may be added to the treated water. At this low concentration, fluoride ion can reduce tooth decay, and also reduce cases of osteoporosis (bone-calcium loss among older adults) and hardening of the arteries.

Post-use cleaning of municipal water occurs at a ***sewage treatment plant.*** The major goal of wastewater treatment is to prevent bacteria and viruses in human waste from infecting the public. However, sewage also contains other undesirable materials, including garbage-disposal scraps, used wash water, slaughterhouse and food-packing plant scraps, as well as organic solvents and waste chemicals from homes, businesses, and industry. Ideally, these should all be removed before "used" water is returned to rivers and streams.

Figure 32 (page 76) summarizes the main steps in sewage treatment. Each step is described below.

- *Screening and grit removal.* Sand and gravel are allowed to settle out; other large objects are removed. Smaller debris is ground up. Solid residues are hauled to an incinerator or landfill for burial.

- *Primary settling.* Floating grease and scum are skimmed off and solids are allowed to settle out as sludge.

- *Aeration.* In one method, sewage filters through sludge held on base-ball-sized rocks (or plastic baffles) in an aeration tank. Air circulates between the rocks; there large numbers of aerobic bacteria digest complex organic substances.

- *Final settling.* More sludge settles out. Most of the sludge is aerated, chlorinated, dried, and sent to an incinerator or landfill. Some sludge may be sent back to the aeration tank.

- *Disinfection.* Chlorine or other substances are added to kill germs.

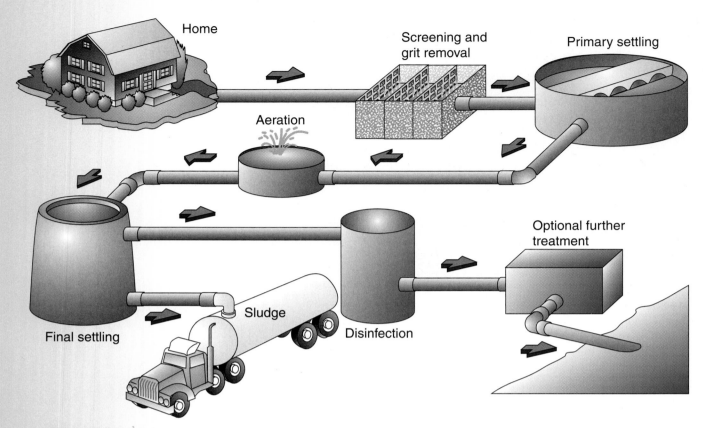

Home

Screening and grit removal

Primary settling

Aeration

Optional further treatment

Final settling

Sludge

Disinfection

Figure 32 Diagram of a municipal sewage treatment plant.

- *Optional further treatment.* The pH of the water may be lowered with carbon dioxide, CO_2. When this gas dissolves in water, carbonic acid (H_2CO_3) is formed, which neutralizes basic compounds in the cleaned-up sewage. In some systems, phosphate ions (PO_4^{3-}) are also removed by precipitation. Heavy-metal ions may be also be removed.

Gas is produced during sludge aeration and digestion. This ***sludge gas*** usually contains about 65% methane (CH_4) and 25% carbon dioxide (CO_2). Methane is a good fuel—it's probably the major component in your community's natural gas supply. In some water-treatment systems the collected methane is burned to heat and dry the sludge residue.

Aerial view of a sewage treatment plant in Jacksonville, Florida.

Chlorine is probably the best known and most common water-treatment substance. It is found not only in community water supplies, but also in swimming pools. Let's examine its use in greater detail.

D.5 CHLORINE IN OUR WATER

The most common cause of illness throughout the world is unhealthful water supplies. Chlorine added to public water supplies has doubtlessly saved countless lives by controlling waterborne diseases. When added to water, chlorine kills disease-producing microorganisms.

In most municipal water treatment systems, chlorine addition (**chlorination**) takes place in one of three ways:

- *Chlorine gas (Cl_2) is bubbled into the water.* Chlorine gas, a nonpolar substance, is not very soluble in water. It does react with water, however, producing a water-soluble, chlorine-containing compound.

- *A water solution of sodium hypochlorite (NaOCl) is added to the water.* (Laundry bleach is also a sodium hypochlorite solution.)

- *Calcium hypochlorite, $Ca(OCl)_2$, is dissolved in the water.* Available both as a powder and small pellets, calcium hypochlorite is often used in swimming pools.

Regardless of how chlorination occurs, chemists believe that chlorine's active form in water is hypochlorous acid, HOCl. This substance forms whenever chlorine, sodium hypochlorite, or calcium hypochlorite dissolves in water.

One potential problem is associated with adding chlorine to municipal water. The following *You Decide* provides the background.

D.6 YOU DECIDE: CHLORINATION AND THMS

Under some conditions, chlorine in water can react with organic compounds produced by decomposing plant and animal matter, forming substances harmful to human health. One group of these substances is known as the trihalomethanes (THMs). One common THM is chloroform ($CHCl_3$), which is carcinogenic. Due to concern over the possible health risks of THMs, the Environmental Protection Agency has placed a limit of 100 ppb (parts per billion) on total THM concentration in municipal water-supply systems. Possible risks associated with THMs must be balanced, of course, against disease-prevention benefits of chlorinated water.

Operators of municipal water-treatment plants can avoid possible THM health risks several ways, but each method has disadvantages:

1. Treatment plant water can be passed through an activated charcoal filter. Activated charcoal can remove most organic compounds from water, including THMs. *Disadvantage:* Charcoal filters are expensive to install and operate.

2. Chlorine could be eliminated altogether. Ozone (O_3) or ultraviolet light could be used to disinfect the water. *Disadvantage:* These methods do not protect the water once it leaves the treatment plant.

Chlorinating tablets are added to swimming pools to rid the water of bacteria and algae.

Treated water can be infected by later addition of bacteria—for example, through faulty water pipes. Also, ozone can pose toxic hazards if not handled and used properly.

3. Pre-chlorination can be eliminated. Chlorine would be added only once, after the water has been filtered and much of the organic material removed. *Disadvantage:* The chlorine added in post-chlorination can still promote the formation of THMs, but to a lesser extent than with pre-chlorination. A decrease in chlorine concentration might also allow bacterial growth in the water.

Your teacher will divide the class into working groups. Your group will be responsible for one of the three alternatives outlined above. Answer these questions:

1. Consider the alternative assigned to your group. Is this choice preferable to standard chlorination procedures? Explain your reasoning.

2. Can you suggest other alternatives beyond the three given above?

Something in the Snake River water was responsible for the fish kill and possible danger to Riverwood residents. Yet, tests for the most common causes of water contamination have not revealed the source of the problem. In fact, the tests have shown that the water is totally safe for human use and contains enough oxygen to support aquatic life. However, the fish kill remains unexplained. Perhaps tomorrow's Riverwood newspaper will finally bring an answer to the mystery!

? PART D: SUMMARY QUESTIONS

1. List and explain the basic steps involved in nature's water purification system, the hydrologic (water) cycle.

2. a. How do municipal water purification and sewage treatment resemble nature's "purification system"?

   b. How are they different from it?

3. a. What advantages does chlorinated drinking water have over untreated water?

   b. What are some disadvantages of using chlorinated water?

EXTENDING YOUR KNOWLEDGE (OPTIONAL)

- During the past century, water-quality standards have become increasingly strict. Discuss several reasons for this trend.

- In recent years many communities have invested in tertiary sewage treatment. Investigate the purpose, design features, advantages, and disadvantages of tertiary treatment. Prepare a written statement that either supports or opposes such water treatment within a community.

- High concentrations of iron(III) ions are undesirable in drinking water. Write a balanced equation showing how these ions might be removed by a sodium ion-exchange resin.

- Prepare a report on the growing controversy over the use and abuse of groundwater supplies within the United States.

E PUTTING IT ALL TOGETHER: FISH KILL— WHO PAYS?

Fish Kill Cause Found; Meeting Tonight

By Karen O'Brien
Staff contributor to *Riverwood News*

The massive fish kill in the Snake River was caused by "gas bubble disease," Mayor Edward Cisko announced at a news conference early today. Because gas bubble disease is noninfectious, humans are at no risk from the water.

Accompanying Mayor Cisko at the conference was Dr. Harold Schmidt of the Environmental Protection Agency. Dr. Schmidt explained that the disease is caused by an excess of air dissolved in the river water. "The excess dissolved air, mostly oxygen and nitrogen, passes through the fish's gills, where it forms gas bubbles. Consequently, less oxygen circulates through the fish's bloodstream. The fish usually die within a few days if the situation is not corrected."

Dr. Schmidt dissected sample fish within a short time after death and found evidence of such gas bubbles, providing positive identification of the disease.

Dr. Schmidt gave assurances that river water containing excess air is not harmful to human health, and that the town's water supply is "fully safe to drink."

Mayor Cisko refused comment on reasons for the water condition, saying, "The cause is still under investigation." But an informed source close to the mayor's office stated that "the most likely cause is the power company's release of water from the dam upstream from Riverwood." The mayor's secretary confirmed that power company officials will meet with the mayor and his staff later today.

Mayor Cisko also invited the public to a special town council meeting scheduled for 8 P.M. tonight in Town Hall. The council will discuss who is responsible for the fish kill, and who should pay the costs associated with the three-day water shutoff. Several area groups plan to make presentations at tonight's meeting.

Editorial:
Attend the Special Council Meeting

Tonight's special town council meeting could result in important decisions affecting all Riverwood citizens. The meeting will address two primary questions: Who is responsible for the fish kill? Who should pay the expenses involved in trucking water to Riverwood during the three-day water shutoff? These questions have financial consequences for all town taxpayers.

Those testifying at tonight's public meeting include power company officials, scientists involved in the river water analy-ses, and engineers from an independent consulting firm familiar with power plant design. Chamber of Commerce members representing Riverwood store owners, representatives from the County Sanitation Commission, and the Riverwood Taxpayers Association will also make presentations.

We urge you to attend this meeting. Many unanswered questions remain. Was the fish kill an "act of nature" or was some human error involved? Was there negligence? Should the town's business com-munity be compensated, at least in part, for financial losses resulting from the fish kill? If so, how should they be compensated, and by whom? Who should pay for the drinking water brought to Riverwood? Can this situation be prevented in the future? If so, at what expense? Who will pay for it?

This newspaper will devote a large portion of the "Letters to the Editor" column in coming days to publish your comments on our community's recent water crisis.

Community council meeting, Austin, Texas.

You will be assigned to one of six "special interest" groups who will testify at the special meeting of the Riverwood Town Council. You will receive some suggestions on information to include in your group's presentation. Consider these suggestions only as a starting point. Identify other points you intend to stress.

Use your planning time to select a group spokesperson and to organize the presentation. Consider preparing written notes. Each group will have two minutes to present its position and one minute for rebuttal. Failure to stay within presentation time limits will result in loss of rebuttal time. Students assigned to the town council group will serve as official timekeepers.

Following the meeting, each group (or, alternatively, each student) will write a letter to the editor of the Riverwood newspaper suggesting answers to questions posed in the editorial reprinted in this section.

Rather than preparing a letter to the editor, your teacher may ask each group to prepare for a simulated television interview. In this case, a television interview team will question one spokesperson from each group. Questions will be based, at least in part, on those raised in the newspaper editorial.

After preparing the letters to the editor or holding the television interviews, viewpoints expressed in the letters or interviews will be compared and discussed by the entire class.

E.1 DIRECTIONS FOR TOWN COUNCIL MEETING

MEETING RULES AND PENALTIES FOR RULE VIOLATIONS

1. The order of presentations is decided by council members and announced at the start of the meeting.

2. Each group will have two minutes for its presentation. Time cards will notify the speaker of time remaining.

3. If a member of another group interrupts a presentation, the offending group will be penalized 30 seconds for each interruption, to a maximum of one minute. If the group has already made its presentation, it will forfeit its rebuttal time.

INFORMATION FOR TOWN COUNCIL MEMBERS (TWO STUDENTS)

Your group is responsible for conducting the meeting in an orderly manner. Be prepared to:

1. Decide and announce the order of presentations at the meeting. Groups presenting factual information should be heard before groups voicing opinions.

2. Explain the meeting rules and the penalties for violating those rules (see above).

3. Recognize each special interest group at its assigned presentation time.

4. Enforce the two-minute presentation time limit. One suggestion: Prepare time cards with "one minute," "30 seconds," and "time is up," written on them. These cards, placed in the speaker's line of sight, can serve as useful warnings.

INFORMATION FOR POWER COMPANY OFFICIALS

The following will help with your presentation:

Normally, only small volumes of water are released from the dam at any particular time. However, releasing large quantities of water from the dam is a standard way to prevent flooding. The potential for flooding was increased due to unusually heavy rains in the area this past summer. The last time such a large volume of water was released from the dam was 30 years ago. A fish kill was reported then but the cause remained unknown. On that occasion, Riverwood and surrounding areas also had experienced an unusually wet summer.

The dam, constructed in the 1930s, had the most current design of that time. Its basic design has not been altered since it was constructed.

INFORMATION FOR SCIENTISTS

The following will help with your presentation:

Gas bubble disease is a noninfectious disease caused by excessive gas dissolved in water. When the quantity of dissolved gases—primarily oxygen and nitrogen—reach a *combined* total of 110–124% of saturation (see page 54), gas bubble disease symptoms can occur in fish. However, such water causes no known harm to humans.

The most dangerous component to fish is excess nitrogen gas dissolved in the water. Fish metabolism can partially reduce the effect of excess oxygen, but there is no way to handle excess nitrogen in blood.

Fish die because the supersaturated gases in the water produce gas bubbles in fish blood and tissues. These bubbles often form in blood vessels of the gills and heart. Blood is unable to circulate throughout the fish's bodies, and death results. Some fish varieties also develop distended (bloated) air bladders. Death can occur from a few hours to several days after gas bubble formation.

A definitive indicator of gas bubble disease is the presence of gas bubbles in the gills of dead fish. However, because some gas bubbles disappear rapidly after fish die, prompt dissection and analysis are required.

Supersaturation of water with oxygen and nitrogen gas may occur near dams and hydroelectric projects as released water forms "froth," trapping large quantities of air. Water at the dam's base may contain oxygen and nitrogen dissolved up to 139% of saturation. Even 90 km (50–60 miles) downstream, gas supersaturation may be as high as 111% of saturation. The Environmental Protection Agency limit for combined oxygen and nitrogen supersaturation in rivers is 110%. Specially designed spillways, providing gradual water release from dams, may substantially lessen or even prevent such gas supersaturation in river water.

INFORMATION FOR ENGINEERS

The following will help with your presentation:

Engineers can predict whether large quantities of air will be trapped in water released from a dam spillway. This involves knowledge of the spillway's physical structure and the water volume that will be released.

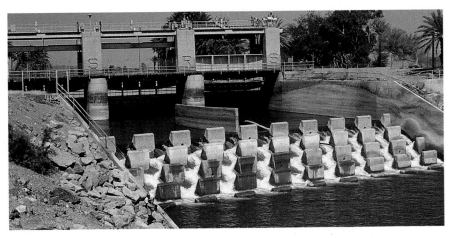

Laguna Dam, Arizona. Sluice gates and retaining rocks on the Colorado River.

The U.S. Corps of Engineers has investigated whether operational or structural changes in dams might reduce the chance of gas supersaturation in released water. Their main goal has been to find ways to reduce the water volume released from spillways and to prevent released water from plunging deeply beneath the river's surface. When water plunges to great depths the increased pressure can force greater amounts of air to dissolve in the water (see page 44).

Three specific suggestions might help Riverwood. The first is to enlarge the reservoir located upstream from the dam. This would provide greater water storage capacity during times of heavy rain and runoff. Water could thus be released from the spillway in smaller quantities, decreasing formation of high levels of gas supersaturation.

The second suggestion is to start a major fish-collecting operation upstream from the dam. Collected fish would be trucked around the dam and released at downstream locations where supersaturation levels are low enough to ensure fish survival.

The third suggestion is to install deflectors on the spillway's downstream side. Spillway deflectors prevent released water from plunging to great depths (see Figure 33). Deflectors release the water at the river surface where the chance of large quantities of air dissolving is reduced.

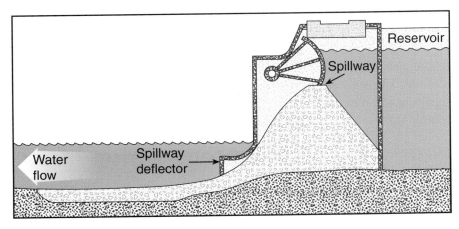

Figure 33 Spillway and deflector.

INFORMATION FOR CHAMBER OF COMMERCE MEMBERS

The following will help with your presentation:

Canceling the annual fishing tournament cost Riverwood merchants a substantial sum. Nearly a thousand out-of-town tournament participants were expected. Many would have rented rooms for at least two nights and eaten at local restaurants and diners. In anticipation of this business, extra help and food supplies were arranged. Fishing and sporting goods stores stockpiled extra fishing supplies. Some businesses have applied for short-term loans to help pay for their extra, unsold inventories.

Local churches and the high school planned family social activities as revenue-makers during the tournament. The school band, for instance, planned a benefit concert for the tournament weekend. The concert would have raised money to send band members to the spring state band competition.

People are likely to remember the fish kill for many years. Tournament organizers predict that fishing competition revenues in Riverwood will be substantially reduced in the future because of this year's adverse publicity. Thus, *total* financial losses resulting from the water emergency may be much higher than most current estimates.

INFORMATION FOR COUNTY SANITATION COMMISSION MEMBERS

The following will help with your presentation:

Analyses of Snake River water after the fish kill (Table 12, page 62) confirmed the safety of Riverwood drinking water according to Environmental Protection Agency guidelines. However, one water contaminant, selenium, significantly increased in concentration since earlier river tests. Despite this increase, selenium's 0.008-ppm level is still below the EPA's 0.01-ppm limit.

The main source of selenium in water is soil runoff. Selenium's soil concentration depends on the soil's geologic origin. Selenium-containing soil near Riverwood came from the debris of prehistoric volcanic eruptions which carried selenium from deep within the Earth's crust to its surface.

Selenium, in trace amounts, is essential to human and animal health. However, if ingested in excessive amounts, its toxic effects are similar to those of arsenic poisoning.

The estimated safe and adequate daily human dose of selenium is 0.05 mg to 0.20 mg. It is contained in foods such as wheat, asparagus, and seafood. The body needs selenium to assist enzymes in protecting and repairing cell membranes. Some authorities think that selenium may—at least in trace amounts—help protect cells against cancer.

INFORMATION FOR RIVERWOOD TAXPAYERS ASSOCIATION MEMBERS

The following will help with your presentation:

Who will pay for the water brought into Riverwood during the water shutoff?

Will taxes be increased to compensate local business people for the financial losses they experienced? (Keep in mind that local merchants themselves are likely to be Riverwood taxpayers!)

If the power company redesigns the dam's spillway, will construction costs be passed on to taxpayers? If so, how?

To what extent (if at all) should information presented by other groups influence your group's opinion regarding who pays? You may find it useful, if possible, to obtain written briefs from the other groups prior to the council meeting. What other sources of information might be useful?

E.2 LOOKING BACK AND LOOKING AHEAD

The Riverwood water mystery is finally solved. It's ironic that the water "contaminants" responsible for the fish kill and for the town's understandable concern are excessive amounts of oxygen and nitrogen gas dissolved in the water. Neither substance qualifies as toxic or dangerous—after all, we live immersed in an atmosphere of these gases.

However, deep sea divers have long feared and respected the hazard known as the "bends"—the formation of nitrogen gas bubbles in a diver's blood if bodily pressure drops too quickly in moving back to the water surface. Thus, even for humans, a substance as seemingly harmless as nitrogen gas can pose life-threatening risks under certain conditions.

Although the analogy is not perfect, the gas bubble disease that killed the fish can be considered a form of the bends. In fact, the same principle that explains "fizzing" in a freshly opened carbonated beverage accounts for both hazardous situations.

Were the analyses of the Snake River water a waste of time or money? Of course not. Even though the tests failed to locate a probable cause for the fish kill, they eliminated several major possibilities, and thus narrowed the field of investigation. In scientific research negative results are often as important and useful as positive ones.

In the case of the Riverwood investigation, one unexpected finding has given chemists an "early warning" of a possible future water-quality problem. The selenium concentration in the Snake River, although not responsible for the fish kill, has increased enough over the past months to merit serious attention and action by authorities. If the fish kill had not happened, would this potential problem have been identified this soon? Possibly. But it's not unusual for a search for a solution to one scientific or technological problem to uncover one or more new problems. In fact, this helps explain why scientific work is considered so challenging and interesting by those who work as scientists and technicians.

Of course, Riverwood and its citizens only exist on the pages of this book. Their water-quality crisis, however, could be real. And the chemical facts, principles, and procedures that clarified their problem and its solution also apply to many experiences in your own residence and community.

Although the fish-kill mystery has been solved, our exploration of chemistry has only just begun. Many issues involving chemistry in the community remain, and water and its chemistry are only one part of a larger story.

The fishing season can resume at Riverwood.

Conserving Chemical Resources

- ▶ Do we have enough resources to meet future human needs?

- ▶ How is chemistry used to extract metals from ores?

- ▶ How can chemistry help create new alternatives for scarce materials?

- ▶ What is being done about the growing problem of waste disposal?

- ▶ How does chemistry explain the similarities and differences among substances?

INTRODUCTION

During a lifetime, a typical U.S. citizen uses an estimated 26 million gallons of water, 52 tons of iron and steel, 6.5 tons of paper, 1,200 barrels of petroleum, 21,000 gallons of gasoline, 50 tons of food, 5 tons of fertilizer, plus a wide variety of other chemical resources. By contrast, members of ancient societies used only relatively tiny fractions of natural resources such as forests, metal ores and minerals, and animals and plants. The chemical resources removed from the earth were miniscule compared to the amounts remaining.

Human longevity and populations grew as humans gained knowledge and skills in farming, metallurgy, crude manufacturing, and medicine. Gradually, people began organizing their discoveries about the natural world, and applied this organized knowledge to invent devices and techniques to make life easier. For example, early metalworkers learned the properties of metals and used them wisely, combining them in proportions that gave the best results. Their work helped give rise to *science*, the gathering, analyzing, and organizing of knowledge about the natural world. Science, in turn, aided ***technology***, applications of science in converting natural resources into goods and services for human needs.

Science and technology have advanced together rapidly. Today, the amount of scientific knowledge doubles in less than a decade. Technology seems to change even more rapidly. Scientific and technological advances have given us stainless steel and polyester fiber, video games and laptop computers, space shuttles and brain scanners, automobile air bags and compact disk players, fast food and diet soft drinks.

Industries that make these products have grown rapidly, and have an increasing appetite for raw materials. In the United States, just 5% of the world's population uses more than half of the world's processed resources. In fact, during the last 40 years, the United States has consumed more fossil fuels (petroleum, natural gas, and coal) and metal ores than were used by all of the world's inhabitants up to that time.

Resources are used in producing many different goods.

These advances in science and technology can be put into perspective by focusing on energy, an essential part of processing Earth's resources. Egyptian pharaohs used the energy of 100,000 laborers to build the Pyramids. In building the Great Wall, Chinese emperors used the energy of two million workers. To keep you warm and clothed, grow and process your food, transport you and your necessities, and entertain you during leisure hours, you have at your personal service enough energy to equal the labor of 200 full-time workers.

But, these services are costly. They have taken vast amounts of resources from Earth. In the process, we have greatly altered it, generating large quantities of waste materials and changing landscapes. The combined effects of mining, processing, using, and discarding materials have caused environmental damage.

We gradually have become more aware of the consequences of using resources. Knowing those consequences, we must learn to balance our withdrawals by returning materials to the environment in useful—or at least not harmful—forms. Only then can the world's resources continue to support its inhabitants.

In this unit, you will explore some chemical principles and issues surrounding our use of resources. You will also become more familiar with some materials that add either richness or discomfort to your life, and you will learn how to help conserve valuable materials.

Human power can be an energy source today as well.

Many of the goods sold here are provided by the work of chemists and chemical engineers.

How much is a penny worth? "One cent" is certainly correct. But then again, maybe it's not. It all depends! Confused? Read on.

CHEMQUANDARY

A PENNY FOR YOUR THOUGHTS

Consider this article:

Zinc Made More Cents

By Karen Chapman
Staff writer for *The Alexandria Post*

After remaining largely unchanged since the first release in 1909, the Lincoln Head penny lost a full 20% of its mass in 1982. The change (no pun intended) was a money-saving measure. Less expensive and lower density zinc was substituted for some of the original copper. Copper prices had risen so much that the copper in a penny was worth more than one cent. The pre-1982 pennies were 95% copper and 5% zinc. Current pennies are 97.6% zinc, coated with a thin copper layer.

When a coin's material value becomes greater than its face value, two undesirable things can happen. First, the U.S. Treasury loses money in manufacturing the coin. When the content of the penny was changed in 1982, it was estimated that the United States would save $25–50 million annually. Second, individuals may remove the coins from circulation, selling them—at a profit—for their metal content.

In light of this information, how much *is* a penny of today really worth?

As in the case of pennies, chemists are often asked to find ways to substitute less costly, more plentiful resources for scarce and expensive ones, without sacrificing key advantages of the original materials. The new pennies are almost as durable as the old pennies, because the hardy copper plating protects the more chemically active zinc from the wear and tear of daily use.

As the ChemQuandary suggested, "worth" of coins has at least two meanings. If the values of the metals in a coin are considered, a penny may be—depending on the current market value for the metals—worth more or less than one cent. As money, of course, a penny is still just a penny.

The U.S. Mint is able to make what is essentially a zinc coin look like a "copper penny." Let's see if you can use some chemistry to alter the appearance of a metallic coin.

▶ To some hobbyists, U.S. pennies (and coins in general) have a third kind of "worth." What is it?

A.1 LABORATORY ACTIVITY: STRIKING IT RICH

GETTING READY

Seeing is believing—or so it is said. In this activity the properties of a metal will appear to change. You will change the appearance of some pennies by heating them with zinc (Zn) metal in a zinc chloride ($ZnCl_2$) solution.

In your laboratory notebook prepare a table like that below, leaving plenty of room for your observations.

Data Table	
Condition	**Appearance**
Untreated penny	
Penny treated with Zn and $ZnCl_2$	
Penny treated with Zn and $ZnCl_2$ and heated in burner flame	

You will use a Bunsen burner to heat the third penny. Before starting the laboratory procedure, examine your burner and identify the parts shown in Figure 1.

Then practice lighting and adjusting the burner as follows:

1. Attach the burner hose to the gas outlet (which should be turned off).
2. Close the air valve and gas valve on the burner.
3. Open the gas outlet valve.
4. Light a match or have your striker ready.
5. Open the burner's gas valve and light the burner immediately by bringing the lighted match to the top of the barrel or sparking the striker.
6. Adjust the burner flame height using the gas valve.
7. Adjust the flame temperature by rotating the air valve. The hottest part of the flame is just above the inner (blue) cone, which is about 2–3 cm high at a relatively high air setting. Cooler flames (lower air settings) have a smaller or undefined inner cone and often include orange-colored regions.
8. Close the gas outlet to extinguish the flame.

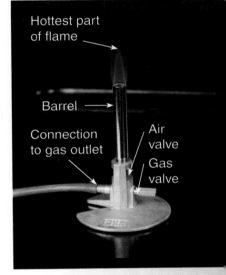

Figure 1 A Bunsen burner.

Hottest part of flame

Barrel

Connection to gas outlet

Air valve

Gas valve

▶ Be sure to keep your head and hair away from the burner flame.

PROCEDURE

1. Obtain three pennies and use steel wool to clean each surface until it is shiny. Record the appearance of the pennies.

2. Assemble a ring stand with a ring clamp and wire gauze. As shown in Figure 2 (page 93), place a 150-mL beaker on the wire gauze and surround the beaker with a second ring clamp so that the beaker cannot tip over.

3. Weigh a 2.0–2.2 g sample of granulated zinc (Zn) or zinc foil. Place it in the beaker.

4. *Caution:* Zinc chloride solution can damage skin. If any of it gets on your skin, immediately wash the affected area with cold tap water. Notify your teacher. Use a graduated cylinder to measure 25 mL of 1 M zinc chloride ($ZnCl_2$) solution. Add the solution to the beaker containing the zinc metal.

5. Cover the beaker with a watch glass. Gently heat the solution until it just begins to bubble, then lower the Bunsen burner flame to continue gentle bubbling. Do not allow the solution to boil vigorously or become heated to dryness.

6. Using forceps or tongs, carefully lower two clean copper pennies into the solution in the beaker. Do not drop the coins into the solution; avoid causing a splash. Put the watch glass on the beaker and keep the solution gently boiling for approximately two to three minutes. You will notice a change in the appearance of the pennies during this time.

7. The third penny is kept out of the solution so that it can act as a **control**—an untreated sample that can be compared with the treated coins.

8. Fill two 250-mL beakers with distilled water.

9. With forceps or tongs, remove the two pennies from the solution. Place them both in one beaker of distilled water. Turn off the burner, but do not discard the zinc chloride solution.

10. Using forceps or tongs, remove the coins from the beaker of water. Rinse them under running tap water, then dry them gently with a paper towel. Set one treated coin aside for later comparisons.

11. Briefly heat the other treated, dried coin in the outer cone of the burner flame, holding it with forceps or tongs as shown in Figure 3 (page 93). Heat the coin only until you observe a change, which will take 10–20 seconds. Do not overheat.

12. Immediately immerse the heated coin in the second beaker of distilled water. Record your observations.

13. Remove the coin from the beaker of water and gently dry the coin with a paper towel.

14. Observe and compare the appearance of the three pennies. Record your observations.

15. When finished, discard the used zinc chloride solution and the used zinc, as directed by your teacher.

16. Wash your hands thoroughly before leaving the laboratory.

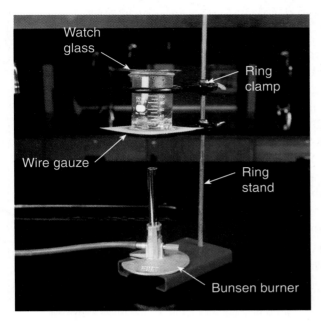

Figure 2 Setup for reaction between copper, zinc, and zinc chloride. The watch glass prevents the solution from splashing out of the beaker.

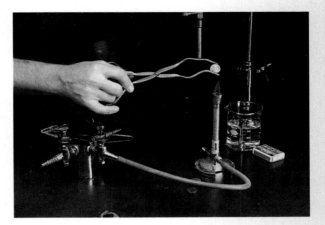

Figure 3 Heating the zinc–zinc chloride treated coin. Using tongs, hold the coin in the *outer* cone of the burner flame.

QUESTIONS

1. a. Compare the colors of the three coins—untreated, heated in zinc chloride solution only, and heated in zinc chloride solution and in a burner flame.

   b. Do the treated coins now look like other metals? If so, explain.

2. If someone says that a precious metal was made in this activity, how can you decide whether it was?

3. Identify two or three practical uses for the metallic changes you observed in this activity.

4. a. What happened to the copper atoms originally present in the treated pennies?

   b. Could the treated pennies be converted back to normal coins? If so, indicate how this could be done.

A.2 USING THINGS UP

In the laboratory activity you just completed, copper seemed to "disappear." Many things we use daily also seem to disappear. For example, fuel in the gasoline tank is depleted as a car speeds along the highway. The ice cream we eat seemingly vanishes. Steel in automobile bodies rusts away.

However, even though the original forms of such materials disappear when they are used, *the atoms composing them remain.*

In the laboratory activity, the copper pennies became coated (plated) with zinc. When heated, copper atoms mixed with zinc atoms to form the mixture we call brass. The original copper and zinc atoms were unchanged, though.

Gasoline burns (reacts with oxygen gas from air) in a car's cylinders, producing energy to propel the vehicle. During burning, carbon and hydrogen atoms from gasoline react with oxygen atoms from air, forming carbon monoxide (CO), carbon dioxide (CO_2), and water (H_2O). These products exit through the car's exhaust system. The original atoms of carbon, hydrogen, and oxygen have not been destroyed, but have been rearranged in new molecules.

As ice cream is digested, atoms from sugars and fats in the ice cream combine with other substances, providing both energy and new compounds for the body. Similarly, even though a steel automobile frame may become pitted by corrosion, the iron atoms can still be found in the rust, a compound of iron and oxygen (Fe_2O_3).

In short, "using things up" means chemically *changing* materials rather than *destroying* them. Nothing is ever really lost in chemical changes if atoms are counted. Sometimes resources can be reclaimed, as in the recovery of aluminum from recycled cans. Other times, the atoms and molecules are so scattered they cannot be efficiently converted. For example, the products of burned gasoline become dispersed in air.

The **law of conservation of matter,** like all scientific laws, summarizes what has been learned by careful observation of nature: *In a chemical reaction, matter is neither created nor destroyed.* Molecules can be converted and decomposed by chemical processes, but atoms are forever. Matter—at the level of individual atoms—is always fully accounted for. Because nature is an exacting bookkeeper, chemical equations must always be balanced. Balanced equations help us represent on paper how nature accounts for all atoms in chemical reactions.

A.3 TRACKING ATOMS

Coal is mostly carbon (C). If carbon completely burns, it combines with oxygen gas (O_2) and produces carbon dioxide (CO_2). Here is how this carbon-burning reaction is written chemically:

$$C(s) \quad + \quad O_2(g) \quad \rightarrow \quad CO_2(g)$$

carbon   and   oxygen   react to   carbon
(coal)   gas   produce   dioxide

► *In chemical equations, (s), (l), and (g) indicate solid, liquid, and gas, respectively. These are the physical states of the substances at the conditions of the reaction.*

Formulas for the two *reactants* (C and O_2) are placed to the left of the arrow ($\rightarrow$); the formula for the *product* (CO_2) appears on the right. The state of each substance is sometimes added as well.

Besides indicating which substances react and are formed, a chemical equation is a "bookkeeping" statement: *All atoms appearing in the products originally came from the reactants.* Producing one molecule of carbon dioxide, CO_2, requires one carbon atom and two oxygen atoms. The carbon atom is furnished by the coal; the two oxygen atoms come from one **diatomic** (two-atom) oxygen molecule, O_2.

When copper (Cu) metal is exposed to the atmosphere, it can take part in several chemical reactions. One is the formation of copper(I) oxide, Cu_2O. This substance is not molecular; it contains no molecules. In contrast, it is an ionic compound composed of Cu^+ and O^{2-} ions. Therefore, a new term is needed. Chemists refer to Cu_2O as a ***formula unit*** of copper(I) oxide.

The chemical statement

$$Cu(s) \quad + \quad O_2(g) \quad \rightarrow \quad Cu_2O(s)$$

copper	oxygen	copper(I)
metal	gas	oxide

▶ *A formula unit is the symbol for an ionic compound.*

is not an equation, because it is not balanced. Although it correctly identifies reactants and products, it fails to meet "bookkeeping" standards. For each two-atom O_2 molecule that reacts, two oxygen atoms must appear on the product side. Because a formula unit of Cu_2O contains only one atom of oxygen, two units of Cu_2O must form:

$$Cu(s) + O_2(g) \rightarrow 2\,Cu_2O(s)$$

▶ *A chemist's "bookkeeping standards" are set by the fact that atoms are neither created nor destroyed in chemical reactions.*

How many copper atoms are needed to produce these two units of Cu_2O? Each Cu_2O unit contains two copper atoms. Thus, four copper atoms are needed to form two units of Cu_2O:

$$4\,Cu(s) + O_2(g) \rightarrow 2\,Cu_2O(s)$$

This is the completed equation for the chemical reaction of copper metal with oxygen gas, producing copper(I) oxide. The numbers placed in front of chemical symbols or formulas are called ***coefficients*** (in this case, 4, 1, and 2 from left to right).

▶ *The coefficient for O_2 is 1 in this equation. In general, when a coefficient is 1, it is not written, but understood to be there.*

An atom inventory can help confirm that a chemical equation is balanced. Just count the atoms of each element on both sides of the equation:

Reactant (left) side	Product (right) side
4 Cu atoms	4 Cu atoms and
2 O atoms (in 1 O_2 molecule)	2 O atoms (in 2 formula units Cu_2O)

▶ *coefficient*
↓
$2\,Cu_2O$
↑
subscript

In the following *Your Turn,* you can practice recognizing and interpreting chemical equations. Then you will observe a chemical reaction in the laboratory in which copper seems (once again!) to disappear. Carefully observing the reaction products, however, will help you decide the fate of reactant atoms.

YOUR TURN

Balanced Equations

For each chemical statement written below,

a. interpret the statement in words,

b. complete an atom inventory of the reactants and products, and

c. decide whether the statement—as written—is a chemical equation.

To help guide your work, the first item is worked out.

1. The reaction between propane, C_3H_8, and oxygen gas, O_2, is a common heat source for campers, recreational vehicle users, and others using liquid propane fuel (LPG). The reaction can be represented by

$$C_3H_8(g) + O_2(g) \rightarrow CO_2(g) + H_2O(g)$$

a. Interpreting this statement in words, we can write: "Propane gas reacts with oxygen gas to give carbon dioxide gas and water vapor."

b. Listing all reactant and product atoms, we find

Reactant side	Product side
3 carbon atoms	1 carbon atom
8 hydrogen atoms	2 hydrogen atoms
2 oxygen atoms	3 oxygen atoms

c. Because the numbers of carbon, hydrogen, and oxygen atoms are different in reactants and products, the original statement is not a chemical equation.

▶ Sometimes heat is included in the equation: $C_3H_8(g) + O_2(g) \rightarrow CO_2(g) + H_2O(g) + heat$

To balance the original statement, first note that each reactant carbon atom appears on the product side in a CO_2 molecule. Because a CO_2 molecule has only one C atom, the original three reactant carbon atoms must appear in *three* CO_2 molecules (3 CO_2). Next consider the fate of hydrogen atoms. The reactant hydrogen atoms appear on the product side as parts of water molecules. Two hydrogen atoms are needed in each water molecule. Therefore, eight hydrogen atoms must be incorporated into *four* H_2O molecules (4 H_2O).

Adding these two coefficients to the original chemical statement balances the carbon and hydrogen atoms:

$$C_3H_8(g) + O_2(g) \rightarrow 3\ CO_2(g) + 4\ H_2O(g)$$

Now consider oxygen. We find *ten* oxygen atoms on the product side—six oxygens in 3 CO_2 (3 × 2 = 6) plus four oxygens in 4 H_2O (4 × 1 = 4). Five pairs of oxygen atoms (5 O_2) will be needed on the reactant side to furnish these ten oxygen atoms. The completed chemical equation must be:

$$C_3H_8(g) + 5\ O_2(g) \rightarrow 3\ CO_2(g) + 4\ H_2O(g)$$

▶ Once correct formulas are written for reactants and products, only coefficients are used to balance the equation. No subscripts can be changed, for example H_2O to H_2O_2.

To check this result, we can complete a new atom inventory:

Reactant side	Product side
3 carbon atoms (in C_3H_8)	3 carbon atoms (in 3 CO_2)
8 hydrogen atoms (in C_3H_8)	8 hydrogen atoms (in 4 H_2O)
10 oxygen atoms (in 5 O_2)	10 oxygen atoms (in 3 CO_2 plus 4 H_2O)

This balanced inventory confirms that a chemical equation has been written. The final equation and atom inventory both suggest that matter has neither been created nor destroyed—demonstrating the law of conservation of matter.

Now try balancing these on your own:

2. The burning of wood or paper:

$$C_6H_{10}O_5(s) + O_2(g) \rightarrow CO_2(g) + H_2O(g)$$

3. The decomposition of nitroglycerin when dynamite explodes:

$$C_3H_5(NO_3)_3(l) \rightarrow N_2(g) + O_2(g) + CO_2(g) + H_2O(g)$$

4. A combination of metallic silver with hydrogen sulfide and oxygen gases in air, forming silver tarnish, Ag_2S (silver sulfide), and water:

$$Ag(s) + H_2S(g) + O_2(g) \rightarrow Ag_2S(s) + H_2O(l)$$

A.4 LABORATORY ACTIVITY: USING UP A METAL

GETTING READY

In this activity you will observe the chemical reaction between nitric acid, HNO_3, and copper metal, Cu. Many metals are chemically attacked by acids. As you will see, such a reaction causes the metal material to deteriorate or be worn away.

PROCEDURE

1. Obtain a 1-cm piece of copper wire and place it into an empty well of a wellplate. Place a white piece of paper beneath the wellplate to better observe any changes during the reaction.
2. Carefully add just enough nitric acid (HNO_3) to the well to cover the copper wire.
3. Observe the reaction for five minutes. Record all changes observed in the wire, the nitric acid solution, and the well contents.
4. Discard the wellplate's contents according to your teacher's directions.
5. Wash your hands thoroughly before leaving the laboratory.

QUESTIONS

1. What happened to the copper wire during the reaction?
2. How do you explain the color that forms in the wellplate?
3. Here is a chemical equation for the reaction you just observed:

$$\text{Cu(s)} + 4\,\text{HNO}_3\text{(aq)} \rightarrow \text{Cu(NO}_3)_2\text{(aq)} + 2\,\text{NO}_2\text{(g)} + 2\,\text{H}_2\text{O(l)}$$

| copper | + | nitric acid | → | copper(II) nitrate | + | nitrogen dioxide | + | water |

CAUTION

8 M nitric acid (HNO_3) will chemically attack skin! If any of the nitric acid is spilled on you, wash the affected area with tap water for several minutes. Notify your teacher.

CAUTION

Do this reaction in a well-ventilated space. Do not breathe the fumes generated by the reaction.

▶ *The (aq) notation—short for aqueous—indicates that the substance is dissolved in water.*

4. Use this equation and your original observations to answer these questions.

    a. You observed several color changes during the reaction. Identify the color of each substance in the chemical equation.

    b. Where did the copper atoms from the wire go?

    c. What elements were in the gas molecules released by the reaction?

    d. Where did these atoms come from?

    e. Where did the atoms in the water produced come from?

A.5 RESOURCES AND WASTE

Resources are simply materials—such as plants, animals, minerals, rocks, even gases—that can be withdrawn from the natural environment to satisfy human needs.

In many ways Earth is like a spaceship. The resources "on board" are all we can count on for our lifelong "trip." Some resources—such as fresh water, air, fertile soil, plants, and animals—can eventually be replenished by natural processes. These are *renewable resources.* If we are careful, nature will help us maintain supplies of these materials. That means neither using renewable resources faster than they can be replenished, nor creating environmental damage by using renewable resources carelessly.

Other materials—such as metals, natural gas, coal, and petroleum—are considered *nonrenewable resources,* because they cannot be replenished (or can be formed only over millions of years). Nonrenewable resources can be "used up," or become so widely dispersed that it is virtually impossible to gather them together again.

As scientific and technological knowledge increase, our society uses resources at ever-increasing rates. To satisfy the needs of just one average U.S. citizen, some 23,000 kg (23 tons) of various resources must be withdrawn from Earth each year.

Waste and resources share a common fate at a landfill.

Using or obtaining a resource often results in producing new, unwanted material. Burning coal generates corrosive gases that enter the atmosphere. Extracting a metal from an ore often leaves behind solid wastes that must be discarded. Even eating meat leaves useless bones and scraps behind.

We also produce waste when we use many common consumer products. Foil wrappers become waste when we use chewing gum. Some products—such as yesterday's newspaper—become waste after they fulfill their initial purpose. Others, such as telephones and computers, become waste when we discard them for newer models.

When we throw such materials away, we do not really get rid of them. The chemical elements found in trash may be less useful to us in that form, but they are still there. They don't go "away," but must be dealt with.

Our society faces increasing problems associated with *waste*—things we no longer want or need. Each person in this country throws "away," on average, about 2 kg (4 lb) of unwanted materials daily. About half of this is paper and other combustibles. Combined, the materials directly discarded by U.S. citizens would fill the New Orleans Superdome from floor to ceiling twice each day! And this does not include the far larger amounts of discards generated in producing the original consumer items.

Waste has earned several names depending on where we deposit it. Materials we gather and throw into cans or garbage disposals are called "trash" or "garbage." "Pollution" is often caused by unwanted, sometimes harmful materials, discarded carelessly.

Many potential discards are actually "resources out of place." For example, used glass bottles can often be sterilized, cleaned, and used again. Aluminum cans and glass bottles can also be reprocessed—or recycled—at a lower cost than making containers from new materials. *Recycling* means reprocessing the materials from manufactured items to make new manufactured items (for example, using glass from old bottles to make new ones). Figure 4 illustrates the general steps involved in a typical recycling process—the recycling of aluminum cans.

▶ *Coal contains sulfur. When coal burns it generates the corrosive gases sulfur dioxide (SO_2) and sulfur trioxide (SO_3).*

▶ *In Japan and Europe most glass bottles are refilled, not discarded.*

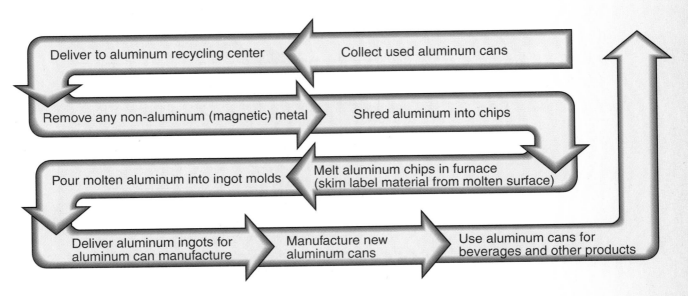

Figure 4 Aluminum can recycling.

U.S. citizens discard a Superdome of trash every 12 hours.

What do you do with wastes generated by your daily activities? You probably throw many of them into wastebaskets or trash containers. But where do they go from there?

A.6 DISPOSING OF THINGS

▶ *About 1/3 of U.S. households have curbside recycling.*

The graph in Figure 5 shows what happens to discards after they leave our trash cans. The top line represents the total tons of solid waste handled by U.S. municipalities each year. The quantities recovered for recycling, or

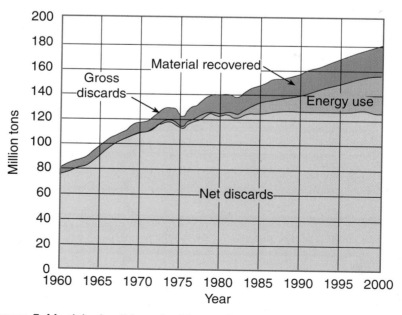

Figure 5 Municipal solid waste. The total quantity discarded (gross discards) and the quantities recovered for recycling or burned to produce useful energy are shown. The remainder (net discards) must be disposed of by landfill, dumping, or incineration.

burned to produce useful energy, are also shown. The rest—net discards—are deposited in landfills and dumps. The graphs have been extended to the year 2000.

Increased attention to waste disposal is vital for us as individuals, communities, nations, and planetary inhabitants. Figure 6 shows the composition of solid waste. You may be surprised to learn that the largest single category of discarded material by volume is paper and paperboard (34%), followed by plastics (20%). Newspapers alone compose 14% of landfill volume.

In this "throw away" society, some two billion razors and blades are discarded each year. American drivers annually generate more than 250 million scrap tires. Our nation's wastes have created a $20-billion-per-year industry that collects, processes, and stores these discarded materials. Waste-related issues are becoming urgent—some 80% of our currently used landfills will become filled and close down within the next 20 years. The "away"—as used in the phrase "throwing away"—will become less clear in coming years!

What are the prospects for addressing these problems? Changing our habits might, of course, alter what happens in the future. Research into new uses of waste is under way. Compacted waste can be coated with asphalt to produce building materials. A process developed by the U.S. Bureau of Mines can convert 1,000 kg of organic waste materials (those with high amounts of carbon and hydrogen) into 250 kg of oil. Crushed glass can replace as much as 30% of the stone and sand used in asphalt for road construction. Since the nation produces about a billion tons of asphalt each year, this could be a significant use for waste glass. Roads have also been built using a mixture made from crushed glass and rubber from scrap tires.

More than 130 waste-to-energy plants currently operate in this country, burning more than 300 million tons of solid waste for energy. Each ton of garbage that serves as "fuel" in these plants produces about a third of the energy released by a similar quantity of coal. Also, some of the residue of burning can be used as deposits for future resource "mining."

However, there are some problems with such plants. Although many organic wastes (such as paper, fabric, and some polymers) that are composed primarily of carbon, hydrogen, and oxygen can be burned to carbon dioxide gas and water vapor, metallic and mineral-based materials are not combustible. And materials that do burn create fly ash that can cause air pollution and generate other solid residues that require disposal.

▶ *The volume of discarded wastes that go into a landfill is more important than the mass of the wastes.*

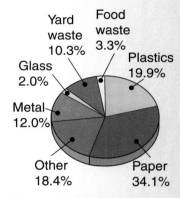

Figure 6 What's in our garbage? (Source: Data from "Characterization of Municipal Solid Waste in the United States: 1990 Update, Executive Summary." June 1990, United States Environmental Protection Agency.)

CHEMQUANDARY

WONDERING ABOUT RESOURCES AND WASTE

Consumers have a choice when asked "Paper or plastic bags?" in the grocery store checkout line. Which choice do you recommend? What information would you need to make and support your choice?

In fact, we rarely think about the variety of natural and manufactured things we use, or about resources involved in making them. What materials do we use daily? What chemical resources are involved in supporting our daily activities? What items do we throw away? What happens to them? It's time to take stock!

A.7 YOU DECIDE: CONSUMING RESOURCES

In this activity you will identify several items you use daily and analyze the resources involved in their use or manufacture.

Assume you use a wooden pencil today. It is probably composed of wood, graphite (a form of carbon produced from wood or other natural materials), and paint. The paint contains natural or synthetic (manufactured) pigments, originally dispersed in a solvent (possibly from petroleum). The pencil's eraser is made of rubber (either synthetic or extracted from a plant), held in place by a metal or plastic collar. Of these materials, wood, graphite, natural rubber, and plant pigments come from renewable resources, while synthetic pigments, solvents, plastics, and metal come from nonrenewable resources.

You will also be asked to identify waste materials involved in using each item. For example, if you used the pencil eraser, you probably just brushed aside the eraser crumbs, which eventually merged with dust on the floor. When the wooden pencil becomes too short, you will probably throw the pencil stub into a wastebasket.

1. List five things you used or used up since yesterday. (Focus on *simple* things, such as toiletries or packaged foods. You may also have used an automobile or portable TV, but these would involve nearly every type of natural resource.)

We can never have more aluminum than was in the Earth at its formation. Unlike aluminum, wood is a renewable resource.

2. For each listed item, answer two questions:

   a. What resources were used to manufacture it?

   b. How did (or will) you dispose of unwanted materials involved in its use?

3. Classify each resource in Question 2a as renewable or nonrenewable.

   Your class will divide into small groups to discuss the remaining questions. A full-class discussion will then highlight major points developed by the groups.

4. Prepare a summary list of all items identified by each group member.

5. a. How many items listed by your group were made from renewable resources?

   b. What percent of all items listed were made from renewable resources?

6. a. Identify four items from your group's summary list that are made mainly of nonrenewable resources.

   b. For each item, try to think of a substitute that would be made mainly of renewable resources.

7. a. List all the waste disposal methods identified by group members in Question 2b.

   b. If everyone in the United States used these same disposal methods, what benefits or problems might arise?

   We use resources in many ways. But how do we decide which resource to select for a particular use? Clearly, we must know the requirements of the intended use, key properties of available materials, and their cost. Two of these concerns involve important applications of chemical knowledge, as we will soon discover.

? PART A: SUMMARY QUESTIONS

1. a. State the law of conservation of matter.

   b. How is such a scientific law different from a law created by the government?

2. Complete atom inventories to decide whether each of these equations is balanced:

   a. The preparation of tin(II) fluoride, an ingredient of some toothpastes (called "stannous fluoride" on some labels):

   $$Sn(s) \ + \ HF(aq) \rightarrow SnF_2(aq) \ + \ H_2(g)$$
   tin   hydrofluoric   tin(II)   hydrogen
   metal   acid   fluoride

   b. The synthesis of carborundum for sandpaper:

   $$SiO_2(s) \ + \ 3\,C(s) \rightarrow SiC(s) \ + \ 2\,CO(g)$$
   silicon   carbon   silicon   carbon
   dioxide   carbide   monoxide
   (sand)   (carborundum)

   c. The reaction of an antacid with stomach acid:

   $$Al(OH)_3(s) + 3\,HCl(aq) \rightarrow AlCl_3(aq) + 3\,H_2O(l)$$
   aluminum   hydrochloric   aluminum   water
   hydroxide   acid   chloride

3. Why are the phrases "using up" and "throwing away" inaccurate from a chemical viewpoint?

4. It has been suggested that Earth is similar to a spaceship (Spaceship Earth).

   a. In what ways is this analogy useful?

   b. In what ways is it misleading?

5. Describe at least two benefits of discarding less and recycling more waste material. Consider the law of conservation of matter and resources available on Spaceship Earth in forming your answer.

6. What advantages does reuse have over recycling?

7. List

   a. four renewable resources, and

   b. four nonrenewable resources.

Recycling takes a lot of space and time, but the savings make it profitable.

B WHY WE USE WHAT WE DO

Every human-produced object, old or new, is made of materials selected for their suitable characteristics or properties. What makes a specific material best for each use? We can begin to answer this question by looking at some properties of materials.

B.1 PROPERTIES MAKE THE DIFFERENCE

The Statue of Liberty, built in the late 1880s, had to be strong yet flexible to withstand the winds blowing across New York Harbor. It had to be made of material that could be shaped and would maintain a pleasing appearance. The materials had to be readily available at reasonable cost. The artist and the architect needed to know the characteristics of available materials, and to select those used to build the statue and its base.

The unique properties of materials make them suitable for specific purposes. These often include *physical properties* such as color, density, or odor—properties that can be determined without altering the chemical makeup of the material. A *physical change,* such as melting or boiling at certain temperatures, may also be important. In a physical change, the identity of the substance is not changed.

The *chemical properties* of a material often also play important roles in its usefulness. Such properties involve transforming the substance into new materials. For example, the reactivity of a substance with acids, or iron's rusting when exposed to moist air are chemical properties. A *chemical change* is a chemical reaction in which the identity of one or more substances changes, and one or more new substances are formed. A chemical

A knowledge of chemistry was essential for the renovation of the Statue of Liberty.

Rescuers free a victim of an auto wreck, an example of a physical change.

Firefighters control a fire, an example of a chemical change.

Why We Use What We Do

change can often be detected by observing the formation of a new gas or solid, a color change, a change in the surface of a solid material, or a temperature change (indicating that heat has been absorbed or given off).

Let's classify some common properties of materials as physical or chemical properties.

YOUR TURN

Properties: Physical or Chemical?

Consider this statement: *Copper compounds are often blue or green.* Does that statement describe a physical or a chemical property? To answer, first consider whether a change in the *identity* of a substance is involved. Has it been chemically changed? If the answer is "no," then it is a physical property; if "yes" it involves a chemical property.

Color is a characteristic physical property of individual chemical compounds. A green copper compound in a jar on the shelf is not undergoing any change in its identity. Color is, therefore, a physical property. Note, however, that a *change* in color often indicates a change in identity and may represent a chemical property. For example, the change of litmus from blue to pink when exposed to acid is a chemical property.

Consider this statement: *Oxygen gas supports the combustion of a fuel.* Does this refer to a physical or chemical property of oxygen? We apply the same key question: Is there a change in the identity of the substance? Fuel combustion—or burning—involves a chemical reaction between the fuel and oxygen gas that changes both reactants. Thus, the statement refers to a chemical property of oxygen gas.

Now, classify each of these statements as describing a physical or a chemical property:

1. Metals, when pure, have a high luster (are shiny and reflect light).
2. Some metals may become dull when exposed to air.
3. Atmospheric nitrogen is a relatively nonreactive element at room temperature, but forms nitrogen oxides at the high temperatures of an automobile engine.
4. Milk turns sour if left too long at room temperature.
5. The hardness of diamonds enables them to be used on drill bits.
6. Metals are typically ductile (can be drawn into wires).
7. Bread dough increases in volume if it is allowed to "rise" before baking.
8. Argon gas, rather than air, is used in light bulbs to retard evaporation of the metal filament.
9. Metals are typically much better conductors of heat and electricity than are nonmetals.

Many problems can arise when materials are being chosen for use. The best substance for a purpose may be either unavailable or too expensive. Or perhaps the substance has an undesirable chemical or physical property that overshadows its desirable properties. What can be done in such situations? Often, a substitute material with most of the important properties of the original substance can be found and used.

That is what happened in the early 1980s when copper became too expensive to be used as the primary metal in pennies. Zinc is about as hard as copper. Zinc's density (7.14 g/mL), although somewhat less, is still near that of copper (8.94 g/mL). It is also readily available and less expensive than copper. For these reasons, zinc was chosen to replace most of the copper in all post-1982 pennies.

A blacksmith modifies properties of a metal sample through a variety of treatments.

However, zinc is chemically more reactive than copper. The zinc-plated pennies that had been made during World War II—known to coin collectors as "white cents" or "steel cents"—had no protective coating and quickly corroded. Post-1982 pennies have been plated with copper to increase their durability and to maintain the coin's familiar appearance.

Whether it's copper or zinc in a penny, or tungsten in a light bulb, each substance has its own specific chemical and physical properties. With millions of substances available, how can we sift through the options to identify the "best" substance for a given need?

Luckily, nature has simplified things. All substances are made of a relatively small number of building blocks—atoms of the different chemical elements. Knowledge of similarities and differences among common elements can greatly simplify the challenge of matching substances to uses. We will learn more about elements in the next section.

B.2 THE CHEMICAL ELEMENTS

Earlier (page 94) we noted that all matter is composed of atoms. One element differs from another because its atoms have a set of properties that differ from those of all other elements. More than 100 chemical elements are known. However, only about one third are important to us on a daily basis. Table 1 (page 108) lists some common elements and their symbols. A listing of all the elements (names and symbols) is given inside the back cover of the book.

Table I	
Common Elements	
Name	**Symbol**
Aluminum	Al
Antimony	Sb
Argon	Ar
Barium	Ba
Beryllium	Be
Bismuth	Bi
Boron	B
Bromine	Br
Cadmium	Cd
Calcium	Ca
Carbon	C
Cesium	Cs
Chlorine	Cl
Chromium	Cr
Cobalt	Co
Copper	Cu
Fluorine	F
Gold	Au
Helium	He
Hydrogen	H
Iodine	I
Iron	Fe
Krypton	Kr
Lead	Pb
Lithium	Li
Magnesium	Mg
Manganese	Mn
Mercury	Hg
Neon	Ne
Nickel	Ni
Nitrogen	N
Oxygen	O
Phosphorus	P
Platinum	Pt
Potassium	K
Silicon	Si
Silver	Ag
Sodium	Na
Sulfur	S
Tin	Sn
Tungsten	W
Uranium	U
Zinc	Zn

The next activity will help you become more familiar with the common elements listed in Table 1.

YOUR TURN

Chemical Elements Crossword Puzzle

Your teacher will distribute a crossword puzzle for you to complete. Puzzle clues, appearing at the close of this unit (page 150), are descriptions of properties and uses of all elements listed in Table 1. Use this table as a guide in completing the puzzle.

Properties of elements vary over wide ranges, as shown by Table 2. Some elements, such as magnesium and calcium, are very similar; others, such as iodine and gold, are quite different. Chemical compounds composed of similar elements often have similar properties.

Table 2	Some Properties of the Elements	
Property	**From**	**To**
Density		
Metallic elements	0.53 g/mL (Li)	22.6 g/mL (Os)
Nonmetallic elements	0.0008 g/mL (H_2)	4.93 g/mL (I_2)
Melting point		
Metallic elements	−33 °C (Hg)	3410 °C (W)
Nonmetallic elements	−249 °C (Ne)	3727 °C (C)
Chemical Reactivity		
Metallic elements	Low (Au)	High (Cs)
Nonmetallic elements	None (He)	High (F_2)

Elements can be grouped or classified in several ways according to similarities and differences in their properties. Two major classes are *metals* and *nonmetals.* Everyday experience has given you some knowledge of the properties of metals. In the next laboratory activity you will have a chance to explore further the properties of metals and nonmetals.

Several elements called *metalloids* are not clearly metals or nonmetals; they have properties that are intermediate. In some properties metalloids are like metals and in others they are like nonmetals.

Every element can be classified as a metal, a nonmetal, or a metalloid. What properties of matter are used for this classification? The next activity will help you find out.

B.3 LABORATORY ACTIVITY: METAL, NONMETAL?

GETTING READY

In this activity you will investigate properties of seven elements in order to classify them as metals, nonmetals, or metalloids. You will examine each for its physical properties of color, luster, and form (for example, is it crystalline, like table salt?). You will attempt to crush each sample with a hammer. This way you can decide whether each element is **malleable** (flattens without shattering when struck) or **brittle** (shatters into pieces). You may also test for electrical conductivity, a physical property. (As an alternative, your teacher may demonstrate this test.)

Next, you will observe differences among these elements' chemical properties. You will find out whether each element reacts with hydrochloric acid, HCl(aq), and with a copper(II) chloride ($CuCl_2$) solution.

Prepare a data table in your notebook, leaving plenty of space to record the properties of the seven element samples, which have been coded with letters *a* to *g*.

Data Table					
Element	Appearance	Result of Crushing	Conductivity (optional)	Reaction with Acid	Reaction with $CuCl_2$(aq)
a. b. c. d. e. f. g.					

PROCEDURE

Physical Properties

1. *Appearance:* Observe and record the appearance of each element. Include physical properties such as color, luster, as well as form.

2. *Crushing:* Gently rap each element sample with a hammer. Decide whether the samples are malleable or brittle.

3. *Conductivity (optional):* If a conductivity apparatus is available, use it to test each sample. Touch both electrodes to the element sample, but don't allow the electrodes to touch each other. If the bulb lights, the sample has allowed electricity to flow through it. Such a material is called a **conductor.** If the bulb fails to light, the material is a **nonconductor** (see Figure 7).

Figure 7 Testing for conductivity.

Chemical Properties

Test each sample for reactivity with acid as described below. The formation of gas bubbles indicates that a chemical reaction has occurred. Also, a change in the appearance of any element sample indicate a chemical reaction has taken place.

1. a. Label seven wells of a clean wellplate or reaction strip a to g.

   b. Place a sample of each element in its appropriate well in the wellplate or reaction strip. The sample should be a 1-cm length of wire or ribbon, or 0.2–0.4 g of solid.

CAUTION

Avoid touching the electrodes with your hands; you can get an electric shock.

c. Add 15–20 drops of 0.5 M HCl solution to each well containing a sample.

d. Observe and record each result. Decide which elements reacted with the hydrochloric acid, and which did not.

e. Discard the wellplate contents as instructed by your teacher.

2. Test each element sample for reactivity with 0.1 M copper(II) chloride ($CuCl_2$) solution as described below.

a. Repeat Steps 1a and 1b.

b. Add 15–20 drops of 0.1 M copper(II) chloride ($CuCl_2$) solution to each well containing a sample.

c. Observe the test tubes for three to five minutes—changes may be slow. Record each result. Decide which elements reacted with the copper(II) chloride, and which did not.

d. Discard the wellplate contents as instructed by your teacher.

3. Wash your hands thoroughly before leaving the laboratory.

QUESTIONS

1. Sort the seven coded elements into two groups, based on similarities in their physical and chemical properties.

2. Which elements could fit into either group, based on certain properties?

3. Consider the following information. Then reclassify each element as a metal, a nonmetal, or a metalloid.

- Metals are elements that have a luster, are malleable, and conduct electricity (physical properties).
- Many metals react with acids and with copper(II) chloride solution (chemical properties).
- Nonmetals are usually dull in appearance, brittle, and do not conduct electricity (physical properties).
- Most nonmetals do not react with acids or with copper(II) chloride solution (chemical properties).
- Elements that have some properties of both metals and nonmetals are metalloids.

> ► *Some metalloids are semiconductor components used to manufacture electronic chips.*

Even though we have narrowed our view from all known substances to slightly more than 100 chemical elements, the quantity of information known about all these elements is still great. How can we manage and conveniently organize our knowledge about these fundamental building blocks of the world's resources? Nature again provides an answer.

B.4 THE PERIODIC TABLE

Scientists search for patterns and regularities in nature. If an underlying pattern can be found, information can be organized to increase its usefulness, clarity, and applicability. The ability of chemists to predict properties of newly developed or even theoretical substances has been greatly enhanced by the discovery of a pattern among the elements. That major discovery has guided the development of new and useful materials.

By the mid-1800s, about 60 elements were known. The five nonmetals known at that time—hydrogen (H), oxygen (O), nitrogen (N), fluorine (F), and chlorine (Cl)—are gases at room temperature. Two other elements are liquids—the metal mercury (Hg) and the nonmetal bromine (Br). The rest are solids, with widely differing properties.

Several scientists of that era tried to devise a classification system that placed elements with similar properties near each other on a chart. Such an arrangement is called a *periodic table.* Dimitri Mendeleev, a Russian chemist, published a periodic table in 1869. We still use a very similar table today. The periodic table has a pattern similar in some respects to a monthly calendar in which weeks repeat on a regular (periodic) seven-day cycle.

The periodic tables of the 1800s were organized according to increasing atomic masses and an element's tendency to combine with atoms of other elements. It was known that atoms of different elements have different masses. For example, hydrogen atoms have the lowest mass of all, oxygen atoms are about 16 times more massive than hydrogen atoms, and sulfur atoms are about twice as massive as oxygen atoms (making them about 32 times more massive than hydrogen atoms). Based on such comparisons, an *atomic mass* was assigned to each element.

Atoms of various elements also differ in how many atoms of another element they combine with. For example, one atom of potassium (K) or cesium (Cs) combines with only one atom of chlorine (Cl), producing the compounds KCl and CsCl. We can represent such one-to-one compounds as ECl (E stands for the Element combining with chlorine). Atoms of other elements may combine with two, or three, or four chlorine atoms, giving sets of compounds with the general formulas ECl_2, ECl_3, and ECl_4. Many elements can be organized in patterns based on their "combining capacity" with oxygen and with chlorine.

In the first periodic tables, elements with similar properties were placed in vertical groups (columns). However, rather than just telling one of the great scientific detective stories, we invite you to follow a similar path.

Dimitri Mendeleev (1834–1907) published the first useful periodic table.

B.5 YOU DECIDE: GROUPING THE ELEMENTS

You will be given a set of 20 element data cards. Each card lists some properties of one of the first 20 elements.

1. Arrange the cards in order of increasing atomic mass.

2. Next, place the cards in a number of different groups. Each group should include elements with similar properties. For example, you might put all elements with boiling points below 0 °C in one group, and in another all elements with boiling points above 0 °C. Or you might examine the formulas of chlorine-containing compounds, and group the elements by the number of chlorine atoms in these formulas.

3. Examine the cards within each group for any patterns. Arrange cards within each group in some logical sequence.

4. Observe how particular properties vary from group to group.

5. Arrange all the card groups into some logical sequence.

Why We Use What We Do

6. Decide on the most reasonable and useful patterns within and among card groups. Then tape the cards onto a sheet of paper to preserve your pattern for classroom discussion.

B.6 THE PATTERN OF ATOMIC NUMBERS

Early periodic tables offered no explanation for similarities found among element properties. The reason for these similarities, which was discovered about 50 years later, serves as the basis for the modern periodic table.

Recall that all atoms are composed of smaller particles, including equal numbers of positively charged protons and negatively charged electrons (page 32). One essential difference discovered later among atoms of different elements is their number of protons, the ***atomic number.*** Every sodium atom contains 11 protons, and so the atomic number of sodium is 11. Each carbon atom contains 6 protons. If the number of protons in an atom is 9, it is a fluorine atom; if 12, it is a magnesium atom. In short, knowing the atomic number (number of protons) allows us to identify which element it represents.

In the modern periodic table, all elements are in sequence according to their atomic numbers. This table also reflects the arrangement of electrons in atoms. As we shall see, electron arrangement is closely related to the properties of atoms.

In the next exercise you will draw graphs to explore the relationship between atomic numbers and the groupings of elements having similar properties in the periodic table.

YOUR TURN

Periodic Variation in Properties

Your teacher will give you the atomic numbers of the 20 elements you arranged in Section B.5. Use these atomic numbers and information on the element cards to prepare the two graphs described below. It will be helpful to label each plotted point with the symbol of the element involved.

1. Label the x axis of Graph 1 with atomic numbers from 1 to 20.

2. Select either the formulas for oxide or chloride compounds. Label the y axis for the number of chlorine or oxygen atoms as follows:

   a. To prepare to plot chloride data, label the y axis 0 for formation of no chloride, 1 for ECl compounds (1 chlorine atom for each E atom), 2 for ECl_2, 3 for ECl_3, and 4 for ECl_4.

   b. To prepare to plot oxide data, label the y axis 0 for formation of no oxide, 0.5 for E_2O (0.5 oxygen atom for each E atom, 1 oxygen atom for 2 E atoms), 1 for EO, 1.5 for E_2O_3 (do you see why?), 2 for EO_2, and 2.5 for E_2O_5.

3. Plot the oxide or chloride data from the element cards.

4. Label the x axis of Graph 2 with the atomic numbers from 1 to 20.

Data graph 1

Data for chemical property (formation of oxides or chlorides)

1   Atomic number   20

Graph 1 Trends in a chemical property.

5. Select either melting point or boiling point data; label the *y* axis as shown in the accompanying figure. Use as much of the space on your graph paper as possible.

6. Plot the data from the element cards. Do not include data for the element with atomic number 6 (carbon)—the *y*-axis value for this element will be quite far off the graph.

7. On each graph, connect adjacent points with straight lines.

Questions

1. a. Does either graph reveal a repeating or cyclic pattern? (*Hint:* Focus on elements in the peaks or valleys.)

   b. Are these graphs consistent with your earlier grouping of the elements? Why or why not?

2. Based on these graphs, why is the chemist's organization of elements called a *periodic* table?

3. When elements are listed in order by their atomic numbers, and grouped according to similar properties, they form seven horizontal rows, called **periods.** This periodic relationship among elements is summarized in the modern periodic table inside the back cover. To become more familiar with the periodic table, locate the 20 elements you classified earlier (page 111). How do their relative positions compare with those shown on your chart?

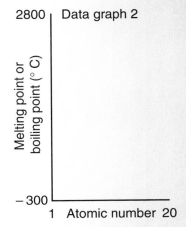

Graph 2 Trends in a physical property.

▶ *The repetition of similar properties at regular intervals when elements are arranged by increasing atomic number is known as the **periodic law.***

▶ *Helium family elements are known as **noble gases.***

▶ *Using his periodic table, Mendeleev correctly predicted the properties of several elements that had not yet been discovered (but were discovered later).*

Each vertical column in the table contains elements with similar properties. These are called **groups** or **families** of elements. For example, the lithium (Li) family consists of the six elements in the first column at the left side of the table. These elements are all highly reactive metals that form ECl chlorides and E_2O oxides. Like sodium chloride (NaCl), all chlorides and oxides of lithium family elements are ionic compounds. By contrast, the helium family, at the right side of the table, consists of very inert elements (of these elements, only xenon and krypton are known to form any compounds).

The arrangement of elements in the periodic table provides an orderly summary of the key characteristics of elements. If we know the major properties of a certain chemical family, we can predict some of the behavior of any element in that family.

YOUR TURN

Mendeleev's Periodic Table

Some properties of an element can be estimated by averaging the properties of the elements located just above and just below the element in question. This is how Mendeleev predicted the properties of elements unknown in his time. He was so certain about his conclusions that he left gaps in his periodic table for missing elements, with predictions of their properties. When these elements were eventually discovered, they fit in exactly as expected. Mendeleev's fame rests largely on the correctness of these predictions.

For example, germanium (Ge) was an element undiscovered when Mendeleev proposed his periodic table. However, in 1871 he predicted the existence of germanium (he named it ekasilicon). Given the information that the boiling points of silicon (Si) and tin (Sn) are 3267 °C and 2603 °C respectively, we can estimate the boiling point of germanium.

The three elements are in the same group in the periodic table. Germanium is preceded by silicon and followed by tin in the group. Taking the average of the boiling points of these elements gives

$$\frac{(3267 \text{ °C}) + (2603 \text{ °C})}{2} = 2935 \text{ °C}$$

When germanium was discovered in 1886, its boiling point was found to be 2834 °C. The estimated boiling point for germanium, 2935 °C, is within 4% of germanium's known boiling point. The periodic table has helped guide us (and Mendeleev) to a useful prediction.

Here is an example of how formulas for chemical compounds can be predicted from periodic table relationships:

Carbon and oxygen form carbon dioxide (CO_2). What formula would you predict for a compound of carbon and sulfur?

The periodic table indicates that sulfur (S) and oxygen (O) are in the same family. Knowing that carbon and oxygen form CO_2, what formula is predicted for the formula of the compound formed by combining carbon and sulfur? Based on CO_2 as the formula for carbon dioxide, a logical prediction would be CS_2 (carbon disulfide)—and it's the correct prediction!

1. The element krypton was not known when Mendeleev developed his periodic table. Given that under similar pressure the boiling point for argon (Ar) is −186 °C, and for xenon (Xe) is −112 °C, estimate the boiling point of krypton.

2. a. Estimate the melting point of rubidium (Rb). The melting points of potassium (K) and cesium (Cs) are 64 °C and 29 °C, respectively.

   b. Would you expect the melting point of sodium (Na) to be higher or lower than that of rubidium (Rb)?

3. Mendeleev knew that silicon tetrachloride ($SiCl_4$) existed. Using his periodic table, he correctly predicted the existence of "ekasilicon," an element one place below silicon in the periodic table. Use your periodic table to write the formula for the compound formed by Mendeleev's "ekasilicon" and chlorine.

Periodic Table of the Elements

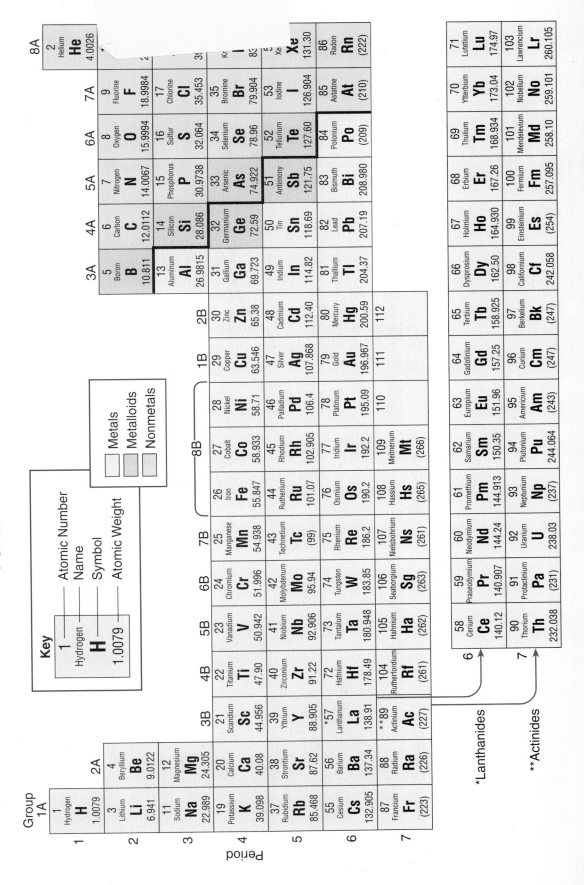

4. Here are formulas for several known compounds: NaI, MgCl$_2$, CaO, Al$_2$O$_3$, and CCl$_4$. Using that information, predict the formula of a compound formed from:

   a. C and F.   d. Ca and Br.

   b. Al and S.   e. Sr and O.

   c. K and Cl.

5. Which family of elements is so chemically inactive that its elements were originally regarded as "inert?"

B.7 CHEMICAL REACTIVITY

Magnesium reacts with oxygen. In fact, it reacts so quickly that we use small samples of magnesium in some fireworks. When heated, magnesium ignites, producing a brief, blinding flash. The equation for the reaction of magnesium with oxygen is written below.

$$2\,Mg(s) \quad + \quad O_2 \quad \rightarrow \quad 2\,MgO(s)$$

magnesium   oxygen   magnesium oxide

   Iron reacts with oxygen also, but generally it does so much more slowly. Rusting is the chemical combination of these two elements. Rusting (a form of corrosion) includes a complex series of reactions requiring water as well as oxygen gas. Cooks know, for example, that dry iron utensils do not rust. Also, automobiles rust faster near an ocean than they do in a desert.

   To a chemist, "rust" is a combination of several iron-containing compounds including iron(III) oxide:

The reaction between oxygen and iron is destroying this truck.

$$2 \text{ Fe(s)} \quad + \quad 3 \text{ O}_2\text{(g)} \quad \rightarrow \quad 2 \text{ Fe}_2\text{O}_3\text{(s)}$$

iron   oxygen   iron(III) oxide

By contrast, gold (Au) does not react with oxygen gas. This is one reason gold is so prized as a material for long-lasting, highly decorative objects (such as jewelry). Gold-plated electrical contacts, such as for automobile air bags, are dependable because no oxides—which are non-conducting—form on the contact surfaces.

By observing how readily certain metals react with oxygen, we learn something about their reactivities. We can rank elements in the relative order of their chemical reactivity. Such a ranking is called an **activity series.** You have already tested the reactivities of some elements with acid and with copper(II) chloride; in the following laboratory activity you will investigate another method for comparing the relative chemical reactivities of metals and their ions.

B.8 LABORATORY ACTIVITY: METAL REACTIVITIES

GETTING READY

In this activity you will observe and compare some chemical reactions of several metallic elements with solutions containing ions of other metals. You will investigate reactions of three metals (copper, magnesium, and zinc) with solutions of ionic compounds that contain metal ions — copper(II) nitrate, magnesium nitrate, zinc nitrate, and silver nitrate.

In your laboratory notebook, prepare a data table like the one below.

Data Table	Relative Reactivities of Metals			
	Solutions			
Metal	$Cu(NO_3)_2$	$Mg(NO_3)_2$	$Zn(NO_3)_2$	$AgNO_3$
Cu	—			
Mg		—		
Zn			—	
Ag	NR	NR	NR	—

We have filled in results for reactions of silver metal, Ag (which is too expensive for this use). NR stands for "no visible reaction." The dashes indicate reaction combinations that are not needed.

PROCEDURE

1. Obtain three 1-cm strips of each of the three metals to be tested. Clean the surface of each metal strip by rubbing it with sandpaper or emery paper. Record observations on the metals' appearance.

2. a. Place a piece of copper in wells A2, A3, and A4.
   b. Place a piece of magnesium in wells B1, B3, and B4.
   c. Place a piece of zinc in wells C1, C2, and C4.

3. a. Add 5 drops of 0.2 M copper(II) nitrate [$Cu(NO_3)_2$] to wells B1 and C1.

   b. Add 5 drops of 0.2 M magnesium nitrate [$Mg(NO_3)_2$] to wells A2 and C2.

   c. Add 5 drops of 0.2 M zinc nitrate [$Zn(NO_3)_2$] to wells A3 and B3.

   d. Add 5 drops of 0.2 M silver nitrate ($AgNO_3$) to wells A4, B4, and C4.

4. Write your observations in the data table. If no reaction is observed, write NR in the data table. Record the observed changes if a reaction occurred.

5. Dispose of your solid and wellplate materials as directed by your instructor.

6. Wash your hands thoroughly before leaving the laboratory.

QUESTIONS

1. Which metal reacted with the most solutions?

2. Which metal reacted with the fewest solutions?

3. List the four metals in order, placing the most reactive metal first (the one that reacted with the most solutions) and proceeding in order to the least reactive metal (the one that reacted with the fewest solutions).

4. Refer to your activity series list in Question 3. Explain to a classmate why the Statue of Liberty was made with copper instead of zinc.

5. a. Which material that you observed in this activity might have been a better choice for the statue than copper?

   b. Why wasn't it chosen?

6. Given your knowledge of relative chemical activity among these four metals,

   a. which metal is most likely to be found in an uncombined or "free" state in nature?

   b. which metal would be least likely to be found uncombined with other elements?

B.9 WHAT DETERMINES PROPERTIES?

▶ *We will look more closely at electron arrangements in the Petroleum unit.*

What causes differences in metals' reactivities—or in other properties that vary from element to element? Recall that atoms of different elements have different numbers of protons (atomic numbers). Therefore, these atoms also have different numbers of electrons. *Many properties of elements are determined largely by the number of electrons in their atoms and how these electrons are arranged.*

A major difference between metals and nonmetals is that *metal atoms lose their outer electrons much more easily than do nonmetal atoms.* Under suitable conditions, one or more of these outer electrons from metals may transfer to other atoms or ions.

For example, in Laboratory Activity B.8, magnesium atoms each transferred two electrons to zinc cations:

$$\text{Mg(s)} \quad + \quad \text{Zn}^{2+}\text{(aq)} \quad \rightarrow \quad \text{Mg}^{2+} \quad + \quad \text{Zn(s)}$$

| magnesium metal | zinc ion | magnesium ion | zinc metal |

▶ The nitrate anions, NO_3^-, remain unchanged in this reaction.

Active metals can give up one or more of their electrons to ions of less-active metals. You observed similar reactions between other pairs of metals and cations.

Some physical properties of metals depend on attractions among atoms. For example, stronger attractions among atoms of a metal result in higher melting points. Magnesium's melting point is 651 °C, while sodium's is 98 °C. Thus, we can infer that the attractions between the atoms in magnesium metal are much stronger than in sodium metal.

Chemical and physical properties of other kinds of substances are also explained by the makeup of their atoms, ions, or molecules and by attractions among these particles. As we pointed out earlier, water's abnormally high melting and boiling points are due to the strong attraction between water molecules.

Understanding properties of atoms is the key to predicting and correlating the behavior of materials. This information, often combined with a bit of imagination, allows chemists to find new uses for materials and to create new chemical compounds to meet specific needs.

▶ Several thousand new compounds are synthesized each year.

B.10 MODIFYING PROPERTIES

Throughout history—first by chance and more recently guided by science—we have greatly extended the array of materials available for human use. Chemists have learned to modify the properties of matter by physically blending or chemically combining two or more substances. Sometimes only slight changes in a material's properties are desired. At other times chemists may create new materials with dramatically different properties.

The black "lead" in a pencil is mainly graphite, a natural form of the element carbon. Pencil lead is available in various levels of hardness. Hard pencil lead, No. 4, produces very light lines on paper. Soft writing lead (such as No. 1) makes very black, easily smudged lines. Pencil lead hardness is controlled by the amount of clay mixed with the graphite. Increasing the quantity of clay produces harder pencil lead (less graphite can be rubbed off onto paper).

The laboratory activity Striking It Rich (page 91) demonstrated how metallic properties can be modified by creating *alloys,* solid mixtures of atoms of different metals. Brass was formed when you heated the zinc-coated penny. Brass is an alloy containing copper and from 10 to 40% zinc. It is harder than copper and has an attractive gold color, unlike either copper or zinc.

Examples of materials designed to meet specific needs are easily found. The properties of certain plastics can be custom made —in some cases without even changing the material's chemical composition. For example, polyethene can be tailored to display relatively soft and pliable

▶ When first discovered, graphite was mistakenly identified as a form of lead, and the name stuck.

properties (as in a squeeze bottle for water), or can be crafted to be hard and brittle—almost glass-like in its behavior.

Penicillin is an antibiotic produced in nature by a mold. The penicillin molecule has been modified by chemists, resulting in a family of "semisynthetic" penicillins with improved or more specific effectiveness.

Such custom-tailoring at the molecular level is possible because of chemical knowledge—knowledge of how the atomic composition of materials affects their observable properties and behavior.

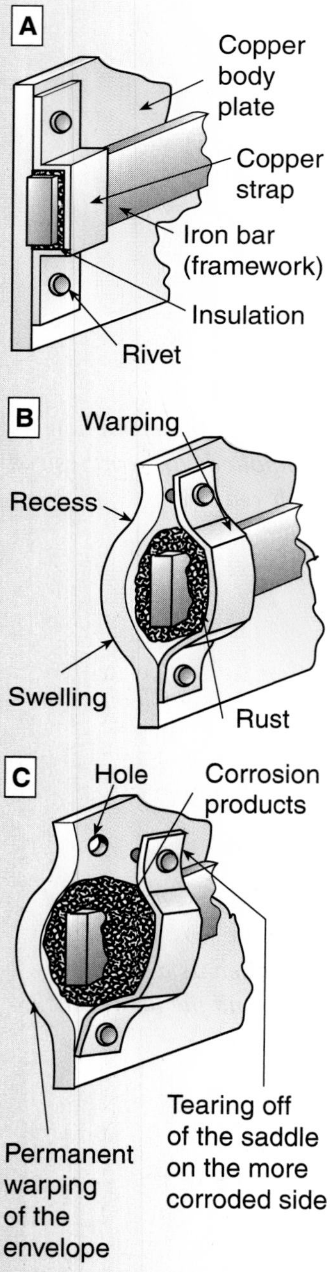

B.11 YOU DECIDE: RESTORING MS. LIBERTY

Corrosion is the chemically produced deterioration of metals. It extracts a costly toll on the U.S. economy and uses up a large portion of the world's resources. The rusting, weakening, and degradation of metals within such diverse settings as automobiles, bridges, ship hulls, gardening utensils, and porch railings costs our economy an estimated $70 billion each year.

Not even the Statue of Liberty—that unique and instantly recognized national symbol built in the 1880s—is immune to corrosion. By the early 1980s serious corrosion within the statue demanded major attention. Several questions faced restoration experts, including what caused the statue to deteriorate, how damage could be repaired, and how further deterioration could be avoided.

The statue's body is composed of 300 individual copper plates. Each plate was originally connected by riveted copper straps to an iron-bar framework, as shown in Figure 8a. When different metals contact each other in the presence of moisture and dissolved ions, the metals react, causing corrosion.

The original statue designers knew such contact would cause corrosion. Therefore, the nineteenth-century builders used shellac-soaked asbestos to separate copper from the iron wherever the two materials met along the iron framework. Over the years, condensed water vapor and rain collected at the junctions. The humid interior of the statue, which was far from waterproof, also hastened corrosion of the iron. The iron bars rusted and swelled. As a result of the swelling, more than 40% of the 450,000 rivets pulled loose. This left holes in the copper and allowed the plates to sag (Figure 8b and c).

In addition to internal structural damage, the statue's copper exterior had been affected by air pollution. Copper metal normally reacts with substances in the air to form a stable, attractive, green coating or *patina,* on the exposed copper surface. This protects the underlying copper. However, acidic pollutants can convert copper patina into a related substance, which is more water soluble. As rainwater dissolves this new substance, more copper metal is exposed and can undergo further corrosion.

In short, three major sources of statue deterioration required attention:

- Corrosion of the iron framework at contact points between iron and copper.
- The humid atmosphere inside the statue, which hastened iron corrosion.
- Removal of the protective patina on the exterior caused by airborne pollution.

Figure 8 A Statue of Liberty body plate–iron framework connection, as originally constructed (*A*). Corrosion of the framework (*B* and *C*).

A

Copper body plate

Copper strap

Iron bar (framework)

Insulation

Rivet

B Warping

Recess

Swelling

Rust

C Hole   Corrosion products

Permanent warping of the envelope

Tearing off of the saddle on the more corroded side

A further problem was created by the multiple coats of paint and tar that had been applied to the statue's interior over the years. The damage due to corrosion could not even be fully assessed—nor repairs made—until the paint and tar were removed.

How would you have tried to solve these restoration problems?

1. Using common experience and your chemistry knowledge, propose one or more solutions to each of the three problems identified above.

2. If substituting materials is part of your proposal, consider factors such as the chemical and physical properties of materials you propose using, their cost, and the need to preserve the Statue of Liberty's design and appearance.

You will share your ideas with the class and compare them with the actual solutions chosen by the restoration committee.

? PART B: SUMMARY QUESTIONS

1. Describe how a chemical or physical property could be used to distinguish
   a. brass from gold.
   b. hydrogen from helium.
   c. tungsten from iron.
   d. a lithium-family metal from silver.

2. What two properties make nonmetals unsuitable for electric wiring?

3. Given the correct formulas Al_2O_3 and $BeCl_2$, predict formulas for compounds containing
   a. Mg and F.
   b. B and S.

4. For medical reasons, people with high blood pressure are advised to limit the amount of sodium ions in their diet. Normal table salt (NaCl) is sometimes replaced by a commercially available substitute, potassium chloride, for seasoning.
   a. Write the formula for potassium chloride.
   b. Why are its properties similar to those of sodium chloride?

5. Decide whether each grouping of three elements belongs to the same chemical family. If not, identify the element that does not belong with the other two:

   a. Sodium, potassium, magnesium
   b. Helium, neon, argon
   c. Oxygen, arsenic, sulfur
   d. Carbon, nitrogen, phosphorus

6. Why is it incorrect to consider any particular metal as chemically "perfect" or "best?"

7. If two different metals are used in construction, why is it desirable to separate them with a nonconducting material?

This gold-plated jewelry will not corrode. This property plus its pleasing appearance, malleability, and ductility make gold an ideal ornamental metal.

CHEMISTRY AT WORK

Preserving the Past . . . for the Future

Mary Striegel is Research Associate at the National Center for Preservation Technology and Training (NCPTT) in Natchitoches, Louisiana. NCPTT is an effort by the National Park Service to advance the practice of historic preservation in the fields of archeology, architecture, landscape architecture, materials conservation, and history.

Mary supervises scientists in the Center's Materials Research Program, which uses field tests to determine how air pollution and acid rain affect historic buildings, outdoor works of art, and various other materials. In laboratory work, the scientists isolate different pollutants in small chambers in order to investigate the way materials and preservation treatments interact. In case studies, they observe the effects of pollutants on artifacts that have been exposed in the environment.

A Case Study of the Effect of Air Pollution

In the "Hiker Project," the Center is monitoring 52 nearly identical bronze statues that were erected throughout the country between 1906 and 1965 to honor veterans of the Spanish-American War.

The statues make an ideal case study of the effects of outdoor airborne pollutants and the effectiveness of preservation techniques. "There's even one statue," explains Mary, "that has been kept indoors since it was erected, which serves as a control."

Variables include a statue's proximity to sources of pollution, the conservation and preservation efforts made by local communities, and other factors. Mary and other scientists monitor the condition of all the statues to study the effects of pollutants.

Based on the results of its research, the Center is developing recommendations for preservation methods. One observation, for example, is that simply hosing down a bronze statue periodically with water may help prevent a great deal of corrosion. The Center will share its results with other organizations that are trying to preserve works of art and architecture.

Take a Science Walk . . .

Explore your neighborhood or community to find outdoor statues showing evidence of corrosion that might be caused by airborne pollutants. Where would be some likely places for finding such effects? Where would they be unlikely?

If you don't have any statues in your neighborhood, look for other metal fixtures such as hinges, railings, metal gutters, and wire fences. Are there patterns to the locations where you find evidence of corrosion? If so, what could explain this?

Photograph courtesy of NCPTT

CONSERVATION IN NATURE AND THE COMMUNITY

You are now prepared to explore the chemical "supplies" aboard Spaceship Earth. Let's consider how nature conserves resources and examine the need for appropriate conservation practices within our society.

C.1 SOURCES OF RESOURCES

Human needs for resources must continue to be met by chemical supplies currently present on Earth. These supplies or resources are often cataloged by where they are found. Table 3 indicates the composition of our planet.

Table 3		Earth's Composition
Layer of Planet	**Thickness (Average)**	**Composition (Decreasing Order of Abundance)**
Atmosphere	100 km	N_2 (78%), O_2 (21%), Ar (0.9%), He + Ne (<0.01%), variable amounts of H_2O, CO_2, etc.
Hydrosphere	5 km	H_2O, and in the oceans that cover some 71% of Earth's surface, approximately 3.5% NaCl and smaller amounts of Mg, S, Ca, and other elements as ions
Lithosphere: Crust	6400 km; Top 40 km	Silicates (compounds formed of metals, Si, and O atoms). Metals include Al, Na, Fe, Ca, Mg, K, and others Coal, oil, and natural gas Carbonates such as $CaCO_3$ Oxides such as Fe_2O_3 Sulfides such as PbS
Mantle Core	40–2900 km 2900 km to the Earth's center	Silicates of Mg and Fe Fe and Ni

The atmosphere, hydrosphere, and outer portion of the lithosphere are sources of *all* resources for *all* human activities. We use nitrogen, oxygen, neon, argon, and a few other gases from the atmosphere. We take water and extract some dissolved minerals from the hydrosphere. However, we rely mainly on the lithosphere, the solid part of Earth, for most chemical resources. For example, that is where we find petroleum and metal-bearing ores.

▶ *An **ore** is a naturally occurring rock or mineral from which it is profitable to recover a metal or other material.*

Our deepest mines barely scratch the surface of Earth's crust. If Earth were an apple, all the lithosphere's resources would be concentrated in the apple skin. From this thin band of soil and rock we obtain almost all the raw materials needed to build homes, automobiles, appliances, computers, videotapes, CDs, and sports equipment—in fact, all manufactured objects.

These chemical supplies are well suited to support a variety of life forms. However, many important resources are not uniformly distributed among nations. There is no connection between the abundance of these resources and either land area or population. Table 4 shows the major countries that produce several commercially important metals.

Table 4	Worldwide Production (%) of Selected Metals		
Metal	**Country (%)**	**Metal**	**Country (%)**
Aluminum	Australia (30) Guinea (13) Jamaica (11) Brazil (10)	Nickel	Former USSR republics (23) Canada (20) New Caledonia (12) Indonesia (9)
Cadmium	Japan (16) United States (9) Canada (9) Belgium (9)	Tin	China (24) Brazil (17) Indonesia (14) Bolivia (8)
Copper	Chile (21) United States (12) China (11) Canada (10)	Zinc	Canada (18) Australia (14) China (9) Peru (8)
Lead	Australia (16) United States (19) Canada (8) Zambia (5)		

(Table 4 data calculated from *World Resources 1994–95*)

Sometimes a particular region is the predominant supplier worldwide of several metals vitally important to industry, including chemical manufacturing. For example, Africa contains the major percentages of the world's known reserves of chromium (80%), cobalt (54%), and manganese (61%), in terms of their dollar values.

Many nations have a surplus of one or more chemical resources but deficiencies in others. Throughout history, unequal resource distribution has motivated great achievements and brutal wars. The development of the United States as a major industrial nation has been largely because of the quantity and diversity of our chemical resources. Yet, in recent years the United States has imported increasing amounts of certain chemical resources. For example, about 75% of the nation's requirements for the metal tin (Sn) are currently met by imported supplies.

In addition to land, which provides most of our chemical resources, ocean waters contain significant dissolved amounts of compounds of nearly 20 metals. Also, solid nodules on the ocean floor contain as much as 24% manganese (Mn), 14% iron (Fe), and trace amounts of copper (Cu), nickel (Ni), and cobalt (Co).

Even if the oceans were to become a major new source of minerals, the total supply of nonrenewable resources "aboard" Earth would remain unchanged. We may be able to postpone, but cannot avoid, the possibility of running out of some nonrenewable resources. The central question still remains: How can we deal wisely with the world's resources? Conservation is the best answer.

C.2 CONSERVATION IS NATURE'S WAY: BALANCING CHEMICAL EQUATIONS

The law of conservation of matter is based on the notion that Earth's basic "stuff"—its atoms—are indestructible. All changes we observe in matter can be interpreted as rearrangements among atoms. Balanced chemical equations represent such changes.

It is often important to know how much of a desired element or compound we can obtain from a natural resource. The answer to such a question usually begins with a chemical equation. Earlier (page 95) you practiced how to recognize properly written chemical equations. Now you will learn how to write such balanced equations for yourself.

As an example, consider the reaction of hydrogen gas with oxygen gas, producing gaseous water. First, write the reactant formula(s) to the left of the arrow and the product formula(s) to the right keeping in mind that hydrogen and oxygen are *diatomic* molecules.

$$H_2(g) + O_2(g) \rightarrow H_2O(g)$$

Check this expression by completing an atom inventory: Two hydrogen atoms appear on the left and two on the right. So, the hydrogen atoms are balanced. However, there are two oxygen atoms on the left and only one on the right. Because oxygen is not balanced, the expression requires additional work.

Here is an *incorrect* way to complete the balanced equation:

$$H_2(g) + O_2(g) \rightarrow H_2O_2(g) \quad \textbf{Incorrect!}$$

Even though this chemical statement satisfies atom-inventory standards (two hydrogen and oxygen atoms on both sides), the expression is wrong. The answer changes the product's identity from water, H_2O, to hydrogen peroxide, H_2O_2. Hydrogen peroxide is *not* produced in this reaction; water is. Therefore, the expression above is incorrect, because it does not represent the correct substances.

The *correct* way to balance an equation is to place appropriate coefficients *before* chemical formulas. Select coefficients so that the atoms of each element become balanced. Here is one way to proceed:

To balance the oxygen atoms, write the coefficient 2 in front of water's formula:

$$H_2(g) + O_2(g) \rightarrow 2\ H_2O(g)$$

Blast furnace at one of the Inland Steel Company facilities.

▶ *Under certain conditions, this reaction can explode violently. When controlled, it can power some types of rockets. Used in fuel cells, it can generate electricity.*

▶ *Subscripts are not changed when balancing chemical equations. Coefficients must be used instead.*

Now two oxygen atoms appear on each side. But, unfortunately, now there are two hydrogen atoms on the left and four on the right—hydrogen is no longer balanced.

We can correct this by placing the coefficient 2 in front of the formula for hydrogen gas:

$$2\,H_2(g) + O_2(g) \rightarrow 2\,H_2O(g)$$

Two pairs of hydrogen atoms contain four hydrogen atoms. The equation is now balanced. When two molecules of hydrogen gas react with one molecule of oxygen gas, two molecules of gaseous water are formed. This is a correct symbolic summary of the reaction. It is based on nature's conservation of atoms.

Here are some additional rules of thumb to help you balance equations:

- Treat polyatomic ions, such as NO_3^- and CO_3^{2-}, as units rather than balancing their atoms individually.

- If water is present in a chemical equation, balance hydrogen and oxygen atoms last.

- Recount all atoms when you believe an equation is balanced—just to be sure!

YOUR TURN

Writing Balanced Chemical Equations

For another example of how a chemical statement can be converted to a chemical equation, consider this:

The reaction of methane (CH_4) with chlorine (Cl_2) occurs in sewage treatment plants and often in chlorinated water supplies. A common product is chloroform ($CHCl_3$). Chloroform is also made by the chemical industry for use in certain pharmaceutical preparations. A chemical statement describing this reaction is

$$\underset{\text{methane}}{CH_4(g)} + \underset{\text{chlorine}}{Cl_2(g)} \rightarrow \underset{\text{chloroform}}{CHCl_3(l)} + \underset{\substack{\text{hydrogen}\\\text{chloride}}}{HCl(g)}$$

To change this expression to a balanced equation, we can be guided by this line of reasoning: One carbon atom appears on each side of the arrow, so the carbon atoms balance. The coefficients for the two compounds containing carbon—CH_4 and $CHCl_3$—can be left as 1's. We will write these coefficients into the expression to help keep track of what we have done; they can be removed later:

$$1\,CH_4(g) + Cl_2(g) \rightarrow 1\,CHCl_3(l) + HCl(g)$$

Four hydrogen atoms are on the left, but only two on the right (one in $CHCl_3$, a second in HCl). To increase the number of hydrogen atoms on the right side, the coefficient for HCl must be adjusted. Because two more

▶ *Chloroform is one of the THMs mentioned in the Water unit, page 77.*

hydrogens are needed on the right side, the number of HCl molecules must be changed from 1 to 3. This gives four hydrogen atoms on the right:

$$1 \ CH_4(g) + Cl_2(g) \rightarrow 1 \ CHCl_3(l) + 3 \ HCl(g)$$

Now both carbon and hydrogen atoms are in balance. How about chlorine? We note there are two chlorine atoms on the left and six on the right side. These six chlorine atoms (3 in $CHCl_3$; 3 in 3 HCl) must have come from three chlorine (Cl_2) molecules. Thus, we use 3 as the Cl_2 coefficient. In keeping with common chemical practice, we remove the two "1" coefficients:

$$CH_4(g) + 3 \ Cl_2(g) \rightarrow CHCl_3(l) + 3 \ HCl(g)$$

Here is an atom inventory for the completed equation:

Reactant side	Product side
1 atom C	1 atom C
4 atoms H	4 H atoms (1 in $CHCl_3$, 3 in HCl)
6 atoms Cl	6 atoms Cl
(in 3 Cl_2 molecules)	(3 in each product)

Copy each of the following expressions onto your own paper and—if needed—balance the expression:

1. Two blast furnace reactions used to obtain iron from its ore:
   a. $C(s) + O_2(g) \rightarrow 2 \ CO(g)$
   b. $Fe_2O_3(s) + CO(g) \rightarrow Fe(l) + 3 \ CO_2(g)$

2. Two reactions in the refining of a copper ore:
   a. $Cu_2S(s) + O_2(g) \rightarrow CuO(s) + SO_2(g)$
   b. $CuO(s) + C(s) \rightarrow Cu(s) + CO_2(g)$

3. Ammonia (NH_3) in the soil reacts continuously with oxygen gas (O_2):

   $$NH_3(g) + O_2(g) \rightarrow NO_2(g) + H_2O(l)$$

4. Ozone, O_3, can decompose to form oxygen gas, O_2:

   $$O_3(g) \rightarrow O_2(g)$$

5. Copper metal reacts with silver nitrate solution to form copper(II) nitrate solution and silver metal:

   $$Cu(s) + AgNO_3(aq) \rightarrow Cu(NO_3)_2(aq) + Ag(s)$$

6. Combustion of gasoline in an automobile engine can be represented by:

   $$C_8H_{18}(l) + O_2(g) \rightarrow CO_2(g) + H_2O(g)$$

There are about 2×10^{25} (20,000,000,000,000,000, 000,000,000) molecules of water in this beaker.

▶ *Even if you connected a million paper clips each second, it would take you 190 million centuries to finish stringing a mole of paper clips.*

▶ *Recall that mol is the symbol for the "mole" unit.*

C.3 ATOM, MOLECULE, AND ION INVENTORY

As your answer to Question 2b above, you obtained the balanced equation

$$2\ CuO(s) + C(s) \rightarrow 2\ Cu(s) + CO_2(g).$$

Here is one interpretation of this equation: *Two formula units of copper(II) oxide and one atom of carbon react to produce two atoms of copper and one molecule of carbon dioxide.* This interpretation—although correct—involves such small quantities of material that a reaction on such a small scale would be completely unnoticed. Such information would not be very useful, for example, to a metal refiner who wants to know how much carbon would be needed to react with a certain amount of copper(II) oxide.

Chemists have devised a counting unit called the **mole** (symbolized mol) that helps solve the refiner's problem. You are familiar with other counting units such as "pair" or "dozen." A pair of water molecules would be two water molecules. One dozen water molecules refers to 12 water molecules. Likewise, one mole of water molecules is 602,000,000,000,000,000,000,000 water molecules. This number—the number of particles (or "things") in one mole—is more conveniently written as 6.02×10^{23}. Either way, this is a very large number!

Suppose you could string a mole of paper clips (6.02×10^{23} paper clips) together and wrap the string around the world. It would circle the world about 400 trillion (4×10^{14}) times.

However, as large as one mole of molecules is, drinking that amount of water would leave you quite thirsty on a hot day. One mole of water is less than one-tenth of a cup of water—only 18 g (or 18 mL) of water. But that is why the mole is so useful in chemistry. It represents a number of atoms, molecules, or formula units large enough to be conveniently weighed or measured in the laboratory. Furthermore, the atomic masses of elements can be used to find the mass of one mole of any substance, a value known as the substance's **molar mass.**

This notion can be developed through some specific examples. Suppose we want to find the molar masses of carbon (C) and copper (Cu). Or, putting it another way, we want to know the mass of one mole of carbon atoms and one mole of copper atoms (6.02×10^{23} atoms in either case).

Rather than counting these atoms onto a laboratory balance (good luck!), we can quickly get the answers from the elements' atomic masses. Here is how:

First find the atomic masses of these elements on the periodic table (located inside the back cover). Carbon's atomic mass is 12.01; copper's is 63.55. If we simply add units of "grams" to these values, we have their molar masses:

$$1\ mol\ C = 12.01\ g \qquad 1\ mol\ Cu = 63.55\ g$$

In brief, the mass (in grams) of one mole of an element's atoms equals the numerical value of the element's atomic mass. As you have seen, this is much more easily done than said!

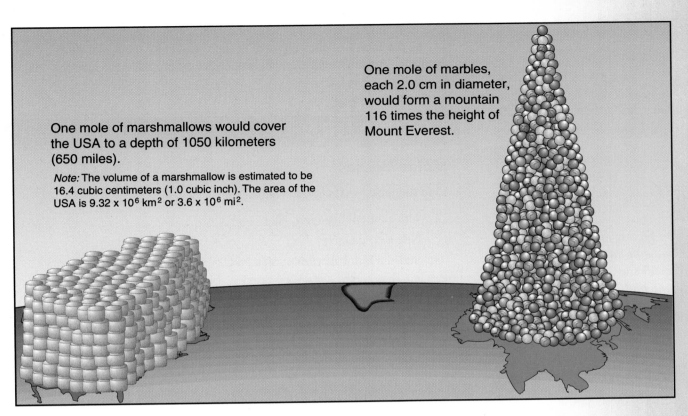

One mole of marshmallows would cover the USA to a depth of 1050 kilometers (650 miles).

Note: The volume of a marshmallow is estimated to be 16.4 cubic centimeters (1.0 cubic inch). The area of the USA is 9.32×10^6 km^2 or 3.6×10^6 mi^2.

One mole of marbles, each 2.0 cm in diameter, would form a mountain 116 times the height of Mount Everest.

The mole concept.

The molar mass of a *compound* is the sum of the molar masses of its component atoms. To illustrate this idea, consider two compounds of interest to the copper metal refiner—copper(II) oxide (CuO) and carbon dioxide (CO_2).

One dozen units of CuO contains one dozen Cu atoms and one dozen O atoms. Likewise, using the chemist's counting unit, one mole of CO_2 contains one mole of C atoms and two moles of O atoms.

Adding the molar mass of copper and that of oxygen atoms gives

$$1 \text{ mol Cu} \times \frac{63.55 \text{ g Cu}}{1 \text{ mol Cu}} = 63.55 \text{ g Cu}$$

$$1 \text{ mol O} \times \frac{16.00 \text{ g O}}{1 \text{ mol O}} = 16.00 \text{ g O}$$

Molar mass of CuO = (63.55 g + 16.00 g) = 79.55 g CuO

The molar mass of carbon dioxide—or any other compound—is found similarly.

$$1 \text{ mol C} \times \frac{12.01 \text{ g C}}{1 \text{ mol C}} = 12.01 \text{ g C}$$

$$2 \text{ mol O} \times \frac{16.00 \text{ g O}}{1 \text{ mol O}} = 32.00 \text{ g O}$$

Molar mass of CO_2 = (12.01 g + 32.00 g) = 44.01 g CO_2

In summary, the molar mass of a *compound* is found by first multiplying the moles of each element in the formula by the molar mass of the element. Then the total element molar masses are added to give the compound's molar mass.

YOUR TURN

Molar Masses

Find the molar mass of each substance:

1. The element nitrogen: N
2. Nitrogen gas: N_2
3. Sodium chloride (table salt): NaCl
4. Sucrose (table sugar): $C_{12}H_{22}O_{11}$
5. Chalcopyrite: $CuFeS_2$
6. Malachite: $Cu_2CO_3(OH)_2$ (*Hint:* This formula shows 5 mol of oxygen atoms)

► *Saying simply "one mole of nitrogen" can be confusing. It could refer to one mole of N atoms (Question 1) or a mole of N_2 molecules (Question 2). Therefore, it is important to specify what particular substance is involved.*

Now we can return to metal refining. The mole "counting unit" makes it easy to find the mass, for example, of carbon dioxide released during refining. The coefficients in a chemical equation show both the numbers of molecules (or formula units) of reactants and products *and* the numbers of moles of these substances:

2 CuO(s)	+	C(s)	→	2 Cu(s)	+	CO_2(g)
2 formula units of CuO		1 atom of C		2 atoms of Cu		1 molecule of CO_2

Thus, for every two moles of CuO that react, one mole of CO_2 will be formed. We already know the molar masses of all four substances in the equation:

2 CuO(s)	+	C(s)	→	2 Cu(s)	+	CO_2(g)
2 mol of CuO		1 mol of C		2 mol of Cu		1 mol of CO_2
159.10 g of CuO (2 × 79.55 g CuO/mol)		12.01 g of C per mole		127.10 g of Cu (2 × 63.55 g Cu/mol)		44.01 g of CO_2 per mole

► *2 mol of CuO is 159.10 g, containing 2 mol of copper atoms and 2 mol of oxygen atoms.*

Our metal refiner now knows that if 159.10 g of CuO is processed by this reaction, 127.10 g of Cu will be produced, along with 44.01 g of CO_2.

Check the table again. Note that the total mass of reactants (159.10 g + 12.01 g = 171.11 g) equals the total mass of products (127.10 g + 44.01 g = 171.11 g). This is a good illustration of the law of conservation of matter—in a chemical reaction matter is conserved; it is neither created nor destroyed.

Thanks to the mole concept, chemical equations allow chemists to account for the masses of all substances involved in chemical reactions. However, monitoring and accounting for resources used in the manufacture of materials and goods is more difficult.

C.4 CONSERVATION MUST BE OUR WAY

If nature always conserves, why do we read at times that a resource is "running out?" In what sense can we "run out" of a resource?

First, remember that nature conserves *atoms,* but not necessarily *molecules.* For example, nature's current production of petroleum molecules is far less than the rate at which society extracts and burns them, converting petroleum's molecules to carbon dioxide, water, and other molecules. We cannot put these smaller molecules together to make petroleum again.

We can deplete a resource—particularly a metal—another way. For profitable mining, an ore must contain some minimum percent of metal. (This limit depends on the metal ore—from as low as 1% for copper, or 0.001% for gold, to as high as 30% for aluminum.) As ores with high metal concentrations are depleted, lower concentration ores are processed. Meanwhile, through use, we disperse the originally concentrated resource. Atoms of the metal, once located in rich deposits in limited parts of the world, become "spread out" over the entire globe. Our economy may eventually become unable to support further extraction of such a metal for general use. For practical purposes, our supply of this resource will be depleted.

Can we avoid this situation? We know that nature conserves automatically at the atomic level. How can we conserve our resources? That is, how can we slow down the rate at which we use them?

You probably are familiar with the environmental "three R's": reduce, reuse, and recycle. In chemistry, these can be slightly revised as the "four R's" of resource conservation and management—rethink, replace, reuse, recycle. As an overall strategy, we should continually *rethink* our personal and societal habits and practices involving resource use. For example, when you respond to a grocery clerk's question "Paper or plastic?" what consequences—if any—does your answer have? Such rethinking can help us to re-examine old assumptions, identify resource-saving strategies, and—at times—uncover new solutions to old problems. Perhaps the most important part of rethinking resource conservation and management is to implement the most direct method—*source reduction.* This simply means decreasing the amount of resources that we use. The less of them we use, the more resources will remain available for future generations.

To *replace* a resource means finding substitute materials with similar properties—preferably materials from renewable resources.

Some manufactured items can be refurbished or repaired for *reuse* rather than sent to a landfill. Unwanted or outgrown clothing can be redistributed to others. Broken equipment can be repaired.

To *recycle* means to gather and collect certain items for reprocessing, allowing the resources present in them to be used again. A growing number of communities realize the wisdom of this aspect of resource

Which 2 of the "4 Rs" are represented here?

management. How is a recycling program planned and organized? Let's consider such an activity in detail.

C.5 YOU DECIDE: RECYCLING DRIVE

Our nation has been called a "throw away" society. We discard about 1.1 million tons of disposable plates and cups each year. That represents enough to offer six picnics annually to the entire world's population! In a more general sense, about 30% of U.S. production of major materials is eventually discarded.

Some critics say that our society overpackages food and consumer items. Packaging materials constitute about a half of all solid wastes in the United States by volume, and nearly a third by mass.

Think about the single-use packaging involved in many fast-food meals. The throw away inventory would include a hamburger container, disposable cup and plastic lid, plastic straw, paper french fry bag, plastic catsup pouch, paper packages for salt and pepper, paper napkin, and—if it is a take-out order—a paper bag to hold the meal!

Some packaging is necessary. Packaging protects ingredients, retains freshness, keeps small items together, and in many instances promotes cleanliness and safety. But even a simple ballpoint pen is often packaged in plastic and cardboard, and is placed in a bag by the cashier for you to take home. (And finally—like some other consumer items—the pen itself might be designed to be discarded rather than refilled or reused.)

What are the alternatives? One is to use less, simply reducing the amount thrown away as waste. Another is to buy in bulk to use less packaging material. Other alternatives are to use items for longer times or to

Consumers can exchange used cans and plastic and glass containers for money at these automatic recycling units.

reuse or recycle them. Many fast-food restaurants have begun using containers made of recycled paper and cardboard instead of plastic.

Recycling requires commitment and effort. What are its benefits? To find answers, let's examine three materials: paper, aluminum, and glass.

Paper. This is an important renewable resource. Because paper is made from tree pulp, new seedlings can be planted to replace trees cut down. However, "renewing" this resource takes time! It takes about 25 years for seedlings to grow to trees large enough to be economically useful. About 17 trees are needed to produce a ton of paper. That ton, large as it sounds, is only enough to supply two citizens with the paper they will use in one year!

Energy is required to make paper from a tree. Less than half as much energy is needed to process recycled paper as is used in making new paper. About 28% of paper is now being recycled.

Aluminum (Al). This is a nonrenewable chemical resource—the total number of aluminum atoms in the world is, for all practical purposes, fixed and unchanging. Once we have consumed this resource, there will be no more.

Aluminum is the most abundant metal in the earth's crust (8%). However, most of this aluminum is locked in silicates, from which it cannot be easily extracted. U.S. demand for aluminum is so great that domestic supplies of the ore (bauxite) do not meet national needs.

We import about 85% of aluminum resources used in this country. Producing aluminum from bauxite ore requires considerable energy. Recycling used aluminum consumes only 5-10% of the energy needed to produce aluminum from its ore. Thanks to organized national efforts, we now recycle over half of our used aluminum cans.

▶ *Aluminum production is discussed in detail in the Industry unit.*

Glass. A simple form of glass is made by melting together, at high temperatures, sand (silicon dioxide, SiO_2), soda ash (sodium carbonate, Na_2CO_3) and limestone (calcium carbonate, $CaCO_3$). All three materials are nonrenewable, but plentiful. It is estimated that about 20 to 25% of each new glass container currently made in the United States is produced from recycled, used glass.

According to recent analyses, glass recycling helps reduce a portion of the energy involved in glass manufacture. If 100% waste glass and recycled glass were used in new containers, energy savings in processing would be about 15%, with an additional 16% savings due to the lack of mining and transportation costs for new materials.

Working in small groups, discuss possible answers to the following questions. Your group's answers will be discussed later by the entire class.

1. Which of the three materials—paper, aluminum, or glass—is most important to recycle

   a. for economic reasons?

   b. for environmental reasons?

   Explain your answers.

2. a. If recycling is important, should the federal government require that certain materials be recycled?

b. If so, identify some materials that should be recycled under such a law.

c. How could such a law be enforced?

3. As individuals, we can replace, reuse, or recycle materials in various ways. For example, we can use both sides of paper for writing, or—when given a choice—purchase beverages in returnable bottles and return them. Identify at least five other ways we, as individuals, can replace, reuse, or recycle certain materials.

4. We plant forests to supply our paper needs. Assume that all printed and typewritten communication is replaced in the future by computer-based electronic networks.

a. What current uses of paper would cease?

b. What occupations or jobs would be eliminated?

c. What occupations or jobs would be created?

d. What kind of chemistry book would you be reading instead of this paper-based one?

e. How else would your daily routine change due to this technological advancement?

? PART C: SUMMARY QUESTIONS

1. a. List and briefly describe the three major physical regions of Earth.

   b. Which region serves as the main "storehouse" of chemical resources used in manufacturing consumer products?

2. Write balanced chemical equations for each of these:

   a. Preparing phosphoric acid (used in making soft drinks, detergents, and other products) from phosphate rock and sulfuric acid:

   $$Ca_3(PO_4)_2(s) + H_2SO_4(aq) \rightarrow H_3PO_4(aq) + CaSO_4(s)$$

   b. Preparing tungsten from its ore:

   $$WO_3(s) + H_2(g) \rightarrow W(s) + H_2O(l)$$

   c. Heating lead sulfide ore in air:

   $$PbS(s) + O_2(g) \rightarrow PbO(s) + SO_2(g)$$

3. Describe an everyday routine activity you would be willing to change in order to reduce problems of solid waste disposal.

4. Find the molar mass of each substance:

   a. The element oxygen, O

   b. Oxygen gas, O_2

   c. Ozone, O_3

   d. Caffeine, $C_8H_{10}N_4O_2$

   e. Lye drain cleaner, NaOH

   f. Aspirin, $C_9H_8O_4$

5. One method to produce chromium metal includes, as the final step, the reaction of chromium(III) oxide with silicon at high temperature:

   $$2\ Cr_2O_3(s) + 3\ Si(s) \rightarrow 4\ Cr(s) + 3\ SiO_2(s)$$

   a. How many moles of each reactant and product are shown in this chemical equation?

   b. What mass (in grams) of each reactant and product is specified by this equation?

   c. Show how this equation illustrates the law of conservation of matter.

D METALS: SOURCES AND REPLACEMENTS

Technological advances and changes in how and where we live can rapidly alter our resource needs. Even though such changes are difficult to predict, we must be prepared to deal with them.

Copper serves as a useful case study of present and projected uses of a vital chemical resource. We will consider sources of our copper supply and how these materials are converted to pure copper. You will produce a sample of metallic copper. Finally, we will look at some possible replacements for this resource.

D.1 COPPER: SOURCES AND USES

Copper is among the most familiar and widely used metals in modern society. Among the elements, it has the second-highest electrical conductivity (silver is first). This property, together with its low cost, corrosion resistance, and ability to be drawn easily into thin wires, has made copper the world's most common metal for electrical wiring. It is also used in the manufacture of brass, bronze, and other copper alloys, a variety of important copper-based compounds, jewelry, and works of art.

Copper resources are widely, but unevenly, distributed throughout the world. The United States has been a major world supplier of copper ore. Canada, Chile, Peru, and Zambia also have significant supplies of copper ore.

Can adequate profit be earned from mining a particular metallic ore at a certain site? The answer depends on several factors:

- the metal's supply–demand status
- type of mining and processing needed to obtain the metal
- amount of useful ore at the site
- distance of the mine from metal refining facility and markets
- percent of metal in the ore.

The first copper-rich ores mined contained 35–88% copper. Such ores are no longer available, but less-rich ores also can be profitable. In fact, it is now possible economically to mine ores with less than 1% copper. Copper ore is chemically processed to produce metallic copper, which is then formed into a variety of useful materials. Figure 9 (page 136) summarizes the copper cycle from sources to uses to waste products.

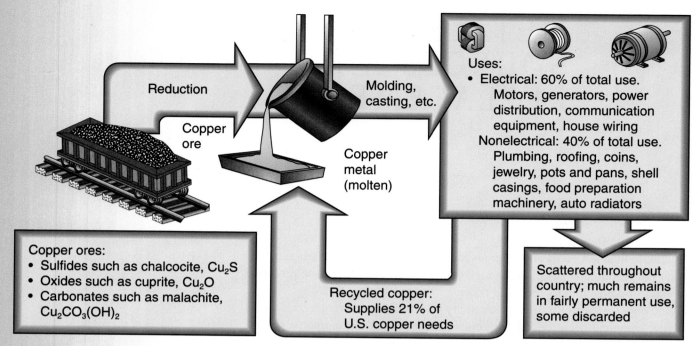

Figure 9 The copper cycle.

Reduction

Copper ore

Molding, casting, etc.

Copper metal (molten)

Uses:
• Electrical: 60% of total use. Motors, generators, power distribution, communication equipment, house wiring
Nonelectrical: 40% of total use. Plumbing, roofing, coins, jewelry, pots and pans, shell casings, food preparation machinery, auto radiators

Copper ores:
• Sulfides such as chalcocite, Cu_2S
• Oxides such as cuprite, Cu_2O
• Carbonates such as malachite, $Cu_2CO_3(OH)_2$

Recycled copper: Supplies 21% of U.S. copper needs

Scattered throughout country; much remains in fairly permanent use, some discarded

Are we doomed to deplete the rich deposits of this valuable resource? Will future developments increase or decrease our need for copper? What copper substitutes are available? This activity will help you answer these questions.

YOUR TURN

Uses of Copper

Some copper properties are listed in Table 5. Let's consider how these properties make copper suitable for uses listed in Figure 9.

Table 5	Properties of Copper
Malleability and ductility	High
Electrical conductivity	High
Thermal conductivity	High
Chemical reactivity	Relatively low
Resistance to corrosion	High
Useful alloys formed	Bronze, brass, etc.
Color and luster	Reddish, shiny

Here is an example: What properties make copper suitable for use in electrical power generators? Certainly copper's high electrical conductivity is essential to this application. Copper's malleability and ductility are also important, making it possible to form copper wires and wrap them in a generator. Corrosion resistance is also a benefit in such large, expensive equipment.

1. Consider the remaining copper uses that are listed in Figure 9 (page 136). For each, identify particular properties that explain copper's appropriateness.

2. a. How would increased recycling of scrap copper affect future availability of this metal?

   b. Is there a limit to the role copper recycling can play?

   c. Why?

3. For each copper use listed below, describe a technological change that could decrease the demand for copper:

   a. Communications

   b. Coins

   c. Power generation

   d. Wiring inside appliances

D.2 EVALUATING AN ORE

How do geologists know how much copper or other metal is present in a particular ore? Some fundamental chemical ideas apply.

A compound's formula indicates the relative number of atoms of each element in the substance. For example, one common commercial source of copper is the mineral chalcocite—copper(I) sulfide, Cu_2S. Its formula tells us that copper(I) sulfide contains two copper atoms for each sulfur atom. The formula can also help us find out how much of the mineral's mass can be converted to pure copper—an important factor in copper mining and production.

Chalcocite.

In short, the percent of metal in an ore helps determine whether a particular deposit is worth mining. Apply your knowledge of chemical composition in completing this exercise.

Mine at Bingham Canyon, Utah—largest human-made hole on Earth.

Percent Composition

Table 6	Copper-containing Minerals
Common Name	**Formula**
Chalcocite	Cu_2S
Chalcopyrite	$CuFeS_2$
Malachite	$Cu_2CO_3(OH)_2$
Azurite	$Cu_3(CO_3)_2(OH)_2$

Some copper-containing minerals are listed in Table 6. Let's find the percent copper in chalcocite, Cu_2S, and in an ore containing 5.0% chalcocite.

From the formula Cu_2S we know that one mole of Cu_2S contains two moles of Cu, or (as shown in the margin) 127.1 g Cu. We also know that the molar mass of Cu_2S is (2×63.55 g) + 32.00 g = 159.2 g. Therefore,

$$\% \ Cu = \frac{\text{Mass of Cu}}{\text{Mass of } Cu_2S} \times 100 =$$

$$\frac{127.1 \text{ g Cu}}{159.2 \text{ g } Cu_2S} \times 100 = 79.84\% \ Cu$$

A similar calculation (try it yourself) indicates that Cu_2S contains 20.16% sulfur. Note that the sum of percent copper and percent sulfur equals 100.00%. (Why?)

To find the percent copper in the ore, assume we have a 100-g sample of the ore. Because the ore is 5.0% chalcocite, 100 g of ore will contain 5.0 g chalcocite; that is, (5.0 g chalcocite/100 g ore) $\times$ 100 g ore = 5.0 g chalcocite. From our first calculation, we know that 79.84% of this mass is copper. Thus

$$5.0 \text{ g } Cu_2S \times 0.7984 = 4.0 \text{ g Cu}$$

If 4.0 g of copper is present in 100 g of ore, the ore contains 4.0% copper, (4.0 g/100 g) $\times$ 100 = 4.0%. This value could also be obtained by using the relationship:

$$5.0 \text{ g } Cu_2S \times \frac{127.1 \text{ g Cu}}{159.2 \text{ g } Cu_2S} = 4.0 \text{ g Cu}$$

1. a. Calculate the percent copper in each of the last three copper minerals listed in Table 6.

   b. Which of the three could be mined most profitably? Assume that each mineral was present at the same concentration in an ore. Also assume that copper could be extracted from each ore at the same cost.

2. Two common iron-containing minerals are hematite (Fe_2O_3) and magnetite (Fe_3O_4). If you had an equal mass (such as 1 kg) of each, which sample would contain the greater amount of iron? Support your answer with calculations.

$$\blacktriangleright 2 \ mol \ Cu \times \frac{63.55 \text{ g } Cu}{1 \ mol \ Cu} =$$

$$127.1 \text{ g } Cu$$

3. Assume that a 100-g sample of iron ore from Site A contains 20 g of Fe_2O_3, while a 100-g sample from Site B contains 15 g of Fe_3O_4.

   a. What is the mass of iron in each sample?

   b. Which ore contains the larger percent of iron?

D.3 METAL REACTIVITY REVISITED

Humans have been described as "tool-making animals." Readily available stone, wood, and natural fibers became the earliest tool materials. A variety of useful implements could be made from such naturally occurring materials. However, the discovery that fire could transform materials in certain rocks into strong, malleable metals triggered a dramatic leap in civilization's growth.

Gold and silver, found as free elements rather than in chemical combination with other elements, were probably the first metals used by humans. These metals were formed into decorative objects. It is estimated that copper has been used for 10,000 years for tools, weapons, utensils, and decorations. Bronze, an alloy of copper and tin, was invented about 3800 B.C. Thus, humans moved from the Stone Age into the Bronze Age.

Eventually early people developed iron metallurgy, the extraction of iron from its ores. This led to the start of the Iron Age more than 3,000 years ago. In time, as humans learned more about chemistry and fire, a variety of metallic ores were transformed into increasingly useful metals.

CHEMQUANDARY

DISCOVERY OF METALS

Copper, gold, and silver are not Earth's most abundant metals. Aluminum, iron, and calcium, for example, are all much more plentiful. Then why were copper, gold, and silver among the first metallic elements discovered?

You explored some chemistry of metals in the laboratory activity Metal Reactivities (page 117). You found, for example, that copper is more reactive than silver, but less reactive than magnesium. A more complete activity series is given in Table 7 (page 140), which includes brief descriptions of common methods for retrieving these metals from their ores.

The most reactive metals are listed first; less reactive metals are closer to the bottom. Reactive metals are more difficult to obtain from their ores than are less reactive metals.

In general, *a more reactive metal will release a less reactive metal from its compounds.* An activity list can be used to predict whether certain reactions can be expected. For example, you found in the laboratory that zinc metal, which is more reactive than copper, will react with copper ions in solution. However, zinc metal will not react with magnesium ions in solution, because zinc is less reactive than magnesium.

Use Table 7 and the periodic table to complete the following.

Table 7	Metal Activity Series		
Element	Metal Ions(s) Found in Ore	Metal Obtained	Reduction Process Used To Obtain the Metal
Lithium	Li^+	$Li(s)$	Electric current passed through the molten salt (electrolysis)
Potassium	K^+	$K(s)$	
Calcium	Ca^{2+}	$Ca(s)$	
Sodium	Na^+	$Na(s)$	
Magnesium	Mg^{2+}	$Mg(s)$	
Aluminum	Al^{3+}	$Al(s)$	
Manganese	Mn^{2+}	$Mn(s)$	Heated with coke (carbon) or carbon monoxide (CO)
Zinc	Zn^{2+}	$Zn(s)$	
Chromium	Cr^{3+}, Cr^{2+}	$Cr(s)$	
Iron	Fe^{3+}, Fe^{2+}	$Fe(s)$	
Lead	Pb^{2+}	$Pb(s)$	Element occurs free (uncombined) or is obtained by heating in air (roasting)
Copper	Cu^{2+}, Cu^+	$Cu(s)$	
Mercury	Hg^{2+}	$Hg(l)$	
Silver	Ag^+	$Ag(s)$	
Platinum	Pt^{2+}	$Pt(s)$	
Gold	Au^{3+}, Au^+	$Au(s)$	

YOUR TURN

Metal Reactivity

1. What trend in metallic reactivity is found from left to right across a horizontal row (period) of the periodic table? (*Hint:* Compare the reactivities of sodium, magnesium, and aluminum. In which part of the periodic table are the most reactive metals found? Which side of the periodic table contains the least reactive metals?)

2. a. Will iron (Fe) metal react with a solution of lead nitrate, $Pb(NO_3)_2$?

   b. Will platinum (Pt) metal react with a lead nitrate solution?

   c. Explain your answers.

3. Compare the three reduction processes described in the activity table.

   a. Which would you expect to require the greatest quantity of energy? Why?

   b. Which is likely to be least expensive? Why?

4. The least reactive metals are easiest to obtain from their ores. Use specific examples from the activity series in your answers to these questions:

   a. Would least-reactive metals necessarily be the cheapest metals?

   b. If not, what other factor(s) influence the market value of a given metal?

CHEMISTRY AT WORK

All That's Gold Doesn't Glitter . . . Or Does It?

Did you ever wonder where the metal comes from that makes up a high school gold ring—or who was responsible for finding the material in the first place? **Sandy Haslem** might have helped. She's a metallurgist with Barrick Goldstrike, a mining company in Elko, Nevada, that finds and recovers gold.

It's not an easy job. The concentration of gold in the earth's crust is only about 15 parts per billion (ppb). But by using a knowledge of chemistry and highly efficient recovery methods, Sandy and her coworkers can extract gold from material that contains as little as 0.47 grams of gold per metric ton (1,000 kg)!

One method of gold recovery is "heap leaching." Material thought to contain gold is stacked on an impermeable pad, and a dilute basic solution containing cyanide ions (CN^-) is dripped down through the heap. A gold cyanide complex, $[Au(CN)_2]^-$, forms with any gold atoms that it contacts:

$$4\,Au + 8\,CN^- + 2\,H_2O + O_2 \rightarrow 4\,[Au(CN)_2]^- + 4\,OH^-$$

Later, the gold is recovered through a process called electrowinning, which uses an electrical current to transfer the gold onto steel wool. Finally, technicians heat the gold-plated steel wool to liquify the gold, and pour the liquid gold into bar molds.

Sandy supervises lab employees who research and develop safe and economical methods for recovering gold. Sandy and her team also monitor a mine's processing operations and make sure that the environment isn't degraded as a result of those operations.

Some Questions To Ponder . . .

Your class ring is not 100% pure gold. Pure gold is very soft, so it is mixed with other metals (copper, silver, nickel, or zinc) to increase its

hardness and durability. The more of the other metal added, the lower the karat value. Pure gold is 24 karat. Your ring is probably 12 karat, or about 50% gold.

Because of the specialized processes it uses, Sandy's company is able to profitably recover gold from fairly low-grade ore—that is, ore containing extremely small traces of gold. A more common ore grade for the gold mining industry is 1.6 grams of gold per metric ton of ore.

1. If a class ring has a mass of 11 g, how many metric tons of ore were processed in order to produce the gold to make the ring?

2. What is the value of the gold in the class ring, if the market value for gold is $14 per gram? (You may want to try using today's actual price for gold, available in the business section of many newspapers).

3. Gold miners in the last century looked for gold nuggets. What property of gold allowed them to be able to find it in this form?

D.4 METALS FROM ORES

The process of converting a combined metal (usually a metal ion) in a mineral into a free metal is called *reduction.* This term has a specific chemical meaning. To convert metal ions to metal atoms, the ions must gain electrons. *Any process in which electrons are gained by a species is called "reduction."* Reduction of one copper(II) ion, for example, requires two electrons:

$$\textbf{\textit{Reduction:}} \quad \underset{\substack{\text{copper(II)} \\ \text{ion}}}{Cu^{2+}} \quad + \quad \underset{\substack{\text{two} \\ \text{electrons}}}{2\,e^-} \quad \rightarrow \quad \underset{\substack{\text{copper} \\ \text{metal}}}{Cu}$$

The reverse process, in which an ion or other species loses electrons, is called **oxidation.** Under the right conditions copper *atoms* can be oxidized:

$$\textbf{\textit{Oxidation:}} \quad \underset{\substack{\text{copper} \\ \text{metal}}}{Cu} \quad \rightarrow \quad \underset{\substack{\text{copper(II)} \\ \text{ion}}}{Cu^{2+}} \quad + \quad \underset{\substack{\text{two} \\ \text{electrons}}}{2\,e^-}$$

▶ *A way to remember this is "**OIL RIG**"; **O**xidation **I**s **L**oss of electrons, **R**eduction **I**s **G**ain of electrons.*

We live in an electrically neutral world. Whenever one species loses electrons another must simultaneously gain them. That is, oxidation and reduction never occur separately. They occur together in what chemists call **oxidation–reduction reactions** or simply **redox reactions.**

You have already observed redox reactions in the laboratory. For example, in the laboratory activity Metal Reactivities (page 117), some metals were oxidized. Here is one oxidation–reduction reaction you observed:

$$\underset{\substack{\text{copper} \\ \text{metal}}}{Cu(s)} \quad + \quad \underset{\substack{\text{silver} \\ \text{ion}}}{2\,Ag^+(aq)} \quad \rightarrow \quad \underset{\substack{\text{copper(II)} \\ \text{ion}}}{Cu^{2+}} \quad + \quad \underset{\substack{\text{silver} \\ \text{metal}}}{2\,Ag(s)}$$

Metallic copper atoms were oxidized (converted to Cu^{2+} ions) and silver ions (Ag^+) from $AgNO_3$ solution were reduced (converted to Ag atoms).

In the same activity you found that copper ions could be recovered from solution as copper metal by reaction with magnesium metal, a more active element than copper. Magnesium atoms were oxidized, copper ions reduced:

$$\underset{\substack{\text{copper(II)} \\ \text{ion}}}{Cu^{2+}} \quad + \quad \underset{\substack{\text{magnesium} \\ \text{metal}}}{Mg(s)} \quad \rightarrow \quad \underset{\substack{\text{copper} \\ \text{metal}}}{Cu(s)} \quad + \quad \underset{\substack{\text{magnesium} \\ \text{ion}}}{Mg^{2+}(aq)}$$

In some circumstances this might be a useful way to obtain copper metal. However, as is always the case, to obtain a desired product, something else is used up—in this case, magnesium, another desirable metal.

How do redox reactions occur? Many metallic elements are found in their minerals as ions because they combine readily with other elements to form ionic compounds. Obtaining a metal from its mineral requires not only energy, but also a way to reduce the metal ions to the metal. A source of electrons, known as a **reducing agent,** must be provided. As Table 7 (page 140) shows, a variety of techniques are used, depending on the metal's reactivity and the availability of inexpensive reducing agents and energy sources. We will consider the methods used.

Molten aluminum metal from electrolytic cells.

Pyrometallurgy is the treatment of metals and their ores by heat, as in a blast furnace. Two of the three reduction processes shown in Table 7 are based on pyrometallurgy—heating an ore in air (roasting) and heating it with an added reducing agent. Carbon (coke) and carbon monoxide are the common reducing agents. A more active metal can be used if neither of these will do the job. Pyrometallurgy is the most important and oldest ore-processing method.

Electrometallurgy involves using an electric current to supply electrons to metal ions, thus reducing them. This process is used when no adequate reducing agents are available or when very high metal purity is needed. Electrometallurgy includes electroplating, electrorefining, and using an electric arc furnace to make steel.

Hydrometallurgy is the treatment of ores and other metal-containing materials by reactants in water solution. You used such a procedure when you compared the reactivity of four metals as directed on page 117. Hydrometallurgy is not commonly used industrially due to the high cost of the more active metal. However, as lower grade ores must be used, it will become economically feasible to use hydrometallurgy and other "wet processes" with minerals that can be dissolved in water.

In the following laboratory activity you will try your hand at electrometallurgy.

D.5 LABORATORY ACTIVITY: PRODUCING COPPER

GETTING READY

In this activity you will produce copper metal from a solution of copper(II) chloride ($CuCl_2$) by electrolysis. A 9-V battery will provide the electric current.

▶ *In the industrial refining of copper, pure copper is obtained by electrometallurgy from less-pure copper produced by pyrometallurgy.*

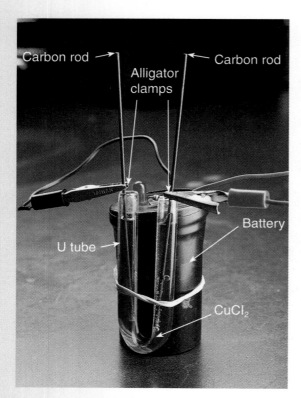

Figure 10 Apparatus for electrolysis of copper(II) chloride solution.

Industrial-scale electrolysis uses large quantities of electrical energy, making it an expensive—although effective—way to obtain or purify metals.

PROCEDURE

1. Obtain two graphite (carbon) rods in the form of pencil lead. (Depending on the size of the electrolysis apparatus you will use, the graphite rods may still be part of two wooden pencils. If so, be sure graphite protrudes from both ends of each pencil so electrical contact can be made.) These graphite rods will serve as the terminals or **electrodes** for the electrolysis process.

2. Set up the apparatus shown in Figure 10. Attach the 9-V battery connector to the battery, but *do not* connect the wire leads to the electrodes or allow the two wires to touch each other.

3. Use a funnel to pour just enough copper(II) chloride ($CuCl_2$) solution into the U-tube so the graphite electrodes can be partially immersed in the solution.

4. Have your teacher approve the setup.

5. Clamp wires to the two graphite terminals.

6. Observe the reaction for approximately five minutes. Record your observations in your notebook.

7. *Cautiously* sniff each electrode.

8. Reverse the wire connections to the electrodes and repeat Step 6.

9. Wash your hands thoroughly before leaving the laboratory.

QUESTIONS

1. Describe changes observed during the electrolysis.

2. The **cathode** is the terminal (electrode) at which reduction occurs.
   a. Which electrode was the cathode in the electrolysis?
   b. What change did you observe at the cathode?

3. The other electrode, the **anode,** is where oxidation occurs.
   a. Which electrode was the anode in the electrolysis?
   b. What change did you observe at the anode?

4. Write a balanced equation for the overall chemical change in this electrolysis reaction. The oxidation product is chlorine gas, $Cl_2(g)$.

D.6 FUTURE MATERIALS

As we continue extracting and using chemical resources from Earth, we are sometimes forced to consider alternatives. One option is to find new materials that can substitute for less-available resources. An ideal substitute satisfies three requirements: Its useful properties should match or exceed those of the original material; it should be plentiful; and, of course, it should be inexpensive. Substitute materials seldom meet these conditions completely. Thus, we must consider the benefits and burdens involved in such substitutions. We will consider the promises and challenges of several representative new materials in this section.

Ceramics are both ancient and new. Clay is one of the most plentiful materials on this planet. It is mainly composed of silicon, oxygen, and aluminum, along with magnesium, sodium, and potassium ions and water molecules. Early humans found that clay mixed with water, then molded and heated, formed useful ceramic products such as pottery and bricks.

In more recent times, researchers found that when other common rock materials are heated to high temperatures, useful "fired" compounds, also called ceramics, can form. Figure 11 (page 146) compares the sources, processing, and products of conventional ceramics with newer, stronger engineering ceramics.

What properties of conventional ceramics made them useful in pottery and bricks? Characteristics such as hardness, rigidity, low chemical reactivity, and resistance to wear were certainly important. The main attractions of ceramics for future use, however, are their high melting points and their strength at high temperatures. They might become, in fact, attractive substitutes in some steel applications. For example, scientists believe that diesel or turbine engines made of ceramics might be able to operate at higher temperatures. Such high-temperature engines would run with increased efficiency, reducing fuel use.

The major problem still facing researchers is that ceramics are brittle. They can fracture if exposed to rapid temperature changes, such as during hot-engine cool-down. Great hope remains, however, for the future of ceramics. They have become vitally important materials.

Plastics have already replaced metals for many uses. These synthetic substances are composed of complex carbon-atom chains and rings with hydrogen and other atoms attached. Plastics generally weigh less and can be designed to be "springy" or resilient in situations where metals might become dented. Plastic bumpers on automobiles are one example.

Plastics can be designed with a wide range of properties, from softness and flexibility to hardness, rigidity, and brittleness. Unfortunately, plastics are made from petroleum, an important nonrenewable resource already in great demand as a fuel (see Petroleum unit).

Optical fibers have already revolutionized communications. Voice or electronic messages can be sent through these thin, specially designed glass tubes of very pure silicon dioxide (SiO_2) as pulses of laser light. As many as 50,000 phone conversations or data transmissions can take place simultaneously in one glass fiber the thickness of a human hair. A typical 72-strand optical fiber ribbon can carry well over a million messages.

Optical fibers are well on their way toward replacing conventional copper wires in phone and data transmission lines. The fiber's larger carrying capacity and noise-free characteristics outweigh their higher initial cost. Some forecasters predict that at least half of all U.S. homes will have optical fiber installed by 2015. A fundamentally new way to send signals has been created.

Optical fibers. Single optical fiber filaments such as these can be as long as 100 km.

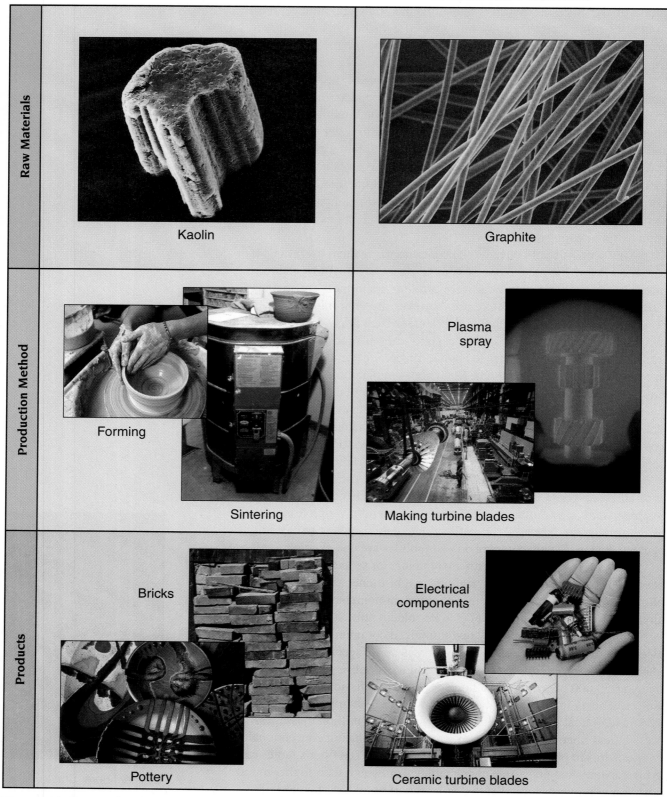

CONVENTIONAL CERAMICS

ENGINEERING CERAMICS

Raw Materials

Kaolin

Graphite

Production Method

Forming

Sintering

Plasma spray

Making turbine blades

Products

Bricks

Pottery

Electrical components

Ceramic turbine blades

Figure 11 Conventional ceramic products such as bricks and pottery are made from clay. Engineering ceramics, which may be higher melting, stronger, or less brittle than conventional ceramics, are produced from various natural and synthetic minerals.

Advanced polymer composites find use already in a variety of important applications, ranging from "Stealth" aircraft and race cars to rocket motor cases, skis, and inline skates. These fiber-reinforced resins have a variety of attractive features, including their low densities, ability to withstand high temperatures, and strength and dimensional stability. Such composites have replaced metals in many applications. Current advanced polymer composite prices can be as high as hundreds of dollars per pound, but such composites deliver unique and highly desirable properties.

By now it should be apparent that chemists, chemical engineers, and materials scientists continue to find new and better alternatives to traditional materials for a variety of applications.

Now it's time to try your hand regarding possible uses of some alternative materials.

YOUR TURN

Alternatives to Metals

1. Select four copper uses from the list found in Figure 9 (page 136). For each, suggest an alternative material that could serve the same purpose. Consider both conventional materials and possible new materials.

2. Suggest some common metallic items that might be replaced by ceramic or plastic versions.

3. Suppose silver became as common and inexpensive as copper. In what uses would silver most likely replace copper? Explain.

? PART D: SUMMARY QUESTIONS

1. Give at least two reasons why estimates of future resource supplies might be highly uncertain.

2. a. Why is metal recycling important?
   b. Does such recycling guarantee we will always have sufficient supplies of a given metal? Why?

3. In selecting a metal to meet a particular need, chemists and engineers consider the properties desired to meet the need.

   a. List three uses of copper metal.
   b. For each, list the property or properties that explain copper's use for that particular application.

4. Give the chemical name and find the percent metal (by mass) in each compound:
   a. Ag_2S
   b. Al_2O_3
   c. $CaCO_3$

5. A 100-g sample of ore contains 10 g of lead(II) sulfate, $PbSO_4$.

   a. What is the percent $PbSO_4$ in this ore?

   b. What is the percent Pb in $PbSO_4$?

   c. Finally, what is the percent Pb in the original ore sample?

6. Consider these two equations. Which reaction is more likely to occur? Why?

   a. $Zn^{2+}(aq) + 2 Ag(s) \rightarrow 2 Ag^+(aq) + Zn(s)$

   b. $2 Ag^+(aq) + Zn(s) \rightarrow Zn^{2+}(aq) + 2 Ag(s)$

7. a. Would it be a good idea to stir a solution of lead nitrate with an iron spoon? Explain.

   b. Write a chemical equation to support your answer.

8. Which two families (groups) of elements contain the most active metals?

9. Three different types of reduction processes are used to process metallic ores (Table 7, page 140). What *similarities* do these different processes share?

10. Consider the following reaction:

$$Zn(s) + Ni^{2+}(aq) \rightarrow Zn^{2+}(aq) + Ni(s)$$

   a. Which reactant has been oxidized? Explain your choice.

   b. Which reactant has been reduced? Explain your choice.

   c. What is the reducing agent in this reaction?

E PUTTING IT ALL TOGETHER: MAKING SENSE OF CENTS

Since antiquity, humans have used various materials as forms of currency. Such "money" in early cultures came from valued materials available locally. The particular materials used have varied enormously with time, region, and culture—sperm whale teeth (Fiji Islands), bricks of pressed tea (China), beetle legs (Mussan Islands), beaver pelts and wampum (North American), and stones and beads (many cultures). Archeological evidence points to the fact that metals have been used in currency for at least 4,000 years.

The first metal pieces, made of silver and gold alloy, specifically stamped as coins appeared about 2,600 years ago. These small and limited numbered coins are a far cry from the more than 20 million coins produced annually by the U.S. Mint.

E.1 WHAT IS MONEY, ANYWAY?

As civilizations evolve, the nature of their currencies changes, for a variety of reasons. For example, in 1982 the percent of copper in a U.S. penny was lowered because the one-cent coin was costing more than one cent to produce. Less expensive zinc replaced almost all of the copper (page 90). Great Britain's sterling silver coin used to be 92.5% silver, but since 1947 has been replaced by a coin containing 75% copper and 25% nickel.

What considerations must be taken into account when changing a currency or designing a new one? That's the question for you and your classmates to consider with the following monetary challenge. A new country has been formed and needs its own currency. The country has brought in you and your classmates as consultants to help to design a new metal currency for this nation.

In developing your design, it is important to remember the general idea of what has been a focus of this chapter—the properties of a material dictate what its uses are. For example, a low-melting material could not be used to make a frying pan or baking dish; those items require a substance with a high melting point. Therefore,

• What characteristics do you want the new currency to have?

• What chemical properties should the new currency have?

• What physical properties would be important for the new currency?

• What other design features are important to people who use the currency?

These questions and others you and your group may raise, plus the concepts in this unit, should help to guide you in selecting what kind of material you will choose for your currency.

Your teacher will assign you to work in a small group. Each group will develop its own composition and design for a new currency, and then share that design with the whole class.

E.2 LOOKING BACK AND LOOKING AHEAD

In ending this *ChemCom* unit, you can stop and take stock of what you have learned thus far. To date, you have "uncovered" some of chemistry's working language (symbols, formulas, and equations), laboratory techniques, laws (law of conservation of matter and the periodic law) and theories (atomic). This knowledge can help you better understand some important societal issues. Central among these issues is the use and abuse of Earth's chemical resources. Water, metals, petroleum, food, air, basic industries, and even personal health are all resources that need wise management to obtain maximum benefits for all people, while minimizing environmental costs.

We have also explored other factors that enter into policy decisions concerning technological problems. Although chemistry is often a crucial ingredient in recognizing and resolving such issues, many problems are too complex for a simple technological "fix." Issues of policy are not usually black or white, but many shades of gray. As voting citizens you will be concerned with a variety of issues that require some scientific understanding. Tough decisions may be needed. The remaining units in *ChemCom* will continue to prepare you for this important role.

Next we will deal with petroleum, such an important nonrenewable chemical resource that it deserves its own unit.

CHEMICAL ELEMENTS CROSSWORD PUZZLE

To be used with *Your Turn* (page 108). Here are clues for the crossword puzzle your teacher will distribute.

DOWN

1. An unreactive, gaseous element that is a product of the nuclear reaction (fusion) of hydrogen atoms. This reaction occurred at the beginning of time and occurs today in stars such as our sun. The second most abundant element in the universe, it is quite rare on Earth. Small concentrations are found in some natural gas deposits. It is used in blimps because of its low density. (Only hydrogen, which is highly flammable, has a lower density.) It is also used in cryogenic (low-temperature) work because it can be compressed to a liquid that has a temperature of -269 °C.

2. A reactive, metallic element. Its compounds are used as a medical "cocktail" to outline the stomach and intestines for X-ray examination. Its compounds also give fireworks green colors.

4. A highly reactive metal. It is used in the manufacture of synthetic rubber and drugs. One of its compounds has been used successfully to treat a certain type of mental illness. It finds limited use in nuclear weapons.

6. A widely distributed nonmetal never found naturally in its elemental state uncombined with any other element. It is an essential component in all cell protoplasm, DNA, and various animal tissues and bones. It is also one of the three main elements in fertilizers.

9. An unreactive gas. In the comic book world, a mineral containing this element could weaken Superman. In the real world, a radioactive form of this element is a by-product of most nuclear explosions; its presence in the atmosphere can indicate whether a nation is testing nuclear weapons.

10. A reactive metal with a high melting point. It is used in manufacturing rocket nose cones because this low-density substance is remarkably strong.

12. A reactive, silver-white metal that is second in abundance to sodium in ocean water. Due to its low density and high strength, its alloys are often used for structural purposes in the transportation industry, as in "mag" wheels. It is also used in fireworks and incendiary bombs because it ignites readily. Some of its compounds, such as Epsom salt and milk of magnesia, have medicinal uses.

14. A component of all living matter and fossil fuels; the black material on a charred candle wick.

18. Nicknamed "quicksilver," it is the only metallic element that is a liquid at room temperature. It is used in thermometers because it expands significantly and uniformly when heated. Its high density makes it a practical substance to use in barometers used to measure atmospheric air pressure. It is a toxic "heavy" metal.

19. The lightest and most abundant element; the fuel of the universe. It is believed that all other elements were originally formed from a series of stellar nuclear reactions beginning with this element. It is found in numerous compounds such as water and in most carbon-containing compounds.

20. A highly reactive metal of low density. It is one of the three main elements found in fertilizer. Its compounds are quite similar to those of sodium, though typically more expensive.

22. A soft, dense metal used in bullets and car batteries. It was once used extensively both in plumbing and in paints. Concern over its biological effects caused a ban on its use for these purposes. It has been phased out as a gasoline additive for the same reason.

25. With the highest melting point of any pure element, it is the filament in ordinary (incandescent) light bulbs. Its one-letter symbol comes from the name wolfram.

27. A metallic element. It is added to steel to increase its strength.

28. A metallic element that serves as the negative pole (electrode) in the common flashlight battery. It is used to plate a protective film on iron objects (as in galvanized buckets). Alloyed with copper, it becomes brass.

29. A metal that is used to make stainless steel. Combined with nickel, it forms nichrome wire which is used in toasters and other devices where high electrical resistance is desired to produce heat.

30. This metal has a relatively low melting point. It is used in fire detection and extinguishing devices as well as in electrical fuses.

31. The most chemically reactive metal. Though it is quite rare, it is used in some photoelectric cells and in atomic clocks, which have far greater accuracy than mechanical or electric clocks.

33. A reddish, lustrous, ductile, malleable metal that occurs in nature in both free and combined states. It forms the body of many statues, including the Statue of Liberty. Other uses include electrical wiring, pennies, and decorative objects.

34. This magnetic, metallic element is used extensively for structural purposes. Outdoor stair railings may be made of this element.

38. A yellow, nonreactive, metallic element that has been highly valued since ancient times for its beauty and durability.

39. A metallic element that is used as a corrosion-resistant coating on the inside of cans used for packaging food, oil, and other substances.

ACROSS

3. A highly reactive, gaseous nonmetal. Its compounds are added to some toothpastes and many urban water supplies to prevent tooth decay.

5. A reactive metal whose compounds make up limestone, chalk, cement, and the bones and teeth of animals. Milk is a good nutritional source of this element.

7. An expensive, silver-white metal used in jewelry. It is also used in some industrial processes to speed up chemical reactions.

8. A solid purple-black nonmetal that changes to a deep purple gas upon heating. An alcohol solution of this element serves as an effective skin disinfectant. A compound of the element is added to sodium chloride (table salt) to prevent goiter.

10. Used in borosilicate (Pyrex®) glass, Boraxo® soap, drill bits, and control rods in nuclear power plants.

11. One of the three magnetic elements, this metal is used in 5-cent pieces and other coins, in electroplating, and in nichrome wire.

13. The most abundant metal in Earth's crust, this silver-white element is characterized by its low density, resistance to corrosion, and high strength. It is used for a variety of structural purposes, such as in airplanes, boats, and cars.

14. A hard, magnetic metal used in the production of steel. A radioactive form of this element is used in cancer treatment.

15. A silver-white, lustrous, radioactive metal. Its oxide is used as fuel in nuclear power plants and in atomic warheads.

16. An unreactive, gaseous element used in advertising signs for the bright reddish-orange glow it produces when an electric current is passed through it.

17. A yellow nonmetal that occurs in both the free and combined states. It is used in making match tips, gunpowder, and vulcanized rubber. Its presence in coal leads to acid rain if it is not removed before the coal is burned.

21. This metal is the best conductor of heat and electricity. Its scarcity prevents it from common use for such purposes. It was used extensively in the past in the manufacture of coins, but has become too expensive. It is used today for fine eating utensils and decorative objects. Some of its compounds are light-sensitive enough to be used in photographic film.

23. A gaseous nonmetal, the most abundant element on Earth. It makes up some 21% of Earth's atmosphere and is essential to most forms of life.

24. The second most abundant element in Earth's crust. It is the principal component of sand and quartz and finds use in solar cells, computer chips, caulking materials, and abrasives.

26. A metallic element used in nuclear power plant control rods and in NiCad rechargeable batteries.

30. A red, highly reactive, fuming liquid with a foul smell. It finds limited use as a disinfectant.

32. An odorless, colorless, unreactive gas used in most incandescent light bulbs.

35. An element with a symbol based on its Latin name. It is used with lead in car batteries.

36. A soft, highly reactive metal. Its compounds include table salt, lye, and baking soda.

37. A gaseous nonmetal that makes up 78% of Earth's atmosphere. Its compounds are important components of proteins, fertilizers, and many explosives.

40. A highly reactive, greenish-yellow gas used as a bleach and water disinfectant. It is a component of table salt.

Petroleum: To Burn? To Build?

- ► What explains petroleum's dual importance as both a fuel and a raw material from which to make many other substances?

- ► How can we make best use of the petroleum still remaining?

- ► What will happen to our way of life when global petroleum supplies become depleted?

- ► Can we find satisfactory alternatives to petroleum?

INTRODUCTION

The first *ChemCom* unit focused on an important renewable resource, water. In the second unit you studied some important nonrenewable resources. These are materials we just "borrow" from the earth: we can extract a metal from its ore, fabricate it, or recycle it, but we do not create any more of it. You should see quite clearly by now that we passengers on Spaceship Earth must rely on resources currently "on board." Nowhere is this more evident than in our energy-hungry world fueled mainly by **petroleum** (Latin *petr-,* rock, + *oleum,* oil).

Petroleum is a vitally important nonrenewable resource. Our society runs on it. As pumped from underground, petroleum is known as crude oil—"black gold." This greenish-brown to black liquid may be as thin as water or as thick as soft tar. Because crude oil cannot be used as is, it is shipped by pipeline, ocean tanker, train, or barge to oil refineries. There it is separated into simpler mixtures—some ready for use and others that require further chemical treatment. The refined petroleum is a mixture of various **hydrocarbons.** These are molecular compounds that contain atoms of only two elements—hydrogen and carbon.

Nearly half of the total U.S. energy needs are met by burning one form or another of petroleum as a fuel. Most petroleum is used in this way. Converted to gasoline, it powers most U.S. automobiles for traveling an

1974 gasoline lines in California. Will this scene be repeated?

average of 10,000 miles each year. Other petroleum-based fuels do many things in this country: provide heat to homes and businesses; deliver energy to generate electricity for many homes and industries; and propel diesel engines and jet aircraft.

What chemistry information can an informed consumer learn from this gasoline pump?

But we would miss a major role of petroleum in society if we think of petroleum only as a fuel, something to be burned. We need to remember its other major use is as a raw material from which a stunning array of familiar and useful products are manufactured—from the compact discs (CDs), sports equipment, clothing, automobile parts, and carpeting we use every day, to crucial prescription drugs, miraculous artificial limbs, and the spacesuits worn by astronauts. This unit's title, "Petroleum: To Burn? To Build?," refers to the two major roles of petroleum in modern society—as fuel (the To Burn role) and as a chemical raw material for synthesizing many things (the To Build role).

Figure 1 shows how much of an average barrel of petroleum is used for burning (as an energy source) and for "molecular building" (as a source of reactants to produce other useful materials) or for nonfuel uses. Astonishingly, 87% of a typical barrel of petroleum is burned outright as fuel. Of the 13% used for nonfuel purposes, about 3% is used for producing other substances (drugs, plastics); the other 10% provides lubricants, road-paving materials, and other miscellaneous nonfuel products. Thus, for every gallon of petroleum going into useful materials, more than 6 gallons are simply burned.

Because each of these roles—burning and building—is vital, important decisions about how to use and save petroleum lie ahead. In this unit we will consider our dependence on petroleum. Because petroleum is nonrenewable and its world-wide use is high and increasing, how long will supplies last? How much should we burn and how much should we save for making new materials? There are some even more basic questions: What in petroleum's chemical composition makes it so valuable? Can alternatives be developed to serve as petroleum substitutes? Or should we reduce our use of this valuable material by changing the way we live?

The answers to these questions will help clarify our options as petroleum supplies dwindle. However, many such questions are not easily answered. Much depends on how seriously our society seeks the answers and faces their consequences. Unfortunately, national interest in these matters tends to vary with the price of crude oil—a price strongly influenced by economic and political conditions.

▶ *One barrel contains 42 gallons of petroleum.*

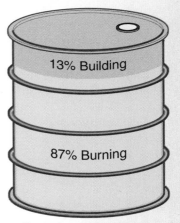

Based on 1994 figures

Figure 1 Petroleum: burning and building.

Twice during the 1970s the normal flow of crude oil imported to the United States was interrupted. As a result, the cost of gasoline rose to almost $2 per gallon. Gasoline rationing occurred, creating long lines at gas pumps. Many individuals and businesses, voluntarily or by state mandate, reduced their use of heating oil by lowering thermostats and improving home insulation. For a few years these events led government and industry to intensify research on alternative energy sources that would lessen our dependency on oil, especially oil from foreign sources. This changed, though, by the mid-1980s. Crude oil prices dropped somewhat in 1986, to as little as $14.50 a barrel (42 gallons), and so gasoline bills didn't leave as large a dent in the pocketbook. Pump prices of 70–90 cents a gallon for gasoline were common. With relatively cheap fuel readily available, interest and government funding for alternative energy sources declined.

The compelling geological and political realities of crude oil remain, however. It is not distributed uniformly around the world; countries that are major sources are sometimes politically unstable. Approximately 60% of the world's known crude oil reserves are located in just five Middle Eastern nations: Iran, Iraq, Kuwait, Saudi Arabia, and the United Arab Emirates (Figure 2).

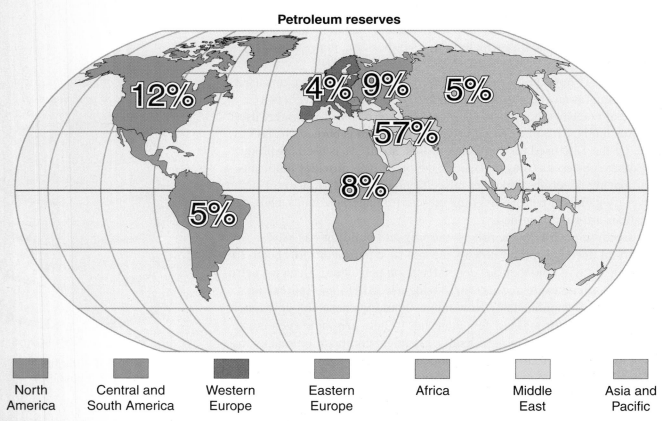

Figure 2 Distribution of world's petroleum reserves.

Political tensions and military turmoil in the Middle East have highlighted the risks in relying on large quantities of imported petroleum. The 1991 Persian Gulf war following Iraq's invasion of Kuwait underscored how quickly major known oil sources can be jeopardized. At the time Iraq overran Kuwait in August 1990, the United States was importing 47% of its required petroleum. In March 1991, when Operation Desert Storm ended, our dependency on foreign oil had dropped to 41%; by April 1994 it had not only risen again, but had actually increased to 52%. Strategic planning for the future must take into account how vulnerable such an alarmingly high dependency on foreign oil makes our nation. In order to diversify its geographical sources of oil, the United States has increased oil imports from Venezuela, Canada, and Mexico to supplement purchases from the Middle East.

Forecasting the future price of crude oil is difficult. Figure 3 shows how the price of crude oil has varied over time. The price may remain stable for years. Or it may move to new heights or depths as political and economic conditions shift. To predict the future of petroleum realistically, we need to look hard at how we use it, how fast it is being used, and approximately how much is still available.

▶ *By contrast, the combined petroleum reserves of the United States, Japan, and Europe equal only about 12% of the world's known supply.*

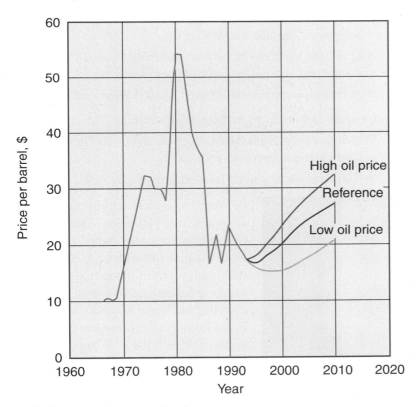

Figure 3 Changes in crude oil prices.

A PETROLEUM IN OUR LIVES

Although we use petroleum mainly as fuel, we also make a remarkable variety of things from it. Let's consider some familiar products that are made directly from petroleum or from petroleum-based substances.

A.1 YOU DECIDE: IT'S A PETROLEUM WORLD

1. Study Figure 4. List 10 items in the figure that are made from petroleum.

2. After you complete the list, your teacher will supply a version of Figure 4 from which all items that are made from petroleum have been removed. Compare the two figures.

   a. Add to your list those items you originally missed. List those items that surprised you as coming from petroleum.

   b. How many petroleum-based products did you overlook?

3. Suppose there were a severe petroleum shortage.

   a. Which five petroleum-based products shown in Figure 4 would you be most willing to do without?

   b. Which five would you be least willing to give up?

Figure 4 Life with petroleum.

Were you surprised by all the items that are made from petroleum? Often we don't realize how important something is until we try to imagine life without it. Petroleum is such a large part of modern life that we tend to take it for granted. We must realize, however, that the world is rapidly using up its limited supply of this nonrenewable resource. That leads to an important question: will we have sufficient petroleum for our needs in the future?

A.2 PETROLEUM IN OUR FUTURE

The United States uses about 18.4 million barrels (773 million gallons) of petroleum *every day*. Much of this is used industrially in manufacturing and shipping, not by individuals. If it all were distributed evenly among all U.S. residents, your daily "share" would be about three gallons.

Petroleum literally powers our modern society. It fuels a large portion of personal, public, and commercial transportation; it heats homes and office buildings; it drives the turbines that generate most of our nation's electricity. (See Figure 5.) Petroleum also is a rich chemical "feedstock," a source of reactants needed to produce plastics, fabrics, pharmaceuticals, and other items that greatly affect our lives. In short, petroleum is a significant and—at least for now—irreplaceable part of our standard of living.

Some petroleum experts predict that world oil production from known petroleum reserves will reach its

Petroleum products in motion.

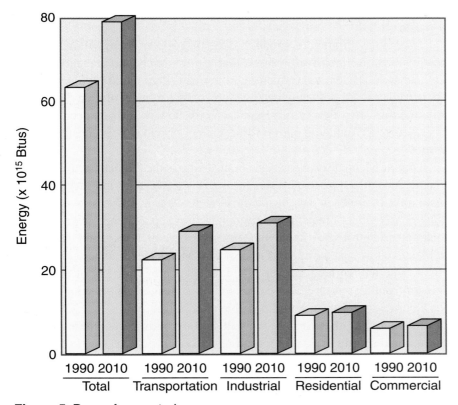

Figure 5 Power from petroleum.

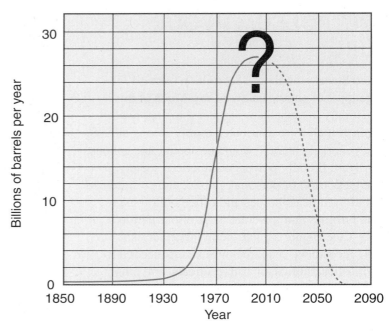

Figure 6 Past production from known petroleum reserves (solid line). Expected production from reserves (dashed line).

peak around the turn of the century (see Figure 6). After that, a steady decline in petroleum production may occur as reserves dwindle. In less than 100 years, petroleum production from presently known reserves may be as low as it was in 1910. In fact, at the end of the 1800s, petroleum was used only in lamps. In those days it served as a replacement for a limited supply of whale oil.

Like other commercial products, the price of petroleum will increase steadily as its availability decreases; that is how supply and demand economics works. If demand for petroleum remains high while supplies diminish, petroleum will gradually become too costly for many of today's common uses. In a time of petroleum scarcity, how would you decide to use this resource? Would you burn it for its energy or make other substances from it?

Dimitri Mendeleev, a Russian chemist who is the "father" of the periodic table, recognized petroleum's value as a raw material for industry. After visiting the oil fields in Russia and the United States in 1879, he warned that using petroleum as a fuel "would be akin to firing up a kitchen stove with bank notes." Petroleum's apparent abundance, ease of storage and handling, and relative low cost have led us to disregard Mendeleev's advice.

A time will come, however, when the choice of "burning or building" will confront all nations. Those decisions will depend on knowing about petroleum's sources, uses, and possible alternatives.

In examining these issues, we will start by finding out where we get our national supply of this important resource.

► *How important do you expect petroleum to be in your life when you are about 40 years old?*

Oil's Well That Ends Well: The Work of a Petroleum Geologist

Susan Landon is an independent petroleum geologist who uses her knowledge of chemistry, geology, geography, biology, and mathematics to analyze rocks and other geologic and geographic features for indications of likely oil and natural gas reserves.

Susan searches for the folded or faulted terrain that often contains petroleum. Hydrocarbons (oil or natural gas) may be trapped in reservoirs in these areas of the earth. Using data about a given geographic area—from satellite images, aerial photographs, topographic maps, and information about the rocks—Susan predicts the amount of oil or natural gas that could have been generated from organic matter in the rocks, and where it might occur.

If signs are good for a substantial petroleum deposit, Susan and her team work to find investors to fund further exploration, which often includes more detailed research, analysis, and drilling. During drilling, Susan and her team examine rock samples and data showing electrical and physical properties of the rocks

encountered. Based on their findings, they recommend either completing or plugging the well.

In Susan's opinion, a liberal arts background with a major in geology or allied science is good preparation for graduate work in geology.

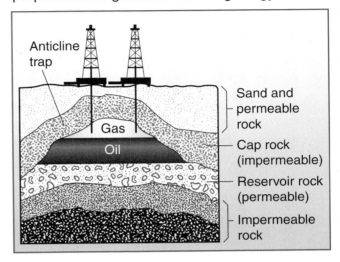

Wells? Now *There's* a Deep Subject!

There are many fancy techniques for *predicting* the presence of underground oil or natural gas, but the only way really to know *for sure* is to drill a well.

To drill a well, you must start with a drilling rig, consisting of a derrick—a vertical metal structure that holds and supports drilling pipe. Rotated by a motor, the drilling pipe ends in a bit that digs and scrapes down through soil and rock. Drilling fluid (usually a special mixture of water and other materials . . . called "mud") is pumped down through the pipe to cool the bit and to carry the chips of rock back to the surface. A geologist can examine the chips and other material to determine the rock type and identify the presence of oil.

Sometimes the oil or natural gas flows to the surface because of underground pressure, but more often a pump must be installed. From the well, the oil or gas is transported by truck or pipeline to refineries. Then it's processed and delivered to commercial and private consumers, who in turn convert it into everything from polypropylene socks to carpeting to household heat.

Photographs courtesy of Susan Landon

A.3 YOU DECIDE: WHO HAS THE OIL?

Offshore drilling rig.

Like many other chemical resources, Earth's petroleum is unevenly distributed. Large amounts are concentrated in relatively small areas. In addition, nations with large reserves do not necessarily consume large amounts of the resource. On the other hand, regions with a high demand for petroleum may be far away from regions with high petroleum reserves. Nations therefore often make exporting and importing arrangements, known as *trade agreements,* to obtain desired goods.

Figure 7 indicates where the world's known supplies of petroleum are located, along with the regional distribution of the world's population. The world's supplies, called *reserves,* can be tapped by available technology at costs consistent with current market prices. Figure 7 also shows the regional distribution of the world's consumption of petroleum. Use the figure to answer these questions:

1. Which region has the most petroleum?
2. Which region has the most petroleum relative to its population?
3. Which region has the least petroleum relative to its population?
4. Which regions consume a greater percentage of the world's supply of petroleum than they possess?

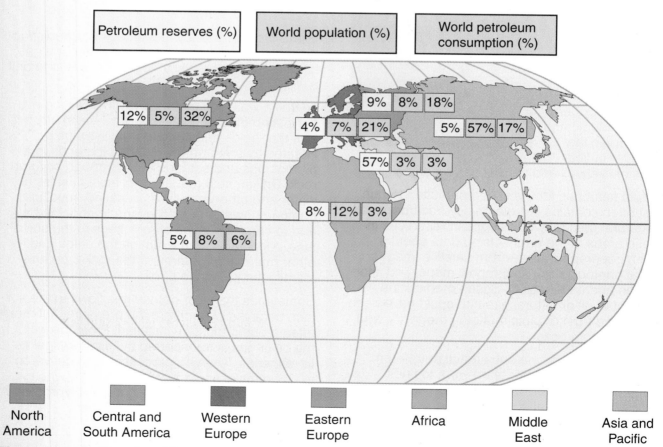

Figure 7 Distribution of world's petroleum reserves, population, and consumption of petroleum.

5. Which regions consume a smaller percentage of the world's supply of petroleum than they possess?

6. Which regions are likely to export petroleum?

7. Which regions are likely to import petroleum?

8. List several pairs of regions that might make petroleum trade agreements with each other. (A given region may be used more than once.)

9. Not all trade agreements that are possible actually exist. What factors might prevent trade relationships between regions?

10. What two regions are the largest petroleum consumers?

11. Why do these two regions consume such large quantities of petroleum?

12. a. Compare North America and the Middle East in terms of (i) population, (ii) petroleum reserves, and (iii) petroleum consumption.

    b. What possible consequences arise from these comparisons?

Thus far, we have recognized how greatly we depend on petroleum. Society's increasing reliance on this nonrenewable, limited resource must be analyzed carefully. However, before we can address such issues, we need a better chemical understanding of petroleum itself. What features make petroleum so valuable for both burning and building? We will tackle this question in the next section.

PART A: SUMMARY QUESTIONS

1. About 13% of the petroleum used in the United States is used for "molecular building" or nonfuel products that have an immense impact on our lives. The remaining 87% of the petroleum is burned as fuel.

    a. The United States uses 18 million barrels of petroleum each day. How many barrels of petroleum—on average—are burned as fuel daily in the United States?

    b. How many barrels of petroleum are used daily in the U.S. for building or nonfuel purposes?

    c. List 10 household items made from petroleum.

    d. What materials could be substituted for these 10 items if petroleum were not available to make them?

2. Suppose petroleum's cost increased by 10%. Assume that the nation's supply of petroleum decreased at the same time—a one-time drop of 2 million barrels daily.

    a. Would you expect the prices of gasoline and petroleum-based products to increase by 10% as well? Explain your answer.

    b. If not, would you expect the increase to be greater or less than 10%? Why?

3. a. What kind of petroleum trade relationship would be expected between North America and the Middle East?

    b. When other world regions become more industrialized and global petroleum supplies decrease, how might the North American–Middle East trade relationship change?

4. One barrel—the traditional unit of volume for petroleum—equals 42 gallons. Assume that each barrel of petroleum provides 21 gallons of gasoline. Your car goes 27 miles on a gallon of gasoline, and you travel 10,000 miles each year. How many barrels of petroleum must be processed to operate your car for a year?

B PETROLEUM: WHAT IS IT? WHAT DO WE DO WITH IT?

Petroleum has been called "black gold." Why is it so valuable? How does it differ from other resources, such as the metallic ores that produce aluminum, iron, or copper?

Petroleum is chemically very different from metal ores. A metallic ore usually contains only the single substance we want to use. Petroleum, on the other hand, is a mixture of hundreds of molecular compounds. These compounds share two desirable chemical properties. First, they are rich in energy that is released when the molecules are burned. Second, if not burned, they can be transformed chemically to produce a wide variety of useful materials.

Petroleum is far more exciting and challenging to chemists than its black, sometimes gooey appearance might suggest. Chemists have learned how to chemically combine small petroleum molecules, the raw materials, into giant, chainlike molecules, producing films, fibers, artificial rubber, and plastics. They have also learned to convert petroleum into perfumes, explosives, and medicines such as aspirin, acetaminophen, and codeine.

What kinds of molecules make petroleum such a chemical treasure? How can the different molecules in this gooey mixture be separated? Industrially, crude oil is separated into its components by *fractional distillation,* a process that uses the differences in boiling points of substances in a mixture.

You can separate a mixture of two substances by a simple *distillation,* a technique you will learn in the following activity. Although fractional distillation is a more complicated process than simple distillation, it is based on the same general principle.

Crude oil and related materials.

B.1 LABORATORY ACTIVITY: SEPARATION BY DISTILLATION

GETTING READY

In this activity you will separate two substances from a mixture by distillation. Distillation is a method for separating liquids according to their boiling points. The substance with the lower boiling point will vaporize and leave the distillation flask and pass through the condenser before the second substance boils. The condensed liquids, called *distillates,* are collected separately.

Heating the mixture to be distilled raises its temperature initially. However, when the first substance of the mixture begins to vaporize from the mixture, the temperature remains steady until the component has been distilled from the mixture. Continued heating will cause the temperature to rise again until the second component begins to distill. At this point, the temperature will again remain steady until the second component distills.

You will take temperature measurements every 30 seconds during the distillation in order to determine the temperature at which each of the two substances distills. You will then compare those temperatures with the boiling points of several compounds listed in Table 1 in order to identify the two substances distilled from the mixture.

Table 1 Properties of Possible Components of Distillation Mixture			
Substance	Formula	Boiling Pt., °C	I_2 Solubility
2-propanol (rubbing alcohol)	C_3H_7OH	82.4	Brown solution
TTE	$C_2Cl_3F_3$	45.8	Violet solution
Water	H_2O	100	Light yellow-brown
Cyclohexane	C_6H_{12}	80.7	Violet solution

Prepare a table in your notebook like that shown below.

Data Table
Observations of starting mixture being distilled:

Distillation:

Time, s	Temperature, °C
0	_____
30	_____
60	_____
90	_____
etc.	

Temperature at which first drop of first distillate enters the beaker: _____

Temperature at which first drop of second distillate enters the beaker: _____

Observations of distillates:

	Distillate 1	Distillate 2
Before iodine added:		
After iodine added:		

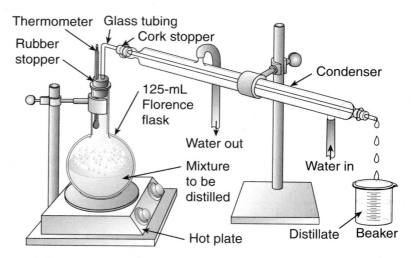

Thermometer Glass tubing
Cork stopper
Rubber stopper
Condenser
125-mL Florence flask
Water out
Mixture to be distilled
Water in
Hot plate
Distillate Beaker

Figure 8 Distillation of a simple mixture.

PROCEDURE

1. Assemble an apparatus like that shown in Figure 8. Label the two beakers "Distillate 1" and "Distillate 2."

2. Using a clean, dry graduated cylinder, measure 50 mL of the distillation mixture and pour it into the flask. Add a boiling chip to the flask.

3. Connect the flask to a condenser as indicated in Figure 8. Attach the hoses to the condenser and water supply appropriately. Put the Distillate 1 beaker at the outlet of the condenser to catch the distillate.

4. Check the apparatus to make sure that all connections are secure so that the distillation apparatus will not leak.

5. Turn on the water to the condenser and begin heating the flask.

6. Record the temperature at 30-second intervals.

7. Note the temperature at which the first drop of the first distillate enters the beaker. When the temperature begins to rise again after the first substance has been distilled, replace the Distillate 1 beaker with the Distillate 2 beaker.

8. Continue to heat the mixture and record the temperature every 30 seconds until the second substance distills.

9. Turn off the heat and allow the apparatus to cool. While the apparatus is cooling, test the relative solubility of iodine in Distillate 1 and Distillate 2. To do this, add a small amount of iodine to each beaker. Record your observations in the data table.

10. Disassemble and clean the distillation apparatus as directed by your teacher.

QUESTIONS

1. Use your data to prepare a graph of time (x axis) vs. temperature (y axis).

2. Using your graph, identify the temperature at which the first substance distilled and the temperature at which the second substance

distilled. How well do these temperatures correlate with the temperatures at which the first drops of distillates were collected?

3. Using the data in Table 1, identify each distillate.

4. In which distillate was iodine more soluble? Explain this.

5. What laboratory tests could you perform to find out whether the liquid left behind in the flask is a mixture or a pure substance?

B.2 PETROLEUM REFINING

Unlike your laboratory mixture, crude oil is a mixture of hundreds of hydrocarbon compounds. Separating such a mixture has been a challenge for chemists and chemical engineers. To meet it, they have applied large-scale fractional distillation to oil refining. The refining process, described in the next section, does not separate each component in crude oil, but it produces several usable, distinctive mixtures called *fractions.*

A modern large petroleum refinery can process up to 400,000 barrels of crude oil per day. The petroleum (crude oil) sent to a refinery is a mixture of hydrocarbons. At the refinery, the mixture is *refined*—that is, the components are separated by fractional distillation into fractions, characteristic mixtures of substances. Each fraction has a certain range of boiling points. Figure 9 illustrates the distillation (fractionation) of crude oil. The crude oil is heated until many of its component substances vaporize. During distillation, the smaller, lighter molecules vaporize first and move

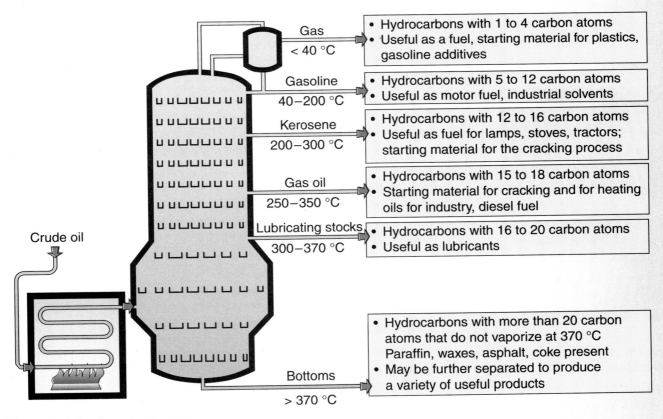

Figure 9 A fractionating tower.

toward the top of the distilling column. Each arrow in Figure 9 indicates the fraction's name and its boiling point range. Higher-boiling fractions contain the largest molecules.

The crude oil is heated to about 400 °C in a furnace. It is then pumped into the base of a distilling column (*fractionating tower*), which is usually more than 30 m (100 ft) tall. The temperature is highest at the bottom of the column and lower at the top. Trays are arranged at appropriate heights inside the column to collect the various fractions.

As the hot crude oil enters the tower, molecules of low-boiling substances in the hot crude oil heat up enough to evaporate from the liquid and rise into cooler portions of the distilling column. As the molecules rise, they cool. Substances with the lowest boiling points do not condense (become liquid); they remain gases, rising to the top of the tower. There they are drawn off separately as petroleum's gaseous fraction.

The remaining fractions condense in the tower and the liquid falls into trays located at various tower heights. These liquid fractions, such as gasoline or kerosene, each have a different boiling point range. Crude oil components with the highest boiling points never become gases; they remain in the liquid state through the entire distillation process. These thick (viscous) liquids—called *bottoms*—drain from the column's base.

You already know the names of some petroleum distillation fractions such as gasoline, kerosene, and lubricating oils. In the following activity you will have the chance to examine samples of some of petroleum's component fractions.

► *The names given to various fractions and their boiling ranges may vary a bit, but crude oil refining always has the general features illustrated.*

B.3 LABORATORY ACTIVITY: VISCOSITY AND DENSITY

GETTING READY

Viscosity is the term for resistance to flow. A material with high viscosity, like honey, flows slowly and with difficulty. A material like water, with a low viscosity, flows readily. In this activity you will determine relative viscosities, which means comparing the viscosity of a substance with that of other substances. Then you can rank the substances comparatively, from the most viscous to the least. You will measure the densities and relative viscosities of several petroleum-based materials, and for comparison, the same properties for water. To determine densities, you will measure the masses and volumes of liquid and solid samples.

In your laboratory notebook, prepare data tables like those on page 171, which you will use for reference.

PROCEDURE

Sample Examination

1. Obtain the six materials listed in your sample examination table.
2. Record the state (solid or liquid) of each sample.
3. Based on Figure 9, which material would you expect to have
    a. the lowest boiling point?
    b. the highest boiling point?

The viscosity of fluid determines how fast it can flow.

Data Table	Sample Examination	
Material	Carbon Atoms per Molecule	State at Room Temperature (Solid, liquid)
Mineral Oil	12–20	
Asphalt	>34	
Kerosene	12–16	
Paraffin wax	>19	
Motor oil	15–18	
Household lubricating oil	14–18	

Data Table	Density Measurements		

Average mass of capped tube: _____

Average mass of bead: _____

Average volume: _____

Liquid	Mass of Capped Tube, Bead, and Liquid (g)	Calculated Mass of Liquid (g)	Calculated Density of Material (g/mL)
Water			
Mineral oil			
Kerosene			
Motor oil			
Household lubricating oil			
Solid	Mass of Sample (g)	Volume Increase (mL)	Calculated Density of Material (g/mL)
Paraffin wax			
Asphalt			

Data Table	Viscosity Measurements	
Material	Average Time (in s) for Bead to Fall	Relative Viscosity
Water		
Mineral oil		
Kerosene		
Motor oil		
Household lubricating oil		

Density Measurement

LIQUID SAMPLES

1. Your teacher will give you average values for the mass of an empty capped tube, the mass of a bead, and the volume of liquid in a sample tube. For convenience, record these values at the top of your Density Measurements data table.

2. Weigh each capped tube containing a petroleum product sample. Also weigh a capped tube containing water. Record the masses.

SOLID SAMPLES

3. Fill a 25-mL graduated cylinder with water to the 10-mL mark.

4. Weigh each solid sample carefully. Record the masses.

5. Carefully drop the first weighed solid sample into the water in the cylinder. Measure the *increase* in volume. Record this value.

6. Repeat Step 5 for the other solid sample.

Density Calculations

LIQUID SAMPLES

1. Calculate and record the mass (in grams) of each liquid sample, using masses obtained in Steps 1 and 2 for the total mass, the capped tube, and the bead. Use this formula:

Sample mass = total mass − (capped tube mass + bead mass)

2. Using your calculated sample masses and the volume of liquid in each capped tube, calculate the density of each sample (in grams per milliliter). Record these values. For example, suppose the sample mass is 40.2 g and the volume is 36.7 mL.

$$\text{Sample density} = \frac{\text{sample mass}}{\text{average volume}} = \frac{40.2\ \text{g}}{36.7\ \text{mL}} = 1.10\ \text{g/mL}$$

SOLID SAMPLES

3. Using the sample masses and increases in water volume found in Step 5, calculate the solid sample densities (in grams per milliliter). The calculation is the same as the one illustrated for liquids in Step 2 of the last calculation except for using the increase in water volume for the volume rather than the liquid volume.

Relative Viscosity

1. Determine the time needed for the bead to fall from top to bottom within the capped tube containing water. Follow this procedure:

   a. Hold the capped tube upright until the bead is at the bottom.

   b. Gently turn the tube horizontally. (The bead will stay at one end.)

   c. Quickly turn the tube upright so the bead is at the top.

   d. Record the number of seconds needed for the bead to fall to the bottom of the tube.

   e. Repeat this procedure three more times. Calculate the average time required for the bead to fall.

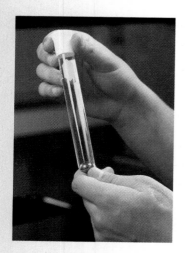

Apparatus for determining relative viscosity.

2. Repeat Step 1 for each liquid petroleum-based sample.

3. Rank your samples in order of relative viscosity, assigning number 1 to the least viscous material (the one through which the bead fell fastest).

4. Wash your hands thoroughly before leaving the laboratory.

QUESTIONS

1. The density of oil plays a major role in oceanic oil spills and in fires. Explain.

2. Propose a generalization, based on your observations in this activity, about the connection between the number of carbon atoms in a molecule and a substance's viscosity.

3. Suppose a classmate suggests that petroleum fractions can be separated at room temperature on the basis of their viscosities.

   a. Do you agree? Explain your answer.

   b. What would be some advantages of such a separation procedure?

CHEMQUANDARY

GASOLINE AND GEOGRAPHY

When shipping gasoline and motor oils to different parts of the nation, petroleum distributors must consider both the ease of evaporation and viscosity of these products. Why must gasoline shipped in winter to a northern state (such as Minnesota) be different from that shipped in summer to a southern state (such as Florida)?

The petroleum fractions you investigated in this section represent a few of the great number of materials present in petroleum. The next activity will provide more detail about the substances obtained from each fraction and what is done with them.

B.4 YOU DECIDE:
CRUDE OIL TO PRODUCTS

The main hydrocarbon fractions in Figure 10 (page 174) are listed in order of increasing boiling points. Low-boiling gases are shown at the top, and high-boiling residues at the bottom. Arrows point to typical refinery products from that fraction, which are sold to users or processors. Some of the ultimate uses of these products are also listed. The box at the bottom illustrates the variety of petroleum-based consumer products.

Your teacher will organize your class into small working groups. Your group will be assigned one of the petroleum fractions shown in Figure 10. Answer these questions regarding your fraction:

1. a. Do any uses of your fraction depend on petroleum's value as an energy source for burning?

   b. If so, name them.

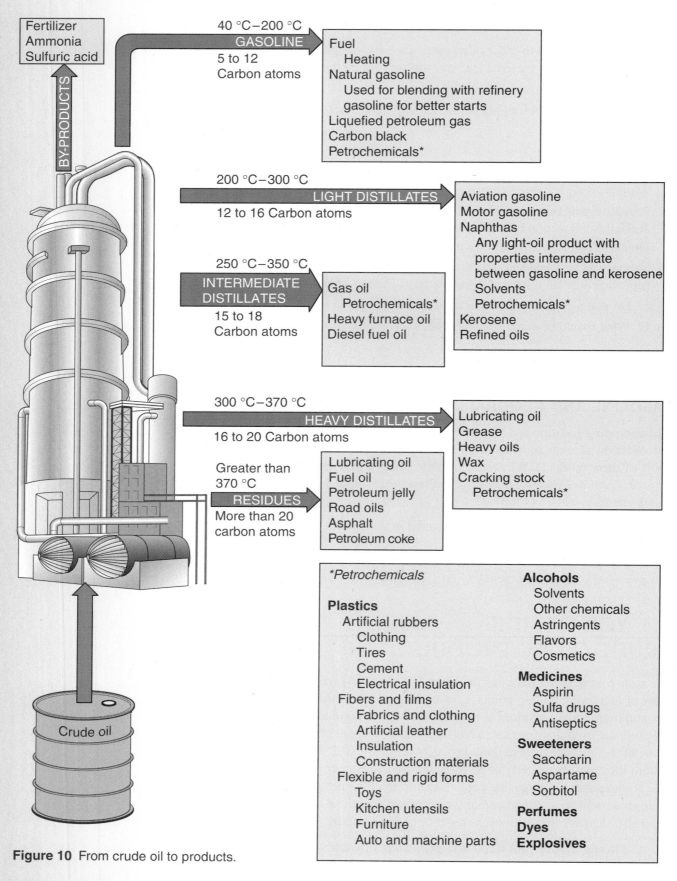

40 °C−200 °C
GASOLINE
5 to 12
Carbon atoms

Fuel
    Heating
Natural gasoline
    Used for blending with refinery
    gasoline for better starts
Liquefied petroleum gas
Carbon black
Petrochemicals*

200 °C−300 °C
LIGHT DISTILLATES
12 to 16 Carbon atoms

Aviation gasoline
Motor gasoline
Naphthas
    Any light-oil product with
    properties intermediate
    between gasoline and kerosene
Solvents
Petrochemicals*
Kerosene
Refined oils

250 °C−350 °C
INTERMEDIATE
DISTILLATES
15 to 18
Carbon atoms

Gas oil
    Petrochemicals*
Heavy furnace oil
Diesel fuel oil

300 °C−370 °C
HEAVY DISTILLATES
16 to 20 Carbon atoms

Lubricating oil
Grease
Heavy oils
Wax
Cracking stock
    Petrochemicals*

Greater than
370 °C
RESIDUES
More than 20
carbon atoms

Lubricating oil
Fuel oil
Petroleum jelly
Road oils
Asphalt
Petroleum coke

Crude oil

Petrochemicals

Plastics
    Artificial rubbers
        Clothing
        Tires
        Cement
        Electrical insulation
    Fibers and films
        Fabrics and clothing
        Artificial leather
        Insulation
        Construction materials
    Flexible and rigid forms
        Toys
        Kitchen utensils
        Furniture
        Auto and machine parts

Alcohols
    Solvents
    Other chemicals
    Astringents
    Flavors
    Cosmetics

Medicines
    Aspirin
    Sulfa drugs
    Antiseptics

Sweeteners
    Saccharin
    Aspartame
    Sorbitol

Perfumes
Dyes
Explosives

Figure 10 From crude oil to products.

2. a. Do any uses of your fraction depend on petroleum's value as a source to make new substances?

   b. If so, name them.

3. Describe how your life might change if your assigned petroleum fraction and its resulting products were totally eliminated during a petroleum shortage.

4. Identify two uses of your fraction that are important enough to continue even during a severe petroleum shortage. Give reasons.

5. Based on your answers to the questions above, which do you believe is more important—burning petroleum to release energy or using it for making derivatives (new materials)? Give reasons for your answer.

The refined petroleum fractions you investigated earlier in this unit were separated from petroleum by distillation. Each fraction has different physical properties, for instance, boiling points. How do chemists explain these differences?

To answer this question, in the following sections we will examine the composition and structure of molecules found in petroleum. We will see again that properties of substances are related to their molecular structures.

B.5 A LOOK AT PETROLEUM MOLECULES

Petroleum's gaseous fraction contains compounds with low boiling points (<40 °C). These small molecules, which contain from one to four carbon atoms, are only slightly attracted to one another or to other molecules in petroleum. Such forces of attraction *between* molecules are called ***intermolecular forces.*** As a result, these small hydrocarbon molecules readily separate from one another and rise through the distillation column as gases.

Petroleum's liquid fractions—including gasoline, kerosene, and heavier oils—consist of molecules having from five to about 20 carbon atoms. Molecules with more than 20 carbon atoms make up the greasy-solid fraction which does not vaporize, even at high distilling temperatures. These greasy, solid compounds have the strongest intermolecular forces of attraction of all petroleum-based substances. That is why they are solids at room temperature.

Complete the following *Your Turn* to learn more about physical properties of hydrocarbons.

▶ *Just as "interstate commerce" means trade between states, intermolecular forces means forces between molecules.*

YOUR TURN

Hydrocarbon Boiling Points

Chemists often gather data about the physical and chemical properties of substances. Although these data can be organized in many ways, the most useful ways uncover trends or patterns among the values. These patterns often trigger attempts to explain the regularities.

The development of the periodic table is a good example of this approach. Recall that you used the properties of neighboring elements in the table to predict a property of an unknown, nearby element. In a similar vein, we can seek patterns among boiling point data for some hydrocarbons. When individual molecules in the liquid state gain enough energy to overcome intermolecular forces, the molecules enter the gaseous state, as in evaporation and boiling.

Answer the following questions about boiling point data given in Table 2.

Table 2	Hydrocarbon Boiling Point
Hydrocarbon	**Boiling Point (°C)**
Butane	−0.5
Decane	174.0
Ethane	−88.6
Heptane	98.4
Hexane	68.7
Methane	−161.7
Nonane	150.8
Octane	125.7
Pentane	36.1
Propane	−42.1

1. a. In what pattern or order are Table 2 data organized?

   b. Is this a useful way to present the information? Why or why not?

2. Assume we want to search for a trend or pattern among these boiling points.

   a. Propose a more useful way to arrange these data.

   b. Reorganize the data table based on your idea.

3. Use your new data table to answer these questions:

   a. Which substance(s) are gases (have already boiled) at room temperature (22 °C)?

   b. Which substance(s) boil between 22 °C (room temperature) and 37 °C (body temperature)?

4. What can you infer about intermolecular attractions in decane compared with those in butane?

5. Intermolecular forces also help explain other liquid properties such as viscosity and freezing points.

   a. Based on their intermolecular attractions, try to rank pentane, octane, and decane in order of increasing viscosity. Assign number 1 to the least viscous ("thinnest") of the three.

   b. Check with your teacher to see if you are correct.

B.6 CHEMICAL BONDING

Hydrocarbons and their derivatives are the focus of a branch of chemistry known as *organic chemistry.* These substances are called organic compounds because early chemists thought living systems—plants or animals—were needed to produce organic compounds. However, chemists have known for more than 150 years how to make many organic compounds without any assistance from living systems. Also, starting materials other than petroleum can also be used to make organic compounds.

In many hydrocarbon molecules, the carbon atoms are joined to each other forming a backbone called a *carbon chain;* the hydrogen atoms are attached to it. Carbon's ability to form chains helps explain why there are so many different hydrocarbons, as you will soon see. Additionally, hydrocarbons can be regarded as "parents" of an even larger number of compounds containing other kinds of atoms attached to strands of carbon atoms.

▶ *The carbon chain forms a framework to which other atoms can be attached.*

ELECTRON SHELLS

To learn how atoms of carbon or other kinds of atoms are held together, we must first examine the arrangement of electrons in atoms. We have seen that atoms are made up of neutrons, protons, and electrons. Neutrons and protons are located in the dense, central region of the atom known as the *nucleus.* Studies have shown that electrons occupy different energy levels in the space surrounding an atomic nucleus. Each energy level (or *shell*) can hold only a certain maximum number of electrons. The first shell of electrons surrounding the nucleus has a capacity of two electrons. The second shell can hold a maximum of eight electrons.

Consider an atom of helium (He), the first noble gas. A helium atom has two protons in its nucleus and two electrons surrounding the nucleus. These two electrons occupy the first, innermost level; this is as many electrons as this level (shell) can hold.

The next noble gas, neon (Ne), has atomic number 10; each of its atoms contains ten protons and ten electrons. Two of these electrons reside in the first level. The remaining eight occupy the second level. Note that each level has reached its capacity, since the maximum number of electrons is two for the first level and eight for the second level.

Helium and neon are each chemically unreactive; that is, their atoms do not combine with those of other atoms (or even each other) to form compounds. Sodium (Na) atoms, with atomic number 11, have just one more electron than neon and are extremely reactive due to that additional electron. Fluorine (F) atoms, with one less electron than neon, also are extremely reactive because of being one electron short of neon. The differences in the reactivities of elements are accounted for by the electron arrangements of their atoms.

A useful guideline to understanding the chemical behavior of many elements is to recognize that *atoms with filled electron levels (shells) are particularly stable,* that is, chemically unreactive. The noble gases are essentially unreactive because their atoms already have filled shells of electrons. All but helium atoms have eight electrons in their outer level;

helium only needs two to reach its maximum of electrons. How does this guideline help explain the stability of the noble gases and the reactivity of the elements sodium and fluorine?

When sodium metal reacts, sodium ions (Na^+) form. The $+1$ electrical charge indicates that each original sodium atom has lost one electron. Na^+ contains eleven protons but only ten electrons, thus the net $+1$ charge. With ten electrons, each sodium ion possesses filled electron shells (two in the first level, eight in the second) like neon atoms. Thus, sodium $+1$ *ions* (unlike sodium atoms) are highly stable; the world's entire natural supply of sodium is found in this ionic form. (By losing an electron, each sodium atom is oxidized—see the Resources unit, page 142.)

By contrast, fluorine atoms react to form fluoride ions (F^-). The -1 electrical charge indicates that each electrically neutral fluorine atom has gained one electron. The F^- ion contains nine protons and ten electrons for a net -1 charge. The ten electrons of a F^- ion give it the same electron distribution as neon. Once again, an element has reacted to attain the special stability associated with filled electron shells. Positive and negative ions combine to form ionic compounds that are held together by electrical attraction between the oppositely charged species. (By gaining an electron, each fluorine atom is reduced—see the Resources unit, page 142.)

In molecular (non-ionic) substances, atoms acquire filled electron shells by sharing electrons rather than by electron loss or gain. Many molecular substances are composed of nonmetals whose atoms do not easily lose electrons. The shared electrons contribute to filling the shells of both atoms.

The hydrogen molecule (H_2) provides a simple example. Each hydrogen atom contains one electron. It is clear that one more electron is needed to fill hydrogen's first shell. Two hydrogen atoms can accomplish this by sharing their single electrons. If each electron is represented by a dot (•), then a hydrogen molecule can be written in this way:

$$H\bullet \; + \; \bullet H \; \rightarrow \; H \vdots H$$

COVALENT BONDS

The chemical bond formed between two atoms that share a pair of electrons is called a ***single covalent bond.*** Through such sharing, both atoms achieve the stability associated with filled electron shells. A carbon atom, atomic number 6, has six electrons—two in the first shell and four in the second shell. Note that four more electrons are needed to fill the second shell to its capacity of eight. This is accomplished through covalent bonding.

Consider the simplest hydrocarbon molecule, methane (CH_4). In this molecule, each hydrogen atom shares its single electron with the carbon atom, as represented here:

$$4\,H\bullet \; + \; \bullet \overset{\displaystyle \bullet}{\underset{\displaystyle \bullet}{C}} \bullet \; \longrightarrow \; H \vdots \overset{\displaystyle H}{\underset{\displaystyle H}{C}} \vdots H$$

Here, as in the formula for hydrogen, the dots represent only the outer-shell electrons for each atom. Such structures are called *electron-dot formulas.* The two electrons in each covalent bond "belong" to both bonded atoms. Dots placed between the symbols of two atoms represent electrons shared by those atoms.

Count the dots surrounding each atom in methane. Notice that by sharing electrons each hydrogen atom in methane has a filled electron energy level (two electrons in the first shell). The carbon atom also has a filled outer shell (eight electrons). Each hydrogen atom is associated with a pair of electrons; the carbon atom has eight electrons (four pairs) around it.

For convenience, each pair of electrons in a covalent bond can be represented by a dash. This is another common way to draw covalently bonded substances called a *structural formula:*

H
· ·
H : C : H
· ·
H

Electron-dot formula
of methane, CH_4

H
|
H — C — H
|
H

Structural formula
of methane, CH_4

Within molecules, atoms are arranged in distinctive three-dimensional patterns. These atomic arrangements help determine the physical and chemical properties of molecules. Although we can draw two-dimensional pictures of molecules on flat paper, assembling three-dimensional models of them gives a more accurate representation. The following activity will give you a chance to assemble such models.

B.7 LABORATORY ACTIVITY: MODELING ALKANES

In this activity you will assemble models of several simple hydrocarbons. The goal is to relate three-dimensional shapes of molecules to the names, formulas, and pictures used to represent molecules on paper.

Pictures of two types of molecular models are shown in Figure 11 (page 180). Most likely, you will use ball-and-stick models. Each ball represents an atom and each stick represents a single covalent bond (shared electron pair) connecting two atoms. Real molecules are not composed of ball-like atoms located at the ends of stick-like bonds. Evidence shows that atoms contact each other, more like those in space-filling models. However, ball-and-stick models are useful because they make it easier to represent the structure and geometry of molecules.

Look again at the electron-dot structure and structural formula for methane (CH_4) above. Methane, the simplest hydrocarbon, is the first member of a series of hydrocarbons known as *alkanes,* which we explore in this activity. Each carbon atom in an alkane forms single covalent bonds with four other atoms. Alkanes are considered *saturated hydrocarbons,* because each carbon atom is bonded to the maximum number of other atoms (four).

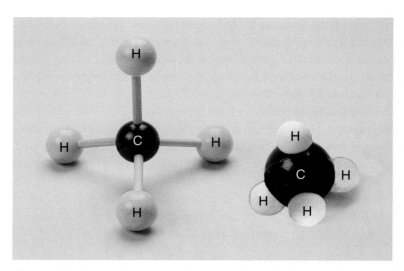

Figure 11 Three-dimensional models: (left) ball-and-stick; (right) space-filling.

PROCEDURE

1. Assemble a model of methane (CH_4). Compare your model with the electron-dot and structural formulas. Note that the angles between atoms are *not* 90°, as you might think from the written formulas. Compare your model with the photographs of models in Figure 11. If you were to build a close-fitting box to surround a CH_4 molecule, the box would be shaped like a triangular pyramid (a pyramid with a triangle as a base), as shown in Figure 12. This shape is called a ***tetrahedron.***

   Here's one way to think about this tetrahedral shape. Assume that the four pairs of electrons—all with negative charges—surrounding the carbon repel one another and stay as far away from one another as possible. This causes the four pairs of shared electrons to arrange themselves so that the bonds point to the corners of a tetrahedron. The angles between the C-H bonds are each 109.5°, a value that has been verified several ways experimentally. The angles are not 90° as they would be if methane were shaped like a flat square.

2. Assemble models of a two-carbon and a three-carbon alkane molecule. Recall that each carbon atom in an alkane is bonded to four other atoms.

   a. How many hydrogen atoms are present in the two-carbon alkane?

   b. How many are present in the three-carbon alkane?

   c. Draw a ball-and-stick model, similar to that in Figure 11, of the three-carbon alkane.

3. a. Draw electron-dot and structural formulas for the two- and three-carbon alkanes.

   b. The molecular formulas for the first two alkanes are CH_4 and C_2H_6. What is the molecular formula of the third?

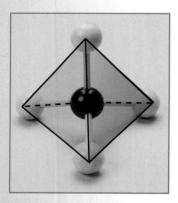

Figure 12 The tetrahedral shape of a methane molecule.

Examine your three-carbon alkane model and the structural formula you drew for it. Note that the middle carbon atom is attached to two hydrogen atoms, but the carbon atom at each end is attached to three hydrogen atoms. This molecule can be represented as $CH_3-CH_2-CH_3$, or $CH_3CH_2CH_3$. Formulas like these provide information about how atoms are arranged in molecules. Thus, for many purposes, such formulas are more useful than shortened molecular formulas such as C_3H_8.

The general molecular formula for any alkane molecule can be written as C_nH_{2n+2}, where n is the number of carbon atoms in the molecule. Thus, even without assembling any models, we can predict the formula of a four-carbon alkane: if $n = 4$, then $2n + 2 = 10$, and the formula is C_4H_{10}.

▶ n *is any positive integer.*

4. Using the general alkane formula, predict molecular formulas for the first ten alkanes. When you are done, compare your molecular formulas with the formulas given in Table 3 to see whether your predictions are correct.

The names for the first ten alkanes are also given in Table 3. These names are composed of a root, followed by -*ane* (designating an alkane). The root refers to the number of carbon atoms in the backbone carbon chain. To a chemist, *meth-* means one carbon atom, *eth-* means two, *prop-* means three, and so on. For those alkanes with five to ten carbons, the root is derived from Greek—*pent-* for five, *hex-* for six, and so on.

5. Write structural formulas for butane and pentane.

6. a. Name the alkanes with these formulas:

    (1) $CH_3CH_2CH_2CH_2CH_2CH_2CH_3$

    (2) $CH_3CH_2CH_2CH_2CH_2CH_2CH_2CH_2CH_3$

    b. Write shorter molecular formulas for the two alkanes listed in Question 6a.

Table 3	Some Members of the Alkane Series		
Name	**Number of Carbons**	**Alkane Molecular Formulas**	
		Short Version	**Long Version**
Methane	1	CH_4	CH_4
Ethane	2	C_2H_6	CH_3CH_3
Propane	3	C_3H_8	$CH_3CH_2CH_3$
Butane	4	C_4H_{10}	$CH_3CH_2CH_2CH_3$
Pentane	5	C_5H_{12}	$CH_3CH_2CH_2CH_2CH_3$
Hexane	6	C_6H_{14}	$CH_3CH_2CH_2CH_2CH_2CH_3$
Heptane	7	C_7H_{16}	$CH_3CH_2CH_2CH_2CH_2CH_2CH_3$
Octane	8	C_8H_{18}	$CH_3CH_2CH_2CH_2CH_2CH_2CH_2CH_3$
Nonane	9	C_9H_{20}	$CH_3CH_2CH_2CH_2CH_2CH_2CH_2CH_2CH_3$
Decane	10	$C_{10}H_{22}$	$CH_3CH_2CH_2CH_2CH_2CH_2CH_2CH_2CH_2CH_3$

7. a. Write a molecular formula for an alkane containing 25 carbon atoms.

   b. Did you decide to write a short version or long version of the formula for this compound? Why?

8. Name the alkane having a molar mass of

   a. 30 g/mol.

   b. 58 g/mol.

YOUR TURN

Alkane Boiling Points: Trends

Prepare a graph of boiling points, using data for the alkanes listed in Table 2 (page 176) and Table 3 (page 181). The *x*-axis scale should range from one to thirteen carbon atoms (even though you will initially plot data for one to ten carbon atoms). The *y*-axis scale should extend from -200 °C to $+250$ °C.

1. From your graph, estimate the average *change* in boiling point (in degrees Celsius) when one carbon atom and two hydrogen atoms ($-CH_2-$) are added to a given alkane chain.

2. The pattern of boiling points among the first 10 alkanes allows you to predict boiling points of other alkanes.

   a. Using your graph, estimate the boiling points of undecane ($C_{11}H_{24}$), dodecane ($C_{12}H_{26}$), and tridecane ($C_{13}H_{28}$). To do this, continue the trend of your graph line by drawing a dashed line from the graph line you drew for the first ten alkanes (this is called *extrapolation*). Then read your predicted boiling points for C_{11}, C_{12}, and C_{13} alkanes on the *y* axis.

   b. Compare your predicted boiling points to actual values provided by your teacher.

3. We have already noted that a substance's boiling point depends (in part) on its intermolecular forces—that is, on attractions among its molecules. For alkanes you have studied, how are these attractions related to the number of carbon atoms in each molecule?

B.8 LABORATORY ACTIVITY: ALKANES REVISITED

GETTING READY

The alkane molecules we have considered so far are **straight-chain alkanes** in which each carbon atom is linked to only one or two other carbon atoms. Many other arrangements of carbon atoms in alkanes are possible. Alkanes in which one or more carbon atoms are linked to three

▶ *A straight carbon chain.*

C—C—C—C—C

or four other carbon atoms are called **branched-chain alkanes.** A branched-chain alkane may have the same number of carbon atoms as a straight-chain alkane.

In this activity you will use ball-and-stick molecular models to investigate some variations in alkane structures that lead to different properties.

▶ *A branched carbon chain.*

C—C—C—C
    |
    C

PROCEDURE

1. Assemble a ball-and-stick model of a molecule with the formula C_4H_{10}. Compare your model with those built by others. How many different arrangements of atoms in the C_4H_{10} molecule can be constructed?

   Molecules having identical molecular formulas but different arrangements of atoms are called *isomers.* By comparing models, convince yourself that there are only two isomers of C_4H_{10}. The formation of isomers helps explain the very large number of compounds that contain carbon chains or rings.

2. a. Draw an electron-dot formula for each C_4H_{10} isomer.

   b. Write a structural formula for each C_4H_{10} isomer.

3. As you might expect, alkanes containing larger numbers of carbon atoms also have greater numbers of isomers. In fact, the number of different isomers increases rapidly as the number of carbon atoms increases. For example, chemists have identified three pentane (C_5H_{12}) isomers. Their structural formulas are shown in Table 4.

Table 4	Alkane Isomers	
Alkane	**Structure**	**Boiling Point (°C)**
C_5H_{12} isomers	CH_3—CH_2—CH_2—CH_2—CH_3	36.1

	CH_3—CH—CH_2—CH_3
	         |
	        CH_3
		27.8

	CH_3
	      |
	CH_3—C—CH_3
	      |
	      CH_3
		9.5

Some C_8H_{18} isomers	CH_3—CH_2—CH_2—CH_2—CH_2—CH_2—CH_2—CH_3	125.6

	CH_3—CH_2—CH_2—CH_2—CH_2—CH—CH_3
	                                    |
	                                CH_3
		117.7

	CH_3
	                      |
	CH_3—CH—CH_2—C—CH_3
	      | |
	     CH_3 CH_3
		99.2

Convince yourself that no other pentane isomer is possible. Assemble other possible isomer models, if you wish, to confirm this statement.

4. Now consider possible isomers for C_6H_{14}.

   a. Working with a partner, draw structural formulas of as many *different* C_6H_{14} isomers as possible. Compare your structures with those drawn by other groups.

   b. How many possible C_6H_{14} isomers were found by your class?

5. Build models of one or more of the possible C_6H_{14} isomers, as assigned by your teacher.

   a. Compare the three-dimensional models built by the class with structures drawn on paper.

   b. Based on your study of the three-dimensional models, how many different C_6H_{14} isomers are possible?

Because each isomer is a different substance, each must have characteristic properties. Let's examine boiling-point data for some alkane isomers.

YOUR TURN

Alkane Boiling Points: Isomers

You have already observed that boiling points of straight-chain alkanes are related to the number of carbon atoms in their molecules. Increased intermolecular attractions are related to the greater molecule–molecule contact possible for larger alkanes. For example, consider the boiling points of some isomers.

1. Boiling points for two sets of isomers are listed in Table 4 (page 183). Within a given series of isomers, how does the boiling point change as the number of carbon side-chains increases (that is, as the amount of branching increases)?

2. Match each of the following boiling points to the appropriate C_7H_{16} isomer: 98.4 °C, 92.0 °C, 79.2 °C.

   a. $CH_3-CH_2-\overset{\displaystyle |}{\underset{\displaystyle CH_3}{CH}}-CH_2-CH_2-CH_3$

   b. $CH_3-CH_2-CH_2-\overset{\displaystyle CH_3}{\underset{\displaystyle CH_3}{\overset{\displaystyle |}{\underset{\displaystyle |}{C}}}}-CH_3$

   c. $CH_3-CH_2-CH_2-CH_2-CH_2-CH_2-CH_3$

3. Here is the formula of a C_8H_{18} isomer:

$$CH_3-CH_2-CH_2-\overset{\overset{\displaystyle CH_3}{|}}{\underset{\underset{\displaystyle CH_3}{|}}{C}}-CH_2-CH_3$$

a. Compare it to each C_8H_{18} isomer listed in Table 4. Predict whether it has a higher or lower boiling point than each of the other C_8H_{18} isomers.

b. Would the C_8H_{18} isomer shown above have a higher or lower boiling point than each of the three C_5H_{12} isomers shown in Table 4?

Chemists and chemical engineers are able to separate the complex mixture known as petroleum into a variety of useful substances. However, petroleum is a nonrenewable and limited resource. This fact poses important questions for our society: How can we best use various petroleum fractions? Should we use them for burning or for building ? If for both, how much should be directed to each use? Most petroleum is now burned as fuel, and we will focus on this use in Part C. What future alternatives might help decrease our dependence on petroleum? All these questions will be explored in the following pages.

Lucas well at Spindletop, Texas, 1901.

? PART B: SUMMARY QUESTIONS

1. a. In what sense is petroleum similar to metallic resources such as copper ore?

   b. In what sense is it different from them?

2. A 35.0-mL sample of octane—a component of gasoline—has a mass of 24.6 g.

   a. What is the density of octane?

   b. What is the mass of a 25.0-mL sample of octane?

3. a. Describe two broad categories of uses for petroleum.

   b. List two examples for each category.

4. Paraffin wax (candle wax) is a mixture of alkanes. A group of 25-carbon alkanes is one component of paraffin. Write the molecular formula for this group of alkanes.

5. a. List two features of molecular structure that determine the relative boiling points of hydrocarbons.

   b. How does each feature influence boiling points?

6. a. Explain what the term isomer means.

   b. Illustrate your explanation by drawing structural formulas for at least three isomers of C_7H_{16}.

7. Suppose two classmates draw you into their conversation about butane. One student says that the two butane isomers each contain 10 hydrogen atoms and 10 covalent bonds. The other student responds that one isomer contains 10 hydrogen atoms and 13 covalent bonds, and the other isomer has 10 hydrogen atoms and 10 covalent bonds. What would you say to set the record straight?

EXTENDING YOUR KNOWLEDGE (OPTIONAL)

- Gasoline's composition varies in different parts of the nation. Does the composition relate to the time of year? If so, what factors help determine the optimum composition of gasoline in various seasons?

- The hydrocarbon boiling points listed in Table 2 (page 176) were measured under normal atmospheric conditions. How would the boiling points change if atmospheric pressure were increased or decreased? (*Hint:* Butane is stored as a liquid in a butane lighter, yet it escapes through the lighter nozzle as gas.)

- When 1,2-ethanediol (ethylene glycol, also known as antifreeze) is dissolved in an automobile's radiator water, it helps protect the water from freezing. This is because the 1,2-ethanediol–water solution has a lower freezing point than that of pure water. Similarly, when an ionic substance such as table salt (NaCl) is dissolved in water, the solution freezes at a lower temperature than pure water does. Why is NaCl a highly undesirable additive for car radiators, but ethylene glycol a suitable additive? (*Hint:* Compare the nature and structure of these two substances.)

1,2-Ethanediol (ethylene glycol).

C PETROLEUM AS AN ENERGY SOURCE

No one knows the exact origin of petroleum. Most evidence indicates that it originated from plants that lived in ancient seas some 500 million years ago. These organisms died and eventually became covered with sediments. Pressure, heat, and microbes converted what was once living matter into petroleum, which became trapped in porous rocks. It is likely that some petroleum is still being formed from sediments of dead organisms. Even so, however, its natural formation is far too slow for us to regard petroleum as a renewable resource.

Human use of petroleum can be traced back almost 5,000 years. Ancient Middle Eastern civilizations used petroleum seeping out of the ground for waterproofing ships and canals and for paving roads. By A.D. 1000, Arabs processed petroleum to obtain kerosene for lighting. Eleventh-century Chinese extracted oil from wells over a half-mile deep. Marco Polo observed and described oil fields in his travels through Persia in the thirteenth century.

The first U.S. oil well was drilled in Titusville, Pennsylvania, in 1859. In the decades since then, our way of living has been greatly altered by petroleum's increasing use. The following activity will help you realize how much life has changed during our recent history of petroleum use.

The John Benninghoff oil farm, Oil Creek, Pennsylvania, 1865.

A portable but inefficient light source from an earlier era.

C.1 YOU DECIDE: THE GOOD OLD DAYS?

Earlier in this century, petroleum was used far less extensively than it is now. To investigate what life was like with less petroleum, you will serve on a team to interview someone old enough to remember this period of our nation's past.

First, however, your class should decide what questions should be included in the interviews. Sample questions are given below. You may use these or develop your own set of 10–12 questions. All interview teams should use the same questions, so that you can compare results later.

SAMPLE INTERVIEW QUESTIONS

1. How would you describe the location where you lived as a child—urban, suburban, or rural?
2. What was the main source of heat in your childhood home?
3. How was this source of heat supplied? Did you obtain the fuel yourself or was it delivered to your home?
4. Considering cleanliness, convenience, and quantity of heat produced, how does that source of heat compare with what we use today?
5. What was the main source of lighting in your childhood home? What source of energy was used to provide the lighting?
6. What, if any, was the main means of public transportation? What provided the energy for this transportation?
7. What was the main source of private transportation? How common was this mode of transportation? What was its source of energy?
8. What fuel was used for cooking?
9. If your family bought your food rather than growing or raising it, how was it packaged?
10. In what kind of container was milk obtained?
11. What kind of soap was used for washing clothes? How did its effectiveness compare to today's soaps and detergents?
12. What were the main fabrics used in clothes? From what were the fabrics made?
13. Were clothes easier or harder to care for than they are now? Please explain.

After deciding what questions to use in the interviews, your class will organize into interview teams. Each team should arrange to interview someone who is more than 70 years old—a team member's grandparent, a neighbor, or a resident of a local retirement center.

Before conducting the interview, each team should practice by interviewing one of its own members. This practice interview will provide current information for comparison and will help you sharpen your interviewing skills. Following each interview, summarize all questions and answers in a written outline or chart.

All interview-team results should be tabulated on a single class poster or bulletin board. The questions can be written in a vertical column; the most common answers can be written alongside each question. Each team can then place a check mark next to the answers they obtained. Interview information that doesn't "fit" the chart can be introduced during the class discussion.

Here are some suggested class discussion topics:

1. Summarize the main differences between earlier years and now regarding

   a. home heating.

   b. lighting.

   c. public transportation.

   d. private transportation.

   e. cooking.

   f. food packaging.

   g. clothes washing.

   h. clothing material.

2. Would a return to the "good old days" be a good thing? Why?

3. If current energy sources become depleted, will we have to return to a lifestyle similar to that of the past? Why?

Would *you* want to live in the "good old days"? Whatever your answer, of course, we still live in the here and now. That means we have to make the best possible use of available energy sources—petroleum being a major one.

To help understand our energy future, we will next investigate energy sources that powered our society over the past century.

C.2 ENERGY: PAST, PRESENT, AND FUTURE

The sun is our planet's primary energy source. Through the process of *photosynthesis,* radiant energy from the sun is stored as chemical energy in green plants. This energy is stored by animals in their own molecules when they eat these plants. The organic molecules found in plants and animals are called *biomolecules.*

Solar energy and energy stored in biomolecules are the basic energy sources for life on our planet. Some of this energy is stored in wood, coal, and petroleum. Since the discovery of fire, human use of these fuels has been a major influence in civilization's development. As your interviews probably suggested, the forms, availability, and cost of energy greatly influence how and even where people live.

In the past, our nation had abundant supplies of inexpensive energy. Until about 1850, wood, water, wind, and animal power satisfied all of our slowly growing energy needs. Wood—then the predominant energy source—was readily available due to the conversion of forests to farmland. Wood served as an energy source for heating, cooking, and lighting. Water, wind, and animal power provided transportation and "fueled" our machinery and industrial processes. In the next activity, you will see just how much all this has changed since then.

YOUR TURN

Fuel Sources over the Years

Figure 13 emphasizes how U.S. energy sources have changed since 1850. There has been a definite shift away from wood, mainly toward fossil fuels—coal, petroleum, and natural gas. Use Figure 13 to answer these questions:

1. a. For how many years after 1850 did wood supply at least 50% of the nation's total energy?
    b. What was the chief mode of travel (other than walking) during that period?

2. a. What factors might explain the decline in wood use after that period?
    b. What energy source increased in importance after wood declined?

3. Compared with other energy sources, only a small quantity of petroleum was used before about 1910. What do you think petroleum's main uses might have been in 1900?

4. Oil became increasingly important about the same time that coal use reached its peak.
    a. When did that occur?
    b. What explains the growing use of petroleum after that date?

5. a. What is the most recent energy source to enter the picture?
    b. What is the major use of this energy source?

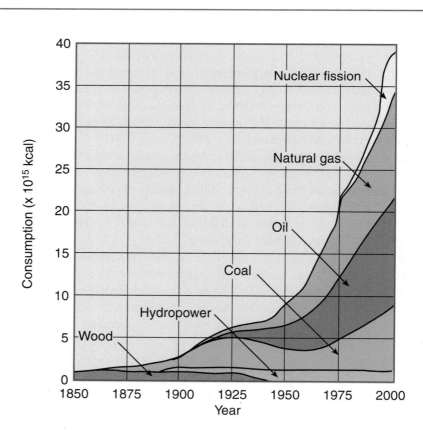

Figure 13 Annual U.S. consumption of energy from various sources (1850–2000). (Adapted from page 39, "Energy and Power" by Chauncey Starr, illustrated by Allen Beechel, September, 1971/Copyrighted by *Scientific American, Inc.* All rights reserved.)

You have just examined the patterns of energy sources used in this country over approximately the past 150 years. During that time, the major sources of energy, as well as the quantity of energy used, have changed. Therefore, we need to recognize not only what various sources have provided energy during this time (see previous *Your Turn*), but also how much energy those sources had to supply to keep up with energy demands.

YOUR TURN

Energy Supplied by Various Fuels

Figure 14 illustrates how the quantities of energy used in the United States from various sources has changed since 1850. For example, the graph indicates that the nation used about 22×10^{18} joules (J) of energy in 1920. Coal was by far the major fuel at the start of the Roaring Twenties.

Use Figure 14 to answer these questions:

1. a. Has U.S. energy consumption since 1850 remained constant, grown steadily, or increased rapidly?

   b. Describe at least two possible reasons for this general trend. Give the approximate time when the reasons might have become significant in terms of energy consumption.

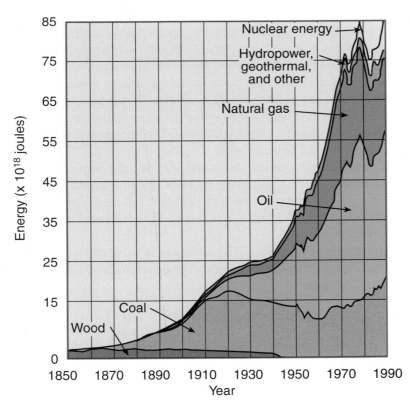

Figure 14 U.S. energy consumption by fuel type, 1850–1990.

2. Between 1970 and 1990:

   a. By what percent did oil consumption change?

   b. What was the percent increase in nuclear energy? How much (in percent) did oil consumption change?

3. Natural gas consumption changed by about 22% between 1970 and 1990.

   a. Using data from the graph, show how to calculate this value.

   b. Did natural gas consumption increase or decrease during this period?

   c. Compare the change in natural gas consumption with that of oil. Offer an explanation.

Changes in supplies and uses of energy are part of our nation's history. But what about the future—your future? How will the energy picture look as you enter the twenty-first century? Are we doomed to run short of energy? In considering these questions, we must first understand how energy is stored in fuels and how we convert this chemical energy into other, more useful forms.

C.3 ENERGY AND FOSSIL FUELS

A coal-burning power plant.

The fossil fuels—petroleum, natural gas, and coal—can be regarded as forms of buried sunshine. Fossil fuels probably originated from biomolecules of prehistoric plants and animals. The stored energy released by burning fossil fuels is energy originally captured from sunlight by prehistoric green plants during photosynthesis. Fossil-fuel energy can be compared to the energy stored in a wind-up toy. That energy was originally supplied to coil a spring in the toy. Most of that energy is stored within the spring in relation to the toy's moving parts. When the spring unwinds, providing energy to the parts, the toy moves. Eventually, the toy "winds down" to a lower-energy, more stable state. The energy that was stored in the spring is released as it unwinds.

Likewise, chemical energy is stored in chemical compounds. When chemical reactions occur, as during the burning of fuels, reactant atoms are reorganized to form products with different and more stable arrangements of their atoms. Some of the stored energy is released in the form of heat and light.

The burning of methane (CH_4) illustrates how we can consider the energy involved in chemical reactions. The equation for the combustion (burning) of methane is

$$CH_4 \ + \ 2\,O_2 \ \rightarrow \ CO_2 \ + \ 2\,H_2O \ + \ \text{Energy}$$

methane gas · oxygen gas · carbon dioxide gas · water

This simple reaction releases a considerable quantity of energy. It is a reaction you might have used in the laboratory to create a Bunsen burner flame.

It is useful to think about this reaction as taking place in separate (and imaginary) *bond-breaking* and *bond-making* steps. Consider the reaction of one methane molecule with two oxygen molecules. As a first step, imagine that all bonds in the reactant molecules are broken, producing separate atoms of carbon, hydrogen, and oxygen. This bond-breaking step is an energy-requiring process—an **endothermic** change; energy must be added (energy appears as a reactant). Figure 15 indicates this change.

▶ An endothermic reaction can proceed only if energy is continuously supplied.

STEP I $\qquad$ Energy + CH_4 + 2 O_2 → C + 4 H + 4 O

The second (and final) imaginary step involves forming the new bonds needed to make the products—one carbon dioxide molecule and two water molecules. The formation of bonds is an energy-releasing process—an **exothermic** change; energy is given off (energy appears as a product).

▶ Once an exothermic reaction begins, it releases energy until the reaction stops.

STEP II $\qquad$ C + 4 H + 4 O → CO_2 + 2 H_2O + Energy

Whether an overall reaction is exothermic or endothermic depends on the quantity of energy added (endothermic process) in the bond-breaking steps compared to the amount of energy given off (exothermic process) in the bond-making steps. If the quantity of energy given off is greater than the quantity added, the *overall* process is exothermic; if more energy is added than is given off, it is endothermic. The energy released in forming bonds in carbon dioxide and water molecule bonds is greater than the energy originally needed to break the C-H bonds in methane and oxygen–oxygen bonds in O_2 gas. For that reason, the overall chemical reaction (Step I + Step II) is exothermic. That is shown by placing energy on the product side of the equation.

$$CH_4 + 2\ O_2 → CO_2 + 2\ H_2O + Energy$$

This very ability to release significant energy per gram when burned makes methane and the other lighter alkanes (C_1 to C_8) valuable as fuels.

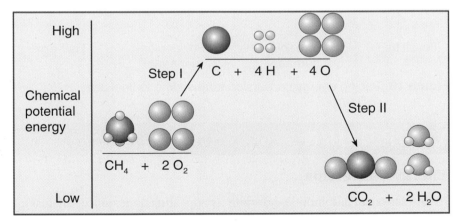

Figure 15 The formation of carbon dioxide and water from methane and oxygen gases. In Step I, bonds are broken, an endothermic process. Step II releases energy (an exothermic process). Because more energy is released in Step II than is required in Step I, the formation of carbon dioxide and water is exothermic.

There is a general principle for energy of reactions: *If a reaction is exothermic, then the reverse reaction is endothermic.* For example, the formation of water from hydrogen and oxygen is exothermic:

$$2 \, H_2 + O_2 \rightarrow 2 \, H_2O + \text{Energy}$$

Therefore, the decomposition of water—the reverse reaction—is endothermic:

$$\text{Energy} + 2 \, H_2O \rightarrow 2 \, H_2 + O_2$$

Scientists and engineers have increased the usefulness of the energy released from burning fuels by developing devices that convert thermal energy into other forms of energy. In fact, much of the energy we use goes through several conversions before reaching us.

Consider the energy-conversion steps needed to dry your hair with a hair dryer (see Figure 16). In the first step, the stored *chemical energy* in a fossil fuel (A) is burned in an electric power plant furnace to produce *thermal energy* (heat) (B). The thermal energy then converts water in a boiler to steam that spins turbines to generate electricity. Thus, the power plant converts thermal energy to *mechanical energy* and then to *electrical energy* (C). When it reaches your hair dryer, the electricity is converted back to thermal energy to dry your hair, and also to mechanical energy (D) as a small fan blade spins to blow the hot air. There is even some conversion to *sound energy*, as any hair-dryer user knows. It is important to recognize that energy is not really consumed. Its form simply changes from thermal to mechanical to electrical, for example. This is summarized by the **law of conservation of energy**—energy is neither created nor destroyed; the total energy of the universe is constant.

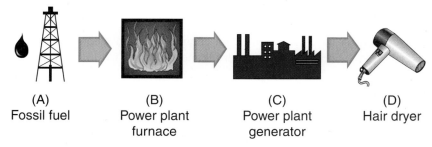

| (A) | (B) | (C) | (D) |
| Fossil fuel | Power plant furnace | Power plant generator | Hair dryer |

Figure 16 Tracing energy conversion from source to final use.

YOUR TURN

Energy Conversion

In one sense, an automobile—despite its appealing appearance—is just a collection of energy converters and energy-using devices. Assume the role of an "energy detective" and follow some typical energy conversions in an automobile. Name the type(s) of energy involved in each step of getting an automobile window defroster to operate when the car is running (Figure 17, page 195).

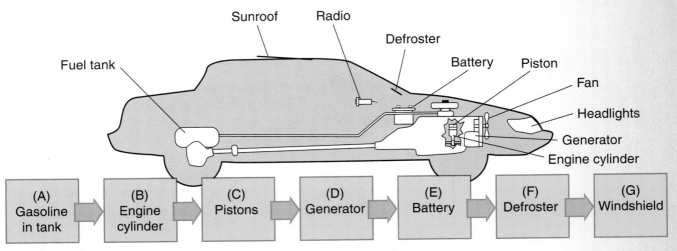

Figure 17 Tracing energy conversions in an automobile.

Although energy-converting devices have certainly increased the usefulness of petroleum and other fuels, some devices have produced new problems. Pollution, for instance, sometimes accompanies energy conversion. More fundamentally, some useful energy is *always* "lost" when energy is converted from one form to another. That is, no energy conversion is totally efficient; some energy always becomes unavailable to do useful work.

Consider an automobile that starts with 100 units of chemical energy in the form of gasoline (Figure 18). Even a well-tuned automobile converts only about 25% of the chemical energy in gasoline to useful mechanical energy (motion). The remaining 75% of gasoline's chemical energy is lost to the surroundings as heat. In the following *Your Turn,* you will see what this means in terms of gasoline consumption and in terms of your pocketbook.

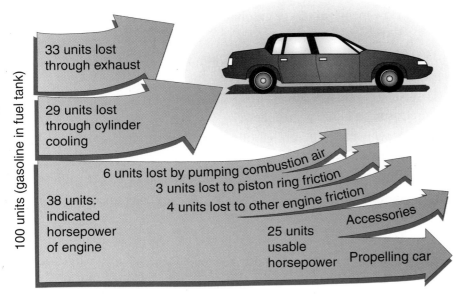

Figure 18 Energy use in an automobile.

Energy Conversion Efficiency

Assume your family drives 225 miles each week and that your car can travel 27.0 miles on a gallon of gasoline. How much gasoline does your car use in one year?

Questions such as this can be answered by using arithmetic. One approach is to attach proper units to all values and then multiply and divide them the same way numbers are processed.

The information given in the sample problem can be expressed this way:

$$\frac{225 \text{ miles}}{1 \text{ week}} \quad \text{and} \quad \frac{27.0 \text{ miles}}{1 \text{ gal}}$$

or, if needed, as inverted expressions:

$$\frac{1 \text{ week}}{225 \text{ miles}} \quad \text{and} \quad \frac{1 \text{ gal}}{27.0 \text{ miles}}$$

The desired answer must have these units:

$$\frac{\text{gal}}{\text{y}}$$

Calculating the desired answer also involves using information you already know—there are 52 weeks in one year, $\frac{52 \text{ weeks}}{1 \text{ y}}$. It also involves using the second unit above in its inverted version so the answer is given in the desired units:

$$\frac{225 \text{ miles}}{1 \text{ week}} \times \frac{1 \text{ gal}}{27.0 \text{ miles}} \times \frac{52 \text{ weeks}}{1 \text{ y}} = 433 \text{ gal/y}$$

Now answer these questions:

1. Assume your 27.0 mpg automobile travels 200 miles each week.
   a. How far will the car travel in one year?
   b. How many gallons of gasoline will the car use during the year's travels?

2. If gasoline costs $1.35 per gallon, how much would you spend on gasoline in one year?

3. Assume your automobile engine uses only 25.0% of the energy released by burning gasoline.
   a. How many gallons of gasoline are wasted each year due to your car's inefficiency?
   b. How much does this wasted gasoline cost at $1.35 per gallon?

4. Suppose you trade the car for one that travels 40.0 miles on a gallon of gasoline.
   a. How much gasoline would this new car use in one year?
   b. Compared with the first car, how many gallons of gasoline would you "save" by driving the more fuel-efficient automobile?

c. How much money would you save on gasoline in one year (assuming you still drive 200 miles per week)?

5. Suppose continued research leads to a car engine with 50.0% efficiency that can travel 50.0 miles on a gallon of gasoline.

   a. In one year, how much gasoline would be saved compared with a car that was 25.0% efficient and averaged 27.0 miles per gallon?

   b. How much money would be saved?

Petroleum supplies are neither limitless nor inexpensive. Thus, increasing energy efficiency is important to maximize the benefits from our remaining petroleum supplies. One way to increase overall energy efficiency is to reduce the number of energy conversions from the fuel to its final use. We can also try to increase the efficiency of our energy-conversion devices.

Unfortunately, devices that convert chemical energy to heat and then to mechanical energy are typically less than 50% efficient. Solar cells (solar energy → electrical energy) and fuel cells (chemical energy → electrical energy) hold promise either for replacing petroleum or for increasing the efficiency of its use.

But what *is* burning? What products form when petroleum-based fuels burn? How much energy is involved? We will investigate these questions next.

C.4 THE CHEMISTRY OF BURNING

You strike a match—a hot, yellow flame appears. You hold it to a candle wick. The candle lights and burns. These events are so common-place you probably don't realize that complex chemical reactions are at work.

Candle-burning involves chemical reactions of the wax (composed of long-chain alkanes) with oxygen gas at elevated temperatures. Many chemical reactions are involved in such burning, or **combustion.** For simplicity, chemists usually consider the overall changes involved. For example, the reaction involved in burning one wax component can be described this way:

$$C_{25}H_{52} \ + \ 38\,O_2 \ \rightarrow \ 25\,CO_2 \ + \ 26\,H_2O \ + \ \text{Energy}$$

wax   oxygen   carbon   water
(alkane)   gas   dioxide gas   vapor

This reaction is exothermic. More energy is given off by the formation of the bonds in the product molecules (carbon dioxide gas and water vapor) than the energy needed to break the bonds in the wax and oxygen gas reactant molecules.

Fuels provide energy as they burn. But how much energy is released? How is the quantity of released energy measured? Find out for yourself in the following activity.

C.5 LABORATORY ACTIVITY: COMBUSTION

GETTING READY

As you know, the boiling points of hydrocarbons are related to the number of carbon atoms in each molecule. Is the quantity of thermal energy released in burning also related to the number of carbon atoms in each hydrocarbon molecule?

The quantity of thermal energy given off when a certain amount of a substance burns is its *heat of combustion.* When one mole of a substance burns, the thermal energy released is called its *molar heat of combustion.*

In this activity you will investigate relationships between the thermal energy released when a hydrocarbon burns, and the structure of the hydrocarbon. You will measure the heat of combustion of a candle (paraffin wax) and compare this quantity with known values for other hydrocarbons.

Prepare a data table similar to the one that follows.

Data Table	
Initial mass of candle + index card	_____ g
Volume of water	_____ mL
Room temperature	_____ °C
Initial temperature of water in can	_____ °C
Final temperature of water in can	_____ °C
Final mass of candle + index card	_____ g

PROCEDURE

1. Prepare the candle by holding a lighted match near its base, so that some melted wax falls onto a 3" × 5" index card. Immediately push the candle into the melted wax and hold it there for a moment to fasten it to the card.

2. Determine the combined mass of the candle and the index card. Record the value.

3. Carefully measure (to the nearest milliliter) 100 mL of chilled water. (The chilled water will be provided by your teacher. It should be 10 to 15 °C colder than room temperature.) Pour the 100-mL sample of chilled water into an empty soft drink can.

4. Set up the apparatus shown in Figure 19 (page 199). (Do not light the candle yet!) Adjust the soft drink can height so the top of the candle wick is about 2 cm from the bottom of the can.

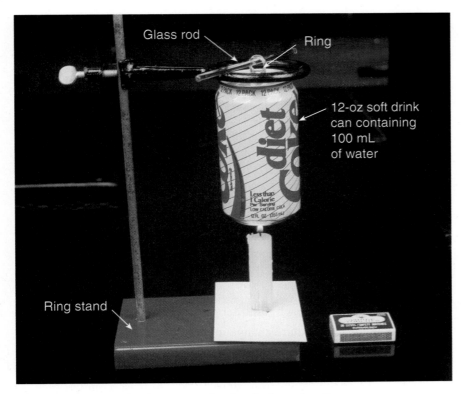

Figure 19 Apparatus for measuring heat of combustion.

5. Measure the room temperature and the water temperature to the nearest 0.2 °C. Record these values.

6. Place the candle under the can of water. Light the candle. Stir the water gently with a stirring rod as it heats.

7. As the candle burns you may need to lower the soft drink can *cautiously* so the flame remains just below the bottom of the can.

8. Continue heating until the temperature rises as far *above* room temperature as it was *below* room temperature at the start. (*Example:* If the water was 15 °C before heating and room temperature is 25 °C, then you would heat the water 10 °C *higher* than room temperature—to 35 °C.)

9. When the desired temperature is reached, extinguish the flame.

10. Continue stirring the water until its temperature stops rising. Record the highest temperature reached by the water.

11. Determine the mass of the cooled candle and index card, including all wax drippings.

12. Wash your hands thoroughly before leaving the laboratory.

CALCULATIONS

A characteristic property of a material is the quantity of heat needed to raise the temperature of one gram of the material by 1 °C. This value is the ***specific heat*** of the material. The specific heat of liquid water, for example, is about 4.2 J/g (joules per gram). Therefore, for each degree

► *The exact value for the specific heat of water is 4.184 joules per gram. However, a rounded-off value of 4.2 J/g is quite useful for this laboratory activity.*

Celsius that liquid water is heated, we know that each gram of the water has absorbed 4.2 J of thermal energy.

Suppose we want to increase the temperature of 10 g of water by 5.0 °C. How much thermal energy is needed?

The answer is reasoned this way: We know it takes 4.2 J to raise the temperature of 1 g of water by 1 °C. However, compared to this standard, we have 10 times more water and we want a 5 times greater rise in temperature. Thus, we need 10×5 or 50 times more thermal energy to accomplish the task. So, 50×4.2 J $= 210$ J. In other words, we need 210 J to heat 10 g of water by 5.0 °C.

The heat of combustion of a substance can be expressed either as the heat released when one *gram* of substance burns or when one *mole* of substance burns (the molar heat of combustion). Let's first find the heat of combustion per gram of paraffin wax.

Use data collected in the laboratory activity to complete the following calculations.

1. Calculate the mass of water heated. (*Hint:* The density of water is 1.0 g per mL—thus, each milliliter of water has a mass of 1.0 g.)

2. Calculate the temperature change in the water. (This equals the final water temperature minus the initial water temperature.)

3. Find the quantity of thermal energy used to heat the water in the soft drink can. Use the values from steps 1 and 2 to reason out the answer, as was illustrated in the sample problem.

4. Find the mass of paraffin burned.

5. Find paraffin's heat of combustion, in units of joules per gram of paraffin (J/g) and also in kJ/g.

► *1 kJ = 1,000 J*

Assume that the thermal energy absorbed by the water equals the thermal energy released by the burning paraffin. Thus, dividing the thermal energy (step 3) by the mass of paraffin burned (step 4) produces the answer:

$$\text{Heat of combustion} = \frac{\text{thermal energy released (step 3)}}{\text{mass of paraffin burned (step 4)}}$$

QUESTIONS

Your teacher will collect your heat of combustion data, expressed in units of kilojoules per gram of paraffin (kJ/g). Use the combined class results for this value to answer the following questions.

1. How does your experimental heat of combustion (in kJ/g) for paraffin wax, $C_{25}H_{52}$, compare with the accepted value for butane, C_4H_{10}? (See Table 5, page 201.)

2. How do the molar heats of combustion (in kJ/mol) for paraffin and butane compare? (*Hint*: To make this comparison, first calculate the thermal energy released when one *mole* of paraffin burns. One mole of paraffin ($C_{25}H_{52}$) has a mass of 352 g. The quantity of thermal energy released will thus be 352 times greater than the thermal energy released by one gram of paraffin.)

3. Explain any differences noted in your answers to questions 1 and 2.

Table 5		Heats of Combustion	
Hydrocarbon	Formula	Heat of Combustion (kJ/g)	Molar Heat of Combustion (kJ/mol)
Methane	CH_4	55.6	890
Ethane	C_2H_6	52.0	1560
Propane	C_3H_8	50.0	2200
Butane	C_4H_{10}	49.3	2859
Pentane	C_5H_{12}	48.8	3510
Hexane	C_6H_{14}	48.2	4141
Heptane	C_7H_{16}	48.2	4817
Octane	C_8H_{18}	47.8	5450

4. Which hydrocarbon—paraffin or butane—is the better fuel? Explain your answer.

5. In calculating heats of combustion, you assumed that all thermal energy from the burning fuel went to heating the water.

   a. Is this a good assumption?

   b. What other laboratory conditions or assumptions might cause errors in your calculated values?

C.6 USING HEATS OF COMBUSTION

With abundant oxygen gas and complete combustion, the burning of a hydrocarbon can be described by the equation

$$\text{Hydrocarbon} + \text{Oxygen gas} \rightarrow \text{Carbon dioxide} + \text{Water} + \text{Thermal energy}$$

Note that energy is written as a product of the reaction, because energy is released when a hydrocarbon burns in a highly exothermic reaction. As you found in the last laboratory activity, this energy can be expressed in terms of kilojoules for each gram (or mole) of fuel burned.

The equation for burning ethane is

$$2\,C_2H_6 + 7\,O_2 \rightarrow 4\,CO_2 + 6\,H_2O + ?\ \text{kJ thermal energy}$$

To complete this equation, we must furnish the quantity of thermal energy involved. A summary of data like that given in Table 5 provides such information. Table 5 indicates that ethane's heat of combustion is 1,560 kJ per mole—burning one mole of ethane releases 1,560 kilojoules of energy.

However, as a part of the chemical equation, thermal energy must be "balanced" just like other reactants and products. The balanced equation represents the burning of *two* moles of ethane ($2\,C_2H_6$). Thus, total thermal energy released will be *twice* that released when one mole of ethane burns:

$$2\ \text{mol} \times 1,560\ \text{kJ/mol} = 3,120\ \text{kJ}$$

▶ *Burning of a hydrocarbon is an exothermic reaction.*

The Lockheed F-117A Stealth Fighter uses high-grade jet fuel processed from petroleum.

The complete combustion equation is

$$2\ C_2H_6 + 7\ O_2 \rightarrow 4\ CO_2 + 6\ H_2O + 3{,}120\ kJ$$

As you found in the previous laboratory activity (and as suggested in Table 5), heats of combustion can also be expressed as energy produced when one gram of hydrocarbon burns (kJ/g). Such values are useful in finding the thermal energy released in burning a particular mass of a fuel.

For example, how much thermal energy would be produced by burning 12 g of octane, C_8H_{18}? Table 5 indicates that when *one* gram of octane burns, 47.8 kJ of energy are released. Burning 12 times more octane will produce 12 times more thermal energy—or

$$12 \times 47.8\ kJ = 574\ kJ$$

The calculation can also be written this way:

$$12.0\ \text{g octane} \times \frac{47.8\ kJ}{1\ \text{g octane}} = 574\ kJ$$

YOUR TURN

Heats of Combustion

To better understand the energy involved in burning hydrocarbon fuels, use Table 5 (page 201) to answer the following questions. The first one is already worked out as an example.

1. How much energy (in kilojoules) is released by completely burning 25.0 mol of hexane, C_6H_{14}?

Table 5 indicates that the molar heat of combustion of hexane is 4,141 kJ. Putting it another way, 4,141 kJ of energy are released when *one* mole of hexane burns. Thus, burning *25 times more* fuel will produce 25 times more energy:

$$25.0 \; \cancel{mol \; C_6H_{14}} \times \frac{4141 \; kJ}{1 \; \cancel{mol \; C_6H_{14}}} = 104{,}000 \; kJ$$

Burning 25.0 mol of hexane would thus release 104,000 kJ of thermal energy.

2. Write a chemical equation, including thermal energy, for the complete combustion of these alkanes:

   a. Methane (main component of natural gas)

   b. Butane (variety of uses, including lighter fluid)

3. Examine the data summarized in Table 5.

   a. How does the trend in heats of combustion for hydrocarbons expressed in kJ/g compare with that expressed in kJ/mol?

   b. Assuming the trend continues to larger hydrocarbons, predict the heat of combustion for decane, $C_{10}H_{22}$, in units of kJ/g and kJ/mol.

   c. Which prediction in question 3b was easier? Why?

4. a. How much thermal energy is produced by burning two moles of octane?

   b. How much thermal energy is produced by burning one gallon of octane? (One gallon of octane has a mass of 2,660 g.)

   c. Suppose your car operates so inefficiently that only 16% of the thermal energy from burning fuel is converted to useful "wheel-turning" (mechanical) energy. How many kilojoules of useful energy would be stored in a 20.0-gallon tank of gasoline? (Assume that octane burning and gasoline burning produce the same results.)

5. The heat of combustion of carbon as coal is 394 kJ per mole.

   a. Write a chemical equation for burning carbon. Include the quantity of thermal energy produced.

   b. Gram for gram, which is the better fuel—carbon or octane? Explain.

   c. In what applications might carbon as coal replace petroleum-based fuel?

   d. Describe one application in which coal would be a poor substitute for a petroleum product.

As automobiles became popular in the United States, gasoline demand increased rapidly. The gasoline fraction normally represents only about 18% of a barrel of crude oil, and so researchers seek ways to increase the quantity of gasoline obtained from crude oil. One key discovery is still in use today: It is possible to alter the structures of some petroleum hydrocarbons so that 47% of a barrel of crude oil can be converted to gasoline. This important chemical technique deserves further attention.

C.7 ALTERING FUELS

As you might expect, not all fractions of hydrocarbons obtained from petroleum are necessarily—at any particular time—in equal demand or use. The market for one petroleum fraction may be much less profitable than another. For example, the invention of electric light bulbs caused the use of kerosene lanterns to decline rapidly in the early 1900s. Petroleum's kerosene fraction, composed of hydrocarbon molecules with 12 to 16 carbon atoms, then became a surplus commodity. On the other hand, automobiles dramatically increased the demand for the gasoline fraction (C_5 to C_{12}) that had been discarded in earlier years (because the gasoline fraction could not be used safely in lanterns—it exploded!).

Chemists and chemical engineers are adept at modifying or altering chemical resources on hand to meet new market opportunities or problems. Such alterations might involve converting less-useful materials to more-useful products, or—as in the case of kerosene in the early 1900s—a low-demand material into high-demand materials.

CRACKING

By 1913, chemists had devised a process for *cracking* molecules in kerosene into smaller (gasoline-sized) molecules by heating the kerosene to 600–700 °C. For example, a 16-carbon molecule might be cracked at 700 °C to produce two 8-carbon molecules:

$$C_{16}H_{34} \rightarrow C_8H_{16} + C_8H_{18}$$

In practice, molecules containing 1 to 14 or more carbon atoms can be produced through cracking. Product molecules containing 5 to 12 carbons are particularly useful components of gasoline, the most important commercial product from refining. Some of the C_1 to C_4 molecules formed in cracking are immediately burned, keeping the temperature high enough for more cracking to occur.

More than a third of all crude oil today undergoes cracking. The process has been improved by adding catalysts, such as aluminum oxide (Al_2O_3). A *catalyst* increases the speed of a chemical reaction (in this case the cracking process) but is not itself used up. Catalytic cracking is more energy efficient because it occurs at a lower temperature—500 °C rather than 700 °C.

Gasoline is sold in a variety of standards (and prices), as you already know. Gasoline composed mainly of straight-chain alkanes, such as hexane (C_6H_{14}), heptane (C_7H_{16}), and octane (C_8H_{18}) burns very rapidly. Such rapid burning causes engine "knocking" and may contribute to engine problems. Branched-chain alkanes burn more satisfactorily in automobile engines; they don't "knock" as much. The structural isomer of octane shown below has excellent combustion properties in automobile engines. (This octane isomer is known chemically as 2,2,4-trimethylpentane. Can you see how the chemical name is related to the structure shown? This compound is frequently called isooctane, its common name.)

▶ Recall that the gasoline fraction obtained from crude oil refining includes hydrocarbons with 5 to 12 carbons per molecule.

The tall catalytic cracking units ("cats") break crude oil down into valuable and useful products.

OCTANE RATINGS

A common reference for gasoline quality is the *octane scale*. This scale assigns the branched-chain hydrocarbon commonly called "isooctane" an *octane number* of 100. Straight-chain heptane (C_7H_{16}), a fuel with very poor engine performance, is assigned an octane number of zero. Other fuels can be rated in comparison with isooctane and heptane. Therefore, a gasoline with the same knocking characteristics as a mixture of 87% isooctane and 13% heptane has an octane number of 87. The higher the octane number of a gasoline, the better its antiknock characteristics. Octane ratings in the high 80s and low 90s (85, 87, 92) are quite common among gasolines today, as any survey of gas pumps will reveal. Assigning an octane number of 100 to isooctane is arbitrary and does not mean that it has the highest possible octane number. In fact, several fuels burn more efficiently in engines than isooctane does, and have octane numbers above 100.

Before 1975, the octane rating of gasoline was increased inexpensively by adding a substance such as tetraethyl lead, $(C_2H_5)_4Pb$, to the fuel. This additive slowed down the burning of straight-chain gasoline molecules, and added about three points to the "leaded" fuel's octane rating. Unfortunately, lead from the treated gasoline was discharged into the atmosphere along with other exhaust products. Due to the harmful effects of lead in the environment, such lead-based gasoline additives are no longer used.

OXYGENATED FUELS

The phase-out of lead-based gasoline additives means alternative octane-boosting supplements are required. A group of additives called *oxygenated fuels* are frequently blended with gasoline to enhance its octane rating. The molecules of these substances contain oxygen in addition to carbon and hydrogen. Although oxygenated fuels deliver less energy per gallon than do regular gasoline hydrocarbons, their economic appeal comes from their ability to increase the octane number of gasoline as well as reduce exhaust-gas pollutants. Two common oxygenated fuels are methanol (methyl alcohol, CH_3OH) and methyl t-butyl ether, MTBE. MTBE has an octane rating of 116 and is added at the refinery; methanol (with octane number 107) is mixed in at the distribution locations. Current annual U.S. production of MTBE is more than 4 billion gallons (about 16 gallons per person), making it one of the top ten industrial chemicals.

Flexible-fuel cars—designed to run on methanol, gasoline, or any mixture of the two—have been demonstrated in several major U.S. cities. A 10% alcohol–90% gasoline blend, sometimes called *gasohol,* can be used without engine adjustments or problems in nearly all automobiles currently in service. In addition to its octane-boosting properties, methanol can be made from natural gas, corn, coal, or wood—a contribution toward conserving petroleum resources.

$$CH_3-O-\underset{\underset{CH_3}{|}}{\overset{\overset{CH_3}{|}}{C}}-CH_3$$

▶ *Methyl t-butyl ether (MTBE)*

YOUR TURN

A Burning Issue

Methanol (CH_3OH) and ethanol (CH_3CH_2OH) are both used as gasoline additives or substitutes. Their heats of combustion are, respectively, 23 kJ/g and 30 kJ/g.

1. Consider the chemical formulas of methanol and ethanol. Gram for gram, why are heats of combustion of methanol and ethanol considerably less than those of any hydrocarbons we have considered? (See Table 5, page 201.) (*Hint*: What would happen if you tried to burn oxygen?)

2. Would you expect automobiles fueled by 10% ethanol–90% gasoline (gasohol) to get higher or lower miles per gallon than comparable cars fueled by 100% gasoline? Why?

Other octane-boosting strategies involve altering the composition of petroleum molecules. This works because branched-chain hydrocarbons burn more satisfactorily than straight-chain hydrocarbons (recall iso-octane's octane number compared to that of heptane). Straight-chain hydrocarbons are converted to branched-chain ones by a process called *isomerization*. Hydrocarbon vapor is heated with a catalyst:

$$CH_3-CH_2-CH_2-CH_2-CH_2-CH_3 \xrightarrow{\text{Catalyst}} \underset{\underset{CH_3}{|}}{CH_3-CH-\underset{\overset{|}{CH_3}}{CH}-CH_3}$$

The branched-chain alkanes produced are blended with gasoline-sized molecules from cracking and distillation, producing high-quality unleaded gasoline.

Although cracked and isomerized molecules improve the burning of gasoline, they also increase its cost. One reason is that extra fuel is used in manufacturing such gasoline.

We have now examined petroleum's use as a fuel. In Part D we will turn to petroleum's "building" role as a source of substances from which to produce an impressive array of useful compounds and materials.

PART C: SUMMARY QUESTIONS

1. Even though petroleum has been used for thousands of years, it became a major energy source only in recent history. List three technological factors that might explain this fact.

2. Compare wood and petroleum as fuels. Identify some advantages and disadvantages of each.

3. From a chemical viewpoint, why is petroleum sometimes considered "buried sunshine"?

4. Complete an energy trace (like that on page 195) to show how energy is transferred from chemical energy stored in an automobile gas tank to the power for opening the automobile's sunroof.

5. Explain "energy efficiency."

6. One gallon of gasoline produces about 132,000 kJ of energy when burned. Assuming that an automobile is 25% efficient in converting this energy into useful work,
   a. How much energy is "wasted" when a gallon of gasoline burns?
   b. Where does this "wasted" energy go?

7. Write a balanced equation for the combustion of propane (bottled gas).

8. Water gas (a 50–50 mixture of CO and H_2) is made by the reaction of coal with steam. Since our nation has substantial coal reserves, water gas might serve as a substitute fuel for natural gas (composed mainly of methane, CH_4). Water gas burns according to this equation:

   $$CO + H_2 + O_2 \rightarrow CO_2 + H_2O + 525 \text{ kJ}$$

   a. How does water gas compare to methane in terms of thermal energy produced?
   b. If a water gas mixture containing 10 mol CO and 10 mol H_2 were completely burned, how much thermal energy would be produced?

9. The combustion of acetylene, C_2H_2 (used in a welder's torch), can be represented as:

$$2\,C_2H_2 + 5\,O_2 \rightarrow 4\,CO_2 + 2\,H_2O + 2512\text{ kJ}$$

   a. What is acetylene's heat of combustion in kilojoules per mole?

   b. For 12 mol of acetylene burned, how much thermal energy is produced?

10. List two factors that would help you decide which hydrocarbon fuel to use in a particular application.

11. a. Explain the meaning of a fuel's octane rating.

   b. A premium gasoline has a 93 octane rating. Explain this in terms of its isooctane and heptane content.

   c. List two ways to increase a fuel's octane rating.

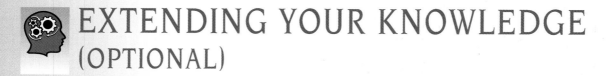# EXTENDING YOUR KNOWLEDGE (OPTIONAL)

Powdered aluminum or magnesium metal release considerable heat when burned. (Their heats of combustion are, respectively, 31 kJ/g and 25 kJ/g.) Why would these highly energetic powdered metals be poor fuel substitutes for petroleum?

D USEFUL MATERIALS FROM PETROLEUM

Just as an architect uses available construction materials to design a building, a chemist—a "molecular architect"—uses available molecules to design new molecules to serve our needs. Architects must know about the structures and properties of common construction materials. Likewise, chemists must understand the structures and properties of their raw materials, the "builder molecules." You will now explore the structures of some common hydrocarbon builder molecules and some products chemists make from them. In fact, you will have the chance to make one product yourself.

D.1 BEYOND ALKANES

Carbon is a versatile building-block atom. It can form bonds to other atoms in several different ways. In alkane molecules, each carbon atom is bonded to four other atoms. Such compounds are called *saturated* because each carbon atom forms only single bonds, using its maximum capacity of bonding to four other atoms. However, in some hydrocarbon molecules, a carbon atom bonds to *three* other atoms—not to four. This series of hydrocarbons is called the *alkenes,* whose first member is ethene, C_2H_4.

If each carbon atom must possess or share eight electrons to fill its outer energy level completely, how is the carbon bonding in alkenes possible? In a single covalent bond, just two electrons are shared between the two bonded atoms. In a *double covalent bond,* however, four electrons are shared between the bonding partners. Compounds containing double bonds are described as *unsaturated* because not all carbon atoms are bonded to their full capacity with four other atoms. Because of their double bonds, alkenes are more chemically reactive than alkanes. Many chemical reactions occur at the double bonds.

Not all builder molecules are hydrocarbons. In addition to carbon and hydrogen, some also contain one or more other elements, such as oxygen, nitrogen, chlorine, or sulfur. One way to think about many of these substances is as "carbon-backbone" hydrocarbons with other elements substituted for one or more hydrogen atoms.

Adding atoms of other elements to hydrocarbon structures drastically changes the chemical reactivity of the molecules. Even molecules composed of the same elements can have quite different properties. The molecules that make up a permanent-press shirt and permanent antifreeze may each contain the same elements—carbon, hydrogen, and oxygen. The dramatic differences in the properties and uses of these substances are in how atoms are arranged in the two molecules. Many of the 10 items from petroleum listed in the first *You Decide* (page 160) are made starting with alkenes as raw materials.

To learn more about this, you will do some atom-arranging in the laboratory.

D.2 LABORATORY ACTIVITY: THE BUILDERS

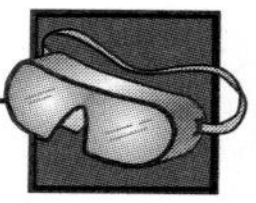

PART 1 ALKENES

1. Examine the electron-dot and structural formulas for ethene, C_2H_4. Confirm that each atom has attained a filled outer shell of electrons:

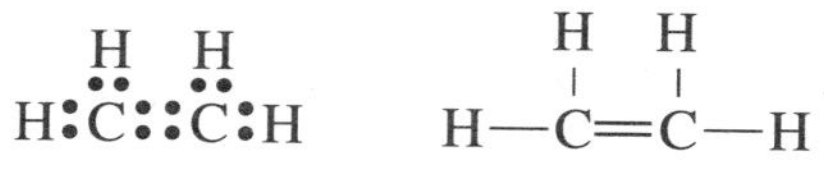

| Electron-dot formula | Structural formula | Molecular formulas |

Alkene names follow a pattern much like that of the alkanes. The first three alkenes are ethene, propene, and butene (there is no one-carbon alkene.) The same root as for alkanes is used to indicate the number of carbon atoms in the molecule's longest carbon chain; each name ends in -*ene*.

2. Examine the molecular formulas of ethene (C_2H_4) and butene (C_4H_8). Recall that the alkane general formula is C_nH_{2n+2}. What general formula for alkenes is suggested by their molecular formulas?

3. Assemble a model of an ethene molecule and one of ethane (C_2H_6). Compare the arrangements of atoms in the two models. Note that although you can rotate the two carbon atoms in ethane about the single bond, the two carbon atoms in ethene resist rotation about the double bond. This is characteristic of all double-bonded atoms in whatever molecule they are found.

4. Build a model of butene (C_4H_8). Compare your model of butene to those made by other class members.

   a. How many different arrangements of atoms in C_4H_8 appear possible? Each arrangement represents a different substance—another example of isomers!

   b. Do the structural formulas in Figure 20 correspond to models built by you or your classmates?

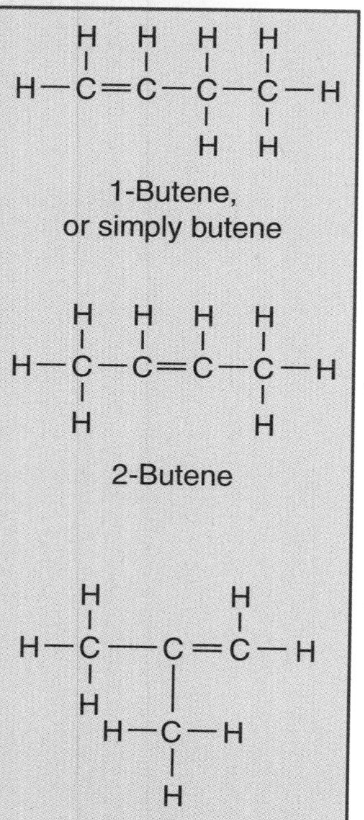

To distinguish isomers of straight-chain alkenes, the longest carbon chain is numbered, with double-bonded carbon atoms receiving the lowest numbers. The isomer name starts with the number assigned to the first double-bonded carbon atom; this explains the names 1-butene and 2-butene. Although the third structure above has a butene molecular formula, it is named as a methyl-substituted propene.

► *Organic compounds are systematically named by internationally agreed-upon rules devised by the International Union of Pure and Applied Chemistry. These rules are known as IUPAC nomenclature.*

► *For historical reasons, ethene is sometimes called ethylene.*

Figure 20 Some isomers of butene.

5. Do each of these pairs represent isomers, or are they the same substance?

   a. $CH_2{=}CH{-}CH_2{-}CH_3$ *or* $CH_3{-}CH_2{-}CH{=}CH_2$

   b. $CH_2{=}\underset{\underset{CH_3}{|}}{C}{-}CH_3$ *or* $CH_3{-}\underset{\overset{||}{CH_2}}{C}{-}CH_3$

6. How many isomers of propene (C_3H_6) are there?

7. Are these two structures isomers or the same substance?

$$\begin{array}{ccc} CH_2 & {-} & CH_2 \\ | & & | \\ CH_2 & & CH_2 \\ \backslash & & / \\ & CH_2 & \end{array} \qquad\qquad \begin{array}{ccc} CH_3 & {-}CH{-} & CH_2 \\ & | & | \\ & CH_2 & {-}CH_2 \end{array}$$

8. Based on your knowledge of molecules with single and double bonds between carbon atoms, assemble a model of a hydrocarbon molecule with a *triple* bond. Your completed model will represent a member of the series known as **alkynes.** Write structural formulas for

   a. ethyne, commonly called acetylene.

   b. 2-butyne.

PART 2   ANOTHER KIND OF ISOMERISM: WHICH SIDE ARE YOU ON?

In Part 1, you learned that the double bond in alkenes prevents the carbons in the double bond from rotating around the bond axis. The double-bonded carbon atoms align in a plane.

$$\text{>}C{=}C\text{<}$$

This inflexible arrangement between the carbon atoms creates the possibility of *cis-trans isomerism.*

As you assembled models in Step 4 above for 2-butene, you might have accidentally come upon this type of isomerism. In *cis-trans* isomerism, two identical groups can be in one of two different molecular positions. Both groups can be on the "same side" of the double bond, or one can be across the double bond from the other.

$$\underset{\textit{cis}\text{-2-butene}}{H{-}\underset{\overset{|}{H}}{\overset{\overset{H}{|}}{C}}{-}\underset{\overset{H}{|}}{C}{=}\underset{\overset{H}{|}}{C}{-}\underset{\overset{|}{H}}{\overset{\overset{H}{|}}{C}}{-}H} \qquad\qquad \underset{\textit{trans}\text{-2-butene}}{H{-}\underset{\overset{|}{H}}{\overset{\overset{H}{|}}{C}}{-}\underset{\overset{H}{|}}{C}{=}\underset{\overset{H}{|}}{C}{-}\underset{\overset{|}{H}}{\overset{\overset{H}{|}}{C}}{-}H}$$

In the case of 2-butene, the hydrogen atoms on the double-bonded carbon atoms are in one of two arrangements, and the two different arrangements represent two different compounds! In one case, the molecule has the two hydrogen atoms on the *same* side of the double bond

Ball-and-stick model of a double bond.

plane. This arrangement is known as the *cis* (same side) isomer, and the compound is called *cis*-2-butene. The other compound has the two hydrogen atoms positioned diagonally across the double-bond plane from each other. This arrangement is the *trans* (across) isomer, and the compound is known as *trans*-2-butene.

Assemble molecular models for compounds with the formula $C_2H_2Cl_2$.

1. a. Is *cis-trans* isomerism possible in this case?

   b. If so, identify the *cis* and the *trans* isomers.

2. One arrangement with the formula $C_2H_2Cl_2$ is the compound 1,1-dichloroethene. In a molecule of this compound, both chlorine atoms are on the same carbon atom. Is *cis-trans* isomerism possible in this case? Explain your answer.

Cis-trans isomerism will be discussed again in the Food unit in relation to fats.

PART 3 COMPOUNDS OF CARBON, HYDROGEN, AND SINGLY BONDED OXYGEN

1. Assemble as many different molecular models as possible using all nine of these atoms:

   2 carbon atoms (each forming four single bonds)
   6 hydrogen atoms (each forming a single bond)
   1 oxygen atom (forming two single bonds)

On paper, draw a structural formula for each compound you have modeled, indicating how the nine atoms are connected. Compare your structures with those made by other students. When you are satisfied that all possible structures have been produced, answer these questions:

   a. How many distinct structures did you identify?

   b. Write their structural formulas.

   c. Are these isomers? Explain.

2. The compounds you may have identified have distinctly different physical and chemical properties.

   a. Recalling that "like dissolves like," which compound should be more soluble in water?

   b. Which should have the higher boiling point?

D.3 MORE BUILDER MOLECULES

So far, we have examined just a small part of the inventory of builder molecules that "chemical architects" have available. Now we will explore two important classes of compounds in which carbon atoms are joined in rings, rather than in chain structures.

Picture what would happen if one hydrogen atom at each end of a hexane molecule is removed, and those two carbon atoms become bonded to each other.

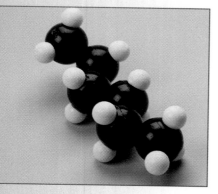

A space-filling model of hexane.

$CH_3CH_2CH_2CH_2CH_2CH_3$

Hexane

$$CH_2$$
$$CH_2 \quad CH_2$$
$$| \qquad |$$
$$CH_2 \quad CH_2$$
$$CH_2$$

Cyclohexane

A space-filling model of cyclohexane.

▶ *The pleasant odor of the first aromatic compounds discovered gave them their descriptive name.*

Cyclohexane is representative of the *cycloalkanes,* saturated hydrocarbons in which carbon atoms are joined in rings. Cyclohexane is a starting material for making nylon.

A very important class of ring compounds begins with benzene (C_6H_6), the simplest member of unsaturated cyclic builder molecules known as *aromatic compounds.* These compounds present a distinctly different chemical personality from those of cycloalkanes and their derivatives. A drawing of a cyclic benzene molecule, based on what you currently know about hydrocarbons, might look like the one below. (In the right hand figure, each "corner" of the hexagonal ring represents a carbon atom with its hydrogen atom.)

or

Chemists who first investigated benzene proposed this and related structures, but a puzzle remained. Carbon–carbon double bonds ($C=C$) such as those depicted above are very reactive. Yet benzene does not react chemically as though it has carbon–carbon double bonds. A new understanding of chemical bonding was needed to explain benzene's puzzling structure.

Substantial evidence now supports the notion that all carbon-carbon bonds in benzene are identical and not well represented by alternating single and double bonds. To represent a benzene molecule on paper, chemists often use this formula instead:

The inner circle represents the equal sharing of electrons among all six carbon atoms. The hexagonal line represents the bonding of six carbon atoms to each other. Each "corner" in the hexagon is the location of one carbon and one hydrogen atom, thus accounting for benzene's formula, C_6H_6.

Only small amounts of aromatic compounds are found in petroleum, but large quantities are produced by fractionation and cracking. Benzene and other aromatic compounds are present in gasoline as octane enhancers; however, their primary use is as chemical builder molecules. Entire chemical industries (dye and drug manufacturing, in particular) have been based on the unique chemistry of aromatic compounds.

A space-filling model of benzene.

Scanning tunneling micrograph of benzene.

D.4 BUILDER MOLECULES CONTAINING OXYGEN

In assembling molecular models with C, H, and O atoms, it is likely that you "discovered" the following type of compounds containing oxygen:

$$CH_3—OH \qquad\qquad CH_3—CH_2—OH$$

Methanol (methyl alcohol)   Ethanol (ethyl alcohol)

Note that each molecule has an OH group attached to a carbon atom. This general structure is characteristic of a class of compounds known as *alcohols.* Alcohols have certain properties in common. The OH is referred to as a *functional group*—an atom or groups of atoms that impart characteristic properties to organic compounds.

If we use the letter R to represent the rest of the molecule (other than the functional group), then the general formula for an alcohol can be written as

$$R—OH$$

Any alcohol

The line indicates a covalent bond between the oxygen of the OH group and another atom in the molecule. In methanol (CH_3OH), the R represents $CH_3—$; in ethanol (CH_3CH_2OH), the R represents $CH_3CH_2—$.

A functional group such as the OH group can be incorporated into an alkane, an alkene, a cycloalkane, an aromatic compound (see examples in margin), or into other structures.

Two other classes of compounds, *carboxylic acids* and *esters,* are versatile and significant builder molecules. Their functional groups each contain two oxygen atoms as are shown below.

$$R—\overset{\overset{\displaystyle O}{\|}}{C}—OH \qquad\qquad\qquad R—\overset{\overset{\displaystyle O}{\|}}{C}—OR$$

Carboxylic acid   Ester

(also written as RCOOH and RCOOR, respectively)

Note that both classes of compounds have one oxygen atom double-bonded to a carbon atom, and a second oxygen atom single-bonded to the *same* carbon atom. Ethanoic acid (acetic acid) and methyl acetate (an ester) are examples of these two types of compounds. Their structural formulas are given in the margin.

Other functional groups include nitrogen, sulfur, or chlorine atoms. When attached to hydrocarbons, these functional groups impart their own characteristic properties to the new molecules. The essential point is that adding functional groups to builder hydrocarbons greatly expands the types of molecules that can be built.

D.5 CREATING NEW OPTIONS: PETROCHEMICALS

Until only about 150 years ago, all objects and materials used by humans were created directly from "found" materials such as wood or stone, or

► *1-Propanol*

$$CH_3CH_2CH_2OH$$

► *Cyclohexanol*

► *Ethanoic acid (acetic acid)*

► *Methyl acetate*

were crafted from metals, glass, and clays. Fibers were cotton, wool, linen, and silk. All medicines and food additives came from natural sources. The only plastics were those made from wood (celluloid) and animal materials (shellac).

Today, many common objects and materials are *synthetic*—created by the chemical industry from oil or natural gas. Such compounds are called *petrochemicals.* They have truly revolutionized how we live and work—what we wear, what we use for our leisure activities (recall the first *Your Turn*), and how we are transported, fed, entertained, and healed. Some petrochemicals, such as detergents, pesticides, pharmaceuticals, and cosmetics, are used directly. Most petrochemicals, however, serve as raw materials in producing other synthetic substances—particularly plastics.

Plastics include paints, fabrics, rubber, insulation materials, foams, glasslike substances, adhesives, molding, and structural materials. Worldwide production of petroleum-based plastics is more than four times that of aluminum products. Over a third of all the fiber and 70% of the rubber, worldwide, are created from petrochemicals.

SPECTRA fibers are the strongest fibers ever made.

BUILDER MOLECULES

What are suitable starting substances for building such petrochemicals? Apart from their ability to burn, alkanes have little chemical reactivity. Few substances can be built directly from them. In contrast, alkenes and aromatics are important builder molecules. The two most important alkenes industrially are ethene (ethylene) and propene (propylene). Aromatic builder molecules such as benzene and styrene can be obtained from petroleum by catalytic cracking and reforming.

The astounding thing is that it takes relatively few different small-molecule compounds (builder molecules) to make several thousand new substances. For simplicity, we will focus on just two common builder molecules—ethene and ethanol—and materials made from ethene.

Ethene (ethylene), because of the high reactivity of its double bond, is readily transformed into many useful products. For example, ethanol (ethyl alcohol) is formed when a water molecule reacts with the double bond of an ethene molecule in the presence of an acid catalyst:

$$
\underset{\text{ethene}}{\begin{array}{c} H \\ \diagdown \\ H \end{array} C = C \begin{array}{c} H \\ \diagup \\ H \end{array}} \;+\; \underset{\text{water}}{H{-}OH} \xrightarrow[\text{catalyst}]{\text{acid}} \underset{\text{ethanol}}{H{-}\overset{\overset{\displaystyle H}{|}}{\underset{\underset{\displaystyle H}{|}}{C}}{-}\overset{\overset{\displaystyle H}{|}}{\underset{\underset{\displaystyle OH}{|}}{C}}{-}H}
$$

Note the OH group characteristic of an alcohol. In this reaction the water molecule "adds" to the double-bonded carbon atoms by placing an H— on one carbon and an —OH group on the other. This type of chemical change is called an *addition reaction.*

Ethanol is used extensively as a solvent in varnishes and perfumes, in preparing many essences, flavors, pharmaceuticals, and in alcoholic beverages. It is also used as a fuel (gasohol).

ADDITION POLYMERS

Ethene undergoes another very important addition reaction with itself. Here ethene serves as a **monomer,** which is a small molecule used to make polyethylene, a **polymer.** A polymer is a large molecule typically composed of 500–20,000 or more repeating units (residues). In polyethene the repeating units are ethene (ethylene) residues. Polyethene is related to ethene in the same way that a long paper-clip chain is related to individual paper clips. The chemical reaction producing polyethene can be written this way:

$$n \; CH_2{=}CH_2 \longrightarrow \quad \overset{\displaystyle H \;\; H \;\; H \;\; H \;\; H}{\underset{\displaystyle H \;\; H \;\; H \;\; H \;\; H}{-C-C-C-C-C-}} \quad \longrightarrow \quad (CH_2CH_2)_n$$

ethene (ethylene), the monomer	the growing polymer chain	polyethene (polyethylene)

Note that ethene may be termed ethylene, and polyethene may be called polyethylene.

Polymers formed in reactions such as this are called—sensibly enough—**addition polymers.** Polyethene (polyethylene), commonly used in bags and packaging, is one of the most important addition polymers. Each year millions of pounds of polyethylene are produced in the United States.

A great variety of addition polymers can be made from monomers that closely resemble ethene as shown in Figure 21 (page 217). One or more of ethene's hydrogen atoms can be replaced by other atoms. These monomers form an array of useful polymers:

$$n \; CH_2{=}CHCl \longrightarrow -CH_2{-}\underset{\displaystyle Cl}{CH}{-}CH_2{-}\underset{\displaystyle Cl}{CH}{-}CH_2{-}\underset{\displaystyle Cl}{CH}-$$

vinyl chloride
(Cl has replaced H)

polyvinyl chloride (PVC), used for shoes, leatherlike jackets, and plastic pipes

$$n \; CH_2{=}CHCN \longrightarrow -CH_2{-}\underset{\displaystyle CN}{CH}{-}CH_2{-}\underset{\displaystyle CN}{CH}{-}CH_2{-}\underset{\displaystyle CN}{CH}-$$

acrylonitrile
(CN has replaced H)

polyacrylonitrile, used in acrylic fiber for clothes and carpets

styrene
(aromatic ring has replaced H)

polystyrene, used in insulation, coffee cups, coolers, toothbrush handles, and combs

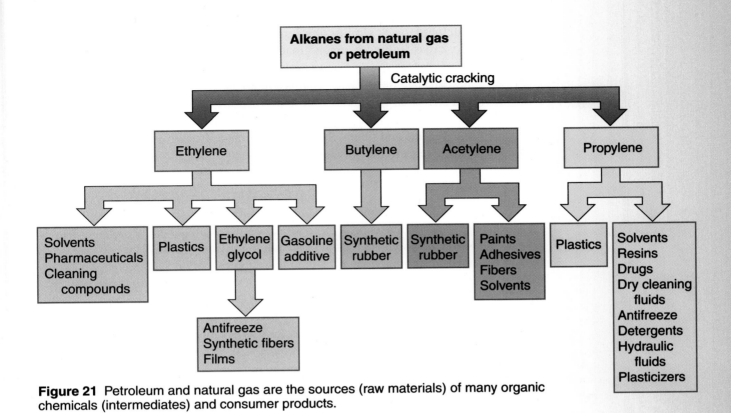

Figure 21 Petroleum and natural gas are the sources (raw materials) of many organic chemicals (intermediates) and consumer products.

The arrangement of covalent bonds in long, stringlike polymer molecules causes them to coil loosely. A collection of polymer molecules (such as those in a piece of rubber or molten plastic) can intertwine, much like strands of cooked noodles or spaghetti. In this form the polymer is flexible and soft.

Polymer *flexibility* can be increased by adding molecules that act as internal lubricants for the polymer chains. Untreated polyvinyl chloride (PVC) is used in rigid pipes and house siding. With added lubricant, polyvinyl chloride becomes flexible enough to be used in raincoats and shoes.

By contrast, polymer *rigidity* can be increased by cross-linking the polymer chains so they no longer readily move or slide (see Figure 22).

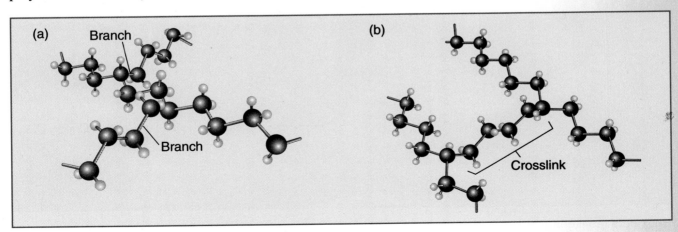

Figure 22 Polyethylene: (a) branched; (b) cross-linked.

Compare the flexibility of a rubber band with that of a tire tread; polymer cross-linking is much greater in the relatively rigid auto tire.

Polymer *strength* and *toughness* can also be controlled. One way is to first arrange the polymer chains so they lie generally in the same direction, as when you comb your hair. Then the chains are stretched so they uncoil. Polymers remaining uncoiled after this treatment make strong, tough films and fibers. Such materials range from the polyethene used in plastic bags to the polyacrylonitrile found in fabrics.

CONDENSATION POLYMERS

Not all polymer molecules are formed by addition reactions. Natural polymers such as proteins, starch, cellulose (in wood and paper), and synthetic polymers including the familiar nylon and polyester are also formed from monomers. But, unlike addition polymers, these polymers are formed with the *loss* of simple molecules such as water from adjacent monomer units. Such reactions are called **condensation reactions** and polymers formed in this way are known as **condensation polymers.** Another very abundant condensation polymer is polyethylene terphthalate (PET). This polyester is most familiar to you as large soft drink containers, its most common use. It is also used in many other applications— as thin film for videotape and as the textile Dacron for clothing and surgical tubing. More than 5 million pounds of PET are produced each year in the United States.

Engineers monitoring the production of Saran Wrap. Sarans are a type of plastic having vinylidine chloride ($CH_2=CCl_2$) as their principal monomer.

CHEMISTRY AT WORK

Manufacturing Safely and Efficiently: It's No Accident

Portia Bass is a Regulatory Affairs Chemist at the E.I. DuPont de Nemours and Company facility in Waynesboro, Virginia. Portia's company makes Lycra®, used in hosiery and clothing; nylon, used in household and automotive carpeting; and Permasep®, a fiber used in the process of sea water desalination.

Portia is responsible for making sure that her company complies with the laws and regulations that protect the environment, her fellow employees, and the community. DuPont must make sure that regulations are being met. Sometimes one word in a regulation can hold up an entire production line, which could be very wasteful and expensive. Portia must have excellent oral and written communication skills in the "languages" of both chemistry and regulations.

After earning degrees in chemistry and analytical chemistry, Portia took additional classes in chemistry, other sciences, and computers. She also benefited from extensive on-the-job

training at DuPont in her current position, as well as in an earlier job monitoring the quality of the plant's production processes.

Stretch Your Imagination . . . About Polymers

Pictured on this page are a few of the products manufactured from materials made by Portia's company. Think about the qualities of the raw materials used in these products. What properties make them useful in the particular products shown here?

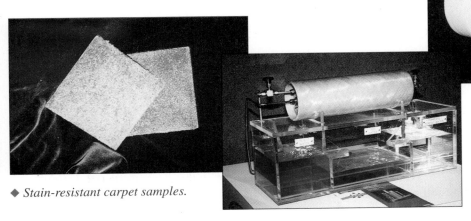

◆ *Nylon thread.*

◆ *Stain-resistant carpet samples.*

◆ *The green container on top contains Permasep®, a semipermeable material used in reverse osmosis processes.*

Photographs courtesy of E.I. DuPont de Nemours and Company

Everyday objects like skateboards are made of different polymers selected for their properties.

Condensation reactions can be used to make small molecules as well as polymers. In the next laboratory activity you will carry out a condensation reaction to make an ester. The reaction is a simple example of how organic compounds can be combined chemically to create new substances. For instance, many synthetic flavorings and perfumes contain esters. And this type of reaction, repeated many times over, produces polyester polymers.

D.6 LABORATORY ACTIVITY: PETROCHEMICALS

GETTING READY

In this activity you will produce a petrochemical by the reaction of an organic acid (an acid derived from a hydrocarbon) with an alcohol to produce an ester that has a pleasing fragrance. Many perfumes contain esters, and the characteristic aromas of many herbs and fruits come from esters in the plants.

For example, an ester called methyl acetate can be produced by the reaction of ethanoic acid (acetic acid) with methanol in the presence of sulfuric acid:

$$CH_3-\overset{\overset{\textstyle O}{\|}}{C}-OH \ + \ H-O-CH_3 \ \xrightarrow{H_2SO_4} \ CH_3-\overset{\overset{\textstyle O}{\|}}{C}-O-CH_3 \ + \ H-OH$$

ethanoic acid   methanol   methyl acetate   water
(acetic acid)

To highlight the interplay of functional groups in the formation of an ester, we can write a general equation for this type of reaction, using the R symbols introduced earlier. (Recall that R stands for the rest of the molecule.)

$$R-\overset{\overset{\textstyle O}{\|}}{C}-OH \ + \ H-O-R \ \xrightarrow{H_2SO_4} \ R-\overset{\overset{\textstyle O}{\|}}{C}-O-R \ + \ H-OH$$

carboxylic   alcohol   ester   water
acid

Note how the acid's and alcohol's functional groups combine to form an ester, with their remaining atoms joining to form a water molecule. The sulfuric acid (H_2SO_4) acts as a catalyst—it allows the reaction to proceed faster, but does not become part of the final product.

You will produce methyl salicylate in this laboratory activity, and will note the characteristic odor of this ester.

PROCEDURE

1. Set up a ring stand, ring, wire screen, and Bunsen burner.
2. Prepare a water bath by adding about 70 mL of tap water to a 150-mL beaker. Add a boiling chip, then place the water bath on the wire screen above the burner.

3. Take a small, clean test tube to the dispensing area. Pour 2 mL of methanol into the tube. Next add 0.5 g of salicylic acid. Then *slowly and carefully add 10 drops of concentrated sulfuric acid to the tube.* As you dispense some of these reagents you might notice their odors. *Do not sniff any reagents directly*—some of them can burn your nasal passages—but record observations of any odors you do notice.

4. Return to your laboratory bench. Place the test tube in the water bath you prepared in step 2. Light and adjust the burner. Start heating the beaker and contents.

5. When the water bath begins to boil, use test tube tongs to move the tube slowly in a small, horizontal circle. Keep the tube in the water, and be sure not to spill the contents. Note any color changes. Continue heating until the water bath has boiled strongly for two minutes. Turn off the burner.

6. If you have not noticed an odor by now, hold the test tube away from you with tongs and wave your hand across the top to waft any vapors toward your nose. Record your observations regarding the odor of the product. Compare your observations with those of other class members.

7. Wash your hands thoroughly before leaving the laboratory.

QUESTIONS

1. a. In a chemistry handbook, find molecular formulas of the acid and alcohol from which you produced methyl salicylate.

   b. Write a chemical equation for the formation of methyl salicylate.

2. Write the formula for amyl acetate, an ester formed from pentanol and ethanoic acid (acetic acid). (See molecular formulas in the margin.) Amyl acetate, with a pearlike odor and flavor, has many uses in products ranging from syrups to paints and shoe polish.)

3. Classify each compound as a carboxylic acid, an alcohol, or an ester:

   a. $CH_3CH_2CH_2CH_2OH$

   b. $CH_3OCOCH_2CH_3$

   c. $CH_3-CH-CH_2-CH_3$
      $\quad\quad\quad\,|$
      $\quad\quad\quad CH_2COOH$

   d. $CH_3-\overset{\overset{\displaystyle O}{\|}}{C}-OH$

CAUTION

Concentrated sulfuric acid must be handled carefully. If any is spilled on you, wash it off with large amounts of flowing water. Report the spill to your teacher.

Carefully waft any vapor toward your nose.

▶ *Pentanol*
   $C_5H_{11}OH$

▶ *Ethanoic acid (acetic acid)*

$$H-\overset{\overset{\displaystyle H}{|}}{\underset{\underset{\displaystyle H}{|}}{C}}-\overset{\overset{\displaystyle O}{\|}}{C}-OH$$

? PART D: SUMMARY QUESTIONS

1. Give at least one specific example (name, formula, and use) of

   a. an industrially important alkene.

   b. an industrially important aromatic compound.

2. Write an equation for the cracking of hexane into two smaller hydrocarbon molecules. (*Hint:* Remember the law of conservation of matter.)

3. a. Write the structural formula for a molecule containing at least two carbon atoms, and representing (i) an alcohol, (ii) an organic acid, or (iii) an ester.

   b. Circle the functional group in each structural formula you wrote.

   c. Name each compound.

4. How does benzene differ from other cyclic hydrocarbons?

5. More than 90% of known chemical compounds are organic (hydrocarbons or substituted hydrocarbons). Identify two characteristics of carbon atoms that help explain the existence of such a large number of carbon-based substances.

6. In your own words describe what is meant by the term polymerization. Use an example of a monomer-to-polymer reaction as part of your answer.

7. Chemical synthesis is one of many branches of chemistry. Try your hand at planning some syntheses in the problems below. One molecule (represented by a question mark) is missing in each equation. Identify the molecule that will complete each equation. (*Hint:* If you are uncertain about the answer, start by completing an atom inventory. Remember that the final equation must be balanced.)

   a. A major type of alkane reaction (other than burning):

   $$CH_3CH_3 + ? \rightarrow CH_3CH_2Cl + HCl$$

   The product, 1-chloroethane (ethyl chloride), was used to make the fuel additive tetraethyllead.

   b. The major way to make 2-propanol (isopropyl alcohol), used as rubbing alcohol:

   $$CH_3-CH=CH_2 + ? \longrightarrow CH_3-CH-CH_3$$
   $$\underset{OH}{|}$$

   c. The conversion of a long-chain organic acid to soap:

   $$CH_3CH_2CH_2CH_2CH_2CH_2-\overset{\overset{O}{\|}}{C}-OH \quad + \quad ? \longrightarrow$$

   $$CH_3CH_2CH_2CH_2CH_2CH_2-\overset{\overset{O}{\|}}{C}-O^- Na^+ \quad + \quad HOH$$

E ALTERNATIVES TO PETROLEUM

Because petroleum is a nonrenewable resource, the total inventory of available petroleum on Earth is decreasing. You know how dependent on petroleum we have become. Fortunately, chemists are already investigating substitutes for petroleum, both to burn and to build.

E.1 ALTERNATIVE ENERGY SOURCES

The way we live—including our homes, agriculture, and industries—requires considerable quantities of energy. In Section C.2 you discovered that the range of energy sources used in the United States has changed over time. As energy demands have accelerated, the nation has relied increasingly on nonrenewable fossil fuels—coal, petroleum, and natural gas. What is the future for fossil fuels and, in particular, for petroleum?

Electricity is necessary for supporting our way of life.

YOUR TURN

Energy Dependency

To address these concerns, we will start by getting a clearer picture of our energy dependency. Once again, examining the facts in an important societal issue involves interpreting numerical data. Consider the graph in Figure 23 (page 224).

1. Which fuel is the nation's most-used energy source?
2. What percentage of our energy needs is met by fossil fuels?
3. What percentage of our energy needs is met by renewable energy sources?

To more fully interpret Figure 23, we should also examine our total supplies of these fuels. Unfortunately, that information is not easy to identify. Estimates vary widely as to the amounts of fossil fuels we have left. For purposes of discussion we will examine data presented in Table 6 (page 224). This represents a somewhat optimistic estimate of extractable supplies of fossil fuels. As you can imagine, the energy values in Table 6 are very large. Because of this, they are expressed in units called quads rather than in joules: 1 quad = 1.05×10^{18} joules.

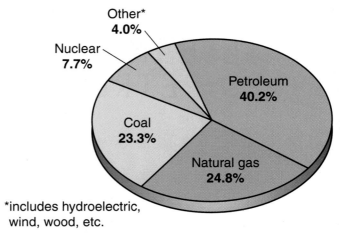

Figure 23 Percent of U.S. energy from various fuel sources.

Table 6	Energy From Reserves
Source	**Energy ($\times 10^{18}$ joules)**
Oil and natural gas liquids	1,000
Natural gas	
Conventional sources	1,000
Unconventional sources	700
Coal	40,300

4. It is projected that the United States will have used an estimated 2,000 quads of energy between 1980 and 2000 (an average of 100 quads yearly).

  a. If we continue to rely on petroleum for 40.2% of our total energy needs, and we use only U.S. petroleum (no imports), how many quads of energy from petroleum will we use from 1996 to the year 2000?

  b. How much of our extractable supply of petroleum would be left by the year 2000?

5. Using data from Figure 23 and Table 6, do calculations like those in Question 4 for:

  a. natural gas (including both conventional and unconventional sources)

  b. coal

Estimates of a fuel's possible lifetime are subject to some serious uncertainties. As supplies of a fuel near exhaustion, our use of that fuel would decline and the fuel would likely become more expensive—thus extending the fuel source's useful lifetime. In addition, as costs increase, economic incentives prompt exploration for new reserves and extraction of more fuel from what were previously cost-ineffective sites (such as low-yield wells and oil shale). Thus, estimates of fuel supplies and future needs are just that—estimates. They are subject to change.

However, despite such limitations, your calculations still illustrate the broad picture: Future petroleum availability presents us with a real problem. We have greater known reserves of coal than petroleum. Can coal serve as a substitute for dwindling petroleum supplies? Are the two fuels interchangeable?

Setting aside possible environmental issues for now, we can explore ways in which petroleum is used as a fuel, and then judge coal's suitability for each major category. Consider Table 7.

Table 7	Petroleum Use (1994)	
End-use Sector	**Million of Barrels/Day**	**Percent of Total Use**
Transportation	10.9	63
Industrial	4.3	25
Residential/commercial	1.4	8
Electric utilities	0.7	4

6. What percent of petroleum is used at fixed-point (stationary) sites?

7. Could coal more readily substitute for petroleum in stationary or mobile uses? Why?

Electricity is also generated from sources other than petroleum—such as hydroelectric (top), coal (left), and nuclear power (right).

The United States is a mobile society. More than 60% of our petroleum is used for transportation, with personal automobiles being the main energy users. Even though efforts to revitalize and improve public transportation systems merit attention, most experts predict we will continue to rely on automobiles into the foreseeable future. (Even energy-conserving mass transit systems must have a fuel source.) What options, then, does chemistry offer us to extend, supplement, and perhaps even replace petroleum as an energy source? The price and availability of crude oil will play a large part in deciding the future of such options.

Petroleum from *tar sands* and *oil shale rock* is an option with some promise. Major deposits of oil shale are located west of the Rocky Mountains. These rocks contain kerogen—partially formed oil. When the rocks are heated, kerogen decomposes to a material quite similar to crude oil. Unfortunately, vast quantities of sand or rock must be processed to recover this fuel. Moreover, current extraction methods use the equivalent of half a barrel of petroleum to produce one barrel of shale oil. Also, very large amounts of water are needed for processing, in an area where water is scarce.

Because our nation's known coal supply is much larger than our known supply of petroleum, another possible alternative to petroleum is a *liquid fuel produced from coal.* The know-how to convert coal to liquid fuel (and also to builder molecules) has been available for decades. This technology was used in Germany more than 50 years ago. U.S. coal-to-liquid-fuel technology is well developed. At present, the cost of mining and converting coal to liquid fuel is considerably greater than the cost of producing the same quantity of fuel from petroleum. However, as petroleum prices increase, obtaining liquid fuel from coal may become a more attractive option.

Certain plants, including some 2000 varieties of the genus *Euphorbia,* capture and store solar energy as hydrocarbon compounds, rather than as carbohydrates. Can these compounds be extracted and used as a petroleum substitute? Future research will provide answers.

Gasoline-burning vehicles can be converted to dual-fuel vehicles which can run on either *natural gas* or gasoline. Natural gas, mainly methane (CH_4), can be compressed and held in high-pressure tanks. This product is commonly known as CNG (compressed natural gas). A refillable CNG tank, capable of powering an automobile 250 miles, can be comfortably installed in a car's trunk. There are more than 70,000 CNG-powered vehicles worldwide. The U.S. government has a project in which 50% of all federal vehicles purchased by 1998 will be alternative-fuel vehicles, including those using natural gas. Using clean-burning natural gas is one strategy for extending the useful lifetime of the world's petroleum reserves.

Other alternative energy sources currently in use or under investigation include hydropower (water power), nuclear fission and fusion, solar energy, wind, and geothermal energies. Despite problems and limitations, these particular options share one positive characteristic: Unlike carbon-based fuels, they do not release carbon dioxide gas to the atmosphere. Carbon dioxide is one of a number of *greenhouse gases*—atmospheric gases that may be contributing to global warming. This is a topic that will be considered in the Air unit.

▶ *A ton of oil shale typically contains the equivalent of 20–80 gallons of oil.*

▶ *About half of U.S. homes are heated by natural gas.*

Prized Petroleum

Whether it is available as crude oil, or from less common, alternative sources such as oil shale or tar sands, petroleum is unquestionably significant to us. In his book *The Prize,* Daniel Yergin summarizes petroleum as a vital resource in these words: "Petroleum remains the motive force of the industrial society and the lifeblood of civilization that it helped create. It also remains an essential element in national power, a major factor in world economies, a critical focus for war and conflict, and a decisive force in international affairs." Answer these questions about Yergin's statement:

1. What does the phrase *motive force* mean?

2. Name three ways in which petroleum has helped to build civilization.

3. Give an example of when petroleum served as a critical focus for war and conflict.

4. The United States is a major factor in world economics, and Yergin says that petroleum, too, is such a factor. In what ways are these two statements related?

5. How does Yergin's statement relate to

   a. the "petroleum—to burn" theme of this unit?

   b. the "petroleum—to build" theme of this unit?

6. Yergin's statement does not specifically mention the options given in Section E.1. In what ways do these options play a part in the role of petroleum as given by Yergin?

We've identified several options for extending petroleum's useful life as an energy source without drastically altering our way of life. More energy-efficient buildings and machines can be constructed, which will lessen our need for petroleum as fuel. We can also use alternative fuels, further reducing the need to "burn" petroleum.

But, what can replace petroleum as a source of chemical building materials?

E.2 BUILDER MOLECULE SOURCES

As petroleum supplies continue to decrease, it is likely that petroleum's use as a fuel will be more and more severely restricted—but petroleum will still be used as a source of petrochemicals. Eventually, because accessible petroleum supplies will someday be exhausted, alternatives to petroleum as a source of builder molecules must also be found.

Coal's use as a chemical raw material is receiving increased attention. In principle, all the carbon compounds now manufactured from petroleum can also be obtained, through appropriate chemical reactions, from coal, water, and air. However, the expense and time involved in opening new coal mines and building conversion plants prevents any rapid conversion to coal. In addition, coal processing is more costly (both financially and environmentally) than comparable processes for petroleum. Thus, making builder molecules from coal would be more expensive—at least until petroleum becomes much more costly than it is now.

Another potential source of builder molecules is *biomass*—plant matter. A major component of biomass is cellulose, found in wood, leaves, and plant stems. Cellulose contains the basic carbon-chain structures needed to build petrochemicals. One possible scenario includes intensive forestry and cultivation of fast-growing plants, as well as use of organic wastes from homes and industries. Ethanol and other builder molecules are already being produced from sugar cane.

Using biotechnology, biomass can be converted into raw materials for petrochemicals. Bioengineering techniques already allow the use of *enzymes* (biochemical catalysts for specific reactions) to produce specific compounds. Biological production methods suitable for large-scale use can provide such key substances as ethanol and acetic acid. These products would be, of course, more costly than petrochemicals, but less so as the cost of petroleum rises.

Significant advances are being made with *biopolymers*—biologically grown polymers and plastics that resemble petroleum-based materials. For example, one British pilot plant uses the bacterium *Alcaligenes eutrophus* to produce about 50 tons annually of a polypropene-like plastic used for films and bags. Part of the appeal of biopolymers is that—unlike petroleum-based polymers—they are fully biodegradable. They may some day serve as a renewable source of plastics—a source not dependent on petroleum supplies.

In a fascinating example of the ultimate in "recycling," researchers at the University of Kentucky have converted plastic milk jugs and soft-drink bottles back into petroleum. In the process, the large polymer chains are broken down into a mixture similar to crude oil. The results of this study are preliminary, but the impact could be significant. For example, just one year's supply of plastic waste in the United States could generate over 80 million barrels of this "crude oil" if the process could be made to operate on a large scale.

We face difficult decisions concerning the use of petroleum for building and burning, but chemistry is providing some options for meeting society's needs. It is likely that, at least for the foreseeable future, most petrochemicals will be manufactured from petroleum, coal, and biomass. How these sources are combined at any given time will be influenced by economics, politics, and available technology.

As petroleum supplies dwindle and become more precious, the cost will rise as well as that of materials made from it. Today, petroleum conservation is the key to buying enough time for replacements to be developed so we can avoid drastic changes to our way of living.

Conservation is also ecologically sound; simply using less petroleum for fuel can help satisfy the basic needs of all global inhabitants, not just of those who are wealthy or have large oil reserves. For example, an increase in auto fuel efficiency to 40 miles per gallon in the United States alone could save 3 million barrels of crude oil daily. Fuel efficiency has gone up, in part, by reducing the weight of cars. Chemistry has played a

▶ *Conservation involves making better use of what we have, using items longer, and recycling what we can (for example, motor oil, plastics, and paper).*

Like petroleum, coal can be used to produce energy or a range of organic chemicals.

major role here by making more durable and less dense plastics available to substitute for heavier metal parts in bumpers, grilles, lights, side panels, batteries, and interiors. Currently, about 10–15% of a car's weight is from polymers, a considerable increase from the 3% used 20 years ago.

? PART E: SUMMARY QUESTIONS

1. Describe two major problems posed by our present use of energy sources.

2. Many experts feel that we should explore ways to use more renewable energy sources such as hydro-, solar, and wind power as replacements for nonrenewable fossil fuels.

   a. Why might this be a wise policy?

   b. For which major energy uses are these renewable sources *least* likely to replace fossil fuels? Why?

   c. Describe how your community might look if it decided to install wind and solar power devices on a large scale.

3. Consider coal, oil shale, and hydrocarbons from plants (biomass) as possible petroleum substitutes. Which do you think holds the greatest promise for the future? Why?

4. Of the two broad uses of petroleum, as a fuel and as a raw material:

   a. Which is likely to be curtailed first as petroleum supplies dwindle?

   b. Give at least two reasons for your choice.

5. What types of compounds do chemists seek as good petroleum substitutes? Why?

6. Some say we are living in the "oil century." What would you say to support or refute this description?

7. U.S. reserves of oil shale are approximately 87 quads, the equivalent of 150×10^9 barrels of oil. Suppose we had to depend on this as our sole source of oil and we used it to maintain our production of 8 million barrels of oil per day.

   a. How many years would this supply last?

   b. There are approximately 249×10^6 persons in the United States, and we use about 24 barrels (about 1,000 gallons) of oil per person per year. At this rate of consumption, how long will the oil shale reserves last?

   c. Why is there a difference between the answers in 7a and 7b?

F PUTTING IT ALL TOGETHER: LIFE WITHOUT GASOLINE

We have used a vast portion of known reserves of many fossil fuels over a relatively short period. Our appetite for energy is enormous and we use up fossil fuels at a prodigious rate. There is little reason to suspect that the rate of fuel consumption will slow down; more likely, it will increase.

F.1 CONFRONTING THE PROSPECTS

How will shifts in supply and demand affect your future? It is worth some serious thinking. Authors have written about their predictions of such events. One noteworthy effort of what life might be like without petroleum for fuel was written by the late biochemist Isaac Asimov, an important writer of science and science fiction. Asked by *Time* magazine in 1977 to describe such a world, Asimov chose to set his prediction 20 years in the future—in the "distant" year of 1997. Here is Asimov's story:

> Anyone older than ten can remember automobiles. They dwindled. At first the price of gasoline climbed—way up. Finally only the well-to-do drove, and that was too clear an indication that they were filthy rich, so any automobile that dared show itself on a city street was overturned and burned. Rationing was introduced to "equalize sacrifice," but every three months the ration was reduced. The cars just vanished and became part of the metal resource.
>
> There are many advantages, if you want to look for them. Our 1997 newspapers continually point them out. The air is cleaner and there seem to be fewer colds. Against most predictions, the crime rate has dropped. With the police car too expensive (and too easy a target), policemen are back on their beats. More important, the streets are full. Legs are king in the cities of 1997, and people walk everywhere far into the night. Even the parks are full, and there is mutual protection in crowds.
>
> As for the winter—well, it is inconvenient to be cold, with most of what furnace fuel is allowed hoarded for the dawn; but sweaters are popular indoor wear and showers are not an everyday luxury. Lukewarm sponge baths will do, and if the air is not always very fragrant in the human vicinity, the automobile fumes are gone.
>
> "The Nightmare of Life without Fuel." *Time,* 1977, April 25, p. 33. Reprinted with permission.

It is clear from today's perspective that real life in 1997 is not so radically different as Asimov's story suggested. (If it were, it's unlikely the book you are now reading would have been shipped from the publisher to your school!) Here are some final issues for you to consider:

▶ *Asimov was a prolific writer who wrote hundreds of fiction and nonfiction books.*

1. Mark Twain said, "It is difficult making predictions, especially those about the future." Asimov predicted many changes in his story.

   a. Which ones have happened? On what basis might he have made such predictions?

   b. What changes have not happened as predicted? How do you account for that?

2. Now it's your turn. Use your imagination *and* what you have learned from this and other chapters to write a story to predict what your "life without petroleum" would be like 20 years from now. Use these questions to guide your writing:

   a. How would your daily life change in such a world?

   b. What change would involve the hardest adjustment for you? Why?

   c. Which would be the easiest adjustment? Why?

3. Asimov was both a scientist and a creative writer. Perhaps his predictions will ultimately prove correct; it might be just a matter of time. Suppose he had chosen to write about a time 100 years into the future, around 2077.

   a. Which of the changes he predicted might happen by that time? Why?

   b. Which changes do you think are not likely to happen, even by then? Why?

   c. Consider the changes *you* predicted to occur within 20 years. How many of them do you think will be realized 100 years from now? Why?

F.2 LOOKING BACK AND LOOKING AHEAD

This unit illustrates once again how chemical knowledge can help us deal with personal concerns and community issues of resource use and excesses.

Clearly, decisions that ignore or attempt to refute natural laws are likely to produce higher "costs" than desired benefits. Chemistry does not provide single correct answers to such major problems, but it does address key questions and gives us options to consider. Such knowledge can help us make better decisions on many difficult issues.

Another complex scientific issue arises in the next unit. This issue has both personal and societal dimensions. You will discover that the Petroleum unit's theme—to build or to burn—applies equally well to food and nutrition. It's food for thought.

Understanding Food

▶ Why do we eat?

▶ How can we become informed decision makers about food choices and diets?

▶ What's wrong with the warning "Don't put chemicals in your mouth"?

INTRODUCTION

▶ *Your body uses chemical reactions to capture, store, and release energy from foods and to build and rebuild itself with substances obtained from food.*

In the previous unit, you learned that petroleum is not only burned as fuel, but is also a raw material in manufacturing many products. Food, the focus of this unit, is also both a building material and a fuel. All of us depend on our diet to provide molecules from which our bodies build new cells, blood, and body tissues. Therefore, what and how much we eat is critically important to our health—we really are what we eat. It may seem obvious, but without a proper diet, the body receives either too little or too much of the raw materials it needs for growth, repair, and replacement. In the United States, guidelines called *Daily Values* are put on food labels to help us make informed choices about healthful eating.

Molecules in food also provide the fuel that is "burned" in the body to provide needed energy for all bodily activities—moving around, maintaining body temperature, your heart's beating, even thinking. Almost any food can provide energy, although foods high in carbohydrates (sugars and starches) provide it more quickly than others. Food requirements for building are much more specific than those for providing energy. The major building-block molecules for body growth and development are proteins and fats. Vitamins and minerals are also absolutely necessary, although in very small quantities—somewhat like specialized parts needed in designing a custom car.

For most cultures, balanced diets developed without any overall strategy. Over time, the great variations in diets among cultures developed from the differences in food crops that could be raised in particular areas. Of these different diets, almost all fulfill basic nutritional requirements. For example, a traditional Mexican diet includes considerable protein from beans and rice; U.S. meals deliver substantial protein from meat, typically beef; Italian menus derive protein mainly from pasta (processed grain) and cheese. Each tradition also has its own recipes for preparing fruits and vegetables, which are sources of vitamins and minerals.

Although foods may appear quite different, once they are digested, their roles in human body chemistry are similar, as we will see in this unit.

Traditional Vietnamese diets include many of the green vegetables available at this open market.

A GETTING STARTED

Food labels are one way we get information about what is in particular foods. In 1994 the U.S. Food and Drug Administration (FDA), with congressional approval, significantly revised the requirements for nutritional labeling. You have seen these labels on packaged foods you use. Perhaps you even read the labels. In this unit we will spend some time discussing the information such labels provide.

For now, let's just take one food, not identified here, and look in detail at its label (shown in Table 1). The label shows the food's nutrient composition and what a 33-g serving provides in terms of meeting the percent of Daily Values. (We will have more to say about Daily Values later.) A *Calorie* is an energy unit.

Table 1	
Nutrition Information 3 items (33 g)	**Percent Daily Value***
Calories 160 Calories from fat 60	

Total fat   7 g
   Saturated fat   1.5 g
   Polyunsaturated fat   0.5 g
   Monounsaturated fat   3 g
	11%
	8%

Cholesterol 0 mg	0%
Sodium 220 mg	9%

Total carbohydrates   23 g
   Dietary fiber   1 g
   Sugars   13 g
	8%
	3%

Protein 2 g	
	Percent Daily Value*
Vitamin A	0%
Vitamin C	0%
Calcium	0%
Iron	4%

*Percent Daily Values are based on a 2,000-Calorie diet. Your daily values may be higher or lower depending on your Calorie needs.

A.1 YOU DECIDE: LABELS AND DAILY VALUES

Table 1 (page 235) contains some very useful information.

1. It is likely that some basic ideas about dietary carbohydrates and minerals were presented in your previous health and biology courses. Use that background to answer the following questions:

   a. Table 1 includes a listing for total carbohydrates. Name some substances that might be included as carbohydrates in this food.

   b. What minerals are listed?

2. Whenever nutrient labeling is used, the label must give at least the percent of Daily Values for total fat and saturated fat; total carbohydrate and dietary fiber; protein; vitamins A and C; and calcium and iron. Does the label in Table 1 fulfill the requirement?

3. Food companies also have the option of listing other vitamins and minerals that are in the food. Which additional ones would you like to have included on this label? Why?

4. Based on what you learned in the Petroleum unit, suggest the major chemical difference between a saturated fat molecule and an unsaturated fat molecule (either monounsaturated or polyunsaturated).

5. Suggest what kind of food is described by Table 1.

A.2 THE FOOD PYRAMID

You've been told (probably more often than you appreciated) to eat the right kinds of foods in the proper amounts. But what *are* the right kinds of foods? What does each kind actually do for us? What are proper amounts?

Variety is the essence of good eating.

In 1991 the Department of Agriculture (USDA) approved the food guide pyramid shown in Figure 1 to help the public make proper dietary choices. The food guide pyramid contains five food groups plus fats, oils, and sweets. It is important to eat foods from all five groups, and the pyramid's shape emphasizes the components of a healthy diet. Daily choices should be made mostly from the bottom section (6–11 servings), some from the middle, and very few from the top of the pyramid. For good health, fats, oils, and sweets at the top should make up only a small portion of our daily diet. That's because these items furnish little in the way of nutrients, though they do provide energy.

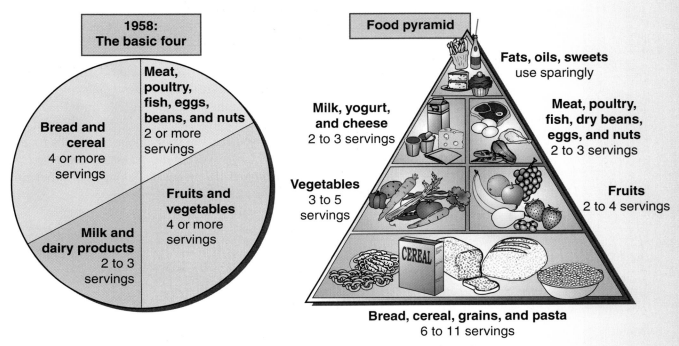

Figure 1 The basic four food groups (1958) and the food guide pyramid (1991).

A.3 YOU DECIDE: KEEPING A FOOD DIARY

Recent USDA surveys reveal that only half of U.S. families have diets rated as "good," and that one-fifth eat "poorly." This next exercise allows you to analyze what you eat in relation to the food guide pyramid.

Keep a complete diary of everything you eat for one day. (You will be assigned a particular day). This information, along with that collected by your classmates, will be used later in this chapter when we consider food nutrients and nutrition.

Record each food item you eat and estimate its quantity. Include all beverages, except plain carbonated water, "no-Calorie" flavored sodas, and tap water. Don't forget snacks! Express your estimates of quantity as volume (such as milliliters, pints, tablespoons, or cups); mass (such as grams, ounces, or pounds); or number (slices, or units such as number of eggs or bananas).

Each time you record a food item, also indicate to which of the five food guide pyramid groups it belongs.

1. Review the information in your food diary.
2. For which food groups did you meet the guidelines?
3. For which food groups did you exceed the guidelines?
4. For those in Item 3 (if any), what specific changes would you make to have your diet meet the guidelines?
5. Why do you think the pyramid gives ranges of recommended servings, such as 6–11 servings from the bread, cereal, grains, and pasta group, rather than a specific number?

In discussions of diets and nutrition, the terms *malnutrition* and *undernourishment* often come up. These conditions are not the same, and it is important to distinguish between them. **Malnutrition** occurs when a diet does not contain all the necessary nutrients, *even though enough food energy is consumed*. This means that the amount of energy from food is not the only consideration for good health. Despite eating enough food (getting sufficient food energy), you can still be malnourished if the food does not provide a balanced diet (supplying all the necessary nutrients). On the other hand, to be **undernourished** means not taking in a sufficient number of calories each day, regardless of what is eaten.

? PART A: SUMMARY QUESTIONS

1. Identify the five major types of dietary nutrients.
   a. Which type is used principally for quick energy?
   b. Which types are used principally for "building"?
2. To which food group in the food guide pyramid does the food described by Table 1 belong?
3. Consider the terms undernourished and malnourished:
   a. Explain the difference between the two terms.
   b. Is it possible for a person to be malnourished but not undernourished?
   c. Can a person be undernourished without being malnourished?
4. Explain the wisdom of the old adage "Variety is the spice of life" concerning one's diet.
5. Is simple "Calorie counting" a wise way to approach dieting? Why or why not?

B FOOD AS ENERGY

All human activity requires "burning" food for energy. The food is not actually burned, of course; its energy is released by a series of chemical reactions in the body during metabolism of the food. In the following sections you will learn how to estimate how much energy is available from different kinds of food, and to estimate the amount of energy it takes to do various activities.

B.1 LABORATORY ACTIVITY: FOOD ENERGY IN A PEANUT

GETTING READY

How do we know how much energy is stored in foods? Chemists can determine this by burning a known amount of a food under controlled conditions and carefully measuring the quantity of thermal energy it releases. This procedure is called *calorimetry* and the measuring device is called a *calorimeter.*

In this experiment you will determine the energy given off by a peanut when it burns, using a setup similar to that used to find the heat of combustion of candle wax (Petroleum unit, page 199). The oil in peanuts burns rapidly when ignited. In a typical calorimeter, the thermal energy released by burning a sample of food—in this case, the peanut—heats a known mass of water. The temperature of the water is measured before and after the peanut is burned, and the thermal energy released by the reaction is then calculated. The procedures and calculations are similar to those you experienced in an earlier calorimetry experiment (Petroleum unit, pages 198–200).

Prepare a data table similar to the one that follows.

▶ *You constructed a simple calorimeter from a soft drink can in the Petroleum unit.*

Data Table		
	Trial 1	**Trial 2**
Brand of peanut		
Mass of peanut, g		
Mass of peanut residue, g		
Mass of peanut burned, g		
Volume of water in can, mL		
Mass of water in can, kg		
Initial temperature of water, °C		
Final temperature of water, °C		

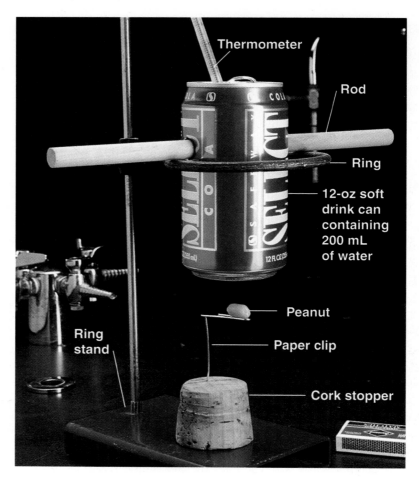

Figure 2 Apparatus for determining the energy released when a peanut burns.

PROCEDURE

1. Make a simple stand for the peanut using a paper clip and cork, as shown in Figure 2.

2. Measure (to the nearest milliliter) 200 mL of room-temperature water in a graduated cylinder. Pour the water into an empty soft drink can.

3. Set up the can and water as shown in Figure 2. Use a thermometer to measure the temperature of the water to the nearest 0.2 °C. Leave the thermometer in the can.

4. Measure the mass of a whole peanut and record this value. Put the peanut on the support stand and place it under the can, so that the peanut is about 2 cm from the bottom of the can.

5. Use a kitchen match to light the peanut directly.

6. *As soon as* the peanut stops burning, carefully stir the water with the thermometer. Measure the final (highest) temperature of the water. Record this value.

7. Allow the peanut residue to cool, and then measure its mass. Record this value.

8. Repeat Steps 3–7 with a new peanut (Trial 2).

CALCULATIONS

If the mass of water and its temperature change are known, the calories needed to change the temperature can be calculated. In this laboratory activity, the energy to heat the water came from the burning peanut. Therefore, the calories it took to heat the water equal the heat given off (calories) by the burning peanut. (*Note:* A small amount of the heat from the peanut was used to heat the can also.)

A *calorie* (with a lowercase *c*) represents the thermal energy needed to raise the temperature of one gram of water one degree Celsius. Thus, water's specific heat, expressed in calorie units, is 1.0 calorie per gram per degree Celsius. Let's put this value to work in a sample problem, to review skills you learned in the Petroleum unit. Suppose you want to heat 250 mL (250 g) of water from room temperature (22 °C) to just under boiling (99 °C). How much thermal energy will be needed?

Here's how the answer can be reasoned out: We know it takes 1.0 calorie to heat 1 g of water by 1 °C. In this case, however, we have *250* times more water to heat, and rather than heating the sample by 1 °C, we need to increase its temperature by 77 °C (99 °C − 22 °C). The total thermal energy needed must be 250 times greater due to the larger mass, and 77 times greater due to the larger temperature change. To include *both* changes, the total energy needed must be 250 × 77, or about 19,000 times greater than the original 1.0 calorie. Thus, the final answer must be 1.0 cal × 19,000, or *19,000 cal.* Because it takes 1,000 calories to equal a kilocalorie (abbreviated kcal or Cal), the answer can also be expressed as *19 kcal or 19 Cal.* The problem solution can also be summarized this way, to show that all the units combine to give the desired unit of "Calories" in the final answer:

▶ *Remember that 1,000 calories is a kilocalorie, 1 kcal.*

$$250 \text{ g} \times 1 \text{ cal/g °C} \times (99 - 22) \text{ °C} = 19,000 \text{ cal or } 19 \text{ kcal} = 19 \text{ Cal}$$

Use the data collected in the laboratory activity to complete the following calculations:

1. Calculate the mass of water heated, first in grams and then in kilograms (kg); 1 kg = 1,000 g. The density of water is 1.0 g/mL. Thus, each milliliter of water has a mass of 1.0 g.

2. Calculate the temperature change in the water. This equals the final (highest) temperature minus the initial temperature.

3. Calculate the number of calories used to heat the water.

4. The food *Calorie* (with an uppercase *C*) is a much larger unit. It equals 1,000 calories, or one kilocalorie. Therefore, one food Calorie would increase the temperature of one kilogram (1,000 g, or one liter) of water by 1 degree Celsius. To distinguish the two kinds of calories, remember the following relationships:

    1,000 calories (cal) = 1 kilocalorie (kcal) = 1 Calorie (Cal)

   a. Calculate the number of Calories used to heat the water.

   b. Determine the number of Calories given off by the burning peanut.

5. Calculate the Calories per gram of peanut burned for each sample; then calculate the *average* Cal/g.

6. Use the data from the label on the peanut package to determine the Cal/g for the peanut you used:

$$\frac{\text{Cal per serving}}{\text{mass (g) of one serving}}$$

7. Use the values you found in Steps 5 and 6 to calculate the percent difference between the label value and the experimental average value:

$$\% \text{ difference} = \left(\frac{\text{Label value} - \text{Experimental average value}}{\text{Label value}} \right) \times 100$$

B.2 FOOD FOR THOUGHT—AND FOR ENERGY

In Laboratory Activity B.1, the food calorie (Cal) was introduced. When nutritionists plan menus, they count Calories (notice the capital C), as does anyone who diets. The capital letter in Calories indicates that these are food Calories, because a Calorie (Cal) is a measure of the energy present in food. For example, a regular serving of french fries (91 g) contains 240 Calories of food energy; a plain baked potato weighs almost twice as much (170 g), but has only 180 Calories.

Where does this food energy come from? The answer is simple: All food energy comes from sunlight. (Recall that sunlight was also the energy stored in petroleum and coal.) By photosynthesis, plants capture solar energy and use it to make large, energy-rich molecules from smaller, simpler ones. Thus, the sun's energy is converted to chemical energy that is stored within the structures of these molecules. We recover some of this stored energy when we metabolize the plants—or eat meat and dairy products from animals that consumed green plants (Figure 3).

What is the energy contained in this serving of french fries?

▶ *All human activities are directly or indirectly powered by solar energy.*

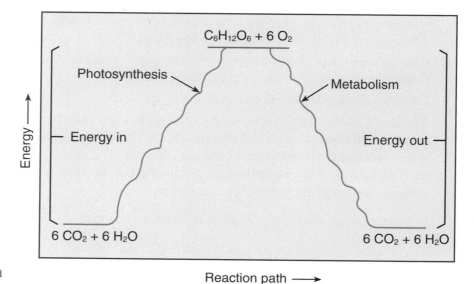

Figure 3 Energy from photosynthesis and metabolism.

From *Chemistry for Health-Related Sciences* by C. T. Sears and C. L. Stanitski, © 1976. Adapted by permission of Prentice-Hall, Inc., Upper Saddle River, NJ.

The energy contents of a wide variety of foods have been determined and are available for diet planning. For example, the table in the Appendix (page 585) includes energy values and nutrient values of common foods.

Now for some additional pencil-and-paper practice with food-based calorimetry.

YOUR TURN

One ounce (28.4 g) of a popular frosted cereal contains 3 teaspoons (12 g) of sugar. When burned, this sugar can heat 860 g of water from 22 °C to 85 °C. How much energy was contained in the three teaspoons of sugar? Based on your answer, how many food Calories are contained in one teaspoon of sugar?

Once again we start with the fundamental water-heating relationship. We know it takes 1.0 cal of thermal energy to heat 1 g of water by 1 °C. In this problem, the burning sugar heated 860 g of water—860 times more than just 1 g. The temperature increase was 63 °C, from 22 °C to 85 °C, 63 times more than the 1 °C used in our water-heating standard, the definition of a calorie. Thus, the total thermal energy involved must have been 860 × 63, or about 54,000 times more than the 1.0 cal standard. Thus, the total energy in three teaspoons of sugar is 1.0 cal × 54,000 = 54,000 cal. In food-energy units, this equals 54 Cal.

Finally, if *three* teaspoons of sugar contain 54 Cal of food energy, then *one* teaspoon of sugar must contain 1/3 as much. Since 54/3 = 18, there are 18 Cal in one teaspoon of sugar.

Now it's your turn.

1. The energy stored in one can of a certain diet drink is capable of heating 160 g of room-temperature (22 °C) water to 60 °C. How many Calories are contained in this diet drink?

2. Suppose you drink six glasses (250 g each) of ice water (0 °C) on a hot summer day.

   a. Assume your body temperature is 37 °C. How many calories of thermal energy does your body use in heating this water to body temperature?

   b. How many Calories?

   c. A serving of french fries contains 240 Cal. How many glasses of ice water would you have to drink to "burn off" the Calories in one serving of french fries?

   d. Based on your answer to Question 2c, is drinking ice water an efficient way to diet?

▶ *The joule is the modernized metric system (SI) unit of energy and is roughly equivalent to the energy it takes to lift 100 g (about the weight of a large egg) one meter. Weight-conscious Americans may one day count joules instead of Calories. This has not yet happened, so we will use the (uppercase) Calorie as the unit of food energy. You can translate (lowercase) calories into joules by using the following approximate relationships:*
1 calorie (cal) = 4.2 joules (J)
1 Calorie (Cal) = 4.2 kJ

What happens to the energy stored in foods we eat? How much body mass is gained if excess food is consumed? The following activity will help you estimate this.

B.3 YOU DECIDE:
ENERGY IN—ENERGY OUT

Table 2 provides estimates of the energy expended in various activities. For example, during 1.0 hour of rollerblading at 12 mph, a 120-lb person expends 630 Cal:

$$\text{Cal expended} = \text{time (min)} \times \text{Cal/min} =$$
$$60 \text{ min} \times 10.5 \text{ Cal/min (from table)} =$$
$$630 \text{ Cal}$$

A 150-lb person expends 702 Cal; a 180-lb individual expends 798 Cal.

Table 2	Energy Equivalents		
Activity	Energy expended, Cal/min: 120 lb	Energy expended, Cal/min: 150 lb	Energy expended, Cal/min: 180 lb
Lying down or sleeping	1.1	1.3	1.6
Sitting	1.3	1.7	2.0
Eating	2.0	2.5	3.1
Studying	1.3	1.7	2.0
Walking, 3.5 mph	4.2	5.3	6.3
Volleyball	4.7	5.9	7.0
Tennis, singles	5.6	7.1	8.5
Jogging, 9 min per mile	10.6	13.2	15.8
Jogging, 7 min per mile	13.7	17.1	20.5
Rollerblading, 8 mph	4.2	5.6	7.0
Rollerblading, 12 mph	10.5	11.7	13.3
Swimming, freestyle	7.8	9.8	11.7
Bowling	3.6	4.5	5.4
Soccer	7.2	9.0	10.8
Basketball, full court	11.6	14.6	17.5
Baseball	3.7	4.7	5.6
Fishing	3.4	4.2	5.0
Stairclimber, 120 steps/min	5.0	6.3	7.6
Skiing, water	6.2	7.8	9.4

1. Use Table 2 to estimate your own daily energy use.

    a. List your typical activities over a 24-hour period and estimate how long each activity takes.

    b. Calculate the total Calories used. Try to estimate energy uses for any activities not given in the table.

c. An average 15- to 18-year-old female engaged in light activity consumes about 2,200 Cal daily. The value for a 15- to 18-year-old male is about 3,000 Cal. How does your own estimated energy use compare with these values?

Compare the total energy you use with the total food energy you consume. Use entries in your food diary and the data in the Appendix (page 585) to calculate the number of Calories you consumed during that day.

2. Let's look at the activity of eating an ice-cream sundae. We will investigate how much exercise it would take to burn off the added Calories, and how much weight you would gain from eating it if you did not exercise.

Two scoops of your favorite ice cream contain 250 Cal; the chocolate topping adds 125 Cal. (You also need to know that 1 lb of body fat contains 4,000 Cal of energy.) Consult Table 2 to answer the following questions.

a. Assume that your regular diet (without the ice-cream sundae) just maintains your current body weight. If you eat the ice-cream sundae,

(1) How many hours of rollerblading would burn it off?

(2) How far would you have to walk at 3.5 mph?

(3) How many hours must you swim?

b. If you choose not to exercise more than usual, how much weight will you gain from the sundae?

c. Now assume that you consume a similar sundae three times each week for 16 weeks. If you do not exercise to burn it off, how much weight will you gain?

d. Of course, another alternative is available—you might decide not to eat the ice-cream sundae. Would you decide to eat the sundae (involving either extra exercise or gaining weight) or not to eat the sundae (involving less pleasure)? Why?

▶ *It would take almost four hours of just sitting to burn off the sundae's Calories.*

These runners are using energy at a rate of about 1600 Cal/h.

3. We have implied that eating an ice-cream sundae will result in weight gain, unless you do additional exercise.

   a. Can you think of a plan that allows you to consume the sundae, do no additional exercise, and still not gain weight?

   b. Explain your plan and be prepared to discuss it in class.

This activity has been based on three possible scenarios:

- If *total energy in* equals *total energy out*, a person will maintain current body weight.
- If *total energy in* is greater than *total energy out,* a person will gain some body weight.
- If *total energy in* is less than *total energy out*, a person will lose some body weight.

Anyone wanting to lose weight must meet the third condition listed above—total energy consumed in food must be less than the total energy used. On the other hand, a person wishing to gain weight must take in more energy than is expended.

Wise dieters know that proper nutrition is not just a question of how much is eaten. What we eat is also critical. Some foods provide more energy than others and some are necessary for their nutrients; some foods provide energy, but not much nutrition. This latter category, which includes soft drinks and alcohol, is at the top of the food guide pyramid. Often such foods are described as furnishing "empty Calories" because these foods have very low nutrient value for the amount eaten.

In the following section, we will examine the major nutrients, beginning with fats. If we "are what we eat," then knowing about such nutrients is important in order to understand what we eat and how it affects our health. Studying the chemical ways in which foods are digested provides an opportunity to use some of the chemistry you have already learned, as well as to learn some new chemical concepts.

B.4 FATS: STORED ENERGY WITH A BAD NAME

The food label shown here raises several questions regarding fat content: (1) Why are there separate listings for total fat, saturated fat, and unsaturated fat? (2) What is the relationship between saturated and unsaturated fats in foods? (3) How much fat is harmful? We will look more closely at these matters as we consider what a fat is.

Unlike carbohydrate and protein, the word *fat* has acquired its own general (and somewhat negative) meaning. To most people, "You're too fat" means "You look overweight." However, from a chemical point of view, *fats* are a major category of biomolecules with special characteristics and functions, just as carbohydrates and proteins are.

Fats are a significant part of our diet. They're present in meat, fish, and poultry; salad dressings and oils; dairy products; and grains. When our bodies take in more food than is needed for energy, much of the excess is converted to fat molecules and stored in the body. When food intake is not large enough to supply the body's energy needs, the body begins to "burn" stored fat.

Fats are composed only of carbon, hydrogen, and oxygen, the same three elements that make up carbohydrates. Fats, however, contain a lower percentage of oxygen than carbohydrates and contain more stored energy. Consequently, each gram of fat contains over twice the energy found in a gram of carbohydrate: 1 g fat is equivalent to 9 Cal, whereas there are 4 Cal for each gram of carbohydrate. Gram-for-gram, you must run more than twice as far or exercise twice as long to "burn up" fat as you do to burn off carbohydrates. For example, there are 11.6 g of fat in a glazed doughnut. Therefore, at 9 Cal/g fat, the doughnut has 11.6 g fat × 9 Cal/g fat, or about 104 Calories from fat. This value, coincidentally, equals the number of Calories from the 26 grams of carbohydrate in the glazed doughnut (26 g carbohydrate × 4 Cal/g carbohydrate = 104 Cal). So, it is not surprising that the body uses fat to store excess food energy efficiently, and that it is difficult to "burn off" excess fat.

Generally, fat molecules are nonpolar and only sparingly soluble in water. This is because they have long hydrocarbon-like portions that are not water soluble. Because of their low water solubility and energy-storing properties, fat molecules are more like hydrocarbons than carbohydrates.

Fats are members of the class of biomolecules called *lipids.* Some lipids are builder molecules that form cell membranes. Others become hormones—chemical messengers that regulate processes in the body.

A typical fat molecule is a combination of a simple three-carbon alcohol called *glycerol* with three fatty acid molecules:

Glycerol + 3 Fatty acids → Fat + 3 Water

The formation of a typical fat is shown in Figure 4.

FATTY ACIDS

Fatty acids are a class of organic compounds made up of a long hydrocarbon chain with a carboxylic acid group (—COOH) at one end. The reaction that produces a fat molecule is similar to one you already completed

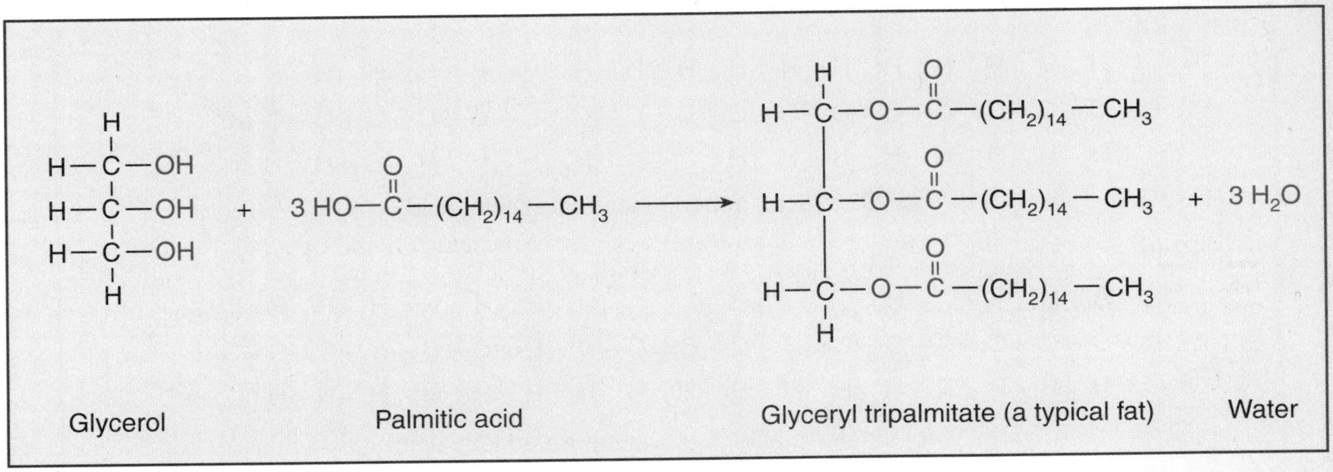

Figure 4 Formation of a typical fat, a triglyceride. Glycerol and three molecules of fatty acid combine in a condensation reaction to form a triester and eliminate three water molecules.

in the laboratory (page 220)—the production of the ester, methyl salicylate. In fat production, glycerol (a molecule containing three —OH groups) reacts with *three* molecules of fatty acid. Each acid molecule forms an ester linkage when it reacts with an —OH group of glycerol, also producing a molecule of water for each ester linkage formed. The main product of this *condensation* reaction is a fat molecule containing three ester groups instead of one. Such a fat is known as a ***triglyceride.*** In general, we will use the terms fat and triglyceride interchangeably.

Recall from the Petroleum unit that hydrocarbons may be saturated (when containing only single carbon–carbon bonds) or unsaturated (when containing double or triple carbon–carbon bonds). Likewise, hydrocarbon chains in fatty acids are either saturated (Figure 5a) or unsaturated (Figure 5b, c). Fats (triglycerides) containing saturated fatty acids are called ***saturated fats;*** triglycerides containing unsaturated fatty acids are called ***unsaturated fats.*** Unsaturated fats contain C=C double bonds, not C≡C triple bonds. The term ***polyunsaturated*** is often used in food advertising in connection with fats. A ***polyunsaturated*** fat contains two or more carbon–carbon double bonds in each fatty acid portion of a triglyceride molecule.

▶ *Saturated fats appear to contribute to heart disease.*

Triglycerides (fats) in butter and other animal fats are nearly all saturated and form solids at room temperature. However, fats from plant sources commonly are unsaturated, containing molecules with several carbon–carbon double bonds. At room temperature these polyunsaturated fats are liquids (oils), such as safflower oil (13% monounsaturated; 78% polyunsaturated fat molecules), corn oil (25% monounsaturated; 62% polyunsaturated), and olive oil (77% monounsaturated; 9% polyunsaturated). Being unsaturated lowers a fat's melting point. These fats are oils

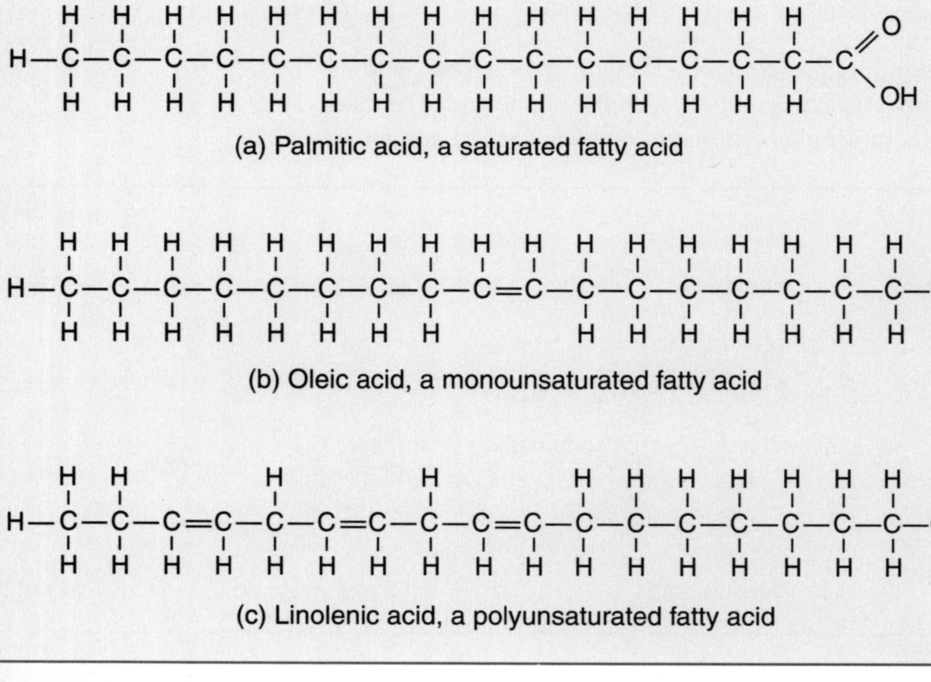

(a) Palmitic acid, a saturated fatty acid

(b) Oleic acid, a monounsaturated fatty acid

(c) Linolenic acid, a polyunsaturated fatty acid

Figure 5 Typical fatty acids.

(liquids) at room temperature. Highly saturated fats (butter, lard, etc.) are solids at room temperature. You also might find it interesting that vegetables and fruits, unlike foods from animals, contain no cholesterol.

Because C=C double bonds in unsaturated fats can undergo addition reactions, but saturated fats cannot, these two types of fats participate differently in body chemistry. Unsaturated fats are much more reactive because of the double bonds. Polyunsaturated fats have become newsworthy because there is increasing evidence that saturated fats may contribute to health problems and some unsaturated fats may not. Saturated fats are associated with formation of plaque (fatlike matter), which can block arteries. The result is a condition known as "hardening of the arteries," or atherosclerosis, a particular threat to coronary (heart) arteries and arteries leading to the brain. If coronary arteries are blocked, a heart attack can result, damaging the heart muscle. If arteries leading to the brain are blocked, a stroke may result, killing brain cells and harming various body functions.

YOUR TURN

Calories from Fat

The label on a package of butter provides the following nutritional data per serving (1 tablespoon, or 14 g):

> Total fat: 10.9 g
> Polyunsaturated fat: 0.4 g
> Saturated fat: 7.2 g
> Calories: 100

We can calculate the percent of polyunsaturated fats in the butter's fat by comparing the mass of polyunsaturated fat with the mass of total fat. A percent value is always the part of interest divided by the whole and multiplied by 100. Here that means 0.4 g/10.9 g × 100 = 3.7%. Thus, approximately 4% of butter's total fat is polyunsaturated.

Now it is your turn.

1. Calculate the percent of saturated fat in the butter's fat.

2. The two percentages for fats do not add up to 100%.
   a. What value does the "missing" percent represent?
   b. How many grams of fat does this represent?

3. Calculate the total percent fat in this serving of butter.

An important consideration when buying foods is the percent of Calories derived from fat and other nutrients. Health professionals recommend that no more than 30% of the total Calories in our diet should come from fats. A direct way to evaluate this is to relate the total number of Calories from fat compared to the total Calories in that serving of food: Calories from fat/total Calories × 100. Remember that 1 g fat = 9 Cal.

4. Let's put this relationship to work.

a. Determine the Calories from fat in 1 tablespoon of butter.

b. Calculate the percent of Calories in butter from fat.

5. A standard serving (14 g) of a commercial margarine contains 10 g of fat and contains 90 Calories. Determine the percent of Calories from fat in this margarine.

A process called *partial hydrogenation* adds hydrogen atoms to some of the double bonds of an oil while allowing some double bonds to remain. By reacting with hydrogen, some of the C=C double bonds are converted to C—C single bonds. Because the number of unsaturated sites is decreased, the partially hydrogenated product is more saturated and the original oil becomes a semisolid. Such partially hydrogenated fats are used in margarines, puddings, and shortening.

In the Petroleum unit we discussed *cis-trans* isomerism. Fatty acids in natural foods have their double bonds arranged typically in the *cis* arrangement (see Figure 6). During hydrogenation, some of the remaining double bonds rearrange to the *trans* form (see Figure 6). Questions have been raised about the nutritional safety of these *trans-fatty acids*. The concern results from studies of diets high in *trans*-fatty acids. The studies show a decrease in blood substances called *HDLs (high-density lipoproteins),* which are associated with lowering the risk of heart disease, and an increase in *LDLs (low-density lipoproteins),* which raise cardiac risk.

Currently, Americans get about 40% of their Calories from fats. The National Research Council, the American Cancer Society, and the American Heart Association recommend that energy from fat make up no more than 30% of the total daily Calories. Further, saturated fats should contribute less than 10% of total Calories. The food guide pyramid reflects these goals for a healthier diet.

Although fat consumption in the United States is decreasing, it is still high compared with recommended levels and with normal fat intake in most other countries. High consumption of fat is a factor in several "modern" diseases, including obesity and atherosclerosis. Most U.S. dietary fat comes from meat, poultry, fish, and dairy products. Deep-fried items, such as french fries, fried chicken, and potato chips, add even more fat to the diet. In addition, when your intake of food is higher than what you burn off with exercise, your body converts excess proteins and carbohydrates to fat for storage.

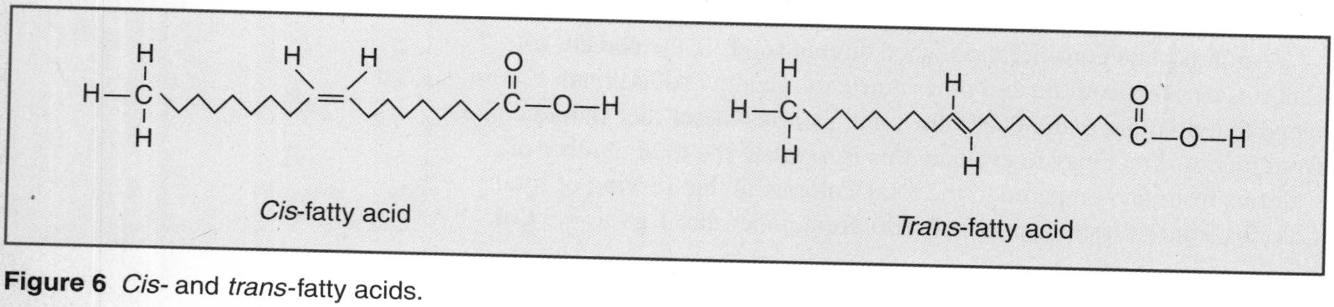

Cis-fatty acid   Trans-fatty acid

Figure 6 *Cis-* and *trans*-fatty acids.

Fats in the Diet

A well-known national fast-food chain offers a special doubleburger that is advertised as a less-fatty alternative to their regular doubleburger sandwich. Here are the nutritional data:

	Serving	Calories	Saturated Fat	Unsaturated Fat
Doubleburger	215 g	500	16 g	10 g
Alternative	206 g	320	5 g	5 g

1. Each doubleburger sandwich has about the same mass. Is the alternative doubleburger less fatty? Answer this question by calculating and comparing the total percent of fat in each of the two sandwiches.

2. Another way to answer Question 1 is to calculate and compare the percent of Calories from fat in each doubleburger.

   a. Does either doubleburger meet the guideline of 30% or less calories from fat?

   b. Does either doubleburger meet the guideline of 10% or less calories from saturated fat?

3. Which doubleburger would likely react with more hydrogen in a partial hydrogenation?

B.5 CARBOHYDRATES: ANOTHER WAY TO COMBINE C, H, AND O

Like fats, carbohydrates are compounds composed of only three elements: carbon, hydrogen, and oxygen. For example, glucose, the key energy-releasing carbohydrate in biological systems, has the formula $C_6H_{12}O_6$.

▶ *Sugars, starch, and cellulose are all carbohydrates.*

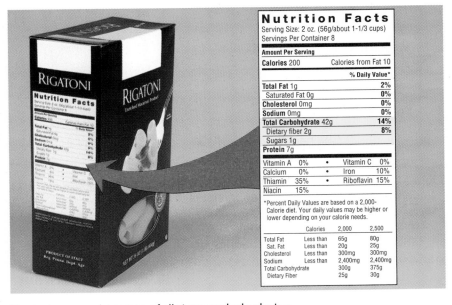

Nutrition Facts
Serving Size: 2 oz. (56g/about 1-1/3 cups)
Servings Per Container 8

Amount Per Serving	
Calories 200	Calories from Fat 10

	% Daily Value*
Total Fat 1g	2%
Saturated Fat 0g	0%
Cholesterol 0mg	0%
Sodium 0mg	0%
Total Carbohydrate 42g	14%
Dietary fiber 2g	8%
Sugars 1g	
Protein 7g	

Vitamin A 0%	•	Vitamin C 0%
Calcium 0%	•	Iron 10%
Thiamin 35%	•	Riboflavin 15%
Niacin 15%		

*Percent Daily Values are based on a 2,000-Calorie diet. Your daily values may be higher or lower depending on your calorie needs.

	Calories	2,000	2,500
Total Fat	Less than	65g	80g
Sat. Fat	Less than	20g	25g
Cholesterol	Less than	300mg	300mg
Sodium	Less than	2,400mg	2,400mg
Total Carbohydrate		300g	375g
Dietary Fiber		25g	30g

Pasta is a good source of dietary carbohydrates.

When such formulas were first discovered, chemists recognized the 2:1 relationship of hydrogen to oxygen in carbohydrates, the same as in water. They were tempted to write the glucose formula as $C_6(H_2O)_6$—implying a chemical combination of carbon and water. So, they invented the term "carbo-hydrate" or water-containing carbon compound. We now know that water molecules are not actually present in carbohydrate molecules, but the name has persisted.

Carbohydrates may be either simple sugars such as glucose or various combinations of two or more simple sugar molecules (Table 3). Simple sugars are called *monosaccharides.* The most common monosaccharide molecules contain five or six carbon atoms bonded together. Glucose (like most other monosaccharides) exists principally in a ring form, but can also occur in a chain form, as shown in Figure 7. Do both forms have the same molecular formula? (Check by counting the atoms.)

Table 3	Carbohydrates		
Classification and Examples	**Composition**	**Formula**	**Common Name of Source**
Monosaccharides		$C_6H_{12}O_6$	
Glucose	—		Blood sugar
Fructose	—		Fruit sugar
Galactose	—		—
Disaccharides	Monosaccharides	$C_{12}H_{22}O_{11}$	
Sucrose	Fructose + glucose		Cane sugar
Lactose	Galactose + glucose		Milk sugar
Maltose	Glucose + glucose		Germinating seeds
Polysaccharides	Glucose polymers	—	
Starch			Plants
Glycogen			Animals
Cellulose			Plant fibers

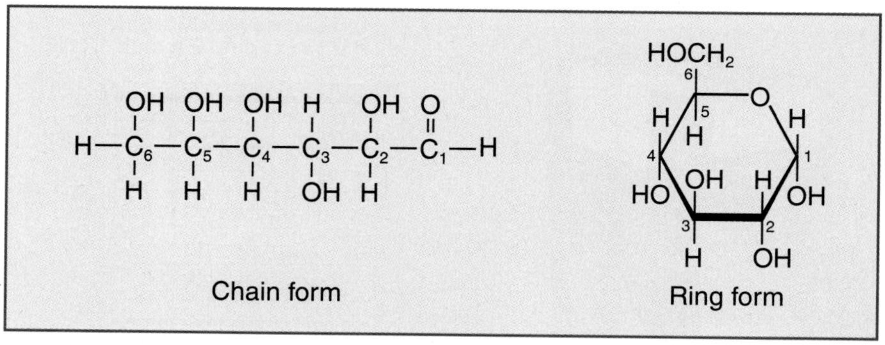

Figure 7 Structural formulas for glucose. The chain and ring forms are interconvertible; the ring form predominates in the body.

Sugars composed of two simple sugar molecules (monosaccharides) are called *disaccharides.* Disaccharides are formed by a condensation reaction between two monosaccharides; this reaction also produces water. Ordinary table sugar, sucrose ($C_{12}H_{22}O_{11}$), is a disaccharide in which the ring forms of glucose and fructose are joined by such a reaction (see Figure 8, page 253). As the molecular structures suggest, monosaccharides

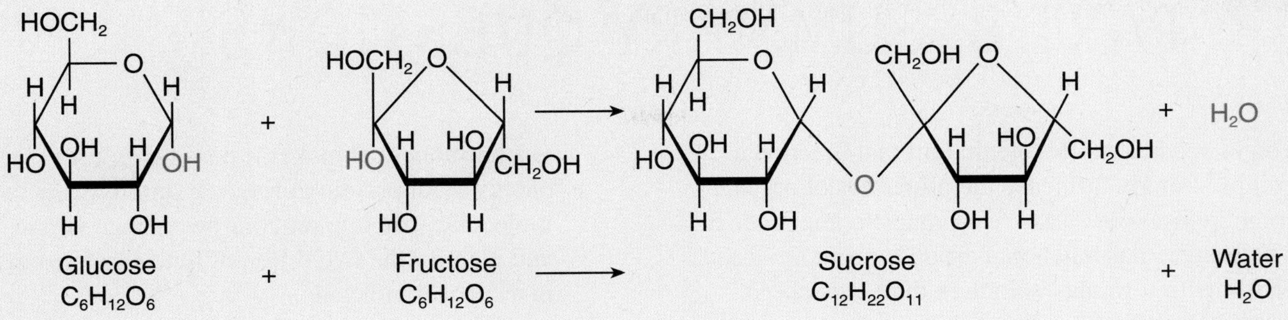

The top figure shows the formation of sucrose:

Glucose $C_6H_{12}O_6$ + **Fructose** $C_6H_{12}O_6$ → **Sucrose** $C_{12}H_{22}O_{11}$ + **Water** H_2O

and disaccharides are polar molecules because of the —OH groups. Thus, these compounds tend to be highly soluble in water, a polar solvent.

The condensation reaction that forms disaccharides can also join many simple sugar units to form polymers from these monosaccharides. Such polymers are called *polysaccharides* (Figure 9). Starch, a major component of grains and many vegetables, is a polysaccharide composed of glucose units. Cellulose, the fibrous or woody material of plants and trees, is another polysaccharide formed from glucose. The types of carbohydrates are summarized in Table 3 (page 252).

Sugars, starch, and fats are the major energy-delivering substances in our diets. Even the smallest muscle twitch or thought requires energy, with each gram of a carbohydrate providing 4 Calories. Nutritionists recommend that about 60% of daily dietary Calories come from carbohydrates. Most of the world's people obtain carbohydrates by eating grains, often consumed as rice, corn bread, wheat tortillas, bread, and pasta. In the United States we tend to eat more wheat breads and potatoes for carbohydrates than people do elsewhere. In all countries, fruits and vegetables also provide carbohydrates. Meats provide a small amount of carbohydrate in the form of glycogen, which is how animals store glucose. On average, each U.S. citizen consumes more than 125 lb (57 kg) of table sugar each year in beverages, breads, and cakes and as a sweetener. A 12-oz, non-diet cola drink contains 9 teaspoons (3 tablespoons) of sugar.

Figure 8 Formation of sucrose. The two shaded —OH groups react, with the elimination of one H_2O molecule.

▶ *Remember the discussion of polyethylene on page 216. A polymer is a large molecule composed of many smaller molecules chemically bonded together.*

▶ *Carbohydrates are all sugars or polymers of sugars.*

▶ *1 g carbohydrate = 4 Cal energy.*

Figure 9 Polysaccharides. Starch and cellulose are both polymers of glucose. They differ in the arrangements of the bonds that join the glucose monomers.

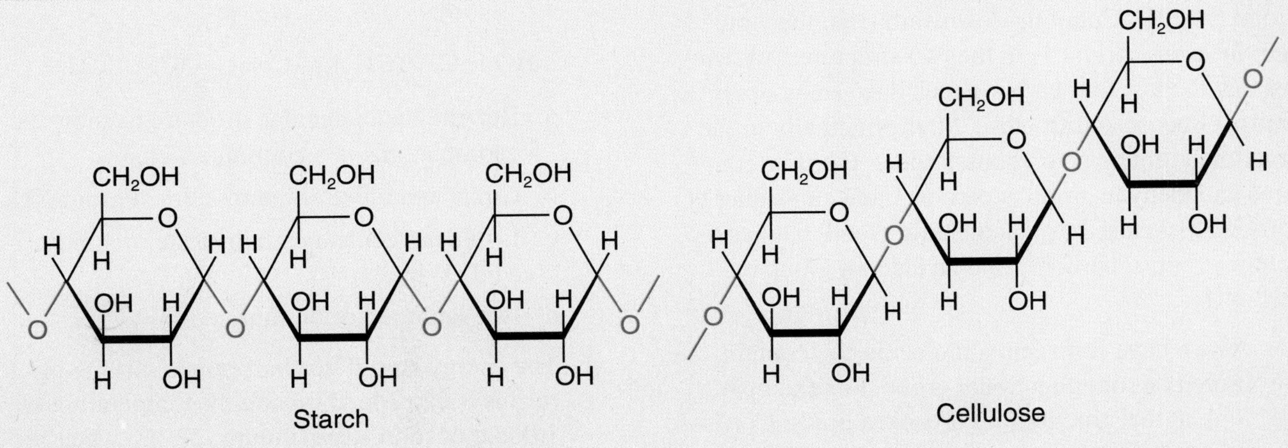

Starch

Cellulose

? PART B: SUMMARY QUESTIONS

You learned in the Petroleum unit that functional groups strongly influence the properties of organic compounds. Some classes of organic compounds in which particular functional groups appear are listed below, with formulas written in their condensed forms. (Each R represents a hydrocarbon segment.)

ROH Alcohol	R—C—R ‖ O Ketone	R—C—O—H ‖ O Carboxylic acid
ROR Ether	R—C—H ‖ O Aldehyde	R—C—O—R ‖ O Ester

Study the straight-chain molecular structures of fructose and glucose given below.

Glucose

$$\begin{array}{c} H \\ | \\ C_1{=}O \\ | \\ H{-}C_2{-}OH \\ | \\ HO{-}C_3{-}H \\ | \\ H{-}C_4{-}OH \\ | \\ H{-}C_5{-}OH \\ | \\ H{-}C_6{-}OH \\ | \\ H \end{array}$$

Fructose

$$\begin{array}{c} H \\ | \\ H{-}C_1{-}OH \\ | \\ C_2{=}O \\ | \\ HO{-}C_3{-}H \\ | \\ H{-}C_4{-}OH \\ | \\ H{-}C_5{-}OH \\ | \\ H{-}C_6{-}OH \\ | \\ H \end{array}$$

Analyzing the two structures, we see that in each case the most abundant functional group is the alcohol group (—OH). Counting downward from the "top" carbon atom (carbon 1) in the two structures, we can see that the last four carbons each have an alcohol group. Glucose and fructose differ principally in the functional groups on carbons 1 and 2. Glucose contains an aldehyde group at carbon 1 and an alcohol at carbon 2; fructose has no aldehyde group, but has a ketone group at carbon 2 and an alcohol group on carbon 1.

1. As we have just seen, a molecule can contain more than one functional group. For example, look at the structure of cortisol (a lipid). Cortisol

is a hormone that makes it possible to use energy from protein during starvation. Copy the molecular structure on your own paper. Circle and identify the two different functional groups in a cortisol molecule.

Cortisol

2. In general, alcohols react with organic (carboxylic) acids to form esters. Using the equation shown in Figure 4 (page 247) as a guide, write an equation (including structures) for the reaction of glycerol with stearic acid, a fatty acid, to form glyceryl tristearate, a fat (triglyceride). Stearic acid has this structural formula:

$$\begin{array}{c} O \\ \| \\ HO{-}C{-}(CH_2)_{16}CH_3 \end{array}$$

3. Copy the following molecular structure on your own paper:

$$\begin{array}{c} O \qquad\quad H \ \ H \\ \| \qquad\qquad | \ \ | \\ HO{-}C{-}(CH_2)_7{-}C{=}C{-}(CH_2)_7CH_3 \end{array}$$

   a. Rewrite the molecular structure to show the carbon atoms in a continuous chain.
   b. Circle and identify the functional group(s).
   c. Is this molecule a carbohydrate or a fatty acid? Why?
   d. Is it saturated or unsaturated? Why?

4. The energy stored in a half-ounce serving of raisins is capable of raising the temperature of 1,000 g of room-temperature (22 °C) water to

62 °C. How many food Calories are contained in that serving of raisins?

5. Use Table 2 (page 244) to explain the following statement: *Breakfast is the most important meal of the day.* (*Hint:* Estimate how many Calories your body uses between dinner at 6 P.M. and breakfast at 8 A.M.)

6. It has been estimated that U.S. citizens carry around about 2.5 billion pounds of excess body fat.

   a. How many Calories of food energy does this represent? (Note that a pound of fat contains about 4,000 Cal of energy.)

   b. Assume an average human needs about 2,500 Cal each day. If the excess energy calculated in Question 6a could somehow be diverted, how many people could be fed for one year?

7. Health professionals currently recommend that no more than 30% of daily Calories come from fat. It might be helpful to get a sense of how much fat that represents.

   a. If you need 2,220 Cal daily and want to meet the 30% guideline, what is the maximum number of grams of fat you should eat daily?

   b. What would the mass of fat be if you needed 3,000 Cal per day?

8. You have just finished athletic practice and you stop for a fast-food meal that consists of a large non-diet cola drink, one hamburger, and a large order of french fries. The following nutritional information applies:

	Quantity	Calories	Carbohydrates	Fat
Hamburger (1)	166 g	410	34 g	20 g
Fries, large	122 g	400	46 g	22 g
Cola	32 oz	300	80 g	—

Let's analyze the meal with respect to some of the guidelines discussed in this section.

   a. Calculate the percent of Calories in the hamburger derived from

   (1) carbohydrates.

   (2) fats.

   b. Calculate the percent of Calories in the french fries derived from

   (1) carbohydrates.

   (2) fats.

   c. Determine the percent of Calories in the cola drink supplied by carbohydrates.

   d. In what ways does this meal meet or fail to meet the 30% guideline for fat Calories and the 60% guideline for carbohydrates?

   e. If you are attempting to lose weight, which parts of this meal should you eat less often or eliminate from your diet?

 # EXTENDING YOUR KNOWLEDGE (OPTIONAL)

Assume you currently consume 3,000 Cal each day and want to lose 30 lb of body fat over the next two months (60 days).

   a. If you decide to lose this weight only through dieting (no extra exercise), how many Calories would you need to omit from your diet each day?

   b. How many food Calories would you still be allowed to consume daily?

   c. Is this a sensible way to lose weight? Why?

C FOODS: THE BUILDER MOLECULES

Your body mass is roughly 60% water and 20% fat. The remaining 20% consists primarily of proteins, carbohydrates, and related compounds, and the major bone elements calcium and phosphorus.

In the United States, recent generations have grown, on average, taller than their parents. Better nutrition is largely responsible for this change. Our genes determine how tall we may grow, but nutrition determines—at least in part—actual height.

In the following sections you will continue to explore food nutrients as builder molecules. You will soon discover the critical importance of protein in building body tissues.

C.1 FOODS AS CHEMICAL REACTANTS

Biochemistry is the branch of chemistry that studies chemical reactions in living systems. Such processes are rarely simple. Consider the body's extraction of energy from disaccharides and polysaccharides. These carbohydrates are broken down by digestion into glucose, $C_6H_{12}O_6$, the primary substance used for energy in most living systems.

The *overall* reaction for extracting energy from glucose by burning is the same in the body as it is on a laboratory bench in the presence of air. Obviously, glucose is not burned with a flame inside the body. Not only would most of the energy escape uselessly as heat, but the resulting temperatures would kill cells. Yet, in a sense, that burning reaction does continuously occur within each human cell, in a series of at least 22 related chemical reactions or steps known as **cellular respiration** (also called *metabolism*).

This process—and nearly every other biochemical reaction—can take place in the body only with assistance from a class of molecules called **enzymes.** Enzymes are catalysts that help to make and break chemical bonds by interacting selectively with certain molecules, just as the correct key opens a particular lock.

▶ *Enzymes are catalysts—compounds that help chemical reactions occur by increasing their rates of reaction.*

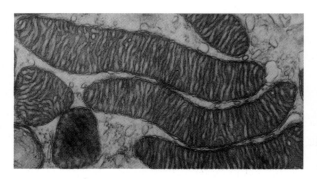

Much cellular respiration takes place in mitochondria. This photo shows details of mitochondria under an electron microscope.

CHEMISTRY AT WORK

Food for Thought

Next time you bite into an ear of corn, consider the amazing "behind the scenes" work that took place to produce the corn before it arrived on your plate.

Sue Adams is one of the many people—whom you will probably never meet—who work to put food on your table. She and her husband John work year-round to run a 1,100 acre farm in Atlanta, Illinois. To protect the environment, they use integrated pest management and practice a technique called *no-till farming*—planting new crops directly in the residue of previously harvested crops. It saves on labor and fuel; more importantly, it reduces loss of valuable topsoil and moisture, and can dramatically reduce the need for materials like fertilizers, pesticides, and herbicides.

In the spring, Sue and John plant half of the fields in corn, and the other half in soybeans. As soon as the crops begin to grow, they scout the fields for evidence of damage by pests and, where necessary, apply special crop protection substances which decompose quickly in sunlight. Just like a chemist in a laboratory, Sue carefully monitors and measures how these substances are used to ensure maximum protection for crops and the environment.

Ever Think of a Satellite as a Farm Tool?

Sue and her husband are on the cutting edge of high-tech agriculture. One example is their use of Global Positioning Satellite (GPS) technology.

GPS employs computers, monitoring units, and radio signals to send, receive, and process information from satellites to study objects on the earth's surface. On the farm, Sue uses GPS technology to plot with great precision which areas of her fields require additional fertilizers or other treatments. She even has a specially equipped all-terrain vehicle which can record and access information while out in the field.

Using GPS data, Sue and John are able to apply fertilizer to only those areas of the farm that need it, minimizing the amount of fertilizer they introduce to the environment. The increased precision has also reduced the costs of fertilizing (materials, equipment, fuel, and labor) by as much as $10 an acre.

Earth and Its Food Supply: What's the Forecast?

In the 1700s, an English economist named Thomas Malthus published his *Essay on the Principle of Population*. The widely read pamphlet argued that unchecked population increases geometrically (1, 2, 4, 8, 16 . . .) while food supplies increase arithmetically (1, 2, 3, 4, 5 . . .). Malthus's theory suggested that one day the world might be unable to feed itself. Indeed, world population has grown significantly. In 1800, world population stood at about 1 billion; today, less than 200 years later, we're close to the 6 billion mark.

In small discussion groups, talk about and answer the following questions.

1. Malthus's prediction has not become reality, at least for a majority of the world's inhabitants. What part of his theory do you think was flawed?

2. The world population by the year 2020 will probably be almost 8 billion people. Why do you suppose there is renewed interest in Malthus's theory?

3. One of the problems with increased population is the total land required for human residence versus that needed for cropland. What new techniques might farmers develop to allow for the space needed by both humans and agriculture?

Photograph courtesy of Ken Kashian, Illinois Farm Bureau

257

C.2 LIMITING REACTANTS

The chemical reactions that take place in the body require the presence of a complete set of ingredients, or *reactants*. In addition, there must be enough reactants to complete the reaction. It's like collecting ingredients to bake a cake. Consider this cake recipe:

2 cups flour	1 1/2 tablespoons baking powder
2 eggs	1 cup water
1 cup sugar	1/3 cup oil

The combination of these ingredients will produce one cake. Now, suppose in the kitchen we find 14 cups flour, 4 eggs, 9 cups sugar, 15 tablespoons baking powder, 10 cups water, and 3 1/3 cups oil. How many cakes can be baked?

Well, 14 cups of flour is enough for 7 cakes (2 cups flour per cake). And there's enough sugar for 9 cakes (1 cup sugar per cake). The supplies of baking powder, water, and oil are sufficient for 10 cakes (confirm this with the recipe). Yet we cannot make 7, 9, or 10 cakes with the available ingredients. Why? Because we only have 4 eggs, just enough for 2 cakes. The egg supply has limited the number of cakes we can bake. The excess quantities of the other reactants (flour, sugar, baking powder, water, oil) simply remain unused. If we want to bake more cakes, we have to find more eggs!

In chemical terms, the eggs in our cake-making example are called the ***limiting reactant*** (or limiting reagent). The limiting reactant is the starting substance (reactant) that is used up first when a chemical reaction occurs. It controls how much (or how little) product can be formed.

YOUR TURN

Limiting Reactants

1. Consider the same cake-making example discussed above, but this time assume that you have 26 eggs.

   a. How many cakes can be made if the other ingredients are present in the same quantities as in the original recipe?

   b. Which ingredient limits the number of cakes you can make this time? (In other words, what is the limiting reactant?)

2. A restaurant prepares carryout lunches. Each completed lunch requires 1 sandwich, 3 pickles, 2 paper napkins, 1 carton of milk, and 1 container. Today's inventory is 60 sandwiches, 102 pickles, 38 napkins, 41 cartons of milk, and 66 containers.

   a. As carryout lunches are prepared, which item will be used up first?

   b. Which item is the "limiting reactant"?

   c. How many complete carryout lunches can be assembled?

3. A booklet is prepared from 2 covers, 3 staples, and 20 pages. Assume we have 60 covers, 120 staples, and 400 pages.

   a. What is the largest number of complete booklets that can be prepared with these supplies?

   b. Which is the "limiting reactant" in this system?

   c. How many of the other two "reactants" will be left over when the booklet preparation process stops?

In chemical reactions, just as in recipes, individual substances react in certain fixed quantities. These relative amounts are indicated in chemical equations by the coefficients. Let's consider the equation for oxidation of glucose. First, the equation can be interpreted in terms of one glucose molecule:

$$C_6H_{12}O_6 \quad + \quad 6\,O_2 \quad \rightarrow \quad 6\,CO_2 \quad + \quad 6\,H_2O \quad + \quad \text{Energy}$$

| 1 glucose molecule | 6 oxygen molecules | 6 carbon dioxide molecules | 6 water molecules | |

Suppose we had 10 glucose molecules and 100 oxygen molecules. Which substance would be the limiting reactant in this reaction? The equation tells us that each glucose molecule requires 6 oxygen molecules. Thus, 10 glucose molecules would require 60 oxygen molecules for complete reaction. That is fine, because 100 oxygen molecules are available—more than we need! The 10 glucose molecules will react with 60 of these oxygen molecules, and 40 oxygen molecules will be left over.

Because the glucose is completely used up first, it is the limiting reactant; there is no excess glucose. Once the 10 glucose molecules are converted, the reaction stops. Additional glucose would be needed to continue it.

The notion of limiting reactants applies equally well to living systems. A shortage of a key nutrient or reactant can severely affect the growth or health of plants or animals. In many biochemical processes, a product from one reaction becomes a reactant for other reactions. If a reaction stops due to a shortage of one substance (the limiting reactant), all reactions following that step will also stop.

Fortunately, in some cases, alternate reaction pathways are available. We have already said that if the body's glucose supply is depleted, glucose metabolism cannot occur. One backup system oxidizes stored body fat in place of glucose. Or, more drastically, under starvation conditions, structural protein can be broken down and used for energy. If dietary glucose is again available, glucose metabolism reactions start up again.

Alternate reaction pathways are not a permanent solution, however. If intake of a nutrient is consistently below what the body requires, that nutrient may become a limiting reactant in vital biochemical processes. The results can easily affect one's health.

▶ *Producing glucose from protein is much less energy efficient than producing glucose from carbohydrates. But your body will use protein in this way if necessary.*

Limiting Reactants: Chemical Reactions

A chemical equation can be interpreted not only in terms of molecules, but also in terms of moles and grams (as discussed on page 130). The glucose oxidation reaction can be rewritten in terms of moles of all reactants and products:

$$C_6H_{12}O_6 + 6\,O_2 \rightarrow 6\,CO_2 + 6\,H_2O + \text{Energy}$$

| 1 mol glucose molecules | 6 mol O_2 molecules | 6 mol carbon dioxide molecules | 6 mol water molecules |

Finally, we can use molar masses to convert these mole values to grams, as follows:

$$C_6H_{12}O_6 + 6\,O_2 \rightarrow 6\,CO_2 + 6\,H_2O + \text{Energy}$$

180 g   (6 × 32 g)   (6 × 44 g)   (6 × 18 g)
= 192 g   = 264 g   = 108 g
(1 mol)   (6 mol)   (6 mol)   (6 mol)

If we needed to burn only 90 g of glucose (half the mass shown above), then only 96 g of oxygen would be needed (half the 192 g shown). If we wanted to burn 90 mg of glucose, 96 mg oxygen would be required to react with all of the glucose. Such relationships can be used to identify the limiting reactant.

Ammonia (NH_3) is an important fertilizer and is used to produce other fertilizers. It is made commercially by the following reaction at high temperature and pressure using an iron catalyst:

$$N_2 + 3\,H_2 \rightarrow 2\,NH_3$$

1. a. How many moles of each reactant and product are indicated by the equation?
   b. How many grams of each reactant and product are indicated by the equation?

2. Assume that a manufacturer has 28 kg of nitrogen gas and 9 kg of hydrogen gas. Also assume that the maximum amount of ammonia is produced from these reactants.
   a. Which reactant will be the limiting reactant?
   b. Which reactant will be left over (i.e., be in excess)?
   c. How much of it will be left over?
   d. How many kilograms of ammonia will be produced?

▶ *Here's an example of how these masses were obtained: One mole of water has a mass of 18.0 g. Six moles of water would have a mass six times larger than this: 6 × 18.0 g water = 108 g water.*

▶ *Algae often pollute lakes and streams that contain too much nitrogen and phosphorus.*

Plants, like animals, must take in essential nutrients to support their growth and metabolism. For example, nutrients for algae must include carbon, nitrogen, and phosphorus. For every 41 g of carbon, algae require 7 g of nitrogen and 1 g of phosphorus. If any one of these elements is in short supply, it becomes a limiting reactant and affects algae growth.

In the following *Your Turn,* you will compare limiting reactants in human and plant systems.

▶ *Trace minerals are present at levels of 1 ppm or less. Higher concentrations may be toxic.*

YOUR TURN

Limiting Reactants in Plants and Humans

Table 4 lists 22 elements currently known to be required in some quantity to support human life.

Table 4	Elements Needed for Human Nutrition*
In major biomolecules	Carbon (C) Hydrogen (H) Oxygen (O) Nitrogen (N) Sulfur (S)
Major minerals	Calcium (Ca) Chlorine (Cl) Magnesium (Mg) Phosphorus (P) Potassium (K) Sodium (Na)
Trace minerals	Chromium (Cr) Cobalt (Co) Copper (Cu) Fluorine (F) Iodine (I) Iron (Fe) Manganese (Mn) Molybdenum (Mo) Nickel (Ni) Selenium (Se) Zinc (Zn)

*Essential in animals and very likely essential in humans.

For comparison, a list of nutrients needed for the growth of common agricultural crops, such as corn and wheat, is given in Table 5 (page 262). Crops cannot grow properly unless they have suitable nutrients in sufficient quantities. That means that farmers must be concerned with limiting reactants in their crop production.

1. Which elements required for plant growth are not essential nutrients for humans?

2. Which elements required for human growth are not essential nutrients for plants?

3. Examine the label on a drugstore product that provides 100% of a person's daily requirement of some minerals. (See Figure 10.)

   a. Are all elements listed in Table 4 included among the ingredients?

   b. If not, which are missing?

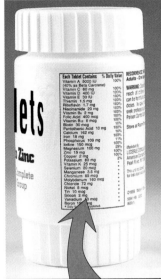

Each Tablet Contains	% Daily Value
Vitamin A 5000 IU (40% as Beta Carotene)	100%
Vitamin C 60 mg	100%
Vitamin D 400 IU	100%
Vitamin E 30 IU	100%
Thiamin 1.5 mg	100%
Riboflavin 1.7 mg	100%
Niacinamide 20 mg	100%
Vitamin B_6 2 mg	100%
Folic Acid 400 mcg	100%
Vitamin B_{12} 6 mcg	100%
Biotin 30 mcg	10%
Pantothenic Acid 10 mg	100%
Calcium 162 mg	16%
Iron 18 mg	100%
Phosphorus 109 mg	11%
Iodine 150 mg	100%
Magnesium 100 mg	25%
Zinc 15 mg	100%
Copper 2 mg	100%
Potassium 80 mg	2%
Vitamin K 25 mcg	*
Selenium 20 mcg	*
Manganese 3.5 mg	*
Chromium 65 mcg	*
Molybdenum 160 mcg	*
Chloride 72 mg	*
Nickel 5 mcg	*
Tin 10 mcg	*
Silicon 2 mg	*
Vanadium 10 mcg	*

Figure 10 The label from a common food supplement.

Table 5	Plant Nutrients Needed by Crops
Basic nutrients	Carbon (C) from CO_2 Oxygen (O) from CO_2 and H_2O Hydrogen (H) from H_2O
Primary nutrients	Nitrogen (N) Phosphorus (P) Potassium (K)
Secondary nutrients	Calcium (Ca) Magnesium (Mg) Sulfur (S)
Micronutrients	Boron (B) Chlorine (Cl) Copper (Cu) Iron (Fe) Manganese (Mn) Molybdenum (Mo) Zinc (Zn)

4. Plants need nutrients, which they get from the soil.

   a. Which plant nutrients are commonly added as fertilizer to increase soil productivity?

   b. These nutrients can also serve as limiting reactants. If they are not available in a nation as added plant nutrients, how would this affect the nation's food production?

Without water and essential nutrients, crops cannot live.

C.3 PROTEINS

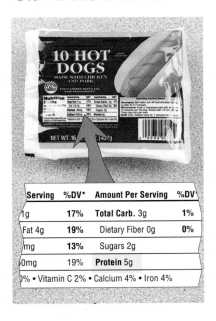

Serving	%DV*	Amount Per Serving	%DV
1g	17%	Total Carb. 3g	1%
Fat 4g	19%	Dietary Fiber 0g	0%
/mg	13%	Sugars 2g	
30mg	19%	**Protein** 5g	
0% • Vitamin C 2% • Calcium 4% • Iron 4%			

Protein has been described as the primary material of life. *Proteins* are the major structural components of living tissue. When you look at another person, everything you see is protein: skin, hair, eyeballs, nails. Inside the body, tissues such as bones, cartilage, tendons, and ligaments all contain protein, as do birds' feathers and the fur, hooves, and horns of animals. In addition, most enzyme molecules that help control chemical reactions in the cell are proteins. Your body contains tens of thousands of different proteins. Protein is constantly needed for new growth and for maintaining existing tissue. For example, red blood cells must be replaced every month. The cells that line the intestinal tract are replaced weekly. Every time we bathe we wash away dead skin cells.

▶ *The word "protein" comes from the Greek word proteios, which means "of prime importance."*

Building body protein.

Table 6 lists just a few major roles of proteins in the human body.

Proteins are polymers built from small molecules called *amino acids.* Each amino acid contains carbon, hydrogen, oxygen, and nitrogen; a few also contain sulfur. Just as monosaccharide molecules are building blocks for more complex carbohydrates, 20 different amino acids are the structural units of all proteins.

Proteins are composed of *very* long chains of amino acids; proteins have molecular weights from 5,000 to several million. Just as the 26 letters of our alphabet combine in different ways to form hundreds of thousands of words, the 20 amino acids can combine in a virtually infinite number of ways to form different proteins.

Table 6	Proteins in the Body	
Type	**Function**	**Examples**

Structural proteins
  Muscle
  Connective tissue
  Chromosomal proteins
  Membranes
	
	Contraction, movement
	Support, protection
	Part of chromosome structure
	Control of influx and outflow, communication
		
		Myosin
		Collagen, keratin
		Histones
		Pore proteins, receptors

Transport proteins	Carriers of gases and other substances	Hemoglobin

Regulatory proteins
  Fluid balance
 	
  Enzymes
  Hormones
	
	Maintenance of pH, water, and salt content of
	  body fluid
	Control of metabolism
	Regulation of body functions
		
		Serum albumin
		
		Proteases
		Insulin

Protective proteins	Antibodies	Gamma globulin

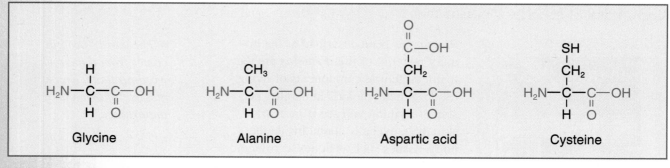

Figure 11 Representative amino acids.

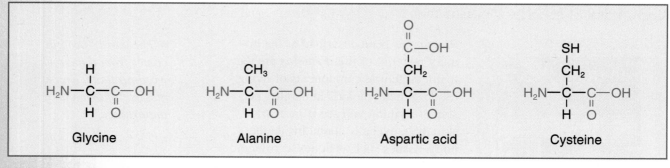

Figure 12 Formation of a dipeptide from two amino acids. All proteins contain amino acids linked in this manner.

All amino acids have similar structural features, as shown in Figure 11. Two functional groups, the amino group (NH_2) and the carboxylic acid group (COOH), are found in every amino acid molecule.

The combining of two amino acid molecules with loss of one water molecule, as illustrated in Figure 12, is a typical condensation reaction. Like starch, nylon, and polyester (see page 218), proteins are condensation polymers.

Test your skill as a "molecular architect" in the following protein-building activity.

YOUR TURN

Molecular Structure of Proteins

1. Draw structures for glycine and alanine on your own paper (see Figure 11).

   a. Circle and identify the functional groups in each molecule.

   b. How are the two molecules alike?

   c. How do the two molecules differ?

2. Proteins are polymers of amino acids. Examine the sample equation in Figure 12 to see how a pair of amino acids join. Notice that the amino group on one amino acid is linked to the carboxylic acid group on another. Each such linkage is called a *peptide linkage*, or peptide bond, and is represented by:

$$-\underset{\underset{O}{\|}}{C}-\underset{\underset{H}{|}}{N}-$$

Because *two* amino acid units join to form a peptide bond, the product is called a *di*peptide. Peptide bond formation is shown in greater detail in Figure 13.

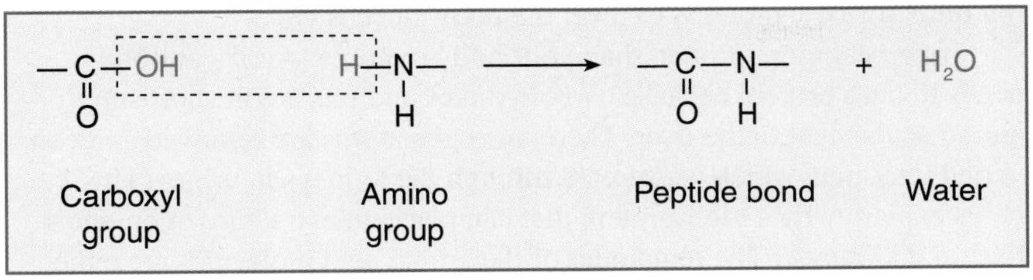

Figure 13 Formation of a peptide bond. Each amino acid can form peptide bonds with two other amino acids.

Because an amino acid contains at least one amino group and one carboxyl group, each acid can form a peptide bond at either or both ends. Using Figures 11, 12, and 13 as models, do the following:

   a. Using structural formulas, write the equation for the reaction between two glycine molecules. Circle the peptide bond in the dipeptide product.

   b. Using structural formulas, write equations for possible reactions between a glycine molecule and an alanine molecule. (*Hint:* Two dipeptide products are possible.)

 3. Examine structural formulas of the two dipeptide products identified in Question 2. Note that each dipeptide still contains a reactive amino group and a reactive carboxylic acid group. That means these dipeptides could react further with other amino acids, forming more peptide linkages.

 4. Consider the following problems, assuming that you have supplies of three different amino acids:

   a. If each type of amino acid can be used only once, how many different *tripeptides* (three amino acids linked together) can be formed? (*Hint:* Represent each type of amino acid by a letter—for instance, a and b and c; a and c and b; and so on). (*Another Hint:* Just as the order of the letters create new words, the order of amino acids in a tripeptide create new and different tripeptides.)

   b. If a given amino acid can appear any number of times in the tripeptide, how many different tripeptides can be formed?

   c. How many tetrapeptides (four amino acids joined together) could be formed from four different amino acids? (For simplicity, assume that each amino acid is used only once per tetrapeptide.)

   d. Given 20 different amino acids and the fact that proteins range in length from about 50 amino acid units to more than 10,000, would the theoretical number of distinct proteins be in the hundreds, thousands, or millions-plus?

   e. How do these insights relate to the uniqueness of each living organism?

When foods containing protein reach your stomach and small intestine, peptide bonds between the amino acids are broken by enzymes. The separate amino acids then travel through the intestinal walls to the bloodstream, to the liver, and then to the rest of the body. There they are building units for new proteins to meet the body's needs.

If you eat more protein than your body requires—or if your body needs to burn protein because carbohydrates and fats are in short supply—amino acids react in the liver. There, nitrogen atoms are removed and converted into urea, which is excreted through the kidneys in urine. (This helps explain why a high-protein diet can place an extra burden on one's liver and kidneys.) The remainder of the amino acid molecules are either converted to glucose and oxidized, or converted to storage fat. One gram of protein furnishes 4 Calories.

The human body can synthesize adequate amounts of 11 of the 20 amino acids. The others, called *essential amino acids,* must be obtained from protein in the diet. If an essential amino acid is in short supply in the diet, it can become a limiting reactant in building any protein containing that amino acid. When this happens, the only way the body can make that protein is by destroying one of its own proteins that contains the limiting amino acid.

Any protein that contains adequate amounts of all essential amino acids is called a *complete protein.* Most animal proteins contain all nine essential amino acids in the needed quantities. Plant proteins and some animal proteins are incomplete; they do not contain adequate amounts of all nine essential amino acids.

Although no single plant protein can provide adequate amounts of all essential amino acids, certain combinations of plant proteins can. Such combinations of foods, which are said to contain *complementary proteins,* form a part of many diets in various parts of the world (see Table 7).

► *The essential amino acids*

Isoleucine
Leucine
Lysine
Methionine
Histidine (for infants)
Phenylalanine
Threonine
Tryptophan
Valine

Table 7	Complementary Proteins
Foods	**Country**
Corn tortillas and dried beans	Mexico
Rice and black-eyed peas	United States
Peanut butter and bread	United States
Rice and bean curd	China and Japan
Rice and lentils	India
Wheat pasta and cheese	Italy

Because your body cannot store protein, it requires a balanced protein diet daily. The recommended amount of protein is 15% of total daily Calories. Too much protein is as harmful as too little, because excess protein causes stress on the liver and kidneys, the organs that metabolize protein. Too much protein also increases the excretion of calcium ions (Ca^{2+}), which are important in nerve transmission and in bone and teeth structures. A protein-heavy diet can even cause dehydration, a problem particularly important to athletes.

How much protein is really needed? The American Dietetic Association recommends that healthy adults take in 0.8 gram of protein daily for every kilogram of body mass. This amounts to 56 g for a 150-lb person and 44 g for a 120-lb individual. In the United States, however, the average daily protein intake is 95 g for 150-lb persons and 65 g for 120-lb individuals. These amounts are 70% and 47% greater than recommended.

For active individuals, the recommended value increases to 1 g of protein per kg of body mass daily. This amount usually varies with the level of physical activity; athletes involved in strenuous exercise and hard training likely need somewhat higher amounts of protein daily.

Table 8	RDAs For Protein		
Age (yr) or Condition	Median Weight (lb)	Median Height (in.)	RDA (g)
Infants			
0–0.5	13	24	13
0.5–1	20	28	14
Males			
11–14	99	62	45
15–18	145	58	58
19–24	160	69	58
25–50	174	69	63
51+	170	68	63
Females			
11–14	101	62	46
15–18	120	64	44
19–24	128	65	46
25–50	138	64	50
51+	143	63	50
Pregnant			60
Nursing			
First 6 mo.			65
Second 6 mo.			62

▶ RDA means U.S. Recommended Dietary Allowance.

▶ A 0–0.5 year infant requires 1.0 g of protein for each pound of body weight (13 g protein/13 lb weight = 1.0 g/lb). By contrast, a 30-year-old female requires only 0.36 g protein per pound (50g/138 lb = 0.36 g/lb).

YOUR TURN

Protein in the Diet

Use information from Table 8 to answer the following questions:

1. How many grams of protein should a person of your age and gender eat each day, according to the RDAs in Table 8? (Assume median weight and height.)

2. Table 8 suggests that, for each pound of body weight, infants actually require more protein than adults. Why should protein values per pound of body weight be highest for infants and become progressively lower as a person ages?

3. Consider a 37-year-old female weighing 125 pounds and of median height.

    a. Would this individual's protein RDA be higher or lower than 50 g?

    b. Why?

4. a. What food do babies consume that provides most of their relatively high protein needs?

    b. Can you find any evidence in Table 8 (page 267) to support your answer?

In 1989 the FDA revised Recommended Dietary Allowances (U.S. RDAs) for all required nutrients, including protein. Additionally, Daily Values were introduced to cover all the nutrients listed on food labels. The Daily Values are used with nutrients and other food components for which no RDA has been established but which are nutritionally important, such as fat and fiber. Food labels now list nutritional information in terms of either a quantity (grams, etc.) or a percent of Daily Values. The Daily Values are calculated based on a daily 2,000-Calorie intake, and must be adjusted accordingly for level of physical activity and the energy demands related to such exercise.

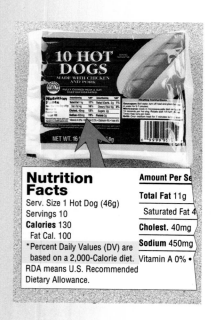

Nutrition Facts

Serv. Size 1 Hot Dog (46g)
Servings 10
Calories 130
  Fat Cal. 100
*Percent Daily Values (DV) are based on a 2,000-Calorie diet. RDA means U.S. Recommended Dietary Allowance.

Amount Per Se

Total Fat 11g
  Saturated Fat 4
Cholest. 40mg
Sodium 450mg
Vitamin A 0%

? PART C: SUMMARY QUESTIONS

1. Explain how the notion of a limiting reactant applies to
   a. human nutrition.
   b. soil nutrients in agriculture.

2. Valine is one of the essential amino acids.
   a. What does the word "essential" mean in reference to valine?
   b. How does the limiting reactant concept apply to valine and the other essential amino acids?

3. Urea, CON_2H_4, has been used to treat sickle cell anemia, a genetic blood disorder. Urea can be prepared by the reaction

   $$CO_2 + 2\,NH_3 \rightarrow CON_2H_4 + H_2O$$

   A reaction mixture contains 34 g of NH_3 and 36 g of carbon dioxide.
   a. Which is the limiting reactant—NH_3 or CO_2?
   b. Which reactant will be left over when the reaction stops?
   c. How much of that reactant will be left?
   d. What is the maximum mass of CON_2H_4 that can be produced?

4. Proteins form part of all structures in a human body, yet a diet of protein alone would be unwise. Why?

5. Diagram the condensation reaction of two alanine molecules to form a dipeptide. (The structural formula of alanine is given on page 264.) Identify the functional groups in the reactants and in the product.

6. a. Calculate the total grams of protein that a 120-pound, 17-year-old female should consume daily (refer to Table 8, page 267).
   b. Use the Appendix (see page 585) to determine how many ounces of chicken would be needed to supply this quantity of protein.
   c. How many cups of whole milk would supply this same amount?

7. a. What is the difference between a complete protein and an incomplete protein?
   b. Discuss their importance in planning a proper diet.

8. Copy the following structures on your own paper.
   a. Circle and identify the functional groups in each molecule.
   b. Which structure represents an amino acid?

   (1) $CH_3-CH_2-C{=}O$
       $\quad\quad\quad\quad\quad\;\;|$
       $\quad\quad\quad\quad\quad\;OH$

   (2) $CH_3-\underset{\underset{H}{|}}{\overset{\overset{NH_2}{|}}{C}}-C{=}O$
       $\quad\quad\quad\quad\quad\quad\quad\;|$
       $\quad\quad\quad\quad\quad\quad\quad OH$

   (3) $H_2N-CH_2-C{=}O$
       $\quad\quad\quad\quad\quad\quad\;|$
       $\quad\quad\quad\quad\quad\quad H$

D OTHER SUBSTANCES IN FOODS

Proteins, carbohydrates, and fats form the major building-block and fuel molecules of life. But other substances—vitamins and minerals—also play a vital role, even though your body requires them only in tiny amounts. Small but essential quantities of vitamins and minerals are supplied by the foods we eat, or by appropriate dietary supplements. What do vitamins and minerals do in the body that is so important?

D.1 VITAMINS

By definition, **vitamins** are biomolecules necessary for growth, reproduction, health, and life. Despite their importance, each vitamin is required in only a small amount. To give you an idea of how "a little bit goes a long way" in terms of vitamins, the total quantity of all vitamins required by a human is about 0.2 g daily.

Vitamins perform very specialized tasks. For example, vitamin D moves calcium ions from your intestines into the bloodstream. Without vitamin D, much of the calcium you ingest would be lost.

The term *vitamin* was coined in the early twentieth century, but even earlier evidence had suggested that chemical substances other than fats, proteins, and carbohydrates are required by the body. For example, scurvy, with symptoms of swollen joints, bleeding gums, and tender skin, was once common among sailors. As early as the 1500s, scurvy was considered a symptom of food deficiency. After the mid-1700s, seafarers learned to carry citrus fruits on long voyages to prevent scurvy. We now know that the disorder is caused by lack of vitamin C. Other health problems are also caused by vitamin deficiencies.

About a dozen different vitamins have been identified. Their existence has been determined and proved by synthesizing them in the laboratory and then testing them in animal diets. Table 9 (page 271) illustrates how vitamins support human life.

Vitamins generally are classified as fat-soluble or water-soluble (see Table 9). Water-soluble vitamins have polar functional groups that allow these vitamins to pass directly into the bloodstream. Because of their solubility, water-soluble vitamins are not stored in the body, and therefore must be ingested regularly. Cooking can wash away or destroy some of them, such as the B vitamins and vitamin C. Your body absorbs fat-soluble vitamins into the blood from the intestine with the assistance of fats in the food you eat. The chemical structures of these vitamins allow them to be stored in body fat, and so it is not necessary to eat them daily. In fact, because they accumulate within the body, they can build up to toxic levels if taken in large quantities.

▶ *Recall that "like dissolves like."*

Table 9	Vitamins	
Vitamin (Name)	**Main Sources**	**Deficiency Condition**
Water-soluble		
B₁ (Thiamin)	Liver, milk, pasta, bread, wheat germ, lima beans, nuts	Beriberi: nausea, severe exhaustion, paralysis
B₂ (Riboflavin)	Red meat, milk, eggs, pasta, bread, beans, dark green vegetables, peas, mushrooms	Severe skin problems
Niacin	Red meat, poultry, enriched or whole grains, beans, peas	Pellagra: weak muscles, no appetite, diarrhea, skin blotches
B₆ (Pyridoxine)	Muscle meats, liver, poultry, fish, whole grains	Depression, nausea, vomiting
B₁₂ (Cobalamin)	Red meat, liver, kidneys, fish, eggs, milk	Pernicious anemia, exhaustion
Folic acid	Kidneys, liver, leafy green vegetables, wheat germ, peas, beans	Anemia
Pantothenic acid	Plants, animals	Anemia
Biotin	Kidneys, liver, egg yolk, yeast, nuts	Dermatitis
C (Ascorbic acid)	Citrus fruits, melon, tomatoes, green pepper, strawberries	Scurvy: tender skin; weak, bleeding gums; swollen joints
Fat-soluble		
A (Retinol)	Liver, eggs, butter, cheese, dark green and deep orange vegetables	Inflamed eye membranes, night blindness, scaling of skin, faulty teeth and bones
D (Calciferol)	Fish-liver oils, fortified milk	Rickets: soft bones
E (Tocopherol)	Liver, wheat germ, whole-grain cereals, margarine, vegetable oil, leafy green vegetables	Breakage of red blood cells in premature infants, oxidation of membranes
K (Menaquinone)	Liver, cabbage, potatoes, peas, leafy green vegetables	Hemorrhage in newborns; anemia

How much is "enough" of each vitamin? It depends on your age and gender. RDAs for vitamins are summarized in Table 10.

▶ *μg = microgram = 10⁻⁶ g. RE = retinol equivalents: 1 retinol equivalent = 1 μg retinol.*

Table 10	RDAs For Selected Vitamins							
Sex and Age	**A (μg RE)**	**D (μg)**	**C (mg)**	**B₁ (mg)**	**B₂ (mg)**	**Niacin (mg)**	**B₁₂ (μg)**	**K (μg)**
Males								
11–14	1000	10	50	1.3	1.5	17	2.0	45
15–18	1000	10	60	1.5	1.8	20	2.0	65
19–24	1000	10	60	1.5	1.7	19	2.0	70
25–50	1000	10	60	1.5	1.7	19	2.0	80
51+	1000	10	60	1.2	1.4	15	2.0	80
Females								
11–14	800	10	50	1.1	1.3	15	2.0	45
15–18	800	10	60	1.1	1.3	15	2.0	55
19–24	800	10	60	1.1	1.3	15	2.0	60
25–50	800	5	60	1.1	1.3	15	2.0	65
51+	800	5	60	1.0	1.2	13	2.0	65

Source: Food and Nutrition Board, National Academy of Sciences—National Research Council, Recommended Dietary Allowances, Revised 1989.

Vitamins in the Diet

1. Carefully planned vegetarian diets are nutritionally balanced. However, people who are vegans (those who do not consume any animal products including eggs or milk) must be especially careful to get enough of two certain vitamins.

   a. Use Table 9 (page 271) to identify these two vitamins, and briefly describe the effect of their absence in the diet.

   b. How might vegans avoid this problem?

2. Complete the following data table about yourself, using data from Tables 10 (page 271) and 11.

Data Table				
Vegetable	**Your RDA**		**No. Servings to Supply your RDA**	
(one-cup serving)	**B$_1$**	**C**	**B$_1$**	**C**
Green peas				
Broccoli				

Table 11	Vitamins B$_1$ and C In Common Vegetables	
	Vitamin (in mg)	
Vegetable (one-cup serving)	**B$_1$ (thiamin)**	**C (ascorbic acid)**
Green peas	0.387	58.4
Lima beans	0.238	17
Broccoli	0.058	82
Potatoes	0.15	30

▶ *For comparison, one cup of orange juice contains 0.223 mg of vitamin B$_1$ and 124 mg of vitamin C.*

   a. Would any of your table entries change if you were a member of the opposite sex? If so, which one(s)?

   b. Based on your completed table, why do you think variety in diet is essential?

   c. Why might malnutrition (because of vitamin deficiencies) be a problem even when people receive adequate supplies of food Calories?

3. Some people believe that to promote vitality and increase resistance to diseases, much larger amounts (megadoses) of certain vitamins should be taken. Of the two broad classes of vitamins, which is more likely to pose a health problem if consumed in large doses? Why?

4. Nutritionists recommend eating fresh fruit rather than canned fruit, and raw or steamed vegetables instead of canned or boiled vegetables.

   a. Which class of vitamins are they concerned about?

   b. Why is their advice sound?

To eat a nutritionally balanced diet, you need to know which foods deliver adequate amounts of the vitamins your body requires. Because nutritionists may not yet know all the nutrients necessary for the best health, it is a good idea to build a diet around a *variety* of foods.

In the following laboratory activity you will find out how the vitamin content, in this case vitamin C, of common foods can be determined.

D.2 LABORATORY ACTIVITY: VITAMIN C

GETTING READY

Vitamin C, also known as ascorbic acid, is a water-soluble vitamin. It is among the least stable vitamins, meaning that it reacts readily with oxygen and so its potency can be lost through exposure to light and heat. In this laboratory activity you will investigate how much vitamin C is in a variety of popular beverages.

This analysis for vitamin C is based on the chemical properties of ascorbic acid (vitamin C) and iodine. Iodine (I_2) solution is capable of oxidizing ascorbic acid, forming the colorless products dehydroascorbic acid, hydrogen ions (H^+), and iodide ions (I^-):

$$I_2 + C_6H_8O_6 \rightarrow C_6H_6O_6 + 2\,H^+ + 2\,I^-$$

iodine   ascorbic   dehydro-   hydrogen   iodide ion
         acid   ascorbic   ion
                 acid

You will perform a ***titration,*** a common procedure used for finding the concentrations or amounts of substances in solutions. You will add a known amount of one reactant slowly from a Beral pipet to another reactant in a wellplate, until just enough for complete reaction has been added. You will recognize the complete reaction by a color change or other highly visible change at the ***endpoint*** of the reaction. Knowing the reaction involved, you can calculate the unknown amount of the second reactant from the known amount of the first reactant.

The titration equipment is illustrated in Figure 14. Complete the procedure as demonstrated by your teacher.

In this analysis, the endpoint is signaled by the reaction of iodine with starch, producing a blue-black product. Starch solution is added to the beverage to be tested, and then an iodine solution is added dropwise from a Beral pipet. As long as ascorbic acid is present, the iodine is quickly converted to iodide ion, and no blue-black iodine–starch product is observed. As soon as all the available ascorbic acid has been oxidized to dehydroascorbic acid, the next drop of added iodine solution will react with the starch, and you will see the blue-black color. The endpoint in the titration, then, is the *first sign of permanent blue-black color* in the beverage-containing well of the wellplate.

You will first complete a titration with a known vitamin C solution. This will allow you to find a useful conversion factor for finding the mass of ascorbic acid that reacts with each drop of iodine solution. You can then calculate the mass (mg) of vitamin C present in a 25-drop sample of

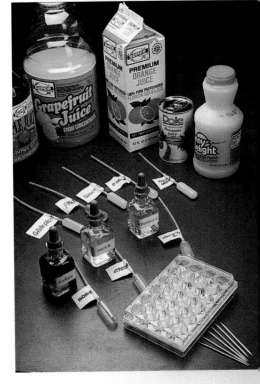

Figure 14 Titration equipment and supplies.

each beverage. This information will allow you to rank the beverages you test in terms of milligrams of vitamin C per drop of beverage.

Prepare a data table like the one below in your laboratory notebook.

Data Table: Drops vit C/mL (Step 1.1) = _____ ; Drops I₂ (Step 1.4) = _____ .				
Beverage	Drops I₂ (Step 2.3) ×	Conversion Factor (Step 1.5) =	mg Vit C per 25 drops of beverage	Rank
1.				
2.				
3.				
4.				
5.				

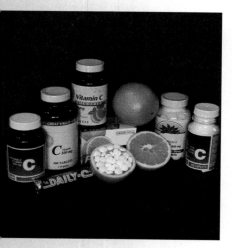

The vitamin C in these bottles is the same compound found in oranges.

PROCEDURE

Part 1. Determining a Conversion Factor

1.1 Determine the number of drops of vitamin C solution in a milliliter by counting the number of drops of vitamin C solution from a Beral pipet that are needed to reach the 1-mL mark of a 10-mL graduated cylinder. Record this value as *drops vit C/mL.*

1.2 Using the same Beral pipet, add 25 drops of vitamin C solution to a well of a clean 24-well wellplate. The vitamin C concentration is 1 mg vit C/mL.

1.3 Add one drop of starch solution to the well.

1.4 With a Beral pipet, add iodine solution one drop at a time to the well containing the vitamin C/starch mixture. Count the number of drops of iodine solution added to reach the endpoint (a "permanent" blue-black color). After each addition of iodine solution, stir the resulting mixture with a toothpick. Continue adding iodine solution dropwise, with stirring, until the solution in the well remains blue-black for at least 20 seconds. If the color fades before 20 seconds, add another drop of iodine solution. (A piece of white paper placed under the wellplate will help you detect the color.) ***Record the number of drops of iodine needed to reach the endpoint.*** *Note:* Colored beverages may not produce a true blue-black endpoint color. For example, red beverages may appear purple at the endpoint.

1.5 The concentration of vitamin C solution used is 1 mg/mL. You used 25 drops of vitamin C solution to react with the iodine solution. You can now use this information to calculate a conversion factor relating milligrams of vitamin C per drop of iodine solution:

$$\frac{25 \text{ drops}}{\text{vit. C}} \times \frac{1 \text{ mL vit C}}{\text{no. drops vit C}} \times \frac{1 \text{ mg vit C}}{1 \text{ mL vit C}} \times \frac{1}{\text{no. drops I}_2}$$

$$\text{(Step 1.1)} \qquad \qquad \text{(Step 1.4)}$$

$$= \underline{\quad\quad} \text{ mg vit C/drop I}_2$$

For example, suppose you found that Step 1.1 required 30 drops of the vitamin C solution to equal 1 mL, and it took 22 drops of iodine solution in Step 1.4. Substituting these values into the conversion factor expression above leads to a calculated value of 0.038 mg vitamin C per drop of iodine solution.

Part 2. Analyzing Beverages for Vitamin C

2.1 Add 25 drops of an assigned beverage to a clean well of a wellplate.

2.2 Add one drop of starch solution to the well.

2.3 With a Beral pipet, add iodine solution one drop at a time to the well containing the beverage/starch mixture. Count the number of drops of iodine solution added to reach the endpoint (a "permanent" blue-black color). After each addition of iodine solution, stir the resulting mixture with a toothpick. Continue adding iodine solution dropwise, with stirring, until the solution in the well remains blue-black for at least 20 seconds. If the color fades before 20 seconds, add another drop of iodine solution. (A piece of white paper placed under the wellplate will help you detect the color.) Record the number of drops of iodine needed to reach the endpoint.

2.4 Repeat Steps 2.1 through 2.3 for each beverage you are assigned to analyze.

2.5 Use the data from Steps 2.3 and 2.4, plus the conversion factor from Step 1.5, to calculate the milligrams of vitamin C per 25 drops of beverage.

2.6 Rank the beverages by number (from 1 to 5; 5 highest) from the highest mg of vitamin C to lowest mg of vitamin C.

2.7. Wash your hands thoroughly before leaving the laboratory.

QUESTIONS

1. Among the beverages tested, were any vitamin C levels

    a. unexpectedly low, in your opinion? If so, explain.

    b. unexpectedly high, in your opinion? If so, explain.

2. What other common foods contain a high level of vitamin C?

Now we are ready to consider the trace nutrients known as minerals. What are they and what do they do in the body?

D.3 MINERALS: AN IMPORTANT PART OF OUR DIET

Minerals are important life-supporting materials in our diet. Some are quite common; others are likely to be found only on laboratory supply shelves.

Some minerals become part of the body's structural molecules. Others help enzymes do their jobs. Still others help in maintaining the health of heart, bones, and teeth. The thyroid gland, for example, uses only a miniscule quantity of iodine (only millionths of a gram) to produce the hormone thyroxine. The rapidly growing field of *bioinorganic chemistry* explores how minerals function within living systems.

Of the more than 100 known elements, only 22 are believed essential to human life. Essential minerals are divided into two categories for convenience: **macrominerals,** or major minerals, and **trace minerals.**

CAUTION

IMPORTANT: Never taste any materials found in the chemistry laboratory.

Iodized salt is an important source of the trace mineral iodine.

As the name suggests, your body contains rather large quantities, at least five grams, of each of the seven macrominerals. Carbon, hydrogen, nitrogen, and oxygen are so widely present both in living systems and in the environment that they are not included in lists of essential minerals. Trace minerals are present in relatively small quantities, less than five grams in an average adult; however, in the diet they are just as important as macrominerals. Any essential mineral—macro or trace—can become a limiting factor (limiting reactant) if it is not present in sufficient quantity.

The minerals, their dietary sources, functions, and deficiency conditions are listed in Table 12. In addition to these, several other minerals—including but not limited to arsenic (As), cadmium (Cd), and tin (Sn)—are known to be needed by other animals. These and perhaps other trace minerals may be essential to human life, as well. The suggestion that the widely known poison arsenic might be an essential mineral probably

Table 12	Minerals	
Mineral	**Source**	**Deficiency Condition**
Macrominerals		
Calcium (Ca)	Canned fish, milk, dairy products	Rickets in children; osteomalacia and osteoporosis in adults
Chlorine (Cl)	Meats, salt-processed foods, table salt	—
Magnesium (Mg)	Seafoods, cereal grains, nuts, dark green vegetables, cocoa	Heart failure due to spasms
Phosphorus (P)	Animal proteins	—
Potassium (K)	Orange juice, bananas, dried fruits, potatoes	Poor nerve function, irregular heartbeat, sudden death during fasting
Sodium (Na)	Meats, salt-processed foods, table salt	Headache, weakness, thirst, poor memory, appetite loss
Sulfur (S)	Proteins	—
Trace minerals		
Chromium (Cr)	Liver, animal and plant tissue	Loss of insulin efficiency with age
Cobalt (Co)	Liver, animal proteins	Anemia
Copper (Cu)	Liver, kidney, egg yolk, whole grains	—
Fluorine (F)	Seafoods, fluoridated drinking water	Dental decay
Iodine (I)	Seafoods, iodized salts	Goiter
Iron (Fe)	Liver, meats, green leafy vegetables, whole grains	Anemia; tiredness and apathy
Manganese (Mn)	Liver, kidney, wheat germ, legumes, nuts, tea	Weight loss, dermatitis
Molybdenum (Mo)	Liver, kidney, whole grains, legumes, leafy vegetables	—
Nickel (Ni)	Seafoods, grains, seeds, beans, vegetables	Cirrhosis of liver, kidney failure, stress
Selenium (Se)	Liver, organ meats, grains, vegetables	Kashan disease (a heart disease found in China)
Zinc (Zn)	Liver, shellfish, meats, wheat germ, legumes	Anemia, stunted growth

surprises you. In fact, however, it is not unusual for substances to be beneficial in low doses, but toxic in higher doses.

Table 13 summarizes the proposed Recommended Dietary Allowances (RDAs) for several macrominerals and trace elements. Use the values given in Table 13 to answer the questions in the following *Your Turn*.

Table 13	RDAs For Selected Minerals					
Sex and Age	Calcium (mg)	Phosphorus (mg)	Magnesium (mg)	Iron (mg)	Zinc (mg)	Iodine (μg)
Males						
11–14	1200	1200	270	12	15	150
15–18	1200	1200	400	12	15	150
19–24	1200	1200	350	10	15	150
25+	800	800	350	10	15	150
Females						
11–14	1200	1200	280	15	12	150
15–18	1200	1200	300	15	12	150
19–24	1200	1200	280	15	12	150
25–50	800	800	280	15	12	150
51+	800	800	280	10	12	150

Source: Food and Nutrition Board, National Academy of Sciences—National Research Council, Recommended Dietary Allowances, Revised 1989.

YOUR TURN

Minerals in the Diet

1. A slice of whole-wheat bread contains 0.8 mg iron. How many slices of whole-wheat bread would you have to eat to meet your daily iron allowance?

2. One cup of whole milk contains 288 mg calcium. How much milk would you have to drink each day to meet your RDA for this mineral?

3. One medium pancake contains about 27 mg calcium and 0.4 mg iron.
   a. Does a pancake provide a greater percent of your RDA for calcium or for iron?
   b. How did you decide?

4. a. How many grams of calcium or phosphorus do you need each year?
   b. Why is this figure so much higher than the RDAs for other essential minerals listed? (*Hint:* Consider their uses.)
   c. List several good sources for each of these two minerals.
   d. Predict the health consequences of a deficiency of these minerals.
   e. Would any particular age group or sex be especially affected by these consequences?

► *"Iodized salt" refers to these products.*

5. Most grocery store table salt has a small amount of potassium iodide (KI) added to the main ingredient, sodium chloride (NaCl).

   a. Why do you think KI is added to the table salt?

   b. If you follow the advice of many heart specialists and do not add salt to your food, what other kinds of food could you use as sources of iodine?

How is the mineral content of a particular food determined? You will explore one method in the following laboratory activity, investigating the relative levels of iron found in foods such as broccoli, spinach, and raisins.

D.4 LABORATORY ACTIVITY: IRON IN FOODS

GETTING READY

Iron in foods is in the form of iron(II) or iron(III) ions. Iron(II) is more readily absorbed from the intestine than is iron(III). Thus, supplements for the treatment of *iron-deficiency anemia* are almost always iron(II) compounds. The most common ingredient is iron(II) sulfate, $FeSO_4$.

This laboratory activity is based on a very sensitive test for the presence of iron ions in solution. In the procedure, all iron in the sample is converted to iron(III) ions. The colorless thiocyanate ion, SCN^-, reacts with iron(III) to form an intensely red ion:

$$\begin{array}{cccc}
Fe^{3+} & + & SCN^- & \rightarrow & Fe(SCN)^{2+} \\
\text{iron(III)} & & \text{thiocyanate} & & \text{iron(III) thiocyanate} \\
\text{ion} & & \text{ion} & & \text{ion (red color)}
\end{array}$$

The intense red color of the iron(III) thiocyanate ion is directly related to the concentration of iron(III) originally present in the solution. This test is so sensitive that iron concentrations so small they must be expressed as parts per million (ppm) produce a noticeable reddish color! Color standards with known concentrations of iron(III) will be available in the laboratory. You will estimate iron (III) concentrations by comparing the colors of test solutions with the color standards.

The samples will be heated to a high temperature to remove the organic portions of the foods, which would interfere with the tests. The organic compounds burn and are driven off as water vapor and carbon dioxide gas. Minerals present (such as iron) remain in the burned ash and are dissolved by an acid solution.

PROCEDURE

1. Record the names of the two food samples you have been assigned. Weigh a 2.5-g sample of each food into a separate porcelain crucible.

2. Place one food-containing crucible on a clay triangle supported by a ring stand. Heat the uncovered crucible with a hot burner flame.

Apparatus for burning food sample.

3. Continue heating until the food sample has turned to a grayish white ash. Do not allow the ash to be blown from the crucible.

4. Remove the burner and allow the crucible to cool on the clay triangle.

5. Begin heating the other food-containing crucible on a second clay triangle/ring stand setup. Continue heating until the sample has turned to ash.

6. Remove the burner and allow the second sample and crucible to cool on the clay triangle.

7. When the first sample has cooled, transfer the entire ash residue to a 50-mL (or larger) beaker. Add 10 mL of 2 M hydrochloric acid, HCl, to the beaker and stir vigorously for one minute. Then add 5 mL of distilled water.

8. Prepare a filtration apparatus, including a ring stand, funnel support, and funnel. Place a piece of filter paper in the funnel and position a test tube under the funnel to collect the filtrate.

9. Pour the mixture from the beaker into the filtration apparatus and collect 5 mL of the filtrate in a test tube. Discard the residue on the filter paper, as well as the remaining solution.

10. Add 5 mL of 0.1 M potassium thiocyanate solution, KSCN, to the test tube containing the filtrate. Seal with the stopper. Gently invert the tube once to mix the solution.

11. Compare the resulting red color to the color standards. It might be helpful to hold white paper behind the tubes while you make the comparison.

12. Record the approximate iron concentration present in your sample, based on your comparisons with the color standards.

13. Repeat Steps 7 through 12 with your other food sample.

14. Obtain iron results from other laboratory teams. Record the food product names and estimated iron levels in the resulting solutions.

15. Wash your hands thoroughly before leaving the laboratory.

▶ The "2 M" (pronounced "two molar") HCl solution contains two moles of HCl in each liter of the solution. Molar concentration—also called molarity—is a convenient way to express the composition of solutions. **Caution:** Be careful not to splash the acid on yourself or others.

▶ "0.1 M potassium thiocyanate" means that each liter of this solution contains 0.1 mol KSCN. A 5-mL sample of this solution thus contains 0.0005 mol KSCN, or 5/1000 as much as the liter of solution.

QUESTIONS

1. The color standards allowed you to estimate the percent of iron in the solutions prepared from your food samples. Do these percentages also apply to the original 2.5-g food samples? Why?

2. Name the foods that are (a) the best and (b) the poorest sources of iron among the food samples you tested.

3. What elements besides iron might be found in the ash?

D.5 FOOD ADDITIVES

Small amounts of vitamins and minerals occur naturally in food. Some foods, especially processed foods such as packaged cookies or frozen entrees, also contain small amounts of *food additives*—substances added to increase the food's nutritive value, storage life, visual appeal, or ease of production.

▶ *Can you guess the
identity of the food product
with this label?*

A typical food label might give the following:

Sugar, bleached flour (enriched with niacin, iron, thiamine, and
riboflavin), semisweet chocolate, animal and/or vegetable shortening,
dextrose, wheat starch, monocalcium phosphate, baking soda, egg
white, modified corn starch, salt, nonfat milk, cellulose gum, soy
lecithin, xanthan gum, mono- and diglycerides, BHA, BHT.

Quite a collection of ingredients! You probably recognize the major
ingredients such as sugar, flour, shortening, and baking soda, and some additives such as vitamins (thiamine, riboflavin) and minerals (iron and monocalcium phosphate). But you probably do not recognize the additives xanthan
gum (an emulsifier) or BHT (butylated hydroxytoluene, a preservative).

Food additives are used for a variety of reasons. Some help preserve
foods or increase their nutritive value. Other additives produce a specific
consistency, taste, or appearance. Intentional food additives have been used
since ancient times. For example, salt has been used for centuries to preserve foods, and spices helped in hot weather to disguise the flavor of food
that was no longer fresh. As people have moved greater distances from
farms, some manufacturers have relied more on food-preservation additives.
Table 14 summarizes some of the major food additive categories. The structural formulas of a few common ones are shown in Figure 15 (page 281).

Table 14	Food Additives	
Additive Type	**Purpose**	**Examples**
Anticaking agents	Keep foods free-flowing	Sodium ferrocyanide
Antioxidants	Prevent fat rancidity	BHA and BHT
Bleaches	Whiten foods (flour, cheese); hasten cheese maturing	Sulfur dioxide, SO_2
Coloring agents	Increase visual appeal	Carotene (natural yellow color); synthetic dyes
Emulsifiers	Improve texture, smoothness; stabilize oil–water mixtures	Cellulose gums, dextrins
Flavoring agents	Add or enhance flavor	Salt, monosodium glutamate (MSG), spices
Humectants	Retain moisture	Glycerin
Leavening agents	Give foods light texture	Baking powder, baking soda
Nutrients	Improve nutritive value	Vitamins, minerals
Preservatives and antimycotic agents (growth inhibitors)	Prevent spoilage, microbial growth	Propionic acid, sorbic acid, benzoic acid, salt
Sweeteners	Impart sweet taste	Sugar (sucrose), dextrin, fructose, aspartame, sorbitol, mannitol

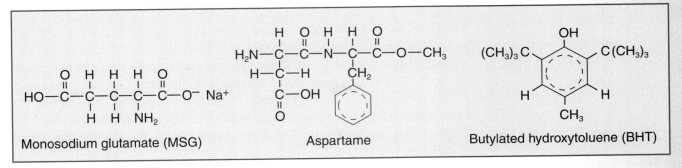

Figure 15 Common food additives.

Why is it not necessarily a good idea to eliminate all additives and preservatives?

Coloring agents are commercially important food additives, especially in processed foods. The following laboratory activity analyzes several commercial food dyes.

D.6 LABORATORY ACTIVITY: FOOD COLORING ANALYSIS

GETTING READY

Many candies contain artificial coloring agents to enhance the product's visual appeal. Colorless candies would be pretty dull! In this laboratory activity, you will analyze the food dyes in two commercial candies, M&M's® and Skittles®, and compare them with the dyes in food coloring.

The separation and identification of the food dyes will be done by *paper chromatography*. This technique uses a solvent (the moving phase) and paper (the stationary phase). Paper chromatography relies on the relative differences in attraction of the dyes for the solvent compared with the paper. The dye that is more attracted to the paper will come out of the solvent and separate out onto the paper, leaving a colored spot, earlier than the dyes that are less attracted to the paper (more attracted to the solvent); they separate out onto the paper later. Individual spots for each dye will appear on the paper. The ratio of the distance from a "starting line" to the spot left by a particular color in a dye compared with the distance moved by the solvent up the paper is known as the R_f value for that color (Figure 16):

$$R_f = \frac{\text{distance to a spot}}{\text{distance moved by solvent}}$$

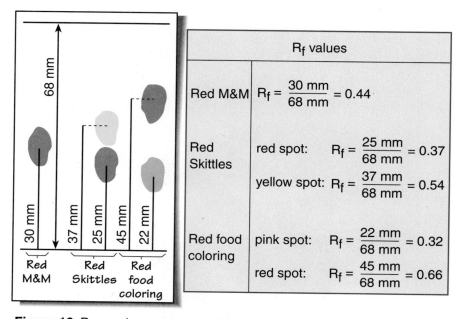

Figure 16 Paper chromatogram with R_f values.

Prepare a data table like the one below in your laboratory notebook.

Data Table					
Sample	Color of initial sample	Dyes observed	Distance to dye spot, cm	Distance solvent moved, cm	R_f value
M&M's					
Skittles					
Food coloring					

PROCEDURE

1. Obtain one M&M, one Skittles, and one food coloring sample, all of the *same* color.

2. Put each candy into a separate well of a wellplate, noting which candy is in each well. Add 5–10 drops of sulfuric acid solution to each of these wells. Stir the mixture in the well with a toothpick, using a clean toothpick for each well, until the color completely

dissolves from the candy. Add 5–10 drops of the food coloring to a different well in the wellplate.

3. Obtain a strip of chromatography paper, handling it only by the edges. With a pencil (*not* a pen), draw a straight line 2 cm from the bottom of the paper. Label the strip as shown in Figure 17.

4. Spot the dyes on the line—the spots should not be large. Use a separate micropipet for each of the three different samples. Place a spot of the M&M's color solution from the micropipet on the line as indicated in Figure 17. Allow the spot to dry and then apply a second spot of the *same* sample on top of the first spot. Using the same two-spot technique, apply spots of the Skittles color solution and the food coloring to the appropriate places on the pencil line as indicated in Figure 17.

5. Pour solvent into the chromatography vessel to the 1-cm line.

6. Lower the spotted chromatography paper into the chromatography vessel until the bottom of the paper rests evenly in the solvent, but so that the spots are above the surface of the solvent. Carefully hold the top of the paper against the inside neck of the vessel and put in the rubber stopper.

7. Allow the solvent to rise past the spots and up the paper until it is about 3 cm from the neck of the vessel. Remove the paper from the vessel and use a pencil to mark the farthest point of travel by the solvent. Dry the paper.

8. Record the colors you observe for each dye sample.

9. Measure the distance (cm) from the initial pencil line where you placed the spots to the center of each dye spot. Record this distance.

10. Calculate the R_f of each dye spot you observed in your samples. Record these values.

Figure 17 Spotting a chromatogram on chromatography paper.

QUESTIONS

1. Why were you instructed to use a pencil rather than a pen to mark the lines on the paper?

2. Which of your dyes, if any, created single spots rather than more than one spot?

3. Which of your dyes has the greatest attraction for the paper?

4. Which of your dyes has the greatest attraction for the solvent?

5. Based on your data:
   a. Do any of your samples have the same dye(s) in them?
   b. What did you use as the basis for answering Question 5a?

6. The candy and food color packages list the dyes they contain. Compare this information with your experimental results.
   a. What similarities exist?
   b. What differences exist?
   c. If differences exist, what are some reasons for them?

Decision making in the grocery store.

FOOD LEGISLATION

Processed food may contain contaminants that indirectly find their way into foods. Typical contaminants are pesticides, mold, antibiotics used to treat animals, insect parts, food packaging materials, or dirt from the processing plant. When we buy food at grocery stores and restaurants, we assume it is completely safe to eat. For the most part, this is true. Nonetheless, in recent years many food additives and contaminants have been suspected of posing hazards to human health.

Food quality in the United States is protected by law. The basis of the law is the federal Food, Drug, and Cosmetic Act of 1938, through which Congress authorized the FDA to monitor the safety, purity, and wholesomeness of food. This act has been amended in response to concerns over pesticide residues, artificial colors and food dyes, and potential *carcinogens* (cancer-causing agents) in foods. Manufacturers must complete tests and provide extensive information on the safety of any proposed food product or additive. A new product must be approved before it is put on the market.

According to the amended federal Food, Drug, and Cosmetic Act, ingredients that manufacturers had been using for a long time prior to the act and that were known not to be hazardous were exempted from testing. These substances, instead of being legally defined as additives, make up the "Generally Recognized as Safe" (GRAS) list. The GRAS list is periodically reviewed in light of new findings and includes items such as salt, sugar, vinegar, vitamins, and some minerals.

Because of the Delaney Clause (added to the act in 1958), every proposed food additive must be tested on laboratory animals (usually mice). The Delaney Clause says that "no additive shall be deemed to be safe if it is found to induce cancer when it is ingested by man or animal." Thus, approval of a proposed additive is denied if it causes cancer in the animals. Some people believe that the Delaney Clause is too strict: they argue that amounts comparable to those that cause cancer in test animals are often vastly more than could ever be encountered in a human diet. As written, the Delaney Clause causes some genuine chemical quandaries. Sodium nitrite ($NaNO_2$), for instance, is a color stabilizer and spoilage inhibitor in many cured meats, such as hot dogs and bologna. Nitrites are particularly effective in inhibiting the growth of the bacterium *Clostridium botulinum,* which produces botulin toxin, cause of the often-fatal disease known as botulism. Thus, nitrites are important for healthful food processing. On the other hand, sodium nitrite may be a carcinogen. In the stomach, nitrites are converted to nitrous acid:

$$NaNO_2 \quad + \quad HCl \quad \rightarrow \quad HNO_2(aq) \quad + \quad NaCl(aq)$$

| sodium | hydrochloric | nitrous | sodium |
| nitrite | acid | acid | chloride |

Nitrous acid can then react with compounds formed during protein digestion, and some of the products formed are very potent carcinogens.

$$HNO_2 + R{-}\underset{\underset{R}{|}}{N}{-}H \longrightarrow R{-}\underset{\underset{R}{|}}{N}{-}N{=}O + H_2O$$

CHEMQUANDARY

NITRITE ADDITIVES

To decide whether nitrites should be used to preserve meats, we can weigh the following risks and benefits:

	Using Nitrites	Eliminating Nitrites
Benefit	Minimizes botulin toxin formation	Removes risk of possible carcinogens
Risk	Increases possibility of carcinogen formation in body	Increases risk of botulin toxin formation

More information and greater knowledge might help us arrive at the best decision. If the choice of nitrites or no nitrites were up to you, what questions would you want answered first?

ARTIFICIAL SWEETENERS

We are a diet-conscious society, and sweeteners have attracted considerable attention. The sweetener aspartame (available under the trade names NutraSweet® and Equal®) is an ingredient in over 160 diet beverages and 3,000 other food products, and is probably the most thoroughly tested FDA-approved food additive. Testing a compound such as aspartame usually takes millions of dollars and years of research.

Aspartame is a chemical combination of two natural amino acids, aspartic acid and phenylalanine. One gram of aspartame contains roughly the same food energy as one gram of table sugar (24 Cal), but aspartame tastes 200 times sweeter. Therefore, much smaller quantities of aspartame are needed to sweeten a product, and so aspartame is used as a "low-Calorie" alternative to sugar that is also safe for diabetics. Annually, thousands of tons of aspartame are used in the United States to sweeten various soft drinks and cold foods. Because aspartame decomposes when heated, it cannot be used directly for cooking. Recently, a version of aspartame has been developed that can be used for baking. Aspartame granules are covered with a protective coating that prevents aspartame from breaking down at high baking temperatures. The coating melts when the baking is done, releasing the sweetener.

Scientific evidence continues to be collected, but no serious alarm regarding aspartame has been issued for the general population. However, aspartame poses a health hazard to phenylketonurics, individuals who cannot properly metabolize phenylalanine. An FDA-required warning "Phenylketonurics: Contains Phenylalanine" on the label of foods containing aspartame points out the potential risk to such people.

▶ *Aspartame was approved by the Food and Drug Administration in 1981 and was introduced commercially in 1983.*

▶ *The structural formula of aspartame is given in Figure 15 (page 281). Aspartame is a dipeptide (page 265).*

By law, processed foods must be labeled with their ingredients listed in decreasing order of quantity; that is, the most abundant ones by weight come first. The original 1938 legislation assigned *standards of identity* to approximately 300 common foods, such as mayonnaise. The law listed the ingredients of these foods, and any food containing them did not have to be labeled with the ingredients again. Food labeling standards changed in 1994, now requiring manufacturers to list the ingredients of all foods, including mayonnaise, in decreasing order of weight.

Persons with specific conditions may have to avoid certain foods and food additives. For instance, diabetics must restrict their intake of sugar. Persons with high blood pressure (hypertension) must avoid salt. Some persons are allergic to certain substances. If such restrictions apply to you, it is essential that you always read food labels. Sometimes a new ingredient will be used in a food you have used safely in the past.

Being aware of what we eat is a wise habit for everyone. In the following activity, you will investigate food labels and consider the additives.

D.7 YOU DECIDE: FOOD ADDITIVE SURVEY

1. As homework, collect the labels from 5 or 6 packaged foods. Select no more than two of the same type of food—for example, no more than two breakfast cereals or two canned soups. Bring your set of labels to school.

2. From the ingredient listings on the packages, select six additives that are not naturally found in foods.

3. Complete a summary table with the following format:

   a. List the six additives in a vertical column along the left side of the page.

   b. Make three vertical columns for each additive with the following headings: "Food Product in Which Found", "Purpose of Additive" (if known), and "Other Information" (regarding the additive).

4. Use Table 14 (page 280), to review the purpose of specific food additives. Then answer this question for each food in your summary table: Which additives do you think should be included in that particular food? Provide reasons.

5. a. Is it possible to purchase the same food product without some of the additives you found?

   b. If so, where?

   c. Is there a difference in price?

   d. If so, which is more expensive?

6. What alternatives can you propose for food additives that are used to prevent spoilage?

CHEMISTRY AT WORK

Keeping Food Supplies Safe for Consumers

State inspectors widen area for corn feed testing

Associated Press

RICHMOND — Inspectors from the Virginia Department of Agriculture expanded their search Thursday for a toxin in corn feed that has killed 18 horses.

Inspectors began taking samples of feed from manufacturers and dealers who mix their own feed in the area south of Interstate 64 and east of Interstate 95. The samples will be tested for Fumonisin B1, a mycotoxin that causes leukoencephalomalacia, also called molfy corn poisoning.

NEWS BRIEFS

SUFFOLK

19TH HORSE DIES: A horse from Suffolk has died and is suspected to be the 19th known victim of "moldy corn poisoning" in Virginia. The horse's owners, who were not identified, could not recall where they purchased the corn,

These actual newspaper headlines illustrate the deadly effects that certain microscopic toxins can have. Here, the consumers of the food were horses, but similar toxins can endanger human consumers as well.

Mary Trucksess is a chemist working to understand and prevent food toxins. She is Research Chemist and Chief of the Bioanalytical Chemistry Branch in the Division of Natural Food Products, Office of Plant and Dairy Food and Beverages. Mary and her fellow chemists at the FDA help ensure the quality of food and drugs (prescription and nonprescription) to protect people from harmful contaminants. Currently Mary is investigating fumonisins—a toxin found mainly in corn, caused by a common soil fungus—which can cause horses to develop a fatal disease of the brain. The FDA is investigating fumonisins to make certain that these toxins do not pose a threat to humans.

Mary's office analyzes other foods, such as peanuts and peanut butter, for aflatoxins. Aflatoxins are naturally occurring toxins that form when certain molds (*Aspergillus flavus* or *Aspergillus parasiticus*) begin to grow on peanuts. If not controlled, these and other toxins could threaten the health of consumers.

Mary also manages and directs other chemists as they conduct research. The chemists in her branch often discuss with her any problems they encounter in their research. She reviews progress reports from the chemists, and recommends different approaches as needed.

For several years, Mary—who holds undergraduate and doctoral degrees in chemistry—has supervised the work of high school honor students who have been selected to intern at the FDA. Mary develops specific projects for the students and trains them in laboratory procedures and data interpretation. Currently, for example, Mary is supervising a high school student who is working on a project to look for the presence of fumonisin in corn tortillas.

? PART D: SUMMARY QUESTIONS

1. Compare water-soluble and fat-soluble vitamins in terms of
   a. daily requirements,
   b. toxicity concerns regarding megadoses, and
   c. cooking precautions.

2. Defend or refute this statement: *Macrominerals are more important than trace minerals.* As part of your answer, give one example of each, including its function and dietary sources.

3. Based on serving sizes indicated in the Appendix (page 585), which food group—fruits and vegetables, dairy products, meats, or grains—is the best source of
   a. calcium?   c. vitamin C?
   b. iron?   d. vitamin A?

4. a. How many cups of orange juice must a 17-year-old drink to meet the RDA for vitamin C? (Consult the Appendix.)

   b. How many cups of whole milk would be needed to meet this RDA?

5. Discuss this statement with other students: *All food additives represent unnatural additions of chemicals to our food and should be prohibited.*

6. Consider the additives listed in Table 14 (page 280).
   a. Which additive type do you think is the least essential? Why?
   b. Which are the most essential? Why?

7. Why is it wise to examine both risks and benefits when considering whether to ban certain food additives?

8. a. Describe the Delaney Clause.
   b. Give one argument in its favor.
   c. Give one argument against the Delaney Clause.

E PUTTING IT ALL TOGETHER: WHICH FOOD IS BETTER?

We began this unit with several questions: Why do we eat? Why is the warning "Don't put chemicals in your mouth" incorrect? These questions have been answered during the study of the unit. You have learned the nutritional importance of carbohydrates, fats, and proteins. You have also examined the relationships between the major nutrients and Calories in various foods, and the importance of vitamins and minerals.

The unit also asks the question "How can we become informed decision makers about food choices and diets?" One way to make informed choices about the food you eat is to read food labels carefully. The labels contain dietary information that allow you to compare several brands of cereals or other food products.

In the following sections you will work with food label data in order to assess the nutritional value of a common dinner.

E.1 FOOD LABELS

The federal government enacted significant changes in food label requirements in 1994. The new labels use more realistic serving sizes and provide more extensive facts about particular nutrients in the labeled food. The label must give at least the percent of Daily Values for total fat, saturated fat, total carbohydrate, and dietary fiber, vitamins A and C, and calcium and iron. Quantities of nutrients in foods are given in terms of mass (grams, milligrams) or as percentages of recommended Daily Values. Keep in mind that the total amount of food you eat in a day is important, not just the percentage of Daily Values in one food or in one meal.

Daily Values are based on current nutrition recommendations for a 2,000-Calorie diet. Near the bottom of the food label is a table showing how much of each of six categories such a diet includes (Figure 18). Each of these amounts represents 100% of the Daily Value for that nutrient.

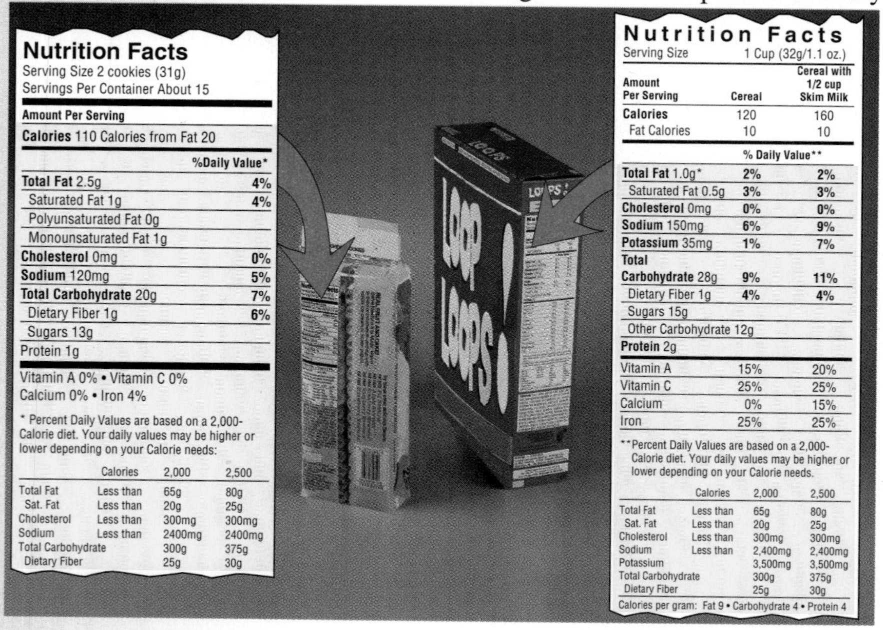

Figure 18 Percent Daily Values: cookies; a cereal.

Consider a food label that says that one serving of the food provides 220 mg of sodium, representing 9% of the Daily Value for sodium. The percent of the Daily Value was calculated by comparing the milligrams of sodium present in one serving with the 2,400 mg of sodium that is the daily recommended maximum (Figure 18). The calculation is just like finding any percentage value:

$$\frac{220 \text{ mg Na}}{2,400 \text{ mg Na}} \times 100 = 9\%$$

Percentages of Daily Values for other nutrients can be calculated in the same way. The number of grams of protein per day can be estimated by multiplying your body weight in kilograms by 0.8, and so mass in kg $\times$ 0.8 = mass of protein (g) needed daily.

Your Daily Values might be higher or lower than the recommended value based on 2,000 Calories, depending on how many Calories you need daily. (See the upcoming *Your Turn*.) That is why many labels, like that in Figure 18, also list maximum amounts for a 2,500-Calorie diet as well. Suppose you know that your level of activity requires 2,200 Calories per day. Your Daily Value for each particular nutrient will be higher than that of someone on a 2,000-Calorie diet by a factor of 2,200/2,000 = 1.10. In our sodium example, your maximum sodium intake daily would be increased from the 2,400 mg recommended for a 2,000-Calorie diet by the 1.10 factor:

Maximum Daily Value of Na = 2,400 mg Na $\times$ 1.10 = 2,640 mg Na

Using this quantity, we can now calculate what percent of your Daily Value for sodium is represented by the 220 mg of sodium in the food example above: (220/2,640) $\times$ 100 = 8% of your Daily Value for sodium.

YOUR TURN

Finding Your Daily Values

To determine your Daily Values, you first need to estimate how many Calories you need daily. You can estimate your daily energy requirement (Calories) using the relationships of 40 Cal/kg of body mass for women, and 45 Cal/kg of body mass for men, where 1 kg = 2.2 lb. For a 165-lb male, for example:

First, converting from pounds to kilograms:

$$165 \text{ lb} \times 1 \text{ kg} / 2.2 \text{ lb} = 75 \text{ kg}$$

Then, finding Calories:

$$75 \text{ kg} \times 45 \text{ Cal/kg} = \text{about } 3,375 \text{ Cal}$$

Now use your required Calories/day and the nutritional information in Figure 18 to answer the following questions about your daily diet:

1. What mass (grams) of total carbohydrate do you need?
2. a. What mass (grams) of total fat should you eat in a day?

Chicken Delight	
Nutritional Information, 1 serving:	
Calories 370	
Calories from fat 70	
Percent Daily Value*	
Total fat 8 g	12 %
Saturated fat 3 g	15 %
Cholesterol 45 mg	15 %
Sodium 470 mg	20 %
Total Carbohydrates 53 g	18 %
Dietary fiber 6 g	23 %
Sugars 35 g	
Protein 23 g	
Vitamin A	30 %
Vitamin C	20 %
Calcium	10 %
Iron	15 %

* Percent Daily Values are based on a 2,000-Calorie diet. Your Daily Values may be higher or lower depending on your Calorie needs.

Chicken Supreme	
Nutritional Information, 1 serving:	
Calories 290	
Calories from fat 130	
Percent Daily Value*	
Total fat 15 g	22 %
Saturated fat 4 g	17 %
Cholesterol 50 mg	17 %
Sodium 900 mg	38 %
Total Carbohydrates 27 g	9 %
Dietary fiber 3 g	12 %
Sugars 3 g	
Protein 14 g	
Vitamin A	6 %
Vitamin C	100 %
Calcium	67 %
Iron	10 %

* Percent Daily Values are based on a 2,000-Calorie diet. Your Daily Values may be higher or lower depending on your Calorie needs.

b. What maximum mass (grams) of saturated fat should you eat in a day?

3. A 113-g serving of a particular food furnishes 23 g of total carbohydrate. Calculate the percent of *your* Daily Value of total carbohydrate that this serving provides.

4. The food in Question 3 also furnishes 10% of the Daily Value for calcium. How many grams of this food would *you* need to eat to provide the same percentage of calcium?

5. Create your own nutrition facts reference list by completing the following table.

Daily Level	
Total fat	About _____ grams or less
Saturated fat	About _____ grams or less
Total carbohydrate	About _____ grams or less
Dietary fiber	About _____ grams or less
Protein (Your mass in kg × 0.8)	About _____ grams or less

E.2 YOU DECIDE: WHAT KIND OF FOOD AM I?

It is early evening, and you decide to pop a quick dinner into the microwave. You check the refrigerator and find that there are only two chicken dinners left: "Chicken Supreme" and "Chicken Delight." (These are fictitious product names, but the following nutritional data are factual.) Which one to choose? Perhaps the nutritional facts will help you to decide.

You may find the values from Question 5 of the previous *Your Turn* to be helpful in answering the following questions:

1. You decide to eat the "Chicken Delight."

   a. In order to avoid exceeding your daily maximum allowance of total fat, how many grams of fat could you have eaten earlier in the day prior to eating this meal?

   b. Suppose it turns out that you consumed 45 g of fat in the meals and snacks before you ate dinner. What percent of your Daily Value for total fat did you eat today, including dinner?

2. A friend comes over, checks the two labels, and tells you that you should have eaten the "Chicken Supreme."

   a. On a nutritional basis, justify to your friend your eating the "Chicken Delight."

   b. On what nutritional basis can your friend claim to be right?

3. Based on your daily energy demand, what percent of *your* Daily Value for iron did your dinner provide?

4. What percent of your protein Daily Value did the "Chicken Delight" provide?

5. One option your friend might consider to reduce fat consumption in the meal is to eat a smaller portion of Chicken Supreme, which has 15 g of total fat per normal serving. What portion of Chicken Supreme should you have eaten to get only 8 g of total fat?

6. Advertisements for vitamin supplements frequently propose that their product will "boost your energy." Make a case in favor of or opposed to this claim.

7. Another friend claims to eat only natural foods because they contain no chemicals. What arguments could you present to show that your friend is wrong?

E.3 LOOKING BACK AND LOOKING AHEAD

As you conclude this *ChemCom* unit, we invite you to take stock of your progress in this chemistry course. You have considered many fundamental aspects of chemistry and chemical resources, as well as personal and social issues surrounding them. In the remaining *ChemCom* units, you will extend your understanding of chemical principles by exploring other important resources and key issues associated with these resources.

Our general goal remains the same—to help you enter into enlightened public discussion of science-related societal issues. Chemistry and you *can* make a difference.

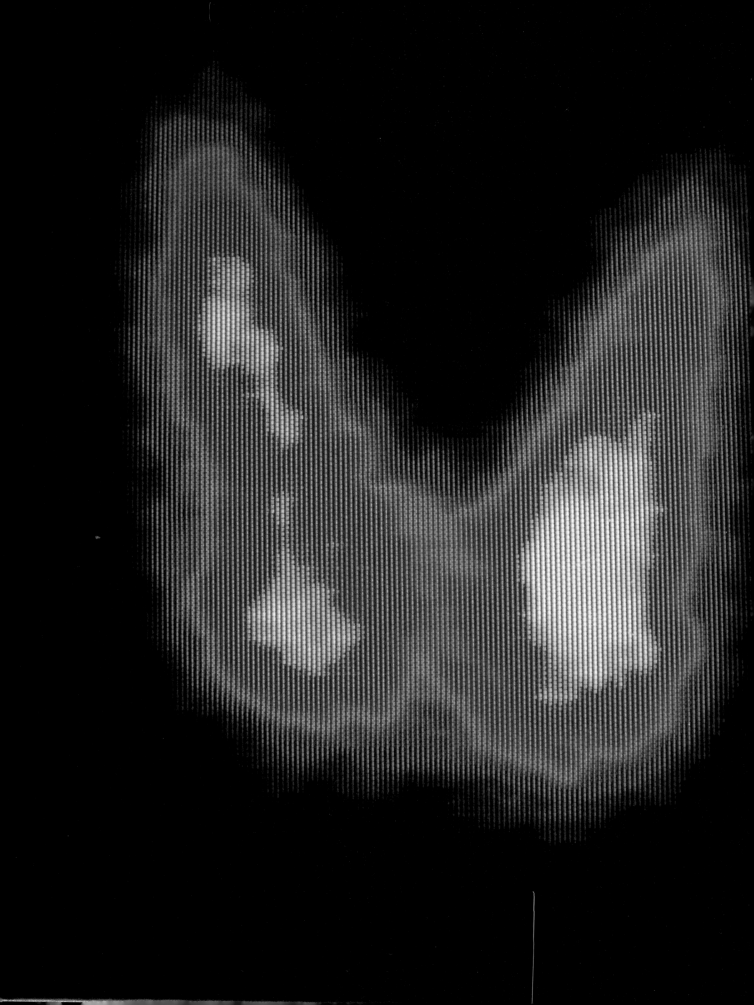

Nuclear Chemistry in Our World

- ▶ Is elimination of all radiation possible? Or even desirable?

- ▶ How do radioactive atoms differ from nonradioactive ones?

- ▶ In what beneficial ways are radioisotopes used?

- ▶ Do the risks of nuclear energy and radioactive emissions outweigh their benefits?

- ▶ As fossil fuels become scarce, what role should nuclear energy play?

So far in your *ChemCom* study, all of the chemistry discussed has involved *chemical changes,* due to rearrangements of outer electrons. We now turn to changes of a much different sort: those associated with the nuclei of atoms. In this unit, we will look closely at a number of processes associated with nuclear changes—nuclear radiation, radioactivity, and nuclear energy—and the science and societal implications behind them.

Nuclear energy itself is not new. Our sun, like all stars, has always run on nuclear power. What *is* new about nuclear energy is how humans use it. When scientists unlocked the secrets of the nucleus, they released the universe's strongest known force and made possible enormous technological benefits. Nuclear science has made important contributions to industry, to biological research, to our energy needs, and especially to medicine. But the production and use of nuclear energy, like all other applications of technology, do involve some risk of accidents or misuse.

Almost every application of nuclear science can have either positive or negative aspects. Are the risks of nuclear technology worth its benefits? Some uses of nuclear technology create greater risks than others; some offer greater benefits than others. Information in this unit will help you as a voting citizen make better decisions on which uses of nuclear technology are worth their associated risks. For example, nuclear radiation can be used to treat cancer, but it can also cause cancer.

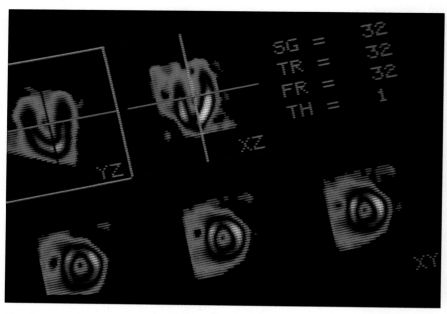

Thallium scan of heart muscles, used to investigate blood flow.

A RADIATION, ENERGY, AND ATOMS

Irradiation means exposing a sample to radiation. However, nuclear radiation is not the only form of radiation. Other forms of radiation, such as ultraviolet and infrared, come from the sun; microwave radiation is the kind that pops popcorn and heats up leftovers. Thus, it is important to recognize that "irradiated" does not mean irradiation by nuclear radiation only.

To many people, anything associated with the term *nuclear* and its companion term, *nuclear radiation*, is cause for alarm, even panic.

Consider the following headlines:

"Is Irradiated Food Safer?"
Ann Arbor (**MI**) *News,* **3/1/92**

"Food Irradiation Makes Florida Debut Amid Persistent Protests Over Safety"
Washington Post, **1/12/92**

Radiation from the sun provides the energy necessary for fruits and vegetables to grow and ripen. In a sense, these foods are irradiated. So, why the concern about irradiating foods, especially if it can make them safer for consumers?

Food irradiation does not mean using solar radiation for growing foods. Rather, it has to do with irradiating various commercial foods, not just fruits and vegetables, in order to reduce their rate of spoiling. The radiation used for this is not sunlight, but a highly energetic form of radiation called **gamma rays.** The gamma rays come from the natural nuclear decomposition of particular types of cobalt or cesium atoms. Gamma radiation does not cause the food to become radioactive, but it destroys bacteria, molds, and yeasts that cause foods to spoil.

In 1963, the Food and Drug Administration (FDA) approved irradiation as a method for preserving some foods; poultry was added to the list in 1990. Worldwide, the process has been endorsed by the United Nations World Health Organization (WHO) and the U.N. Food and Agriculture Organization (FAO). Food irradiation is used in more than 30 countries for over 40 different foods, and is relatively common in European countries, Canada, and Mexico. In the United States and some other countries, foods preserved by irradiation must carry labels with the international logo for irradiated food.

▶ *Improvements in food preservation can lead to larger food supplies—a particular benefit for developing nations.*

Food irradiation logo

NON - IRRADIATED -

IRRADIATED - (0.2 M RAD)

STRAWBERRIES -
15 DAYS STORAGE 38°F (4°C)

An illustration of the effectiveness of gamma radiation in the preservation of strawberries. Those on the right were irradiated. Those on the left were not. The photograph, taken after 15 days of storage, shows that the irradiated berries remained firm, fresh, and free of mold.

A.1 YOU DECIDE: SHOULD YOU EAT IRRADIATED FOOD?

Food irradiation has created controversy in the United States. In 1992, irradiated strawberries went on the market in this country, the first widely sold food to be preserved in this way. A few states have either banned or declared a moratorium on selling irradiated foods.

1. Assume that you go to your local supermarket and see packages of chicken for sale, labeled with the irradiated food logo indicating that they have been preserved by gamma radiation. You hesitate, as some questions form in your mind. List five questions you would like to have answered about food irradiation to help you decide whether to buy the chicken.

2. A company proposes to put a large-scale food irradiation plant in your community. A public hearing will be held on this matter. Make a list of five questions you would want answered at the hearing.

3. Identify two ways in which nuclear radiation is now used in your community.

Frequently, the words radiation and radioactivity are used as though they were the same; they aren't. We now turn to finding out more about nuclear and other types of radiation, as well as radioactivity. In the following sections you will find many of the answers to questions you have probably just raised.

A.2 KINDS OF RADIATION

Electromagnetic radiation is given off by objects, and the range of energies given off by such radiation is called the *electromagnetic spectrum.* Figure 1 shows the major types of electromagnetic radiation in order from low to high energy. Radio waves and microwaves are at the low-energy end of the spectrum; X-rays and cosmic rays are at the high-energy end of the spectrum. Near the center is the familiar visible spectrum, the only range of energies we can see with our eyes. It is remarkable and humbling to realize that, in the vast sweep of the full electromagnetic spectrum, our eyes can detect only this rather narrow, select band of energies.

Each type of electromagnetic radiation has several properties in common:

- It is a form of energy and has no mass.
- It travels at the speed of light.
- It can travel through a vacuum; unlike sound or ocean waves, its movement does not depend on a medium such as air or water to "carry" it.
- It is emitted by atoms as their nuclei decay, or after the atoms are energized in processes such as heating the tungsten filament in a light bulb or lighting the fuse that explodes the compounds in fireworks.
- It moves through space as packets (bundles) of energy called *photons.* Each photon has a certain energy that is characteristic of its type of radiation (visible, ultraviolet, gamma, and so on). The photon's energy determines the radiation's effect on living things and other types of matter.

▶ *"Nuclei" is the plural of "nucleus."*

Radiation is classified as either ionizing or non-ionizing radiation. *Non-ionizing radiation* has much lower energy than ionizing radiation. Electromagnetic radiation in the visible and lower-energy regions is non-ionizing radiation. This form of radiation transfers its energy to matter, "exciting" the molecules to vibrate or to move their electrons to higher energy levels. Chemical reactions sometimes occur as this energy is transferred to molecules. The reactions caused by microwave ovens in cooking food are one example. Non-ionizing radiation is also involved in radio and TV reception and electric light bulbs. Intense microwave and infrared radiation can be lethal, causing burns. Excessive exposure to non-ionizing radiation can be harmful. A sunburn, for example, results from overexposure to non-ionizing radiation from the sun.

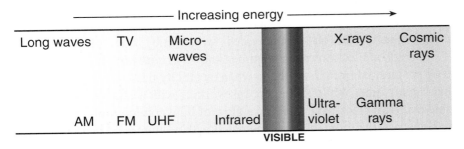

Figure 1 The electromagnetic spectrum.

▶ *Ionizing radiation has more energy than non-ionizing radiation.*

▶ *These fragments are often in the form of ions—hence the name "ionizing radiation."*

Ultraviolet radiation from the sun can be harmful.

▶ *Fluorescent minerals glow in the dark when illuminated by ultraviolet radiation.*

Ionizing radiation, which includes all nuclear radiation and other high-energy electromagnetic radiation (ultraviolet, X-rays, cosmic rays, and gamma rays), has the greatest energy and potential for harm. Energy from such radiation can cause ions to form when electrons are ejected from molecules. These ions are highly reactive molecular fragments that can cause serious damage to living systems. Because ultraviolet (UV) radiation can cause skin cancer, sustained exposure to UV radiation should be, at best, limited.

The difference in the effects of non-ionizing vs. ionizing radiation is something like the difference between being sprayed with a fine mist from a sprinkler and hit with a blast of water from a hose. The body may absorb more energy (water) from the mist (or from non-ionizing radiation), but with less damage, because the impact is spread over a large area. In the case of the hose water (or ionizing radiation), the energy is more focused and is often more harmful.

Nuclear radiation originates from nuclear changes in atoms. Generally, the atoms remain unchanged chemically because in *chemical* changes the atomic number (number of protons) doesn't change. Thus, in chemical reactions an atom of aluminum always remains aluminum (13 protons), and an iron atom is always iron (26 protons).

However, some atoms with unstable nuclei can spontaneously change. Substances that change spontaneously in this way are called **radioactive.** Usually these radioactive atoms change to produce an atom of a different element (one with a different number of protons), an emitted particle, and released energy. Because the nucleus undergoes change, the process is referred to as **radioactive decay.** The emitted particles and energy together make up **nuclear radiation.**

CHEMQUANDARY

ALWAYS HARMFUL?

How true is the statement: "All radiation is harmful and should be avoided?"

A.3 THE GREAT DISCOVERY

Scientists have long been interested in light and other types of radiation. However, nuclear radiation was hard to detect. When and how did scientists first identify it?

The history of our understanding of ionizing radiation began in 1895. In his studies of fluorescence, the German physicist W. K. Roentgen found that certain minerals glowed—or fluoresced—when hit by beams of electrons.

Beams of electrons are sometimes known as **cathode rays,** because they are emitted from the cathode when electricity passes through an evacuated glass tube. While Roentgen was working with a cathode-ray tube shielded with black paper, he saw a glow of light coming from a piece of paper across the room. The paper, coated with a fluorescent material, would be expected to glow when exposed to radiation, but it was not in the path of the electrons from the tube. Roentgen concluded that there must be some unseen radiation coming from the tube and passing through the paper. He named the mysterious rays **X-rays,** X representing the unknown.

Further experiments showed that although X-rays penetrate many materials, they cannot pass through dense materials such as lead and bone. X-rays are now known to be high-energy electromagnetic radiation.

Scientists soon realized how useful X-rays could be in medicine, and the first X-ray photograph ever taken was of Roentgen's wife's hand. Figures 2 and 3 show some modern X-ray pictures.

Roentgen's discovery excited other scientists. Soon hundreds of researchers, including the French physicist Henri Becquerel, were studying these new rays.

Becquerel wondered if fluorescent minerals might emit X-rays. In 1896, he placed a fluorescent mineral, which happened to contain uranium, in sunlight. He then wrapped a photographic plate in black paper, and placed it next to the mineral. If the mineral did emit X-rays as it fluoresced, the film would darken even though it was shielded from sunlight. To Becquerel's delight, the developed plate darkened—the film had been exposed. He stored the covered photographic plates and mineral sample in a drawer.

▶ *Modern X-ray devices are based on an electron beam striking a heavy-metal target such as silver.*

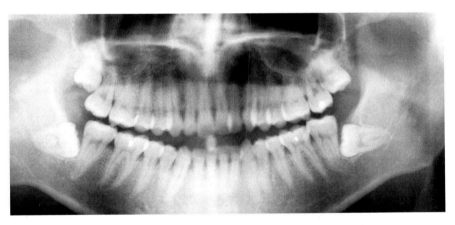

Figure 2 A dental X-ray photograph of a human jaw. Such X-rays are useful for detecting cavities and other dental problems.

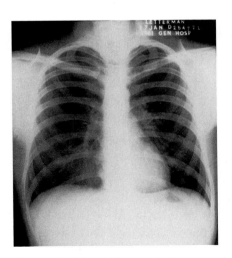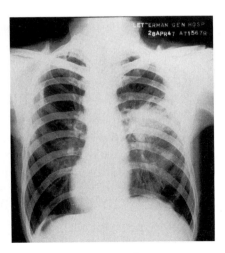

Figure 3 Chest X-rays. Left, normal lungs; right, cancerous lungs.

Following that initial success, skies were cloudy for several days, preventing Becquerel from doing more experiments. He decided to develop the plates on the chance that the mineral fluorescence might have persisted, causing some fogging of the plates. But when he took the plates from the drawer, he was astounded. Instead of a faint fogging, the plates had been strongly exposed.

A fluorescent mineral in a darkened desk drawer could not cause such a high level of exposure. In fact, there was no satisfactory explanation for Becquerel's observations. He interrupted his X-ray work to study the mysterious rays apparently given off by the uranium mineral. These rays proved to be more energetic and possess much greater penetrating power than did X-rays. Becquerel had discovered radioactivity.

Becquerel suggested to Marie Curie, his graduate student at the time, that she attempt to isolate the radioactive component of pitchblende, a uranium ore, as her Ph.D. research. Her preliminary work on this was successful. Her physicist husband, Pierre Curie, left his research work to join her on the pitchblende project. Working together, Marie and Pierre Curie discovered that the radioactivity level of pitchblende was four to five times greater than expected from its known uranium content. The Curies suspected some other radioactive element was also present. After processing more than 1,000 kg (more than a ton) of pitchblende, they isolated tiny amounts (milligrams) of two previously unknown radioactive elements—polonium (Po) and radium (Ra).

Further research revealed that all uranium compounds are radioactive. Interest in radioactivity continued to grow among scientists. Some realized that a better understanding of these rays could provide clues to the structure of atoms. One scientist who was particularly successful in penetrating the atom's mysteries was Ernest Rutherford.

CHEMQUANDARY

SCIENTIFIC DISCOVERIES

What do the events described below have in common with Becquerel's discovery of radioactivity?

1. Charles Goodyear was experimenting with natural rubber (a sticky material that melts when heated and cracks when cold) when he accidentally allowed a mixture of rubber and sulfur to touch a hot stove top. He noted that the rubber–sulfur mixture did not melt. Vulcanization, the process that makes rubber more widely useful by modifying its properties, resulted from this observation.

2. While growing bacteria in a dish, Alexander Fleming noted some mold in the dish. Though bacteria filled most of the dish, none grew in the area surrounding the mold. He later discovered penicillin based on this observation.

3. Roy Plunkett withdrew gaseous tetrafluoroethene ($F_2C=CF_2$) from a storage cylinder, but the gas flow stopped long before the cylinder was empty. He cut open the cylinder and discovered a white solid. Today we know this solid as Teflon® (polytetrafluoroethene).

4. James Schlatter, a research chemist working to develop an anti-ulcer drug, inadvertently got some of the substance on his fingers. When he later licked his fingers to pick up a piece of paper, his fingers tasted very sweet, and he correctly linked the sweetness back to the anti-ulcer drug. Instead of finding an anti-ulcer drug, he discovered aspartame (NutraSweet®), now the dominant commercial artificial sweetener.

A.4 NUCLEAR RADIATION

In 1899, Ernest Rutherford showed that radioactivity includes two different types of rays, which he named **alpha rays** and **beta rays.** He placed thin sheets of aluminum in the pathway of radiation from uranium. Beta rays penetrated more layers of aluminum sheets than alpha rays did. Shortly afterward, a third kind of radiation produced by radioactive elements was discovered. This radiation was called **gamma rays.**

In order to learn about their electrical properties, the three kinds of radiation were passed through magnetic fields. Scientists already knew that when charged particles move through a magnetic field they are deflected by the magnetic force. The path of positively charged particles is deflected in one direction, and the path of negatively charged particles in the opposite direction. Neutral particles and electromagnetic radiation are not deflected by magnetic fields (Figure 4).

Experiments with nuclear radiation in magnetic fields indicated that alpha rays are composed of positively charged particles and beta rays are made up of negatively charged particles. Gamma rays are not deflected by a magnetic field; they were later identified as high-energy electromagnetic radiation, similar to X-rays.

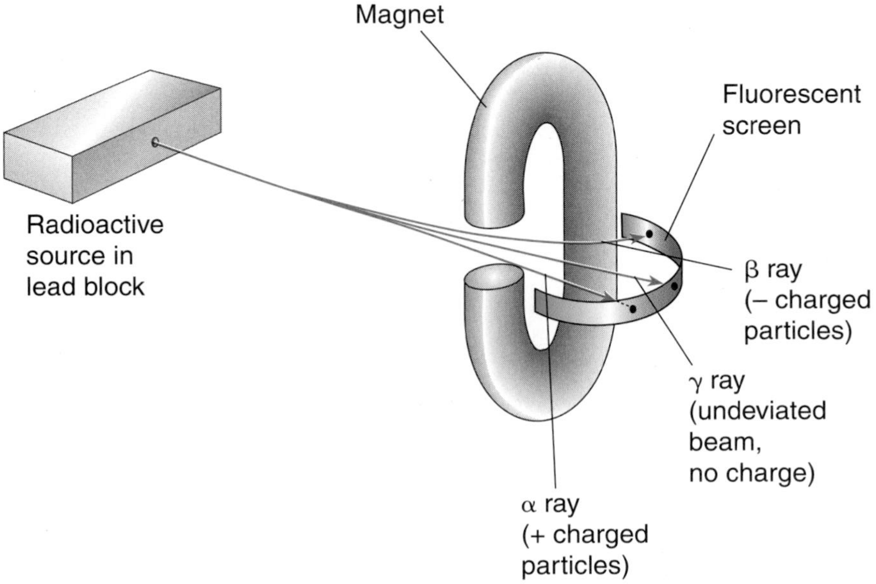

▶ *Alpha (α), beta (β), and gamma (γ) are the first three letters of the Greek alphabet.*

▶ *Figure 4 shows an experiment in which all three types of radiation might be detected.*

Figure 4 Lead block experiment, showing behavior of alpha (α) rays, beta (β) rays, and gamma (γ) rays in a magnetic field.

The nature of radioactivity was thus established; as often happens in science, a new discovery toppled an old theory—in this case, that atoms were the smallest, most fundamental unit of matter. Once alpha, beta, and gamma rays were identified, scientists became convinced that atoms must be composed of still smaller particles.

A.5 LABORATORY ACTIVITY: THE BLACK BOX

GETTING READY

Experiments like those of Roentgen, Becquerel, and Rutherford illustrate that indirect evidence may be essential for exploring properties of an object we cannot see or touch. In this activity you will try to identify objects in sealed boxes. In many ways this activity resembles the work of scientists in determining the nature of the atom—a more fundamental "sealed box."

PROCEDURE

Two sealed boxes, numbered 1 and 2, are at your laboratory bench. Each box contains three objects, all different from one another and from those in the other box.

1. Gently shake, rotate, or manipulate one of the boxes. From your observations, try to determine the size of each object, its general shape, and the material from which it is made. Record your observations, designating the three objects as A, B, and C.

2. Compare your observations and ideas about the three objects with those of other team members. What conclusions can you and your team reach? Can you identify the objects?

Rutherford conducted many important experiments on the nature of alpha rays. Here he is seen in his laboratory at McGill University in Montreal, Canada.

3. Repeat Steps 1 and 2 with the second box.

4. Make your final decisions about all the objects in Boxes 1 and 2. Identify each object by name and a sketch.

QUESTIONS

1. Which of your senses did you use to collect the data?

2. In what ways does this activity resemble efforts of scientists in exploring atomic and molecular structure?

3. Name some theories about the nature of the world that are based primarily on indirect evidence.

From the results of another experiment, Rutherford proposed a fundamental atomic model that remains useful even today. To do so, he developed an ingenious, indirect way to "look" at atoms.

A.6 GOLD FOIL EXPERIMENT

Prior to Rutherford's experiment, the arrangement of electrons and positively charged particles in atoms had been explained in several ways. In the most accepted model, an atom was viewed as a solid mass of positively charged materials, with the negatively charged electrons embedded within, like peanuts in a candy bar.

Hans Geiger and Ernest Marsden, working in Rutherford's laboratory in Manchester, England, focused a beam of alpha particles—the heaviest of the three types of radioactive emissions—at a sheet of gold foil 0.00004 cm (about 2,000 atoms) thick. The scientists surrounded the metal sheet with a screen coated with zinc sulfide (Figure 5). The zinc sulfide-coated screen emitted a small flash of light where each alpha particle struck it. By observing the tiny light flashes, the researchers were able to figure out the paths of the alpha particles interacting with the gold foil.

▶ *In the late 1800s this was known as the "plum pudding model," in reference to the distribution of raisins within this popular English dish.*

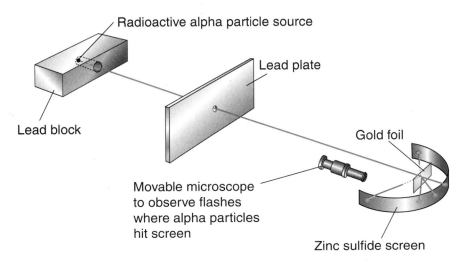

Figure 5 The alpha particle scattering experiment.

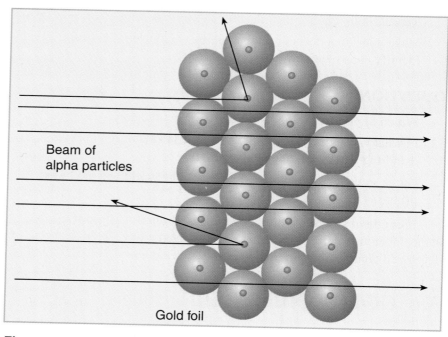

Figure 6 Result of the alpha particle scattering experiment. Most alpha particles passed through the foil; a few were deflected.

Rutherford expected that alpha particles would scatter as they were deflected by the gold atoms, producing a pattern similar to spray from a nozzle. He was in for quite a surprise.

First of all, most alpha particles passed straight through the gold foil as if nothing were there (Figure 6). This implied that most of the volume taken up by atoms is essentially empty space. But what surprised Rutherford even more was that a few alpha particles—about one in every 20,000—bounced back toward the source. Rutherford described his astonishment this way: "It was about as incredible as if you had fired a 15-inch shell at a piece of tissue paper and it came back and hit you."

What these few rebounding alpha particles encountered must have been relatively small, because most alpha particles missed it. But whatever they hit must also have been very dense and electrically charged.

From these results, Rutherford developed the model of the nuclear atom. He named the tiny, massive, positively charged region at the center of the atom the nucleus. He envisioned that electrons orbited around the nucleus, somewhat like planets orbit the sun.

▶ *Here, "massive" means heavy. A massive object may have a small volume.*

A.7 ARCHITECTURE OF ATOMS

Since Rutherford's time, our understanding of atomic structure has expanded and, in some ways, changed. Rutherford's image of a central, massive nucleus surrounded mostly by empty space is still accepted. From subsequent research we now also know that, although useful, the idea of "orbiting electrons" is too simplified. Rather, each electron is believed to occupy a specific region in which it spends most of its time. We can identify the general region, but not the location, of the electron at a given instant.

Further research revealed that the nucleus is composed of two types of particles—**neutrons,** which are electrically neutral, and **protons,** with a positive charge. A neutron and a proton each have about the same mass, 1.7×10^{-24} g. Although this mass is extremely small, it is still much larger than the mass of an electron. As shown in Table 1, a mole (6.02×10^{23}) of electrons has a mass of 0.0005 g. The same number of either protons or neutrons would have a mass of about 1 g. In other words, a proton or a neutron is about 2,000 times more massive than an electron. Protons and neutrons account for most of the mass of the universe.

▶ *Particles smaller than an atom are described as "subatomic particles."*

Table 1	Three Important Subatomic Particles		
Particle	**Location**	**Charge**	**Molar Mass (g/mol)**
Proton	Nucleus	1+	1
Neutron	Nucleus	0	1
Electron	Outside of nucleus	1−	0.0005

▶ *Protons—positive charge Neutrons—no charge Electrons—negative charge*

The diameter of a typical atom is about 10^{-8} cm, but an average nuclear diameter is 10^{-12} cm, only one ten-thousandth ($10^{-12}/10^{-8}$) as big as the diameter of the entire atom. Looking at this another way, the nucleus occupies only about one trillionth (10^{-12}) of the volume of an atom. As a comparison, imagine that a billiard ball represents an atom's nucleus. On that scale, electrons surrounding this billiard-ball nucleus would occupy space extending more than a kilometer (0.6 mile) away in all directions.

As you learned in Section A.2, each atom of the same element has the same number of protons in its nucleus, and each element has a unique number of protons. This number, called the **atomic number**, identifies the element. For example, because each carbon atom nucleus contains six protons, carbon has an atomic number of 6; no other element has that atomic number.

▶ *In most periodic tables, the atomic number is given above the element's symbol.*

However, all atoms of the same element do not necessarily have the same number of neutrons in their nuclei. Atoms of the same element having different numbers of neutrons are called **isotopes** of that element. Naturally occurring carbon atoms, each containing six protons, may have six, seven, or even eight neutrons. The composition of these three carbon isotopes is summarized in Table 2. The carbon isotope with eight neutrons has an unstable nucleus—it is radioactive. Such isotopes are called **radioisotopes.**

Isotopes are distinguished by their different mass numbers. The **mass number** represents the total number of protons and neutrons in an atom of a given isotope. The three carbon isotopes in Table 2 have mass numbers

Table 2	Three Carbon Isotopes			
Name	**Total Protons (Atomic Number)**	**Total Neutrons**	**Mass Number**	**Total Electrons Outside Nucleus**
Carbon-12	6	6	12	6
Carbon-13	6	7	13	6
Carbon-14	6	8	14	6

of 12, 13, and 14. To specify a particular isotope, the atomic number and the mass number are added to the symbol for the element. For example, an isotope of strontium (Sr) with an atomic number of 38 and a mass number of 90 is written this way:

$$^{90}_{38}\text{Sr}$$

A 2+ ion of strontium-90 would be shown as $^{90}_{38}\text{Sr}^{2+}$. To name an isotope in words, just follow the element's name with a hyphen and then the mass number. The isotope depicted above is called strontium-90.

The symbols, names, and nuclear composition of some isotopes are summarized in Table 3.

► The atomic number is a lower left subscript; the mass number is an upper left superscript.

Table 3		Some Common Isotopes			
Symbol	Name	Total Protons (Atomic number)	Total neutrons	Mass number	Total electrons
$^{7}_{3}\text{Li}$	Lithium-7	3	4	7	3
$^{67}_{31}\text{Ga}$	Gallium-67	31	36	67	31
$^{201}_{81}\text{Tl}$	Thallium-201	81	120	201	81
$^{208}_{82}\text{Pb}$	Lead-208	82	126	208	82
$^{208}_{82}\text{Pb}^{2+}$	Lead-208, 2+ ion	82	126	208	80

YOUR TURN

Isotope Notation

Suppose you know that one product of a certain nuclear reaction is an isotope containing 85 protons and 120 neutrons. It therefore has a mass number of 205 (85 protons + 120 neutrons = mass number of 205). What is the name of this element?

$$^{205}_{85}?$$

Consulting the listing of elements, we see that astatine (At) is element 85.

$$^{205}_{85}\text{At}$$

1. Prepare a summary chart similar to Table 3 for the following six isotopes. (Consult the periodic table for missing information.)

   a. $^{12}_{?}\text{C}$ c. $^{16}_{?}\text{O}$ e. $^{202}_{?}\text{Hg}$

   b. $^{14}_{7}?$ d. $^{24}_{12}?^{2+}$ f. $^{238}_{92}?$

2. What relationship do you note between total protons and total neutrons

   a. for atoms of lighter elements?

   b. for those of heavier elements?

A.8 LABORATORY ACTIVITY: ISOTOPIC PENNIES

GETTING READY

You found earlier (Resources unit, page 90) that pre-1982 and post-1982 pennies have different compositions. As you might suspect, they also have different masses. In this activity, a mixture of pre- and post-1982 pennies will represent the atoms of a naturally occurring mixture of two isotopes of the imaginary element "coinium." Using the pennies will help you to learn one way that scientists can determine the relative amounts of different isotopes present in a sample of an element.

You will be given a sealed container that holds a mixture of 10 pre-1982 and post-1982 pennies. Your container might hold any particular atomic mixture of the two "isotopes." Your task is to determine the isotopic composition of the element coinium *without* opening the container.

To illustrate how this can be done, let's consider a mixture of heavy billiard balls and lightweight Ping-Pong® balls. Say you are given a 10-ball mixture of billiard balls and Ping-Pong balls in a sealed box. It is clear that 10 billiard balls would weigh much more than 10 Ping-Pong balls. Thus, the heavier the mixture of 10 balls is, the more billiard balls you must have present.

An obvious—but important—notion is that the mass of the entire mixture equals the sum of the masses of all the billiard balls and the masses of all the Ping-Pong balls. That idea can be expressed this way:

Total mass of balls =
(Number of billiard balls × Mass of one billiard ball) +
(Number of Ping-Pong balls × Mass of one Ping-Pong ball)

Now let's get back to the pennies. Following the billiard ball and Ping-Pong ball example, this relationship applies:

Total mass of pennies =
(Number of pre-1982 pennies × Mass of one pre-1982 penny)
+ (Number of post-1982 pennies × Mass of one post-1982 penny)

Now we are ready to complete the penny calculations. For starters, we know there are 10 pennies in the container. So, if we let x equal the total number of pre-1982 pennies, then $(10 - x)$ = number of post-1982 pennies. Further, the mass of all the pre-1982 pennies equals the number of pre-1982 pennies (x) multiplied by the mass of 1 pre-1982 penny. Likewise, the total mass of the post-1982 pennies equals the number of post-1982 pennies ($10 - x$) times the mass of 1 post-1982 penny.

► *The total mass of pennies can be found by subtracting the mass of the empty container from the mass of the sealed container containing pennies.*

To reduce the number of words, we can write the pennies' relationship this way:

Total mass of pennies =
$$(x \times \text{mass of pre-1982 penny}) + [(10 - x) \times \text{mass of post-1982 penny}]$$

Our goal is to find the value of x—the number of pre-1982 pennies in the container. Once that value is known, we will have figured out the composition of the 10-penny mixture—without opening the container! However, to solve for x, we first need to know the values of the three masses in the equation. That's what the following procedure is designed to do.

PROCEDURE

1. Your teacher will give you a pre-1982 penny, a post-1982 penny, and a sealed container with a mixture of 10 pre- and post-1982 pennies, and will tell you the mass of the empty container. Record the code number of your sealed container.

2. Find the mass of the pre-1982 penny and then the mass of the post-1982 penny.

3. Find the mass of the sealed container of pennies.

4. Find the total mass of the 10 pennies. (*Hint:* Use data from Steps 1 and 3.)

5. Calculate the values of x—the number of pre-1982 pennies, and $(10 - x)$—the number of post-1982 pennies.

6. Calculate the percent composition of the element "coinium" from your data.

QUESTIONS

1. What property of the element "coinium" is different in its pre- and post-1982 forms?

2. a. In what ways is the penny mixture a good analogy or model for actual element isotopes?

   b. In what ways is the analogy misleading or incorrect?

3. Name at least one other familiar item that could serve as a model for isotopes.

A.9 ISOTOPES IN NATURE

Most elements in nature are mixtures of isotopes. Some isotopes of an element may be radioactive, while others are not. From a chemical viewpoint, this does not make any real difference. All isotopes of an element behave virtually the same chemically—they have the same electron arrangement, but they differ slightly in mass.

The proportions of an element's naturally occurring isotopes are usually the same everywhere on the Earth. An element's accepted atomic and molar masses, as shown in the periodic table, represent averages based on the relative abundances of its isotopes. The following *Your Turn* illustrates this idea.

Weighing sample on an electronic balance.

Molar Mass and Isotope Abundance

To calculate the molar mass of an element, it is useful to use the concept of a *weighted average*. You can calculate the weighted average mass for the coins in the element "coinium" from your laboratory activity.

Suppose you found that $x = 4$, that is, your mixture contained four pre-1982 pennies and six post-1982 pennies. As decimal fractions, the composition of your mixture was 0.4 pre-1982 pennies and 0.6 post-1982 pennies. (Notice that the decimal fractions must add up to equal 1.) These fractions, together with the mass of each type of penny (the two isotopes of "coinium"), can be used in the following equation to calculate the weighted average mass of any penny:

Avg. mass of a penny =
(Fraction, pre-1982 pennies)(Mass, pre-1982 penny) +
(Fraction, post-1982 pennies)(Mass, post-1982 penny)

1. Try the average mass calculation for your actual "coinium" mixture.

2. Now calculate the average mass of a penny in your mixture another way: Divide the mass of your total penny sample by 10.

3. Compare the average masses you calculated in Questions 1 and 2. These results should convince you that *either* calculation leads to the same result. If not, consult your teacher.

Now let's do an example for an actual isotopic mixture.

Naturally occurring copper consists of 69.1% copper-63 and 30.9% copper-65. The molar masses of the pure isotopes are

copper-63   62.93 g/mol
copper-65   64.93 g/mol

Using these data, calculate the molar mass of naturally occurring copper. The equation for finding average molar masses is the same as the earlier equation you used for "coinium":

Molar mass =
(Fraction of isotope 1)(Molar mass of isotope 1) +
(Fraction of isotope 2)(Molar mass of isotope 2) + . . .

Because there are only two naturally occurring isotopes for copper, Cu-63 and Cu-65, the average molar mass of copper is found this way:

Molar mass of copper = (0.691)(62.93 g/mol) + (0.309)(64.93 g/mol)
= 63.5 g/mol.

This is the value for copper given in the periodic table (rounded to the nearest 0.1 unit).

4. Naturally occurring uranium (U) is a mixture of three isotopes:

	Molar Mass (g/mol)	% Natural Abundance
Uranium-238	238.1	99.28%
Uranium-235	235.0	0.71%
Uranium-234	234.0	0.0054%

a. Is the molar mass of naturally occurring uranium closer to 238, 235, or 234? Why?

b. Calculate the molar mass of naturally occurring uranium.

Marie Curie originally thought that only heavy elements were radioactive. It is true that naturally occurring radioisotopes (radioactive isotopes) are more common among the heavy elements. For example, isotopes of elements with atomic numbers greater than 83 (bismuth) are all radioactive. However, many natural radioactive isotopes are also found among lighter elements and it is even possible to create a radioactive isotope of any element. Table 4 lists some naturally occurring radioisotopes and their isotopic abundances.

Table 4	Some Natural Radioisotopes	
Name	**Symbol**	**Relative Isotopic Abundance (%)**
Hydrogen-3	$^{3}_{1}H$	0.00013
Carbon-14	$^{14}_{6}C$	Trace
Potassium-40	$^{40}_{19}K$	0.0012
Rubidium-87	$^{87}_{37}Rb$	27.8
Indium-115	$^{115}_{49}In$	95.8
Lanthanum-138	$^{138}_{57}La$	0.089
Neodymium-144	$^{144}_{60}Nd$	23.9
Samarium-147	$^{147}_{62}Sm$	15.1
Lutetium-176	$^{176}_{71}Lu$	2.60
Rhenium-187	$^{187}_{75}Re$	62.9
Platinum-190	$^{190}_{78}Pt$	0.012
Thorium-232	$^{232}_{90}Th$	100
Uranium-235	$^{235}_{92}U$	0.72
Uranium-238	$^{238}_{92}U$	99.28

The history of science is full of discoveries that build upon earlier discoveries. The discovery of radioactivity was such an event, beginning with the work of Roentgen, Becquerel, and Rutherford, and leading in turn to new knowledge of atomic structure.

PART A: SUMMARY QUESTIONS

1. List two identifying characteristics and two examples of electromagnetic radiation.

2. The discovery of X-rays led to investigations of the atomic nucleus. Describe historical events that connect these two research topics.

3. Describe how Becquerel was able to make distinctions among fluorescence, X-rays, and natural radioactivity.

4. Describe experimental evidence supporting the idea that radioactive sources emit charged particles as well as electromagnetic radiation (gamma rays).

5. Describe experimental evidence supporting each of these statements:
   a. An atom is mainly empty space.
   b. An atom contains a tiny, yet relatively massive, positively charged center.

6. Approximately how many electrons would it take to equal the mass of one proton?

7. Complete this table:

Data Table					
Symbol	Name	Atomic Number	Total Protons	Mass Number	Total Neutrons
?	?	6	?	12	?
$^{60}_{27}$?	?	?	?	?	?
$^{207}_{?}$Pb	?	?	?	?	?

8. Potassium (K) occurs as a mixture of three isotopes:

Isotope	Molar Mass (g/mol)
Potassium-39	38.964
Potassium-40	39.964
Potassium-41	40.962

a. If the molar mass of potassium is 39.098 g/mol, which isotope is the most abundant?

b. Explain your answer.

9. Neon (Ne) is found in the Earth's atmosphere at concentrations of about 1 part in 65,000. All three naturally occurring neon isotopes are useful in neon signs and other applications. Vital statistics for neon's isotopes are in Table 5.

Table 5	Neon Isotopes	
Isotope	Molar Mass (g/mol)	% Natural Abundance
Ne-20	19.992	90.51
Ne-21	20.994	0.27
Ne-22	21.991	9.22
		100.00

a. Do you expect that neon's molar mass is closer to 20, 21, or 22? Why?

b. Calculate the molar mass of neon.

10. Almost all boron (B) atoms are found in one of two forms, boron-10 and boron-11. Both isotopes behave alike chemically. They are both useful in fireworks (for green color), in the antiseptic boric acid, and in heat-resistant glass. Only boron-10, however, is useful as a control material in nuclear reactors, as a radiation shield, and in instruments used to detect neutrons. The molar mass of boron is 10.81 g/mol; which boron isotope must be more abundant in nature?

11. What is the minimum information needed to identify a particular isotope?

 # EXTENDING YOUR KNOWLEDGE (OPTIONAL)

Choose one of these subatomic particles and investigate how scientists determined its existence and properties: proton, neutron, electron, neutrino, quark, pi meson, positron, or gluon. What are the practical results (if any) of such scientific studies?

Radiation, Energy, and Atoms

B RADIOACTIVE DECAY

Some 350 isotopes of 90 elements are found in our solar system. About 70 of these isotopes are radioactive. Almost 1,600 additional isotopes have been made in the laboratory. For elements with atomic numbers of 83 or less, the natural abundance of radioactive isotopes is quite low compared with the abundance of stable isotopes.

The radiation emitted by naturally occurring radioisotopes is the source of natural radioactivity in our environment. You may be surprised to learn that a relatively constant level of *natural* radioactivity, called **background radiation**, is always present around you. Background radiation is present in the walls of your room, the building materials (such as brick and stone) in your school and home, in air, land, and sea—even within your own body.

Radioactive isotopes decay spontaneously, giving off alpha or beta particles and gamma radiation. The kind and intensity of radiation emitted help determine the medical and industrial applications of radioisotopes. In addition, the three types of radiation pose their own distinct hazards to human health.

Because nuclear radiation cannot be detected by human senses, various devices have been developed to detect it and measure its intensity. You will work with one type of radiation detector in the following laboratory activity.

B.1 LABORATORY ACTIVITY: α, β, AND γ RAYS

GETTING READY

One early device for detecting radioactivity was the *Geiger–Mueller counter* (Figure 7, page 315). It produces an electrical signal when its detector is hit by particles coming from a radioactive source. In this activity or teacher demonstration, you will use a modern-day counter to compare the penetrating ability of alpha, beta, and gamma radiation through cardboard, glass, and lead.

When ionizing radiation enters the counter's detecting tube, or *probe*, ions are formed and generate the electric current flowing in the tube. Most radiation counters register the current as both audible clicks and a meter reading. The unit of measure on the meter, counts per minute (cpm), indicates the intensity of the radiation.

An initial reading of background radiation must be taken to establish a baseline reference before readings are taken from a known radioactive

An employee being tested for radiation exposure.

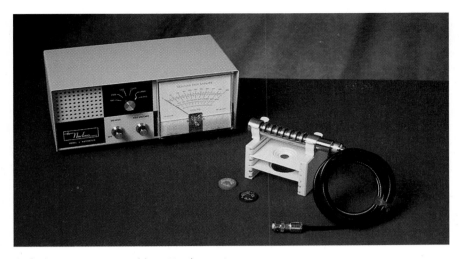

A Geiger counter and its attachments.

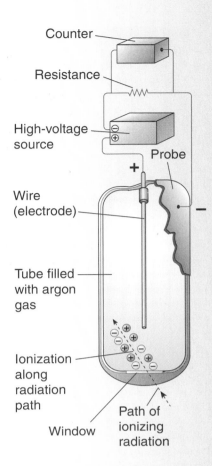

Figure 7 Diagram of a Geiger–Mueller counter showing how radiation entering the probe causes ionization of gaseous atoms or molecules.

From Chemistry—The Study of Matter and Its Changes by J. Brady and J. Holum, © 1993 John Wiley & Sons, Inc. Adapted by permission of John Wiley & Sons, Inc.

source. This background count is then subtracted from each reading taken from the radioactive source.

The radioactive materials in this activity pose no danger to you. Nuclear materials are strictly regulated by state and federal laws. The sources in this activity emit only very small quantities of radiation, and using them requires no special license. Nevertheless, you will handle all radioactive samples with the same care that would be required for licensed materials. You will wear rubber or plastic gloves. Do not allow the radiation counter to come in direct contact with the radioactive material. Check your hands with a radiation monitor before you leave the laboratory.

Prepare a data table like the one below.

Data Table				
		Counts per Minute (cpm)		
Radiation	**No Shielding**	**Cardboard**	**Glass**	**Lead**
Gamma				
Beta				
Alpha				

PROCEDURE

Part 1: Penetrating Ability

1. Set up the apparatus shown in Figure 8, page 316.

2. Turn on the counter; allow it to warm for at least 3 minutes. Determine the intensity of the background radiation by counting clicks for one minute in the absence of any other radioactive sources. Record this background radiation in counts per minute (cpm).

3. Put on your gloves. Using forceps, place a gamma-ray source on the ruler at a point where it produces a high reading on the meter (Figure 8, page 316). Watch the meter for 30 seconds and estimate the average cpm detected during this period. Subtract the background reading from this value and record the result.

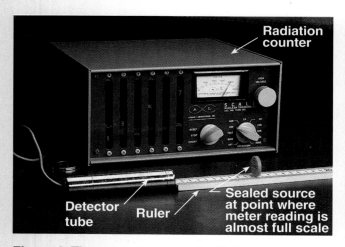

Figure 8 The apparatus setup for Part 1.

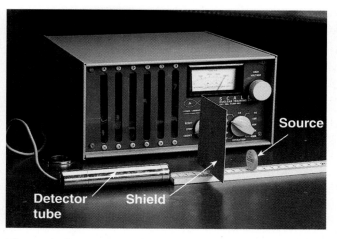

Figure 9 Place the shield between the probe and the radiation source.

4. Without moving the radiation source, place a piece of cardboard (index card) between the probe and the source, as shown in Figure 9.

5. Again watch the meter for 30 seconds. Correct the average reading for background radiation and record the result.

6. Repeat Steps 4 and 5 using a glass plate.

7. Repeat Steps 4 and 5 using a lead plate.

8. Repeat Steps 3 through 7 using a beta-particle source.

9. Repeat Steps 3 through 7 using an alpha-particle source.

QUESTIONS

1. Analyze your results from Steps 4 through 9. Which shielding materials were effective in reducing the intensity of each type of radiation?

2. How do the three types of radiation you tested compare in their penetrating ability?

3. Of the shielding materials tested, which do you conclude
   a. is the most effective in blocking radiation?
   b. is the least effective?

4. What properties of a material do you think determine its radiation-shielding ability?

Part 2: *Effect of Distance on Intensity*

1. Prepare a data table containing two columns: one to record distance from the probe, the other to record radiation intensity (cpm) values.

2. Place a radioactive source designated by your teacher at the point on the ruler that produces nearly a full-scale reading (usually a distance of about 5 cm).

3. Record a corrected average reading over 30 seconds.

4. Move the source so the distance from the probe is doubled.

5. Record a corrected average reading over 30 seconds.

6. Move the source twice more, so the original distance is first tripled, then quadrupled, recording a corrected reading after each move. (For example, if you started at 5 cm, you would take readings at 5, 10, 15, and 20 cm.)

7. Prepare a graph, plotting the corrected cpm on the *y* axis, and the distance from source to probe (in cm) on the *x* axis.

QUESTIONS

Analyze the graph you prepared in Step 7.

1. By what whole-number factor did the intensity of radiation (measured in counts per minute) decrease when the initial distance was doubled?

2. Did this same whole-number factor apply when the distance was doubled again?

3. Try stating the mathematical relationship between distance and intensity using this factor and Figure 10.

► *A factor is a number by which another number is multiplied or divided to become a new number.*

Part 3: Shielding Effects

1. Use forceps to place a source, designated by your teacher, on the ruler at a distance that produces nearly a full-scale reading.

2. Take an average reading over 15 seconds, correcting for background radiation.

3. Place one glass plate between the source and probe. Take an average reading over 15 seconds, correcting for background radiation.

4. Place a second glass plate between the source and probe. Take an average reading over 15 seconds, correcting for background radiation.

5. Repeat Steps 3 and 4, using lead sheets rather than glass plates.

6. Wash your hands thoroughly before leaving the laboratory.

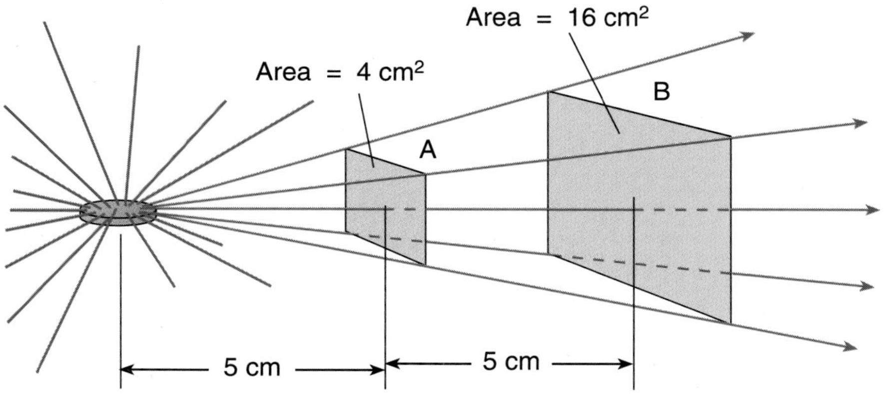

Figure 10 The relationship between distance from source and intensity of radiation. Intensity is counts per minute in a given area. Note how the same amount of radiation spreads over a larger area as the distance from the source increases.

QUESTIONS

1. How effective was doubling the shield thickness in blocking radiation intensity
    a. for glass?
    b. for lead?

2. When you have a dental X-ray, the rest of your body is shielded with a special blanket. What material would be a good choice for this blanket? Why?

3. Which type of radiation, from a source outside the body, is likely to be most dangerous to living organisms? Why?

You have found that the three kinds of radiation differ greatly in their penetrating abilities. Why is this so? What *are* alpha and beta particles? Where do they come from? We will address these questions in the next section.

Highly radioactive materials are handled safely at Amersham International.

B.2 NATURAL RADIOACTIVE DECAY

An alpha particle is composed of two protons and two neutrons. It is the nucleus of a helium-4 atom: $^4_2\text{He}^{2+}$. Alpha radiation (alpha particle emission) is given off by radioactive isotopes of some elements with atomic numbers higher than 83. An alpha particle is nearly 8,000 times heavier than a beta particle. The alpha particle's large mass can cause great damage, but only over very short distances. Because alpha particles are very powerful tissue-damaging agents once inside the body, alpha emitters in air or food are particularly dangerous to human life. Fortunately, alpha particles are easily blocked; as you noted in Laboratory Activity B.1, alpha particles are stopped by a few centimeters of air. An alpha-emitter can even be held safely for a short time in your hand, because skin is thick enough to act as a barrier.

Figure 11 illustrates a radium-226 nucleus emitting an alpha particle. The radium nucleus loses two protons, so its atomic number drops from 88 to 86. An isotope of a different element, radon, is formed by the decrease of 2 in the atomic number of radium. Radium-226 also loses two neutrons, so its mass number drops by 4 to 222 (2 protons and 2 neutrons are lost) producing radon-222. The decay process can be represented by this equation:

$$^{226}_{88}\text{Ra} \quad \rightarrow \quad ^4_2\text{He} \quad + \quad ^{222}_{86}\text{Rn}$$
$$\text{radium-226} \quad \rightarrow \quad \text{alpha particle} \quad + \quad \text{radon-222}$$

Atoms of two elements—helium and radon—have been formed from one atom of radium. Note that atoms are not necessarily conserved in nuclear reactions, as they are in chemical reactions. Atoms of different elements can appear on both sides of a nuclear equation.

Mass numbers and atomic numbers, however, *are* conserved in nuclear reactions. In the equation above, the sum of mass numbers of the reactants equals that of the products (226 = 4 + 222). Also, the sum of atomic numbers of the reactants equals that of the products (88 = 2 + 86). Both of these relations hold true for all nuclear reactions.

Beta particles are fast-moving electrons emitted from the nucleus during radioactive decay. Since they are less massive than alpha particles and travel at very high velocities, beta particles have much greater penetrating ability than do alpha particles. On the other hand, beta particles are not as damaging to living tissue.

During beta decay, a neutron changes into a proton and an electron. The proton remains in the nucleus, but the electron (a beta particle) is ejected at high speed. A third particle, an antineutrino, is also released. This equation describes the process:

$$^1_0\text{n} \quad \rightarrow \quad ^1_1\text{p} \quad + \quad ^{0}_{-1}\text{e} \quad + \quad \text{antineutrino}$$
$$\text{neutron} \quad \text{proton} \quad \text{beta particle}$$
$$\text{(electron)}$$

A beta particle is assigned a mass number of 0 and an "atomic number" (nuclear charge) of −1. Notice that the overall result of beta emission is that a neutron is converted into a proton.

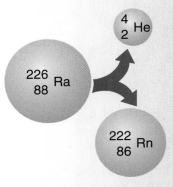

Figure 11 Alpha particle emission from radium-226. The mass number decreases by 4, (2p + 2n), and the atomic number decreases by 2, (2p).

▶ *Ionic charges, such as the 2+ of the alpha particle, are usually not included in nuclear symbols.*

▶ *Note in the equation that n, p, and e are used as the symbols, respectively, for a neutron, a proton, and a beta particle—an electron. It is also acceptable to symbolize an electron (beta particle) by the Greek letter beta: $^{0}_{-1}\beta$.*

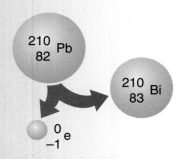

Figure 12 Beta decay of lead-210 leads to bismuth-210.

▶ *Atomic number balance:*
82 = 83 + (−1)

Figure 12 shows beta decay in the nucleus of lead-210, in which the nucleus loses one neutron but gains one proton. Thus, the mass number remains unchanged at 210, but the atomic number increases to 83. The new nucleus formed is that of bismuth-210.

$$\underset{\text{lead-210}}{^{210}_{82}\text{Pb}} \rightarrow \underset{\text{bismuth-210}}{^{210}_{83}\text{Bi}} + \underset{\substack{\text{beta particle} \\ \text{(electron)}}}{^{0}_{-1}\text{e}}$$

Once again, note that the sum of all mass numbers remains constant during this nuclear reaction (210 on each side of the equation). The sum of atomic numbers (the nuclear charge) remains constant as well (82 on each side).

Alpha and beta decay (emission) often leave nuclei in an energetically excited state. This type of excited state, described as metastable, is designated by the symbol m. For example, ^{99m}Tc represents a technetium isotope in a metastable, excited state. Energy from isotopes in such excited states is released as gamma rays—high-energy electromagnetic radiation having as much or more energy per photon as X-rays. Because gamma rays have neither mass nor charge, their emission does not change the mass or charge balance in a nuclear equation. Table 6 summarizes general information about natural radioactive decay.

Table 6	Changes Resulting From Nuclear Decay			
Type	Symbol	Change in Atomic Number	Change in Neutrons	Change in Mass Number
Alpha	$^{4}_{2}\text{He}$	Decreased by 2	Decreased by 2	Decreased by 4
Beta	$^{0}_{-1}\text{e}$	Increased by 1	Decreased by 1	No change
Gamma	$^{0}_{0}\gamma$	No change	No change	No change

Of these three forms of nuclear radiation, gamma rays are the most penetrating. But, under some circumstances, they are the least tissue-damaging over comparable distances. Tissue damage is related to the extent of ionization created by the radiation, expressed as the number of ionizations within each unit of tissue. *Alpha particles cause considerable damage over short range, but protecting against them is easy. Beta and gamma radiation do less damage over longer range, but it is more difficult to protect against them* (see Figure 13).

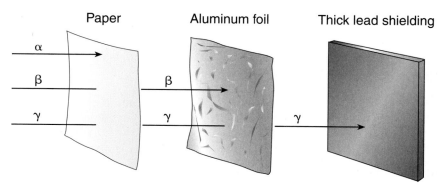

Figure 13 The relative penetrating powers of alpha (α), beta (β), and gamma (γ) radiation. Gamma rays are the most penetrating, alpha particles the least.

New isotopes produced by radioactive decay may also be radioactive, and therefore undergo further nuclear decay. Uranium (U) and thorium (Th) are the "parents" (reactants) in three natural decay series that begin with U-238, U-235, and Th-232, respectively. Each series ends with formation of a stable isotope of lead (Pb). The decay series starting with uranium-238 contains 14 steps, as shown in Figure 14.

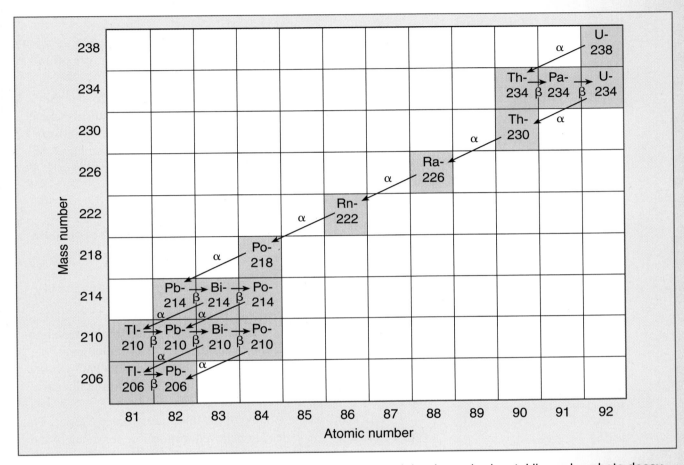

Figure 14 The uranium-238 decay series. Diagonal lines show alpha decay, horizontal lines show beta decay. Here's how to "read" this chart: Locate radon-222 (Rn-222). The arrow to the left shows that this isotope decays to polonium-218 (Po-218) by alpha (α) emission. This nuclear equation applies: $^{226}_{88}\text{Ra} \rightarrow {}^{222}_{86}\text{Rn} + {}^{4}_{2}\text{He}$.

YOUR TURN

Nuclear Balancing Act

The key to balancing nuclear equations is recognizing that both atomic numbers and mass numbers are conserved. Use information in Table 6 (page 320) to complete the following exercise.

Cobalt-60 is a common source of ionizing radiation for medical therapy. Complete this equation for the beta decay of cobalt-60:

$$^{60}_{27}\text{Co} \rightarrow {}^{0}_{-1}\text{e} + \text{?}$$

Beta emission causes no change in mass number, as noted in Table 6; therefore the new isotope will also have mass number 60. We can write the unknown product as ^{60}?. Because the atomic number increases by 1 during beta emission, the new isotope will have atomic number 28, one more than cobalt. The periodic table shows that nickel has an atomic number of 28. The final equation is

$$^{60}_{27}\text{Co} \rightarrow {}^{0}_{-1}\text{e} + {}^{60}_{28}\text{Ni}$$

1. Write the appropriate symbol for the type of radiation given off in each of these reactions:

   a. The following decay process allows archaeologists to date the remains of ancient biological materials. Living organisms take in carbon-14 and maintain a relatively constant amount of it over their lifetime. After death, no more carbon-14 is taken in, so the amount gradually decreases due to decay.

   $$^{14}_{6}\text{C} \rightarrow {}^{14}_{7}\text{N} + ?$$

   b. The following decay process takes place in some types of household smoke detectors.

   $$^{241}_{95}\text{Am} \rightarrow {}^{237}_{93}\text{Np} + ?$$

2. Thorium (Th) occurs in nature as three isotopes: Th-232, Th-230, and Th-228. The first of these is the most abundant. Thorium's radiation intensity is quite low; its compounds can be used without great danger if kept outside the body. In fact, thorium oxide (ThO_2) was widely used in gas mantles in Europe and America during the gas-lighting era to speed combustion of the gas.

   Th-232 is the parent isotope of the third natural decay series. This series and the U-238 series are believed responsible for much of the heat generated inside the earth. (The U-235 series contribution is negligible, since the natural abundance of U-235 is quite low.)

   Complete these equations, representing the first steps in the Th-232 decay series, by identifying the missing items A, B, C, and D. Each code letter represents a particular isotope or a type of radioactive emission in the decay series from Th-232 to Rn-220. For example, in the first equation, Th-232 decays to form Ra-228 by emitting radiation called A. What is A?

   a. $^{232}_{90}\text{Th} \rightarrow \text{A} + {}^{228}_{88}\text{Ra}$ d. $^{228}_{90}\text{Th} \rightarrow {}^{4}_{2}\text{He} + \text{D}$

   b. $^{228}_{88}\text{Ra} \rightarrow {}^{0}_{-1}\text{e} + \text{B}$ e. $\text{D} \rightarrow {}^{4}_{2}\text{He} + {}^{220}_{86}\text{Rn}$

   c. $\text{B} \rightarrow {}^{0}_{-1}\text{e} + \text{C}$

You have just learned that radioisotopes emit radiation. When you investigated the differences among alpha, beta, and gamma radiation (Laboratory Activity B.1), you measured radiation directly in counts per minute (cpm). It can also be measured as the quantity of radiation *absorbed* by the body. The extent of cellular damage caused by absorption of such radiation depends on the quantity absorbed, and that brings us to the topic of the next several sections.

B.3 IONIZING RADIATION—HOW MUCH IS SAFE?

The basic unit for measuring radiation effects on humans is the **rem** (*R*oentgen *e*quivalent *m*an). The rem measures the ability of radiation—regardless of type or activity (cpm)—to cause ionization in human tissue. Normal exposures are so small that the dosage is stated in *millirems (mrem)*; 1 rem = 1,000 mrem. One millirem of any type of radiation produces essentially the same biological effects, whether the radiation is composed of alpha particles, beta particles, or gamma rays. In general, each 1-mrem dose received increases an individual's risk of dying from cancer by one chance in 4 million.

> ▶ *The rem is a term derived as the abbreviation for **R**oentgen equivalent **m**an.*

Each type of radiation can cause damage in different ways. The two main factors that determine tissue damage are the radiation density (the number of ionizations within the same area) and the dose (the quantity of radiation received).

X-rays and gamma rays—ionizing forms of electromagnetic radiation—penetrate deeply into the body. However, they actually cause less damage than alpha or beta particles, if the same amount of tissue is considered. Alpha particles generally have five to fifty times the total energy of beta particles or gamma rays. Although alpha particles are easily stopped external to the body, they can cause extensive damage over a very short distance (about 0.025 mm) if they reach the lungs or bloodstream. Due to their relatively large mass and slower velocities, the alpha particles give up a significant amount of energy within a small area. When traveling through human tissue, alpha particles act more like bulldozers than sports cars.

> ▶ *Recall that an alpha particle is a helium nucleus with a molar mass of 4.0 g/mol.*

B.4 RADIATION DAMAGE: NOW AND LATER

Ionizing radiation breaks bonds in molecules and thus tears molecules apart. At low radiation levels, only comparatively few molecules are harmed, and the body's systems can usually repair the damage. However, the higher the dose (or quantity of radiation) received, the more molecules are affected. Damage to proteins and nucleic acids is of greatest concern because of their importance in body structures and functions, including those passed on to offspring. Proteins form much of the body's soft tissue structure and make up enzymes, molecules that control the rates of bodily processes. When many protein molecules are torn apart within a small region, too few functioning molecules may be left to permit the body to heal itself in a reasonable time.

> ▶ *See the Food unit for more background on proteins and enzymes.*

Nucleic acids in DNA can be damaged in two different ways. Minor damage causes **mutations**—changes in the structure of DNA that may

Radioactive Decay

323

result in the synthesis of altered proteins. Most mutations kill the cell. If the cell is a sperm or ovum, the change may lead to a birth defect. Some mutations can lead to cancer, a disease in which cell growth and metabolism are out of control. If the DNA in many body cells is severely changed, the immediate effect is that proteins can no longer be synthesized to replace damaged ones. Death follows.

Table 7 lists the factors that determine the extent of biological damage from radiation. Table 8 shows the biological effects of *large* dosages of radiation; note that the values in Table 8 are given in rems, not millirems.

Table 7	Biological Damage from Radiation
Factor	**Effect**
Dose	Most scientists assume that an increase in radiation dose produces a proportional increase in risk.
Exposure time	The more a given dose is spread out over time, the less harm it does.
Area exposed	The larger the body area exposed to a given radiation dose, the greater the damage.
Tissue type	Rapidly dividing cells, such as blood cells and sex cells, are more susceptible to radiation damage than are slowly dividing or non-dividing cells, such as nerve cells. Fetuses and children are more susceptible to radiation damage than are adults.

Table 8	Radiation Effects
Dose (rem)	**Effect**
0–25	No immediate observable effects.
25–50	Small decreases in white blood cell count, causing lowered resistance to infections.
50–100	Marked decrease in white blood cell count. Development of lesions.
100–200	Radiation sickness—nausea, vomiting, hair loss. Blood cells die.
200–300	Hemorrhaging, ulcers, deaths.
300–500	Acute radiation sickness. Fifty percent die within a few weeks.
>700	One hundred percent die.

Large radiation doses have drastic effects. Conclusive evidence that high radiation doses cause cancer has been gathered from uranium miners, accident victims, and nuclear bomb victims in World War II. Some of the most useful case studies were based on workers who used radium compounds to paint numbers on watch dials that would glow in the dark.

▶ *DNA molecules control cell reproduction and synthesis of cellular proteins.*

▶ *The radium dial markings glowed in the dark. Modern glow-in-the-dark watches are based on safe, nonradioactive materials.*

The workers used their tongues to smooth the tips of their paint brushes, and thus unknowingly swallowed radioactive radium compounds.

Leukemia, cancer of the white blood cells, is the most rapidly developing and common cancer associated with radiation. Other forms of cancer may also appear after exposure, as well as anemia, heart-related problems, and cataracts (opaque spots on the lens inside the eye).

Considerable controversy continues regarding whether very low radiation doses can cause cancer. Most data on cancer have come from human exposure to high doses, with mathematical projection back to lower doses. Few studies have directly linked low radiation doses with cancer development. Although most scientists agree that natural levels of radiation are safe for most people, some authorities argue that *any* increase in radiation above natural levels increases the probability for cancer. This is similar to the reasoning that led to food additive regulations for carcinogens (Food unit, page 284).

How much radiation do you experience each year? What are its sources? Do you have any control over the radiation? We will deal with these questions next.

B.5 EXPOSURE TO RADIATION

Everyone receives background radiation at low levels from natural and human sources. Among natural sources are these:

- Cosmic rays—exceedingly high-energy particles that bombard the Earth from outer space
- Radioisotopes released from rocks, soil, and groundwater—uranium (U-238 and U-235), thorium (Th-232), and their decay products; other natural radioisotopes
- Radioisotopes in the atmosphere—radon (Rn-222) and its decay products, including polonium (Po-210)

Human sources of background radiation include:

- Fallout from nuclear weapons testing
- Increased exposure to cosmic radiation during air travel
- Radioisotopes released into the environment from nuclear power generation and other nuclear technologies

Some of the radiation we experience comes from sources within our bodies (see Figure 15). On average, people living in the United States receive above 360 mrem per person each year, about 300 mrem (82%) of which comes from *natural* sources. Figure 16 (page 326) shows the percent from each source.

What radiation level is considered safe? It is clearly in your best interests to avoid unnecessary radiation exposure. The federal government's present background radiation standard for the general public is 500 mrem (0.5 rem) per year for any individual. An average of 360 mrem is below this radiation level. The federal government's standard for an individual's maximum safe exposure in the workplace is 5,000 mrem (5 rem) annually.

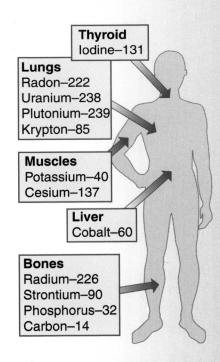

Thyroid
Iodine–131

Lungs
Radon–222
Uranium–238
Plutonium–239
Krypton–85

Muscles
Potassium–40
Cesium–137

Liver
Cobalt–60

Bones
Radium–226
Strontium–90
Phosphorus–32
Carbon–14

Figure 15 All living things contain some radioactive isotopes, including these, which are found in specific parts of the human body.

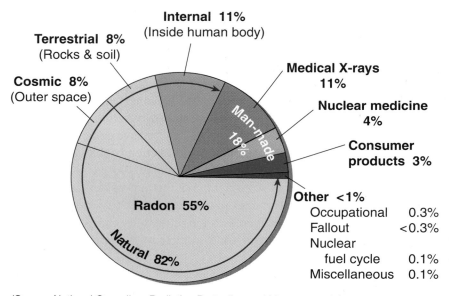

(Source: National Council on Radiation Protection and Measurements)

Figure 16 U.S. sources of ionizing radiation.

CHEMQUANDARY

RADIATION EXPOSURE STANDARDS

Why are exposure-level standards different for some individuals than for the general public? Why are the standards different for those who are occupationally exposed?

YOUR TURN

Radiation from Foods

Many of the foods you eat contain radioactive atoms, contributing about 20 mrems per year. These radioactive atoms and their nonradioactive counterparts occur naturally. They come from the soils in which the foods are grown. Potassium is a mineral important for health (see Food unit, page 276). Potassium-40 (K-40) is a naturally occurring radioisotope of potassium and, thus, is incorporated into our foods. Table 9 (page 327) is a listing of the potassium content of some foods, along with the radio-activity of potassium-40 in those foods, measured in disintegrations per second (dps).

1. List the foods from Table 9 (page 327) you have eaten this week that contain potassium. Take into account the number of helpings of each food eaten during this time.

2. What is the total number of disintegrations per second of K-40 from the food you ate in this time?

Table 9 Potassium Content and K-40 Activity of Selected Foods			
Food	**Quantity**	**Potassium, mg**	**K-40, dps**
Hamburger	4 oz	960	29
Chicken, fried	1/4 chicken	240	7
Hot dog	2 regular	200	6
French fries	3.5 oz	650	19.5
Corn	1 ear	200	6
Banana	1 small	370	9
Apple	1 medium	160	7
Milk, whole	1 cup	370	11
Cola beverage	12 oz	13	0.4
Corn flakes	1 oz	14	0.4
White bread	1 slice	30	0.9
Peanut butter	1 tablespoon	100	3
Egg	1 large	65	2
Oatmeal	1 cup	130	4

3. There are about 60 dps in your body due to K-40 per kilogram of body mass $\left(\dfrac{60 \text{ dps}}{\text{kg body mass}}\right)$.

▶ *1 kg = 2.2 lb*

   a. Calculate the number of dps due to potassium-40 in your body.
   b. Compare this value to what you obtained in Question 2. How do you account for any difference?

4. Suppose you decided to reduce your annual internal exposure to radiation by trying to eliminate all potassium from your diet. Why would this be unwise?

Although we cannot avoid all radiation sources—particularly those from the Earth and its atmosphere—we can make choices about others.

- We can decide whether to have diagnostic X-rays.
- We can decide whether to undergo medical tests that use radioactive isotopes.
- We can choose whether to fly, how often, and how far.
- We can choose where to live and, if need be, adjust our home environment.

B.6 RADON IN HOMES

The gaseous element radon (Rn)—the heaviest of the noble gases—has always appeared as a component of the Earth's atmosphere. It is a decay product of uranium. In the 1980s the public first learned about unusually high concentrations of radioactive radon gas in a relatively small number of U.S. homes.

▶ *You can locate this decay product as Rn-222 in the decay series chart shown earlier on page 321.*

Radioactive Decay

Radon is produced as uranium-238 decays in the soil and in building materials. Some radon produced in the soil dissolves in groundwater. Many houses have foundation and basement floor cracks that permit radon from rocks, soil, and water to seep in. In a tightly sealed house, the radon gas does not have much chance to escape, and radon is now a problem in some areas because of changes in the way we build and use homes. In older homes, outdoor air enters through doors, windows, and the gaps around them, thus diluting radon or removing it from the house. But air conditioning in new buildings decreases the need to open windows. To conserve energy, many new homes are built more air-tight than older homes were. The net result is that indoor air has little chance to mix freely with outdoor air and radon levels may reach high levels. Remedies for high radon levels in homes include increased ventilation, sealing cracks in floors, and removing radon from groundwater. Relatively inexpensive radon test kits are available for home use.

The real threat of radon gas occurs after it is inhaled. Radon decays to produce, in succession, radioactive isotopes of polonium (Po), bismuth (Bi), and lead (Pb). Thus, if radon gas is inhaled, it enters the body and, through radioactive decay, is transformed to these toxic heavy-metal ions that cannot be exhaled as gas. These radioactive heavy metal ions also emit potentially damaging alpha particles within the body. In homes with abnormally high radon gas levels, inhaled dust can also carry traces of the same heavy-metal isotopes deposited by decaying radon. Estimates indicate that about 6% of homes in the United States have radon levels higher than the exposure level recommended by the EPA. It is estimated that around 10% of deaths from lung cancer annually in the United States are due to the effects of radon gas. These figures, although sobering, should be kept in perspective, however. For example, more than 10 times as many people die each year from lung cancer attributed to cigarette smoking.

Radon is only one of *many* sources of background radiation to which each of us is exposed. Complete the following activity to estimate how much radiation you receive.

▶ *Smoking itself is a significant risk factor for lung cancer. In addition, heavy smokers are considerably more susceptible to the effects of radon.*

YOUR TURN

Your Annual Radiation Dose

On a separate sheet of paper list the numbers and letters found in the following table. Then fill in the blanks in column 2 with suitable values. When you finish, add the quantities to estimate your annual radiation dose.

Source of Radiation   **Quantity per Year**

1. Location of your community.

   a. Cosmic radiation at sea level. (Cosmic radiation is radiation emitted by stars across the universe. Much of this is

Source of Radiation	Quantity per Year
deflected by the Earth's atmosphere and ionosphere.)	(U.S. average) 26 mrem

b. Add an additional radiation value based on your community's elevation above sea level:

    1,000 m (3,300 ft) above sea
        level = 10 mrem

    2,000 m (6,600 ft) above sea
        level = 30 mrem

    3,000 m (9.900 ft) above sea
        level = 81 mrem

    (Estimate mrem value for
    intermediate elevations.)   _____mrem

2. House construction material (building materials contain a very small percent of radioisotopes). Brick or concrete, 70 mrem; wood, 30 mrem.   _____mrem

3. Ground; radiation from rocks and soil.   (U.S. average) 26 mrem

4. Food, water, and air.   (U.S. average) 28 mrem

5. Fallout from prior nuclear weapons testing.   (U.S. average) 4 mrem

6. Medical and dental X-rays.

    a. Chest X-rays (10 mrem per X-ray)   _____mrem

    b. Gastrointestinal tract X-rays (200 mrem per X-ray)   _____mrem

    c. Dental X-rays (10 mrem per X-ray)   _____mrem

7. Air travel (increases exposure to cosmic radiation; 1 mrem per 1,000 miles flown).   _____mrem

8. Power plants. If your home is within five miles of a nuclear or coal-fired power plant, add 0.3 mrem.   _____mrem

Total   **_____mrem**

QUESTIONS

1. How does your annual radiation dose compare with the 360-mrem average (page 325)?

2. a. Could you reduce your radiation exposure by changing your lifestyle? Explain.

    b. Would you want to make those changes? Why?

    We will consider nuclear decay and radiation again in Section C when we examine rates of nuclear decay and methods used to detect radiation.

1. Name the type of radioactive radiation released in each case:

   a. A radioactive iodine (I) isotope decays to form a xenon (Xe) isotope with a higher atomic number but the same mass number as the iodine.

   b. Technetium-99m decays, yet its atomic number and mass number both remain unchanged.

   c. A thorium isotope decays to form a radium isotope that has a lower atomic number and mass number than the original thorium.

2. Complete these equations:

   a. This reaction represents the decay mode of the medically important, synthetic radioisotope Co-60.

   $$^{60}_{27}\text{Co} \rightarrow ? + ^{0}_{-1}\text{e}$$

   b. About 0.01% of naturally occurring atoms of this element are radioactive. It is one of the main radioisotopes in your body and decays as shown below.

   $$? \rightarrow ^{40}_{20}\text{Ca} + ^{0}_{-1}\text{e}$$

   c. Radiation from radium-226 has been used in cancer therapy. Radium-226 decays by the reaction shown below. Identify the radiation emitted.

   $$^{226}_{88}\text{Ra} \rightarrow ^{222}_{86}\text{Rn} + ?$$

3. From a human health standpoint, how does ionizing radiation differ from non-ionizing radiation?

4. Name the common unit for measuring the biological effects of radiation.

5. Explain why alpha emitters are fairly harmless outside the human body, but quite dangerous inside it.

6. List four factors that determine the extent of biological damage caused by radiation.

7. Radiation is more destructive to rapidly dividing cells than to those multiplying more slowly. How can this fact be regarded as a

   a. benefit? (That is, how can it be put to a positive use?)

   b. risk?

8. a. Briefly discuss three problems encountered in assessing the risks of radiation exposure at low dose levels.

   b. In the absence of better data, how do you think we should regulate allowable radiation exposures at low dose levels?

9. Consider the following three radiation levels for humans. Explain why the values are different.

   a. The minimum radiation dose at which immediate, observable effects have been detected.

   b. The maximum annual dose allowed for radiation workers.

   c. The radiation standard set for an average person.

C RADIOACTIVITY: NATURAL AND ARTIFICIAL

Another aspect to consider about background radiation is the fact that different radioactive isotopes decay and emit radiation at differing rates. Thus, different radioisotopes have different lifetimes; a given radioisotope has a specific lifetime. Scientists have devised convenient ways to measure and report how fast various radioisotopes decay. In the next section, you will learn about these decay rates, which help determine how useful or hazardous a radioisotope may be.

C.1 HALF-LIFE: A RADIOACTIVE CLOCK

How long does it take for a sample of radioactive material to decay? Knowing the answer to this question allows scientists to predict the total time a radioisotope used in medical diagnosis will remain active within the body, to plan how long hazardous nuclear wastes must be stored, and to estimate the ages of ancient organisms, of civilizations, and of the world itself.

The rate of decay of radioisotopes is measured in half-lives. One *half-life* is the time it takes for one-half the atoms in a sample of radioactive material to decay. Table 10 lists the half-lives of several radioisotopes.

Table 10	Half-Lives of Selected Radioisotopes	
Hydrogen-3	$_1^3H \rightarrow \, _2^3He + \, _{-1}^0e$	12.3 y
Carbon-14	$_6^{14}C \rightarrow \, _7^{14}N + \, _{-1}^0e$	5.73×10^3 y
Phosphorus-32	$_{15}^{32}P \rightarrow \, _{16}^{32}S + \, _{-1}^0e$	14.3 d
Potassium-40	$_{19}^{40}K \rightarrow \, _{20}^{40}Ca + \, _{-1}^0e$	1.28×10^9 y
Radon-222	$_{86}^{222}Rn \rightarrow \, _{84}^{218}Po + \, _2^4He$	3.28 d

Consider carbon-14. From Table 10 we see that this isotope has a half-life of 5,730 years. If we start with 50 billion atoms of carbon-14, in 5,730 years 25 billion atoms will have decayed to nitrogen-14, leaving 25 billion atoms of the original carbon-14. During the next 5,730 years, one-half of the remaining 25 billion atoms will decay, leaving 12.5 billion atoms of carbon-14. And so on.

Half-lives vary greatly for different radioisotopes. For example, the half-life of polonium-212 is 3×10^{-7} seconds; that of uranium-238 is 4.5 billion years.

After 10 half-lives, only about 1/1,000th or 0.1% of the original radioisotope is still left to decay. (Verify this statement with your own

▶ *After 10 half-lives (57,300 years), 50 million C-14 atoms will still be present.*

The ages of excavated artifacts can be determined by measuring the extent of decay of radioactive isotopes in the organic artifacts.

calculations.) That means the isotope's activity has dropped to 0.1% of its initial level of intensity. In many cases this is considered a safe level; in others it is not. Since there is no way to speed up the rate of radioactive decay, radioactive waste disposal is a challenging problem, especially for radioisotopes with very long half-lives. We will look into this dilemma in a later section.

C.2 LABORATORY ACTIVITY: UNDERSTANDING HALF-LIFE

GETTING READY

In this activity you will use pennies to simulate the process of radioactive decay. The pennies will help you discover the relationship between the passage of time and the number of radioactive nuclei that decay.

Suppose all the pennies are atoms of an element called coinium. Further, a heads-up penny represents an atom of the radioactive isotope—let's call it headsium—of coinium. The product of this isotope's decay is a tails-up penny—the isotope tailsium.

You will be given 80 pennies and a container. Placing all the pennies heads up in the container will represent the "starting" composition of our radioisotope. Each shake of the container will represent one half-life period. During this period a certain number of headsium nuclei will decay to give tailsium (that is, some pennies will flip over).

PROCEDURE

In your notebook, construct a data table like the following. The first data entry has been made for you.

Data Table		
Number of Half-Lives	Number of Tailsium (Decayed) "Atoms"	Number of Headsium (Undecayed) "Atoms"
0	0	80
1		
2		
3		
4		

1. Place the 80 pennies heads up in the box.
2. Close the box and shake it vigorously.
3. Open the box. Remove from the box all atoms of *decayed* headsium (coins that have turned over). Record the number of *decayed* headsium and *undecayed* headsium atoms at the end of this first half-life.
4. Repeat Steps 2 and 3 three more times. At this point you will have simulated four half-lives. You should have five numbers for headsium

in your final data summary, representing the number of *undecayed* headsium atoms (coins that have not turned over) remaining after zero, one, two, three, and four half-lives.

5. Following your teacher's instructions, pool the class data by adding the total number of *decayed* atoms (coins that have turned over) for the whole class after the first half-life, the second half-life, and so on.

6. Using your own data and class pooled data, prepare a graph by plotting the number of half-lives on the *x* axis and the number of *undecayed* atoms remaining for each half life on the *y* axis. Plot and label two graph lines—one for your own data, and one for pooled class data.

QUESTIONS

1. a. Describe the appearance of your two graph lines. Are they straight or curved?

   b. Which set of data—your own or the pooled class data—provided the more convincing demonstration of the notion of half-life? Why?

2. How many undecayed headsium nuclei would remain out of a sample of 600 headsium nuclei after three half-lives?

3. If 175 headsium nuclei remain out of an original sample of 2,800 nuclei, how many half-lives must have passed?

4. Name one similarity and one difference between your simulation and actual radioactive decay. (*Hint:* Why was it advisable to pool class data?)

5. How could you modify this simulation to demonstrate that different isotopes have different half-lives?

6. a. How many half-lives would it take for 1 mol of any radioactive atoms (6.02×10^{23} atoms) to decay to 6.25% (0.376×10^{23} atoms) of the original number of atoms?

   b. Would any of the original radioactive atoms still remain

      (1) after 10 half-lives?

      (2) after 100 half-lives?

7. a. In this simulation is there any way to predict when a particular penny will "decay?"

   b. If you could follow the fate of an individual atom in a sample of radioactive material, could you predict when it would decay? Why or why not?

8. What other ways can be used to model the concept of half-life?

The curves you constructed in this activity apply to the decay of any radioactive isotope, with one important difference: the half-life is unique for each isotope. The half-life of one isotope may be millions of years; a different isotope may have a half-life of only fractions of a second.

The following exercise will give you practice applying the half-life concept.

YOUR TURN

Half-Lives

1. Suppose you were given $1,000 and told you could spend one-half of it in the first year, one-half of the balance in the second year, and so on. (One year thus corresponds to a half-life in this analogy.)

   a. If you spent the maximum allowed each year, at the end of what year would you have $31.25 left?

   b. How much would be left after 10 half-lives (that is, 10 years)?

2. Cobalt-60 is a radioisotope used as a source of ionizing radiation in cancer treatment; the radiation it emits is effective in killing rapidly dividing cancer cells. Cobalt-60 has a half-life of five years.

   a. If a hospital starts with a 1,000-mg supply, how many milligrams of Co-60 would it need to purchase after 10 years to replenish the original supply? Does the answer depend on how frequently cobalt isotope radiation was used to treat patients? Why or why not?

   b. How many half-lives would it take for the supply of cobalt-60 to dwindle to

      (1) less than 10% of the original?
      (2) less than 1% of the original?
      (3) less than 0.1% of the original?

3. Even though the activity of a sample is only 0.1% of its original value after 10 half-lives, theoretically it would take virtually forever for a reasonably sized sample of radioactive material to decay completely. To help you understand this concept, consider this analogy:

   A person is attempting to reach a telephone booth that is 512 m away. Assume that the person covers half this distance (256 m) in the first minute, half the remaining distance (128 m) during the second minute, half the remainder (64 m) in the third minute, and so on. In other words, the half-life for this moving process is 1 minute.

   a. If the individual never covers more than half the remaining distance in each one-minute interval, how long—at least in theory—will it take the person to reach the phone booth?

   b. How many half-lives will be needed to get within 25 cm of the booth?

4. Strontium-90 is one of many radioactive wastes from nuclear weapons. This isotope is especially dangerous if it enters the food supply. Strontium behaves chemically like calcium; the two elements belong to the same chemical family. Thus, rather than passing through the body, strontium-90 is incorporated into calcium-based material such as bone. In 1963 the United States, the (then) Soviet

Union, and several other countries signed a nuclear test ban treaty that ended most above-ground weapons testing. Some strontium-90 released in previous above-ground nuclear tests still remains in the environment, however.

a. Sr-90 has a half-life of nearly 29 years. Track the decay of Sr-90 that was present in the atmosphere in 1963 by following these instructions:

   (1) Using 1963 as year zero (when 100% of released Sr-90 was present), identify the year that represents one, two, three, etc., half-lives. Stop when you reach the year 2100.

   (2) Calculate the percent of original 1963 Sr-90 radioactivity present at the end of each half-life year.

   (3) Prepare a graph, plotting the percent of original 1963 Sr-90 radioactivity on the y axis, and the years 1963 to 2100 on the x axis.

   (4) Connect the data points with a smooth curve.

b. What percent of Sr-90 formed in 1963 still remains today?

c. What percent will remain in the year 2100?

C.3 BENEFITS OF RADIOISOTOPES

Various nuclear technologies make use of the ionizing radiation given off as certain radioisotopes decay. Such applications include tracer studies, where the objective is to detect the presence of the isotope, as well as irradiation, where the radiation is used as an energy source.

Knowing that certain elements collect in specific parts of the body—for instance, calcium in bones and teeth—physicians can investigate a given part of the body, by using an appropriate radioisotope as a tracer. In a tracer study, radioisotopes with short half-lives are put into a patient's body for diagnostic purposes. Such radioisotopes, called **tracers**, can track cellular abnormalities, locate damaged areas, and aid in therapy.

Tracers have properties that make them ideally suited for this task. First, radioisotopes behave the same—both chemically and biologically—as stable isotopes of the element. Physicians know that certain elements collect in specific parts of the body, for example calcium in bones and teeth. To investigate a part of the body, an appropriate radioisotope is used. A solution of a tracer isotope is supplied to the body, or a biologically active compound—synthesized to contain a radioactive tracer element—is fed to or injected into the patient (see Figure 17, page 336). A nuclear radiation detection system then allows the physician to track the location of this tracer in the body.

An example of using a radioisotope tracer is diagnosing problems of the thyroid gland, located in the neck. A patient simply drinks a solution containing a radioactive iodine (I-123) tracer. The physician then uses a radiation detection system to monitor the rate at which the tracer is taken

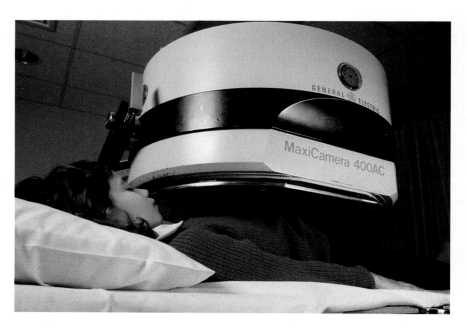

Figure 17 Injection of a radioisotope makes it possible to produce a non-invasive image of the abdomen.

▶ *The "m" in Tc-99m indicates that the isotope is metastable—it readily changes to a more stable form of the same isotope, releasing gamma rays in the process.*

up by the thyroid. A healthy thyroid will incorporate a known amount of iodine. An overactive or underactive thyroid will take up more or less iodine, respectively. The physician then compares the measured rate of I-123 uptake in the patient to the normal rate for someone of the same age, gender, and weight and takes appropriate action.

Technetium-99m (Tc-99m), a synthetic radioisotope, is the most widely used radioisotope in medicine. It has replaced exploratory surgery as a way to locate tumors in the brain, thyroid, and kidneys. Tumors are areas of runaway cell growth, and technetium concentrates where cell growth is fastest. A bank of radiation detectors around the patient's head can pinpoint a brain tumor's precise location. Phosphorus-32 can be used in a similar way to detect bone cancers.

In some cancer treatments, the diseased area is exposed to ionizing radiation to kill cancerous cells. For thyroid cancer, the patient receives a concentrated internal dose of radioiodine which concentrates in the thyroid. In other cancer treatments, an external beam of ionizing radiation (from cobalt-60) may be directed at the cancerous spot. Such irradiation treatments must be administered with great care—high radiation doses can also damage or kill normal cells.

The tracer and cell-killing properties of radioisotopes are used to diagnose and treat other forms of cancers. Other medical applications include the use of radiosodium (Na-24) to search for circulatory system abnormalities and radioxenon (Xe-133) to help search for lung embolisms (blood clots) and abnormalities. Table 11 (page 337) outlines other medical uses for radioisotopes.

Because we cannot see, hear, feel, taste, or smell radioactivity, we must use special detection devices to indicate its presence. You have already seen a radiation counter used, or used one yourself. This and other types of radiation detectors are explained in the next section.

Table 11		Selected Medical Radioisotopes	
Radioisotopes	Half-Life	Use
Used as tracers		
Technetium-99m	6.02 h	Measure cardiac output; locate strokes and brain and bone tumors.
Gallium-67	78 h	Diagnosis of Hodgkin's disease
Iron-59	45.1 d	Determine rate of red blood cell formation (these contain iron); anemia assessment
Chromium-51	27.8 d	Determine blood volume and lifespan of red blood cells
Hydrogen-3 (tritium)	12.3 y	Determine volume of body's water; assess vitamin D usage in body
Thallium-201	74 h	Cardiac assessment
Iodine-123	13.3 h	Thyroid function diagnosis
Used for irradiation therapy		
Cesium-137	30.0 y	Treat shallow tumors (external source)
Phosphorus-32	14.3 d	Treat leukemia, a bone cancer affecting white blood cells (internal source)
Iodine-131	8 d	Treat thyroid cancer (external source)
Cobalt-60	5.3 y	Treat shallow tumors (external source)
Yttrium-90	64 h	Treat pituitary gland cancer internally with ceramic beads

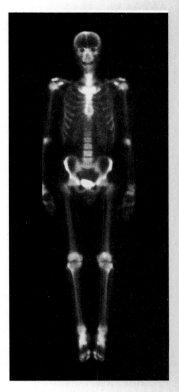

Bone scan.

C.4 RADIATION DETECTORS

One way to detect radioactive decay is to observe the results of radiation interacting with matter. In the radiation counter you used in Laboratory Activity B.1, for example, argon gas is ionized by entering radiation. The ionized gas conducts an electric current, and an electric signal is generated as the ions and components of radiation pass through the probe.

You learned earlier in this unit that radioactivity will expose photographic film. Workers who handle radioisotopes wear film badges or other detection devices to measure their exposure. If they get a dose in excess of federal limits, they are temporarily reassigned to jobs that minimize exposure to radiation.

Devices called *scintillation counters* detect entering radiation as light emitted by the excited atoms of a solid. The scintillation counter probe pictured in Figure 18 has a sodium iodide (NaI) detector, which emits light when ionizing radiation strikes it.

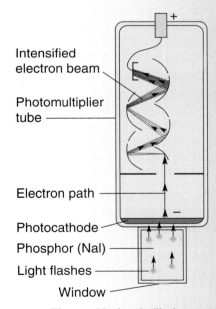

Intensified electron beam

Photomultiplier tube

Electron path

Photocathode

Phosphor (NaI)

Light flashes

Window

+

−

Figure 18 A scintillation counter probe. Ionizing radiation causes flashes of light (scintillations) in the detector (NaI crystal). Each light flash is converted into an electron pulse that is increased many times as it moves through the photomultiplier tube.

Radioactivity: Natural and Artificial

CHEMISTRY AT WORK

High-tech Soldiers in the War Against Cancer

Beverly **Buck** works in radiation oncology, the field of medicine that uses radiation to treat patients who have tumors.

Beverly is a radiation therapist at the Joint Center for Radiation Therapy in Boston. She also serves as Education and Development Coordinator making sure patients are properly treated, and oversees the systems that monitor the radiation equipment. Beverly and other radiation therapists at the Center deliver the actual dosages of radiation prescribed by each patient's radiation oncologist.

Radiation therapy is one weapon among several in the fight against cancer, and is often used together with chemotherapy. Radiation therapy is associated with risks; high-energy radiation can genetically alter or kill normal body cells, but it can also kill cancer cells. With any cancer treatment, the patient, patient's family, and the oncologist physician must carefully weigh the potential benefits and risks.

While radiation therapists must understand the science and technology behind radiation oncology, it's also critical to focus on each patient. Most of their patients are going through major crises in their lives, and so radiation therapists try to recognize and address patient fears and emotions.

To become a radiation therapist, a person must go through an educational program lasting at least two years. Students learn about nuclear chemistry, anatomy, physiology, radiation physics, radiobiology, radiation safety, pathology, oncology, and patient care methods.

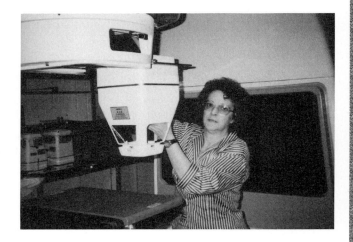

An Activity

There is likely a health facility near where you live that employs radiation therapists. If you were interested in applying for a job opening there, what type of training would you need, and where could you get such training? Is there any type of state license or certification you would need? If so, what are the requirements to get licensed or certified? What other skills or abilities do you think would be useful for the job?

Photograph courtesy of Beverly Buck

Solid-state detectors, which monitor the movement of electrons through silicon and other semiconductors, are the primary detectors used today for detecting and measuring radioactivity.

Ionizing radiation can also be detected in a cloud chamber. You will try out this detection method in the following activity.

C.5 LABORATORY ACTIVITY: CLOUD CHAMBERS

GETTING READY

A *cloud chamber* is a glass container filled with air that is saturated with water or other vapor, like the air on a humid day. If cooled, the air inside will become supersaturated. (Recall from the Water unit, page 43, that this is an unstable condition.) If a radioactive source is placed near a cloud chamber filled with supersaturated air, the radiation will ionize the air inside as it passes through the chamber. Vapor condenses on the ions formed, leaving a white trail behind each passing radioactive emission, revealing the path of the particle or ray. Figure 19 (page 340) is a photograph taken of particle tracks under similar conditions.

The cloud chamber you will use consists of a small plastic container and a felt band moistened with 2-propanol (isopropyl alcohol). The alcohol evaporates faster than water and saturates air more readily. The cloud chamber will be chilled with dry ice to promote supersaturation and cloud formation.

▶ *Cloud chamber trails resemble the "vapor" trails from high-flying aircraft.*

▶ *The temperature of dry ice, solid carbon dioxide, is −78 °C.*

PROCEDURE

1. Fully moisten the felt band inside the cloud chamber with alcohol. Also place a small quantity of alcohol on the container bottom.

2. Using gloves and forceps, quickly place the radioactive source on the chamber bottom. Replace the lid.

3. To cool the chamber, embed it in crushed dry ice. Be sure the chamber remains level.

4. Leave the chamber on the dry ice for three to five minutes.

5. Dim or turn off the room lights. Focus the light source through the container at an oblique angle (not straight down) so that the chamber base is illuminated. (If you do not observe any vapor trails, try shining the light through the container's side, instead.)

6. Observe the air in the chamber near the radioactive source. Record your observations.

QUESTIONS

1. What differences—if any—did you observe among the tracks?

2. Which type of radiation do you think would make the most visible tracks? Why?

3. What is the purpose of the dry ice?

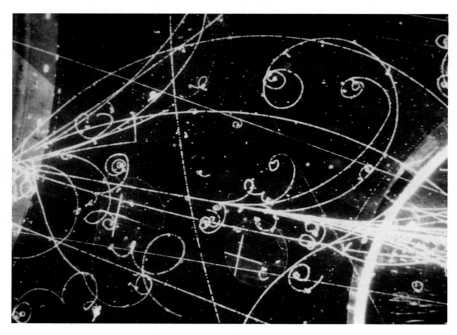

Figure 19 A photograph of particle tracks.

You have seen that the radiation observed in a cloud chamber is emitted from an unstable, radioactive isotope that is eventually converted to a stable, nonradioactive isotope. Do you think it might be possible to reverse the process, converting a stable isotope into an unstable, radioactive isotope? You will find the answer to this question in the next section.

C.6 ARTIFICIAL RADIOACTIVITY

In 1919, to find out whether he could make radioactive elements in the laboratory, Ernest Rutherford enclosed nitrogen gas in a glass tube and bombarded it with alpha particles. After analyzing the gas remaining, he found that some nitrogen had been converted to an isotope of oxygen, according to this equation:

$$\underset{\text{helium-4}}{^4_2\text{He}} \quad + \quad \underset{\text{nitrogen-14}}{^{14}_7\text{N}} \quad \rightarrow \quad \underset{\text{oxygen-17}}{^{17}_8\text{O}} \quad + \quad \underset{\text{hydrogen-1}}{^1_1\text{H}}$$

Rutherford had produced the first synthetic or artificial ***transmutation*** of the elements—the conversion of one element to another. However, the new element, O-17, was not radioactive.

Rutherford continued his work, but he was limited by the moderate energies of the alpha particles he had available. In later years other scientists, using improved tools, would reach Rutherford's goal of creating synthetic *radioactive* nuclei.

By 1930, *particle accelerators* were developed that could produce the highly energetic particles needed for additional bombardment reactions. The first radioactive artificial isotope was produced in 1934 by the French physicists Frédéric and Irène Joliot-Curie (the son-in-law and daughter of Pierre and Marie Curie). They bombarded aluminum with alpha particles, producing radioactive phosphorus-30:

$$_{13}^{27}\text{Al} + _{2}^{4}\text{He} \rightarrow _{15}^{30}\text{P} + _{0}^{1}\text{n}$$

Since then, many transformations of one element to another element have been completed; in addition, new radioactive isotopes of various elements have been synthesized. Many of the important diagnostic radioisotopes noted in Table 11 (page 337) are synthetic. Tc-99m, for example, is both a synthetic element and a radioisotope.

In addition, a number of new elements have been synthesized in nuclear reactions. From 1940 to 1961, Glenn Seaborg and his co-workers at the University of California at Berkeley discovered ten new elements—atomic numbers 94 to 103—a prodigious feat. None of these ten elements occurs naturally. All are made by high-energy bombardments of heavy nuclei with various particles. For example, alpha bombardment of plutonium-239 produced curium-242:

$$_{94}^{239}\text{Pu} + _{2}^{4}\text{He} \rightarrow _{96}^{242}\text{Cm} + _{0}^{1}\text{n}$$

Bombarding Pu-239 with neutrons yielded americium-241, a radioisotope now used in home smoke detectors:

$$_{94}^{239}\text{Pu} + 2\,_{0}^{1}\text{n} \rightarrow _{95}^{241}\text{Am} + _{-1}^{0}\text{e}$$

Seaborg's work was recognized through his receiving the 1951 Nobel prize in chemistry. Albert Ghioso, a colleague of Seaborg's, has led the way in producing new elements beyond lawrencium (element 103). One

▶ *In 1994–96, a research group in Darmstadt, Germany reported the formation of several atoms of elements 110, 111, and 112, all formed by high-energy nuclear collisions. The reports need to be confirmed.*

Codiscoverers of Element 106, Lawrence Berkeley Laboratory, 1974.
Left to right: Matti Nurmia, Jose R. Alonso, Albert Ghiorso,
E. Kenneth Hulet, Carol T. Alonso, Ronald W. Lougheed, Glenn Seaborg,
Joachim M. Nitschke.

Glenn Seaborg being honored for discovering elements 94–103.

such element is 106, produced by bombarding a californium-249 target with a beam of oxygen-18 nuclei to produce the 263 isotope of element 106. To honor Seaborg's work, some nuclear chemists have proposed naming element 106 seaborgium, with the symbol Sg. This symbol is supported by the American Chemical Society, but it has not yet been approved by the International Union of Pure and Applied Chemistry, the official body for naming elements. In fact, there is continuing controversy regarding the naming of elements 104 to 109.

YOUR TURN

Bombardment Reactions

Every bombardment reaction involves four particles:

- The *target nucleus* is the stable isotope that is bombarded.
- The *projectile* (bullet) is the particle fired at the target nucleus.
- The *product* is the heavy nucleus produced in the reaction.
- The *ejected particle* is the light nucleus or particle emitted from the reaction. More than one ejected particle may be released (see the Cm-246 example below).

For example, consider the Joliot-Curies' production of the first synthetic radioactive isotope. The four types of particles involved are identified below:

target nucleus	projectile		product nucleus	ejected particle
$^{27}_{13}\text{Al}$	$+$ $^{4}_{2}\text{He}$	$\rightarrow$	$^{30}_{15}\text{P}$	$+$ $^{1}_{0}\text{n}$
aluminum-27	alpha particle		phosphorus-30	neutron

Nobelium (No) was produced by bombarding curium (Cm) with nuclei of a light element. We can identify this element by completing this equation:

$$^{246}_{96}\text{Cm} + \text{?} \rightarrow\ ^{254}_{102}\text{No} + 4\,^{1}_{0}\text{n}$$

As you learned earlier, completing nuclear equations involves balancing atomic and mass numbers. Because the sum of the product atomic numbers is 102, the projectile must have the atomic number 6; ($96 + 6 = 102$), a carbon atom. The total mass number of products is 258, indicating that the projectile must have been carbon-12; ($246 + 12 = 258$). The completed equation is

target nucleus		projectile		product nucleus		ejected particles
$^{246}_{96}\text{Cm}$	$+$	$^{12}_{6}\text{C}$	$\rightarrow$	$^{254}_{102}\text{No}$	$+$	$4\,^{1}_{0}\text{n}$

For the following items, complete the equations by supplying missing numbers or symbols. Name each particle. Then identify the target nucleus, projectile, product nucleus, and ejected particle.

1. In this reaction, a naturally occurring nonradioactive isotope is converted into a medically useful, radioactive form of the same element.

$$^{59}_{27}\text{?} + \,^{?}_{?}\text{n} \rightarrow\ ^{60}_{?}\text{?}$$

2. Until its synthesis in 1937, technetium (Tc) existed only as an unfilled gap in the periodic table; all its isotopes are radioactive. Any technetium originally on the Earth has disintegrated. Technetium, the first new element produced artificially, is now used extensively in industry and medicine. Each year, for example, millions of bone scans are obtained through the use of technetium.

$$^{96}_{42}\text{?} + \,^{?}_{?}\text{H} \rightarrow\ ^{97}_{43}\text{?} + \,^{1}_{0}\text{?}$$

3. In 1992, the GSI research group in Darmstadt, Germany created element 109 by the bombardment of Bi-209 nuclei. The name Meitnerium (symbol Mt) has been proposed for the new element to honor Lise Meitner, an Austrian physicist who first proposed the concept of nuclear fission (see Section D.1).

$$^{58}_{?}\text{?} + \,^{209}_{83}\text{?} \rightarrow\ ^{?}_{109}\text{Mt} + \,^{1}_{0}\text{?}$$

Not only does the ability to transform one element into another give us powerful technological capabilities, but it also has changed the way we view elements. Within the last 50 years, 17 confirmed *transuranium* elements (those with atomic numbers greater than uranium's number of 92) have been added to the periodic table. For that reason the periodic table has expanded to include a new series of elements, the *actinide series.*

CHEMQUANDARY

TRANSMUTATION OF ELEMENTS

Ancient alchemists searched in vain for ways to transform lead or iron into gold (transmutation of the elements). Has transmutation now become a reality? From what you know about nuclear changes, do you think that lead or iron could be changed into gold? If you do, write equations for the reactions.

? PART C: SUMMARY QUESTIONS

1. Suppose you have a sample containing 800 nuclei of a radioactive isotope. After one hour, only 50 of the original nuclei remain. What is the half-life of this isotope?

2. You have 400 μg (micrograms) of a radioisotope with a half-life of five minutes. How much will be left after 30 minutes?

3. Gold (Au) exists primarily as one natural isotope, Au-197. A variety of synthetic radioisotopes of gold, with mass numbers from 177 to 203, have been produced. Suppose you have a 100-mg sample of pure Au-191, which has a half-life of 3.4 hours.
   a. Make a graph of its decay curve.
   b. Estimate how much Au-191 will remain after
      (1) 10 hours.
      (2) 24 hours.
      (3) 34 hours.
   c. List two reasons why synthetic gold isotopes would not be a good substitute for natural gold in jewelry.

4. Define the term tracer as it applies to medical uses of radioisotopes.

5. a. Describe two medical uses—one internal and one external—for radioisotopes.
   b. Why is it beneficial for radioisotopes used internally to have short half-lives and those used externally to have long half-lives?

6. Americium-241, a synthetic radioisotope with a half-life of 450 years, finds practical use in household smoke detectors. Why would an alpha emitter such as americium-241, with a relatively long half-life, be suitable for such an application?

7. Describe three ways that ionizing radiation can be detected.

8. List the three types of radiation given off by radioactive sources in order of increasing penetrating power.

9. The following two equations represent the synthesis of two transuranium radioisotopes by Seaborg and co-workers. For each, name the target nucleus, the projectile, the product nucleus, and the ejected particle(s).
   a. $^{242}_{96}\text{Cm} + ? \rightarrow ^{245}_{?}\text{Cf} + ^{1}_{0}\text{n}$
   b. $^{238}_{92}\text{U} + ^{1}_{0}\text{n} \rightarrow ^{239}_{93}? + ?$

10. The equations from Question 9 indicate that curium-242 (half-life = 163 days) forms Cf-245 (half-life of 44 minutes), and U-238 (half-life = 4.5×10^9 years) forms Np-239 (half-life of 2.4 days). Which radioisotope in Question 9 would be more useful as an energy source for a space probe? Why?

EXTENDING YOUR KNOWLEDGE (OPTIONAL)

- Neutrinos are fundamental particles with very little mass (much less than an electron), and no charge. Look into the discovery of the neutrino. It is a story of great scientific conviction and persistence. Would you expect an instrument such as a cloud chamber to show neutrino trails? Explain your answer.

- A variety of charged particles (such as alpha and beta particles), uncharged particles (neutrons), and gamma rays have been used as projectiles in nuclear bombardment research. What are some advantages and disadvantages of each? How are the velocities of these nuclear projectiles controlled? How are they "aimed?" Topics you may wish to explore include electrostatic generators, cyclotrons, and linear accelerators. Also of interest is the role played by nuclear reactors in synthesizing radioisotopes.

- U-235 is the parent isotope for the decay series ending with Pb-207. The entire series can be represented as follows:

$$^{235}_{92}U \rightarrow [\text{many steps}] \rightarrow {}^{207}_{82}Pb + ? \, {}^{4}_{2}He + ? \, {}^{0}_{-1}e$$

Based on overall changes in atomic numbers and mass numbers from the parent isotope to the final stable product, how many total alpha and beta particles must be emitted during the decay of one atom of U-235?

- Look into the use of radioactive isotopes for dating artifacts and rocks, and for judging the authenticity of paintings.

D NUCLEAR ENERGY: POWERING THE UNIVERSE

In the 1930s a bombardment reaction involving uranium-238 unlocked a new energy source and led to the development of both nuclear power and nuclear weapons. This was the beginning of the nuclear age. How did scientists first unleash the enormous energy of the atom, and how has nuclear energy been used for generating electricity? Nuclear energy provides our focus in Part D.

D.1 SPLITTING THE ATOM

In the early years of World War II, the German scientists Otto Hahn and Fritz Strassman bombarded uranium with neutrons in the hope of creating a more massive nucleus. Much to their surprise, they found that one reaction product was barium—an atom with only about half the atomic weight (approximately 137) of uranium (235).

Austrian physicist Lise Meitner.

The first to understand what had happened was an Austrian physicist, Lise Meitner, who had worked with Strassman and Hahn earlier. She suggested that neutron bombardment had split the uranium atom into two parts that were nearly equal in size. Other scientists quickly verified Meitner's explanation. The world had witnessed its first *nuclear fission reaction.*

Hahn and Strassman had actually triggered an array of related fission reactions. One reaction that might have produced the barium is this:

$$^{235}_{92}U \ + \ ^{1}_{0}n \ \rightarrow \ ^{140}_{56}Ba \ + \ ^{93}_{36}Kr \ + \ 3\,^{1}_{0}n$$
$$\text{U-235} \quad \text{neutron} \quad \text{barium-140} \quad \text{krypton-93} \quad \text{neutrons}$$

Scientists soon found that the uranium-235 nucleus can *fission* (split) into numerous combinations of products, but usually into a heavier element (such as barium) and a less-massive element (such as krypton). Here is another example:

$$^{235}_{92}U \ + \ ^{1}_{0}n \ \rightarrow \ ^{143}_{54}Xe \ + \ ^{90}_{38}Sr \ + \ 3\,^{1}_{0}n$$
$$\text{U-235} \quad \text{neutron} \quad \text{xenon-143} \quad \text{strontium-90} \quad \text{neutrons}$$

Sculptor Henry Moore created this work to commemorate the birth of the nuclear age.

Uranium-235 is the only naturally occurring isotope that undergoes fission with slow (thermal) neutrons. Many synthetic nuclei—in particular uranium-233, plutonium-239, and californium-252—also fission under neutron bombardment.

D.2 THE STRONG FORCE

The fission of uranium-235 and other fissionable isotopes releases at least a million times more energy than that produced in any chemical reaction. This is what makes nuclear explosions so devastating and nuclear energy so powerful.

The source of nuclear energy lies in the force that holds protons and neutrons together in the nucleus. This force, called the **strong force,** is fundamentally different from—and a thousand times stronger than—the electrical forces that hold atoms and ions together in chemical compounds. The strong force has a very short range, extending only across the atomic nucleus.

How is the strong force related to the energy released in nuclear reactions? Recall what you learned about the energy from petroleum and food. In chemical reactions, when chemical bonds are stronger in products than in reactants, energy is released—often as heat. In nuclear reactions, the strong force can be stronger in product nuclei than in those of reactants, and this, too, releases energy. But this energy is so much greater than that from chemical reactions that something more must be involved.

In chemical and physical changes, energy may be converted from one form to another, but no overall energy loss or gain occurs. Nor does mass increase or decrease in chemical reactions.

In nuclear reactions, though, it's not just mass or just energy that is conserved, but the two together. Mass and energy are related according to Einstein's famous equation, $E = mc^2$. The energy released (E) is equal to the mass lost (m) multiplied by the speed of light squared (c^2). The complete conversion of just one gram of nuclear matter to energy would release energy equal to that produced by burning 700,000 gallons of octane fuel!

When changes take place in atomic nuclei, energy and mass are interconverted in accord with the Einstein equation. In nuclear fission, the mass of the fission products is slightly less than the mass of the atom that originally split. The mass loss is so small that it does not affect the mass numbers of reactants or products. Even so, the conversion of these small quantities of mass into energy accounts for the vast power of nuclear energy.

D.3 CHAIN REACTIONS

Another important result of fission is the emission of two or three neutrons from the fragments of each nucleus. These neutrons can keep the reaction going by splitting more nuclei. The result is a **chain reaction.**

Recall, however, that most of an atom is empty space. The probability that a neutron from a fission reaction will split another nucleus

▶ *More than 200 radio-isotopes have been identified among the products of nuclear fission reactions. These products have half-lives ranging from a fraction of a second to millions of years. Such a diverse mixture greatly increases the difficulty of waste disposal from nuclear reactors.*

▶ *These observations are the basis for the law of conservation of energy and the law of conservation of matter.*

Uranium metal is refined and shipped in a "derby," or 167 kg.

depends on the amount of fissionable material available. Unless a certain *critical mass* (minimum quantity) of fissionable material is present, the neutrons cannot encounter enough nuclei to sustain the chain reaction.

When a critical mass of fissionable material is present, a chain reaction occurs (Figure 20). Recognition that such reactions were possible and could be utilized in military weapons came shortly after the first fission reaction was observed. Germany and the United States soon initiated projects to build "atomic bombs." U.S. aircraft dropped atomic bombs on Hiroshima and Nagasaki in Japan in 1945, near the end of World War II.

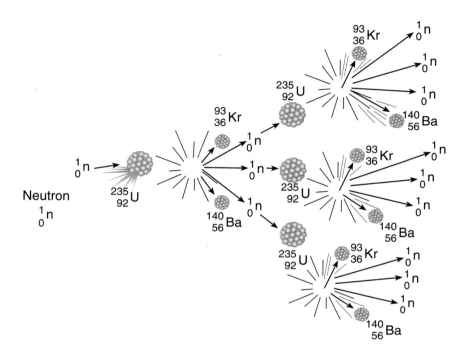

Figure 20 A nuclear chain reaction. The reaction is initiated (left) by a neutron colliding with a uranium-235 nucleus. The reaction continues, in a chain-like fashion, as long as the emitted neutrons encounter other fissionable atoms. The products of each individual fission can vary.

The following activity will give you a clearer picture of the dynamic nature of a chain reaction.

D.4 YOU DECIDE: THE DOMINO EFFECT

1. Obtain a set of dominoes or similarly shaped objects. Set them up as shown in Figure 21 (page 349). Arrange the dominoes so that when each one falls, it will knock over others in succession. The more dominoes you use, the more dramatic the effect.

Figure 21 The "domino" effect. Dominoes arranged to simulate a chain reaction.

2. Knock over the lead domino.

   a. Observe the effect on the other dominoes.

   b. Would you call this an example of an *expanding* (out of control) or *limited* (controlled) chain reaction? Why?

3. Set up the dominoes again. This time, set up only a very few dominoes the same way as in the first arrangement. Arrange the others so that when they fall, they will knock over only *one* other domino or none at all. Make a sketch of your design.

4. Knock over the lead domino. Observe the reaction. Is it an expanding or a limited chain reaction? Why?

5. Which of the two domino-tumbling arrangements is a better model of a nuclear bomb explosion? Why?

6. Imagine that you had the dominoes arranged as in Step 3.

   a. What could you do to stop the reaction once it began?

   b. Do you think this method would be equally useful in stopping the reaction you observed in Step 2? Why?

7. a. In what way is this domino simulation like a nuclear chain reaction?

   b. How is it different?

   c. Can you think of another way to model a nuclear chain reaction?

By controlling the fission rate in a chain reaction, nuclear engineers make it possible to use the released heat to generate electricity. In the following section you will learn how this is done.

D.5 NUCLEAR POWER PLANTS

The first nuclear reactors were designed solely to produce plutonium-239 for use in World War II atomic weapons. Today, 111 commercial nuclear reactors (like that in Figure 22) produce electricity in the United States

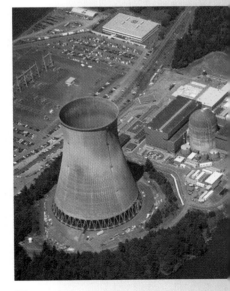

Figure 22 Cooling tower and containment building in the Trojan nuclear power plant. The nuclear reactor is in the domed containment building.

Nuclear Energy: Powering the Universe

(Figure 23). In 1992, such nuclear power plants met about 6% of the nation's total energy needs, and generated nearly 22% of U.S. electrical power. In 1991, 15 states generated over 25% of their electricity from nuclear power plants, and six states generated 50% of their electricity that way (Figure 24, page 351). On a global scale, an estimated 426 nuclear reactors in over 30 nations produce one out of every six units of world electricity. Nearly 100 new nuclear power reactors (none in the United States) were under construction worldwide in 1995.

Most power plants generate electricity by producing heat to boil water. The resulting steam spins the turbines of giant generators, producing electrical energy. A nuclear power plant operates the same way. But coal-, oil-, and natural gas-powered generators use the heat of combustion of fossil fuels to boil water; nuclear power plants use the heat of nuclear fission reactions.

The essential parts of a nuclear power plant, diagrammed in Figure 25 (page 351), are fuel rods, control rods, the moderator, the generator, and the cooling system.

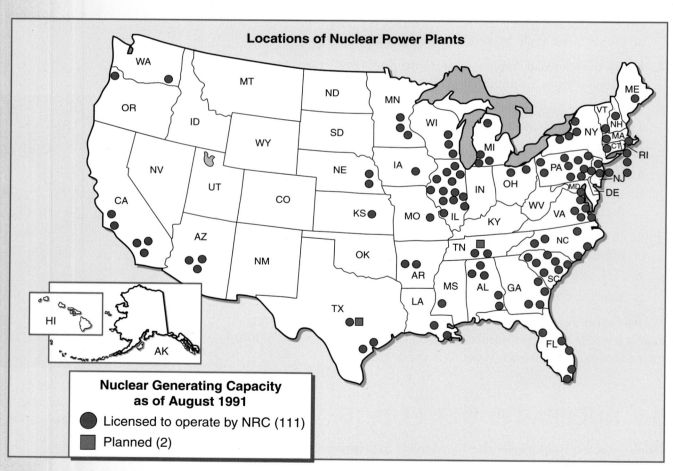

Figure 23 Locations of U.S. nuclear power plants.

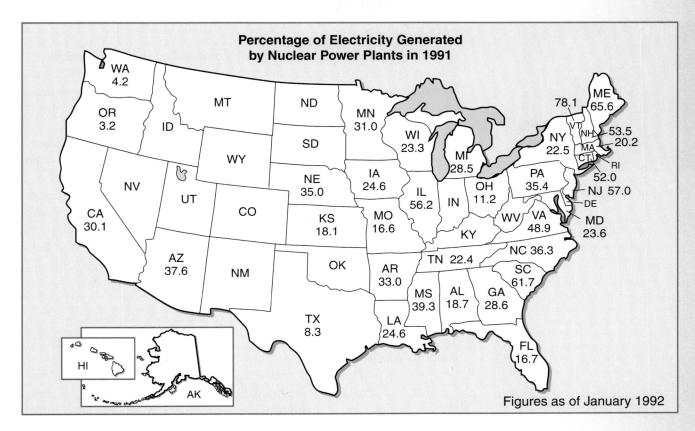

Figure 24 Percentage of electricity generated by nuclear power plants.

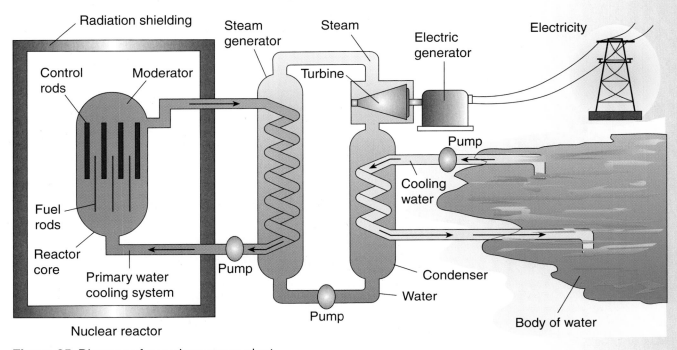

Figure 25 Diagram of a nuclear power plant.

Nuclear fuel rods.

▶ *In contrast, weapons-grade uranium usually contains 90% or more of the fissionable U-235 isotope.*

FUEL RODS

Coal-fired power plants consume tons of coal daily. In contrast, the fuel for a nuclear reactor occupies a fraction of the volume needed for coal, and is loaded in only about once each year. Nuclear reactor fuel is in the form of pellets about the size and shape of short pieces of chalk. One uranium fuel pellet has the energy equivalent of 1 ton of coal or 126 gallons of petroleum. As many as 10 million fuel pellets may be used in one nuclear power plant. These pellets are arranged in long, narrow steel cylinders—the fuel rods. The fission chain reaction takes place inside these rods.

The fuel pellets contain uranium dioxide, UO_2. Most of this uranium is the nonfissionable uranium-238 isotope. Only 0.7% of natural uranium is U-235, the fissionable isotope. In a nuclear reactor, this quantity of U-235 has been "enriched" to about 3%, still only a small fraction of fissionable material. Nevertheless, it is enough to sustain a chain reaction, but far from enough to cause a nuclear explosion.

CONTROL RODS

The nuclei of certain elements, such as boron or cadmium, can absorb neutrons very efficiently. Such materials are placed in control rods. As neutrons are needed to trigger nuclear fission, their absorption by the non-fissionable control rods reduces the number of neutrons available to cause fission. The rate of the chain reaction is controlled by moving the control rods up or down between the fuel rods (see Figure 26). Or, the reaction can be terminated altogether by dropping the control rods all the way down between the fuel rods.

Figure 26 (a) Control rod drive shafts.

(b) Close-up of fuel rod assembly.

MODERATOR

In addition to fuel rods and control rods, the core of a nuclear reactor contains a moderator, which slows down high-speed neutrons so they can be more efficiently absorbed by fuel rods to enhance the probability of fissions. *Heavy water* (symbolized D_2O, where D is the hydrogen-2 isotope), regular water (termed *light water* by nuclear engineers), and graphite (carbon) are the three most common moderator materials.

GENERATOR

In all commercial nuclear reactors, the fuel and control rods are surrounded by a system of circulating water. In simpler reactors, the fuel rods boil this water, and the resulting steam spins the turbines of the electrical generator. In another type of reactor, this water is superheated under pressure, and does not boil. Instead, it circulates through a heat exchanger, where it boils the water in a *second* cooling loop. This is the type of reactor shown in Figure 25 (page 351).

COOLING SYSTEM

The steam that has moved past the turbines is cooled by water taken from a nearby body of water. This cooling water, now heated, flows back to the environment. The cooled steam condenses to liquid water and recirculates inside the generator.

So much heat is generated that some also must be released into the air. The largest and most prominent feature of most nuclear power plants is a tall, gracefully tapered concrete cylinder—the cooling tower, where excess thermal energy is exhausted. Many people mistakenly assume that the cooling tower is the nuclear reactor.

► *The white plumes seen rising from such cooling towers are condensed steam, not smoke.*

The reactor is designed to prevent the escape of radioactive material, should some malfunction cause radioactive material—including cooling water—to be released within the reactor itself. The core of a nuclear reactor is surrounded by concrete walls 2 to 4 meters thick. Further protection is provided by the reactor's enclosure in a building having thick walls of steel-reinforced concrete, designed to withstand a chemical explosion or an earthquake. The reactor building is capped with a domed roof that can withstand significant internal pressure. The Chernobyl reactor building in the Ukraine did not have such a domed containment roof. In the Chernobyl incident, fire from burning graphite (a moderator) generated gas pressure sufficient to blast away the 100-ton steel plate covering the reactor building (see Section E.3).

Nuclear fission is not the only source of nuclear energy. In the next section you will learn about the nuclear reaction that fuels the stars (and, indirectly, all living matter).

D.6 NUCLEAR FUSION

In addition to releasing energy by splitting massive nuclei (fission), we can also generate large quantities of nuclear energy by *fusing*—or combining—nuclei. *Nuclear fusion* involves forcing two relatively small nuclei to combine into a new, heavier atom. As with fission, the energy released by nuclear fusion can be enormous, again because of the conversion of mass into energy. Gram for gram, nuclear fusion liberates even more energy than nuclear fission—between three to ten times as much.

Nuclear fusion powers the sun and other stars. Scientists believe that the sun formed when a huge quantity of interstellar gas, mostly hydrogen, condensed under the force of gravity. When gravitational pressure became great enough to heat the gas to about 15 million degrees Celsius, hydrogen atoms began fusing into helium. The sun began to shine, releasing the energy that drives our solar system. Scientists estimate that the sun, believed to be about 4.5 billion years old, is about halfway through its life cycle.

The following nuclear equation represents the sum of several important nuclear fusion reactions that occur in the sun. The ejected particles are *positrons,* particles with the mass of an electron, but carrying a positive charge.

$$4\,{}^{1}_{1}\text{H} \quad \rightarrow \quad {}^{4}_{2}\text{He} \quad + \quad 2\,{}^{0}_{+1}\text{e}$$
$$\text{hydrogen-1} \qquad \text{helium-4} \qquad \text{positrons}$$

How much energy does a nuclear fusion reaction produce? By adding the masses of products in the above equation and subtracting the masses of reactants, we find that 0.0069005 g of mass is lost when one mole of hydrogen atoms undergo fusion to produce helium. Using Einstein's equation, $E = mc^2$, we can calculate that the energy released through the fusion of one gram (1 mol) of hydrogen atoms is 6.2×10^8 kJ. Here is one way to put that very large quantity of energy in perspective: The nuclear energy released from the fusion of one gram of hydrogen-1 equals the thermal energy released by burning nearly 5,000 gallons of gasoline or 20 tons of coal.

Nuclear fusion reactions have been used in the design of powerful weapons. The hydrogen bomb—also known as a thermonuclear device—is based on a fusion reaction that is ignited by a small atomic (fission) bomb.

Can we also harness the energy of nuclear fusion for beneficial purposes, such as producing electricity? That remains to be seen, but scientists have spent more than several decades pursuing this possibility (Figure 27).

The challenge of maintaining the incredibly high temperatures (at least 100 million degrees) needed to sustain controlled nuclear fusion has been difficult to meet. The nuclear fusion facility at Princeton University has achieved temperatures sufficient to initiate fusion, but so far the quantity of energy put into the process is more than the energy released.

When scientists succeed in controlling nuclear fusion in the laboratory, there is still no guarantee that fusion reactions will become a practical source of energy. Low-mass isotopes to fuel such reactors are plentiful and inexpensive, but confinement of the reaction could be very expensive. Furthermore, although the fusion reaction itself produces less radioactive waste than nuclear fission, capturing and shielding the heat of the reaction could generate just as much radioactive waste as is produced by fission power plants.

In splitting and fusing atoms, we have unleashed the energy that fuels the universe. Much good has come of it. But we have also raised scientific, social, and ethical questions that have not been answered. Along with great benefits, there are also great risks. How much risk is worth the potentially great benefit? We will explore this issue next.

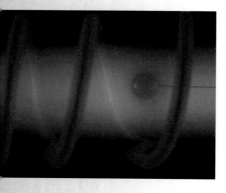

Figure 27 Pellet containing deuterium and tritium in laser path for laser fusion.

► *Bombardment of the containment walls by neutrons released by nuclear fusion would produce large quantities of radioactive materials.*

CHEMISTRY *AT WORK*

Nuclear Chemistry at Full Speed Ahead

As a Petty Officer in the Reactor Laboratory Department aboard the aircraft carrier *USS Dwight D. Eisenhower*, **Bryant Cleveland**'s job is to monitor and adjust water circulating systems in the ship's nuclear reactor. Water on one side of the system (which is kept separate from all other water on the ship) keeps the nuclear reactor cool, and transfers heat energy to a secondary system, which converts the heat into electrical and steam energy for the ship.

"One key to safe and efficient operation," explains Bryant, "is to use only 'Grade A' water—that is, highly demineralized, with a pH level between 6.0 and 8.0 and with less than 0.1 ppm of chlorine." Using Grade A water helps protect the systems' tanks, pipes, and other channels from corrosion.

Whenever technicians need to add substances to the coolant system, they must first make room by discharging water into holding containers on the ship. They filter the water being discharged to trap certain radioactive isotopes, maintaining close watch on isotope levels before and after discharge. Bryant and his shipmates work in shifts all day and night to sample, analyze, and maintain proper chemistry throughout the reactor's coolant systems.

In high school, Bryant took chemistry and science courses. After high school, he joined the U.S. Navy, where he completed studies in algebra, physics, and chemistry, as well as in highly specialized areas such as heat transfer and fluid flow. He also received training on various reactor systems.

◆ *Sights aboard the USS Dwight D. Eisenhower.*

Photographs courtesy of *USS Dwight D. Eisenhower* Photo Lab

355

1. a. Name three isotopes that can undergo neutron-induced fission.

   b. How is fission different from the radioactive decay that Ernest Rutherford explained?

2. Complete this equation for a nuclear fission reaction:

$$^{235}_{92}U + ?n \rightarrow ^{87}_{?}Br + ^{146}_{57}? + ?$$

3. a. What is a nuclear chain reaction?

   b. Under what conditions will a chain reaction occur?

   c. What is the difference between an expanding chain reaction and a controlled one?

4. Why is it impossible for a nuclear power plant to become a nuclear bomb?

5. a. In what ways is a nuclear power plant similar to a facility that burns fossil fuel?

   b. In what ways is it different?

   c. Describe one advantage and one disadvantage of each type of power plant (nuclear and conventional fuel).

6. Explain how the system that controls the chain reaction in a nuclear power plant works.

7. Explain the difference between nuclear fission and nuclear fusion.

8. Explain how both of these statements can be true:

   a. Nuclear fusion has not been used for an energy source in the world.

   b. Nuclear fusion is the number one energy source in the world.

EXTENDING YOUR KNOWLEDGE (OPTIONAL)

- Investigate the *breeder reactor* (a nuclear reactor that produces more nuclear fuel than it uses).

- Investigate possible answers to the question, How and where did the elements that make up our bodies, the earth, and the universe originate?

E LIVING WITH BENEFITS AND RISKS

The nineteenth century poet William Wordsworth used the phrase, "Weighing the mischief with the promised gain . . ." in his assessment of technological advancement (in this case, the railroad). Since the dawn of civilization, however, people have accepted the risks of new technologies in order to reap the benefits. Fire, one of civilization's earliest tools, gave people the ability to cook, keep warm, and forge tools from metals. Of course, when fire is out of control, it can destroy property and life. Every technology offers its benefits at a price.

Some people oppose new technologies in general, arguing that the benefits are not worth the risks. For example, many argued against the introduction of trains and automobiles. More recently, nuclear power has stirred a higher level of intense, continuing opposition and concerns than have most previous technologies.

In the following sections, you will examine how to balance for yourself the benefits and risks of any new technology. One must weigh the potential benefits of the technology against the harm or threats it poses to individuals, society, and the environment.

One way to define an "acceptable" new technology is one having a relatively low probability of harm—that is, having benefits that outweigh dangerous risks. Unfortunately, risk–benefit analysis—weighing what Wordsworth called "mischief" against "promised gain"—is far from an

Every form of transportation has its risks!

exact science. For instance, some technologies present high risks immediately; others may present chronic, low-level risks for years or even decades. Many risks are impossible to measure with certainty. Some potential risks can be controlled by an individual, while others must be controlled by government. In short, it is very difficult to compare or "weigh" various risks. Still, decisions must be made.

Before considering the pros and cons of nuclear power, take a few minutes to weigh the benefits and risks of different modes of travel to complete a journey. This will help sharpen your ability to think in risk–benefit terms.

▶ *In many cases, electing not to make a decision is, in fact, an actual decision, one with its own risks and benefits.*

E.1 YOU DECIDE: THE SAFEST JOURNEY

Suppose you want to visit a friend who lives 500 miles (800 km) away using the safest means of transportation. Insurance companies publish reliable statistics on the safety of different methods of travel. Using Table 12, answer these questions:

Table 12	Risk of Travel
Mode of Travel	**Distance at which one person in a million will suffer accidental death***
Bicycle	10 miles
Automobile	100 miles
Scheduled airline	1000 miles
Train	1200 miles
Bus	2800 miles

*On average, chance of death is increased by 0.000001.

1. Assume there is a direct relationship between distance traveled and chance of accidental death. (That is, assume that doubling distance would double your risk of accidental death.)

    a. What is the risk factor value for traveling 500 miles by each mode of travel in the table? For example, Table 12 shows that the risk factor (chance of accidental death) for biking increases by 0.000001 for each 10 miles. Therefore, the bike-riding risk factor in visiting your friend would be

    $$500 \text{ mi.} \times \frac{0.000001 \text{ risk factor}}{10 \text{ mi.}} = 0.00005 \text{ risk factor}$$

    b. Which is the safest mode of travel (the one with the smallest risk factor)?

    c. Which is the least safe?

2. The results obtained in Question 1 might surprise many people. Why?

3. a. List the benefits of each mode of transportation.

   b. In your view, do the benefits of riskier ways to travel outweigh their increased risks? Explain your reasoning.

4. Do you think these same statistics will be true 25 years from now? Why?

5. What factor(s), beyond the risk to personal safety, would you include in your risk–benefit analysis before deciding how to travel?

6. a. Which mode of travel would you choose? Why?

   b. Would someone else's similar risk–benefit analysis always lead to the same decision as yours? Why or why not? (Putting it another way, is there *one* "best" way to travel?)

CHEMQUANDARY

RISK-FREE TRAVEL?

Is there any way to visit your friend that would be completely risk-free? Would it be safer not to visit your friend at all?

Some uses of nuclear energy and radioactivity—compact power sources, radioisotopes for diagnosing and treating disease, smoke detectors, and many research and development applications—are clearly worth their risks. Other uses, such as stockpiling nuclear weapons, are far more controversial.

Our goal is to provide sufficient understanding to help you weigh the benefits and risks of new applications of nuclear technology now and in the future. We will begin by examining some benefits of radioisotopes.

In Section C.3 we discussed the uses of radioisotopes for medical diagnosis and therapy; these uses offer significant benefits at low risk. Over the past 50 years, radioisotopes have also been used in numerous industrial and research applications, such as tracers. For example, an oil company can use a particular tracer in its gasoline to monitor the progress of its own gasoline through a pipeline containing gasoline from other oil companies. The tracer indicates to the company when its gasoline arrives at a specific pumping station. In another use, an environmental researcher studying the effects of commercial fertilizer runoff into streams can use a phosphorus-32 tracer to track the rate and location of the fertilizer runoff, because fertilizers contain phosphorus.

Radioisotopes are also used as irradiation sources for industrial processes, including sterilizing medical disposables such as injection needles, surgical masks, and gowns. Irradiation is also used to seal plastic containers. The irradiation procedure is relatively simple: An irradiation chamber contains pellets of a gamma-emitting radioisotope, typically cobalt-60. The material to be irradiated is placed on a conveyer belt that moves past the radioisotope at a specified rate.

We began this unit by having you consider the case of irradiated foods. By killing microorganisms that cause spoilage (Figure 28), irradiation extends the time foods remain fresh. As pointed out in Section A.1, food irradiation has been approved for use in over 30 nations for more than 40 different foods. It has been used to preserve astronauts' food for space missions. There are about 160 industrial gamma-irradiation facilities worldwide. Even so, there are risk–benefit concerns associated with producing and buying irradiated foods. These concerns lead us to question whether irradiation has made these foods highly radioactive. Are they safe to eat?

The answer to the first question is clear: Irradiation does not make foods radioactive. Answering the question of food safety, however, is not quite as straightforward. One concern sometimes raised regarding food irradiation is that the energetic, ionizing radiation might break and rearrange chemical bonds in irradiated food. This could possibly create potentially harmful species called unique radiolytic products (URP), unique in that they are formed only by irradiation. Other species produced by irradiation are also formed during regular processing or cooking of foods. Research has not convinced most food irradiation specialists that URPs are formed. At low radiation levels, the potential threat of URPs is regarded by many authorities as insignificant.

Even if consumers decide irradiated foods are safe to use, they may not buy them because they cost more than their nonirradiated counterparts. Ultimately, consumer choices may be made based on cash registers rather than rems.

The benefits of increasing shelf life and reducing spoilage must be weighed against URPs and other such risks. It is especially important to recognize that eating spoiled foods is highly risky. Food-borne illnesses cost more than $10 billion each year in medical care and lost productivity.

Figure 28 Irradiated potatoes (right) remain fresh after two weeks. Non-irradiated potatoes (left) are moldy and sprouting roots after the same period of time.

Risks associated with alternative food-preservation methods, those that do not use radiation, must also be kept in mind.

YOUR TURN

Assessing Risks and Benefits

1. Review the questions you raised in Section A.1 of this unit regarding food irradiation.

   a. What questions can you now answer based on the material in this unit?

   b. What questions remain unanswered?

2. Make two lists—one of benefits, the other of risks associated with irradiated foods. Based on those lists, draft a letter to the editor of the *Ann Arbor* (MI) *News* or the *Washington Post* (see page 297) stating your position about irradiated foods.

We turn now to another nuclear risk–benefit issue that is even more controversial and persistent than food irradiation. Eventually, all uses of nuclear technology contribute to the unresolved problem of what is to be done with radioactive waste.

E.2 NUCLEAR WASTE: PANDORA'S BOX

Imagine this situation. You live in a home that was once clean and comfortable. But you have a major problem: You cannot throw away your garbage. The city forbids this because it has not decided what to do with the garbage. Your family has compacted and wrapped garbage as well as possible for about 40 years. You are running out of room. Some bundles leak and are creating a health hazard. What can be done?

The U.S. nuclear power industry, the nuclear weapons industry, and medical and research facilities share a similar problem. Spent (used) nuclear fuel and radioactive waste products have been accumulating for about 40 years. Some of the material is still highly radioactive, and some of it—even initially—had only a low level of radioactivity. It is uncertain where these materials, regardless of radioactivity levels, will be permanently stored—or how soon. Let's take a closer look at the problem.

Nuclear wastes can be broadly classified into two categories: high-level and low-level. High-level nuclear wastes are either (1) the products of nuclear fission, such as those generated in a nuclear reactor, or (2) transuranics, products formed by neutron absorption by the original U-235 fuel. For example, plutonium-239 is a transuranic material. Low-level nuclear wastes have only low levels of radioactivity. These wastes include used laboratory protective clothing, diagnostic radioisotopes, and air filters from nuclear power plants. Figure 29 (page 362) illustrates the composition of radioactive wastes in this country.

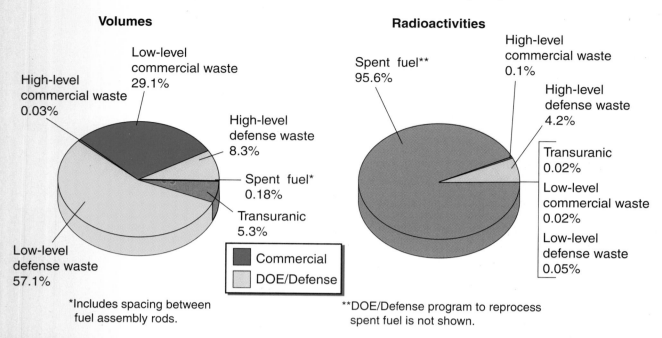

Radioactive Wastes: Volumes and Radioactivities (1990)

Volumes

High-level commercial waste 0.03%

Low-level commercial waste 29.1%

High-level defense waste 8.3%

Spent fuel* 0.18%

Transuranic 5.3%

Low-level defense waste 57.1%

■ Commercial
□ DOE/Defense

*Includes spacing between fuel assembly rods.

Radioactivities

Spent fuel** 95.6%

High-level commercial waste 0.1%

High-level defense waste 4.2%

Transuranic 0.02%

Low-level commercial waste 0.02%

Low-level defense waste 0.05%

**DOE/Defense program to reprocess spent fuel is not shown.

Figure 29 Volumes and radioactivities of radioactive wastes.

▶ *Plutonium could be extracted from spent fuel rods of commercial nuclear reactors. However, such extraction is illegal in the United States, because the extracted plutonium could be used to produce nuclear weapons.*

Figure 29 emphasizes the importance of distinguishing waste volume from radioactivity level when considering nuclear wastes; high volume does not mean high radioactivity. For example, Figure 29 shows that defense efforts produce the largest volume of radioactive wastes (57.1%), but this waste accounts for less than 0.1% of the total radioactivity of nuclear wastes. On the other hand, spent fuel from nuclear reactors occupies less than 1% of the volume of all radioactive wastes, yet accounts for almost 96% of the total radioactivity. Moreover, high-level nuclear waste species also can have extended half-lives, some thousands of years.

YOUR TURN

High- and Low-Level Waste Disposal

Use Figure 29 to answer the following questions.

1. Approximately what percent of nuclear waste is
   a. low-level waste?
   b. high-level, including transuranic wastes?

2. What source accounts for the greatest volume of high-level nuclear waste?

3. What two sources account for the majority of low-level nuclear waste?

4. Should high-level or low-level nuclear wastes receive greater attention? Explain your answer.

Because high-level and low-level nuclear wastes have different characteristics, they are disposed of differently. Low-level wastes can be put into sealed containers and buried in lined trenches 20 feet deep. Low-level military nuclear waste is disposed of at federal government sites maintained by the Department of Energy. Since 1993, each state is responsible for disposal of its own low-level commercial nuclear wastes. Groups of states in several regions have formed *compacts;* a disposal site in one state will be used for disposal of low-level wastes from all compact members.

High-level nuclear waste disposal requires a much different approach. Spent nuclear fuels present the greatest challenge. To get a sense of how much high-level radioactive waste there is, consider that a commercial nuclear reactor (power plant) produces about 30 tons of spent fuel each year. This means that roughly 3,300 tons of waste are generated each year by the 111 commercial nuclear power plants operating in the United States. Approximately 22,000 tons of spent fuel are in storage in 34 states. Eventually, the components of the nuclear power plant reactors themselves will become nuclear waste. By the year 2000, 64 of the 111 U.S. reactors, most licensed for 40 years, will be more than 20 years old.

This country's nuclear weapons program has contributed about 20 million liters of additional stored waste. The volume of military waste is much greater than the amount of commercial waste. This waste is in the form of a sludge—the waste product of extracting plutonium from spent fuel rods of military reactors. (Nuclear weapons are created from the plutonium.) As radioisotopes in the waste decay, they emit radiation and thermal energy. In fact, without cooling, such wastes can become hot enough to boil. This makes military waste containment extremely challenging.

Nevada hazardous waste disposal cell.

Table 13 Some Isotopes in Spent Fuel Rods	
Isotope	**Half-life**
Plutonium-239	24,000 y
Strontium-90	27.7 y
Barium-140	12.8 y

Because of nuclear disarmament treaties, by the year 2003 the United States will need to remove from nuclear weapons as much as 410 tons of highly enriched uranium and about 50 tons of plutonium.

Up to a third of a nuclear reactor's fuel rods must be replaced each year. This is necessary because the uranium-235 fuel becomes depleted as fissions occur and because accumulated fission products interfere with the fission process. The spent fuel rods are highly radioactive, with some isotopes remaining active for many thousands of years. Table 13 lists the half-lives of a few isotopes produced by nuclear fission. All of these and hundreds more are found in spent fuel rods.

By federal law, reactor waste must be stored on site, usually in nuclear waste storage tanks, until a permanent depository is created. Available storage space on site, however, is limited. Final disposal of high-level radioactive waste is the responsibility of the federal government, but permanent long-term disposal sites for high-level wastes have not yet been opened. Legal negotiations and development work are in process to develop such a site at Yucca Mountain, Nevada. If this site turns out be suitable, high-level nuclear wastes would be stored deep underground there.

The method of long-term radioactive waste disposal favored by the U.S. government (and by many other nations) is *mined geologic disposal*, illustrated in Figure 30 (page 365). The radioactive waste would be buried at least 1,000 m below the earth in vaults that would presumably remain undisturbed permanently.

To prepare the waste for burial, it would first be allowed to cool for 30 years by putting the spent fuel rods in very large tanks of water. Over

A 22-ton spent nuclear fuel shipping cask rests virtually undamaged atop this demolished tractor-trailer rig bed after an 84-mph impact into a 10-foot thick concrete target.

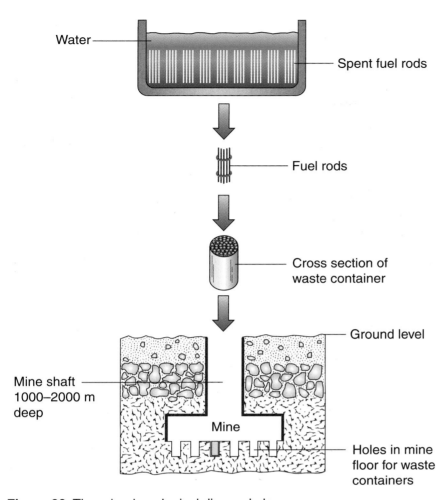

Figure 30 The mined geological disposal plan.

time, the radioisotopes would decay, lowering the level of radioactivity to a point where the materials could be handled safely. Such on-site nuclear waste storage would be only a temporary measure; the tanks would require too much maintenance to be safe for longer periods of time.

Next, cooled radioactive wastes would be locked inside leakproof packages. The currently favored method is to transform the wastes into a type of ceramic. This is *vitrification.* Although the ceramic material would still be highly radioactive, the waste would be much less likely to leak or leach into the environment because of the glasslike "envelope." The vitrified radioactive wastes would be sealed in special containers made of glass, stainless steel, or concrete.

The Savannah River vitrification plant under construction in South Carolina will be the world's largest vitrification facility. Vitrification has been used in France for over a decade, where 75% of that nation's electrical energy is produced by nuclear reactors.

Nowhere in the world, however, has nuclear waste yet been permanently buried. The problem in every country is to find politically and socially acceptable sites. In Japan, where land mass is at a premium, deep ocean burial has been considered.

▶ *This cooling process for existing wastes is already taking place at each nuclear reactor site.*

Some geologic sites formerly assumed to be stable enough for radioactive waste disposal were later discovered to be unsafe. For example, some plutonium was buried at Maxey Flats, Kentucky, in a rocky formation that geologists believed would remain stable for thousands of years. Within 10 years, some buried plutonium had moved dozens of meters away.

What are the current plans to resolve our long-term nuclear waste disposal problems? By U.S. law, the President has selected two sites for permanent radioactive waste disposal from among options provided by the Department of Energy. These sites, located in regions of presumed geologic stability, are the Waste Isolation Pilot Plant (WIPP) near Carlsbad, New Mexico, and at Yucca Mountain, Nevada, a volcanic ridge 100 miles northwest of Las Vegas. Both proposals have created considerable debate and controversy. Layers of legal, environmental, and development work still stand in the way of a final okay for either site. It will probably be at least the year 2010 before a suitable site is ready to receive high-level nuclear wastes for long-term storage.

Nuclear waste disposal is a pressing issue. However, another possible hazard of nuclear technology—a serious accident at a nuclear power plant—must be considered in any thorough risk–benefit analysis.

▶ *The NIMBY syndrome— Not In My Back Yard—is also at work.*

E.3 CATASTROPHIC RISK: A PLANT ACCIDENT

Although a nuclear power plant cannot become an atomic bomb, accidents—through malfunction, earthquake, or human error—could release large quantities of radioactive matter into the environment. Since the late 1970s we have seen major nuclear power plant accidents at Three Mile

Three Mile Island nuclear power plant.

Island in the United States and at Chernobyl in Ukraine (formerly part of the Soviet Union).

The 1979 accident at the Three Mile Island (TMI) nuclear power plant at Middletown, Pennsylvania (near Harrisburg), was the most serious U.S. accident to date in a commercial nuclear power plant. The incident, resulting from an unanticipated and interrelated series of mechanical and operator failures, did not involve any loss of human life. Nonetheless, the accident—together with the evidence and testimony generated in subsequent public hearings—contributed to an increased level of apprehension and "anti-nuclear" sentiment in much of the U.S. general public.

Nuclear reactor unit 2 at TMI lost its flow of cooling water on the morning of March 28, 1979. Cooling water was lost due to a number of mechanical and human errors, triggered initially by an automatic shutoff of two water pump valves in the secondary cooling system. Even though the nuclear reactor automatically shut down, enough thermal energy was released to heat the core to over 5,000 °C, causing about half of it (62 tons) to melt and fall to the reactor's bottom. About 32,000 gallons of radioactive cooling water surrounding the core eventually poured from an open safety relief valve. Alarmed by exaggerated and inaccurate reports of radiation release and the possibility of a hydrogen gas bubble explosion (which did not occur), nearly 200,000 individuals within 50 miles of the plant evacuated the area.

The actual release of radioactivity to the atmosphere was slight at TMI. No individual received more than about 50 mrem, about one-fourth that of a gastrointestinal tract X-ray.

Yet, although the tragedy of complete plant meltdown was avoided at TMI, and the main containment structure remained intact, the TMI experience had a lasting impact on millions of U.S. citizens who remember the uncertainty and fear associated with the accident.

June 3, 1986, at the Chernobyl power station after the disaster.

Because of sentiments questioning the safety of nuclear power due to TMI, a moratorium was declared on the design and construction of any new nuclear power plants in the United States. However, it has been suggested that an overabundance of nuclear power plants already existed due to over-optimistic projections by utility companies of the need for electrical energy.

The Chernobyl disaster has remained the object of considerable study, debate, and concern. It is important, however, to keep the TMI and Chernobyl incidents in perspective. They are not equivalent; the Chernobyl accident was far more serious. More than 2 million times more radioactivity was released into the atmosphere at Chernobyl than at TMI. This release led to significant acute radiation sickness and death among people at the Chernobyl plant and its vicinity. (TMI had no such tragic results.) Also, long-lived plutonium isotopes were released at Chernobyl, but not at TMI.

What happened at Chernobyl? Reactor 4 at the Chernobyl Nuclear Power Station was a large, water-cooled, graphite-moderated nuclear power reactor. On April 26, 1986—some two years after startup—the reactor ran dangerously out of control, heating the cooling water to the boiling point. Pressure from the resulting steam burst cooling pipes and, in a powerful but *non-nuclear* explosion, blew the 1,000-ton roof off the reactor. Radioactive debris flew nearly a mile skyward, fires were triggered (including the burning of graphite moderator material), and molten nuclear fuel spilled from control rods.

Chernobyl's radioactive cloud—including radioactive isotopes released from still-hot debris for days after—eventually spread across the Northern Hemisphere. Although 97% of the radioactive material fell in Europe, lower levels of radioactive material were detected in nations ranging from China to Canada. Over 200 acute cases of radiation sickness were reported initially, including 31 deaths. Hundreds of thousands of people were evacuated from high-radiation areas within the Soviet Union. Officials estimate that cleanup costs will exceed $100 billion by the turn of the century.

Experts note that Chernobyl's reactor design included features known to be unnecessarily risky. *Positive feedback*—in which increased water-boiling caused increased fission and heat, which, in turn, caused more water-boiling—was a consequence of the way the reactor was built. An inherent flaw in its design created problems when the reactor was run at less than one-fifth of its full power. Initial reactor instability was caused by operating changes made for a special reactor experiment during which the reactor was run under reduced power. In addition, the Chernobyl reactor was not protected by a confinement vessel (a steel-reinforced concrete outer shell intended to isolate and confine accidentally released radioactive materials). Such confinement vessels are a standard part of most nuclear reactors, including all those operating in the United States.

The Chernobyl disaster has contributed to heightened public apprehension and concern about nuclear power. In the former U.S.S.R. alone, over 30 nuclear power reactor orders have been cancelled or postponed since 1986.

Studies of such accidents will help us understand the potential risks of nuclear power. One mid-1970s study, made by scientists from the

National Academy of Sciences (NAS), suggested that if a nuclear power plant accident occurred, the chance of dying—either from high radiation doses at the time of the accident or from lower doses suffered from resulting fallout—would be a little greater than the chance of dying in a plane crash. (Keep in mind that airplane travel is very safe.) On the other hand, though risk assessment calculations for airplane travel are certain, such calculations for nuclear accidents are uncertain. The NAS study results could mean that nuclear plants are either much safer, or much more dangerous, than air travel. The reason for such great uncertainty is that scientists must attempt a much more complicated calculation for nuclear accidents based on much less experience and data.

Is nuclear power worth the risk? It is our responsibility to decide what we want for ourselves and our nation, and to make our wishes known to those who govern. We must keep in mind, as well, that nuclear power plants reduce the large amounts of carbon dioxide generated by burning coal in coal-fired power plants, contributing to global warming. Such coal-fired plants also release radioisotopes in their smoke.

In a 1992 public-opinion poll, 73% of respondents said that nuclear energy should play a "very important role" in America's future energy needs; only 22% responded that this issue was "not too important" or "only somewhat important." What are your current impressions? Do you feel better prepared to approach these difficult issues after reading this unit? We hope so.

PART E: SUMMARY QUESTIONS

1. Explain the nuclear waste disposal problem, using the concepts of radioactive decay, half-life, and radiation shielding.

2. Classify the following as either low-level or high-level nuclear waste:

   a. 500 gloves used by a technician in a nuclear medicine laboratory

   b. a barium isotope solution extracted from a spent fuel rod

   c. a 0.05-mg sample of element 106

   d. vials of iodine-123 from a hospital's diagnostic clinic

3. Even if nuclear power plants were phased out, we would still need to deal with existing wastes, plus the much greater wastes associated with nuclear weapon production.

   a. Briefly describe one possible solution to this problem.

   b. Identify at least one technological and one practical political problem associated with your solution.

4. If you were living near Yucca Mountain, Nevada, what questions would you want answered regarding the possible siting of the nuclear waste depository near your community?

5. In a 1993 essay, John Ahearne, a physicist and former chairman of the U.S. Nuclear Regulatory Commission, wrote that "Even though nuclear power is currently important, very important in many states, I doubt that it will play a significant role in meeting America's future energy needs." What factors might have led him to that conclusion?

PUTTING IT ALL TOGETHER: LONG-TERM NUCLEAR WASTE STORAGE

Long-term disposal of high-level nuclear wastes, such as spent fuel rods, is needed. Remember that radioisotopes in these fuel rods have half-lives ranging from about 10 years (barium-140, 12.8 y) to several millennia (uranium-238, 4.5 billion y; plutonium-239, 24,000 y). Also, see Table 13 (page 364).

The EPA has insisted that mined geological disposal of high-level nuclear wastes must be able to isolate such wastes from their "accessible environment" for 10,000 years. In other words, even after the waste canisters degrade, the geological formation must be able to contain the nuclear waste for at least 10 millennia. To put those 10,000 years into a perspective of some human constructions, consider that the earliest Egyptian pyramids are about 5,000 years old, England's Stonehenge is around 4,500 years old, and the Great Wall of China is a relative youngster at 2,200 years.

F.1 A SIGN FOR THE TIMES

Long-term burial of high-level nuclear waste gives you an interesting opportunity to put to use what you have learned about nuclear phenomena and about materials in the Resources unit and Petroleum unit. Assume that the United States constructs a long-term, high-level nuclear waste depository dug deep into a geologically stable site.

Working in groups assigned by your teacher, design markers with symbols to be placed around this site. Keep in mind that visitors and workers at the site may not necessarily share a common language, and that languages are likely to change over 10 millennia. Even the names and symbols for the chemical elements could change during that time! Therefore, such markers should have markings that can be easily interpreted about what is at the site by future Earth dwellers, and should be designed to last until at least the year 12,000. Thus, it will be important for you to consider what materials should be used for the markers and what size they might be.

By the way, this is more than just a theoretical exercise. The Department of Energy asked a team of experts to design such markers.

F.2 LOOKING BACK

Thomas Jefferson, third president of the United States, wrote in 1820: "I know of no safe depository of the ultimate powers of the society but the people themselves; and if we think them not enlightened enough to exercise their control with a wholesome discretion, the remedy is not to take it from them, but to inform their discretion by education."

Responsible citizenship with regard to nuclear energy policies is certainly part of informed discretion. As we stated earlier, the purpose of this unit is not to teach you everything known about nuclear chemistry or to provide the "right" answer to any particular nuclear issue. Our aim has been to give you a foundation for further exploration of this important topic—an issue that involves scientific principles, technology, and societal concerns. With this foundation, you will be better prepared to influence our nation's future polices regarding nuclear energy, in its many applications.

This wall of the power-plant building of the Atomic Energy Commission's National Reactor Testing Station in Idaho records the first known production of useful electric power from atomic energy. The future of the industry will depend on decisions made by an informed citizenry.

Living in a Sea of Air: Chemistry and the Atmosphere

- ► Is *global warming a growing threat? If so, what can we do about it?*

- ► What impact have human activities had on air quality?

- ► What are the possible implications of ozone depletion in the stratosphere?

- ► What are the key benefits and burdens of fossil-fuel burning?

- ► What substances are important in indoor air quality compared with those for outdoor air quality?

- ► What is meant by "acid rain?" How does it affect us?

- ► How can chemistry help us establish and maintain good air quality?

INTRODUCTION

Take a deep breath and exhale slowly. The air that moves so easily in and out of your lungs is our next topic of study.

We live *on* the Earth's surface, but *in* its atmosphere. Air surrounds us just as water surrounds aquatic life. Like the Earth's crust and bodies of water, the atmosphere serves both as a source for obtaining chemical resources and as a sink for discarding waste products. We use some atmospheric gases when we breathe, burn fuels, and carry out various industrial processes. Humans, other living organisms, and natural events also add gases, liquids, and solid particles to the atmosphere. These added materials may have no effect, or they may disrupt local environments—or even change global conditions.

Because human activities can sometimes lower air quality, important questions arise: Should air be regarded as a free resource? How clean should air be? What chemical methods are available to clean polluted air? What does it cost to maintain clean air? What are the financial and environmental costs of polluted air? Who should be responsible for air pollution control? Citizens will continue debating such questions at local, state, national, and international levels in the years ahead because they impact everyone. We cannot easily escape from air pollution, even indoors.

Finding acceptable answers to these complex questions means understanding the chemistry of gases that make up the atmosphere. We need to know about the atmosphere's composition and structure, general properties of its gases, its processes influencing climate, and how natural processes renew the atmosphere through recycling. We will also consider the quality of air we breathe indoors.

This unit explores basic chemistry related to these topics and examines current pollution control efforts. It offers another opportunity to sharpen your chemical knowledge and decision-making skills.

The photo shows the pollution-soiled Milwaukee Veterans Hospital partway through a chemical restoration.

A AIR—THIS MARVELOUS CANOPY

It's said that we don't appreciate the value of things until we are forced to do without them or pay for them. This is certainly true of the Earth's atmosphere. Except for astronauts, patients with respiratory diseases, and undersea divers, people seldom think about the sea of colorless, odorless gases surrounding them—that is, as long as it remains odorless, tasteless, invisible, and available. If it is present and not too polluted, we tend to take for granted the approximately 14 kg of air we each breathe daily. Such an attitude is not wise, as we will soon see.

How much do you know about the atmosphere and current problems involving air quality? Test your knowledge by completing the following exercise.

A.1 YOU DECIDE: THE FLUID WE LIVE IN

On a separate sheet of paper indicate whether you think each numbered statement below is true (T) or false (F), or whether it's too unfamiliar to you to judge (U). Then, for each statement you believe is true, write a sentence describing a practical consequence or application of the fact. Reword each false statement to make it true.

Do not worry about your score. You will not be graded on this exercise; it's intended to start you thinking about the wonderful fluid in which we live.

▶ *Fluids include liquids and gases.*

1. You could live nearly a month without food, and a few days without water, but you would survive for only a few minutes without air.

2. The volume of a given sample of air (or any gas) depends on its pressure and temperature.

3. Air and other gases are weightless.

4. The atmosphere exerts nearly 15 pounds of force on each square inch of your body.

5. The components of the atmosphere vary widely at different locations on Earth.

6. The atmosphere acts as a filter to prevent some harmful radiation from reaching the Earth's surface.

7. In the part of the atmosphere nearest the Earth, air temperature usually increases as altitude increases.

8. Minor air components such as water vapor and carbon dioxide play major roles in the atmosphere.

9. Two of the ten most-produced industrial chemicals are obtained from the atmosphere.

10. Ozone is regarded as a pollutant in the lower atmosphere, but as an essential component of the upper atmosphere.

11. Clean, unpolluted air is a pure substance.

12. Air pollution first occurred during the Industrial Revolution.

13. No human deaths have ever been directly attributed to air pollution.

14. Natural events, such as volcanic eruptions and forest fires, can produce significant air pollution.

15. Destruction of materials and crops by air pollution involves a significant economic loss for our nation.

16. The main source of air pollution is industrial activity.

17. The "greenhouse effect" is a natural warming effect that may become harmful because of excessive burning of fossil fuels.

18. In recent years, rain in industrialized nations has become less acidic.

19. Most air pollution caused by humans originates with burning fuels.

20. Pollution control has not improved overall air quality.

When you are finished, your teacher will give you answers to these items, but will not elaborate on them. However, each item will be discussed at some point in this unit.

The next section includes some demonstrations and at-home activities that illustrate the nature of the air you breathe.

A.2 LABORATORY DEMONSTRATION: GASES

Because atmospheric gases are generally colorless, odorless, and tasteless, you might incorrectly conclude that they are "nothing." However, gases have definite physical and chemical properties, just as solids and liquids have. Your teacher will perform demonstrations that illustrate some of these properties. The demonstrations will help you to answer four major questions about air.

- Demonstrations 1–3:   Is air really matter?
- Demonstrations 4–8:   What is air pressure all about?
- Demonstrations 9–10:   Why does air sometimes have an odor?
- Demonstrations 11–12:   Is air heavy? Will it burn?

For each demonstration, complete each of these four steps:

1. Before the demonstration, try to predict what will happen.

2. Carefully observe the demonstration. Explain any differences between your prediction and the actual outcome.

3. Try to identify the properties of gases that are being demonstrated. Try to decide why gases exhibit those properties.

4. Think of one practical application of each demonstrated gas property.

Complete these additional activities at home, recording your results and explanations in your notebook:

1. Put the end of a straw in a glass of water, and hold another straw outside the glass. Place the ends of both straws in your mouth and try to drink the water through the first straw. Explain what happens.

2. Punch a hole the diameter of a straw through the screw-on lid of a clean, empty jar. Insert a straw and seal the connection with clay, putty, or wax. Fill the jar to the brim with water, then screw on the top and try to drink the water through the straw. Explain your observations.

3. Based on your observations, what makes it possible to drink liquid through a straw?

The properties of the elusive stuff called air make it vitally important. Let's examine one aspect of air's importance.

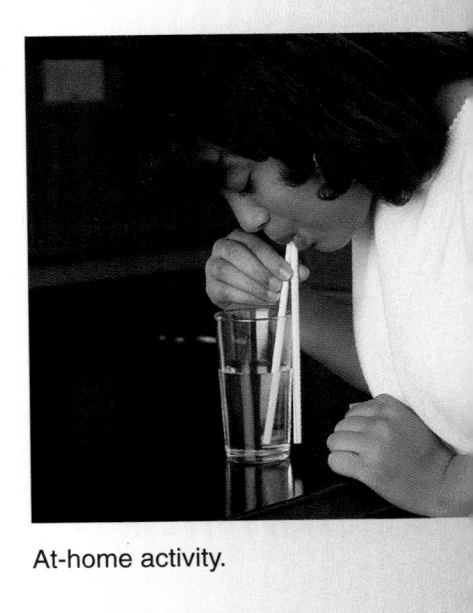

At-home activity.

A.3 AIR: THE BREATH OF LIFE

Seen from the moon, the Earth's atmosphere blends with its water and land masses, presenting a picture of considerable beauty. Other planets also may possess exotic beauty, but their environments would be hostile to life as we know it. By contrast, our world supports millions of species of living organisms, from one-celled amoebas to redwood trees and elephants.

In the Water unit you considered water's key role in supporting life on the Earth. Although our present atmosphere was formed much later than the world's waters, air also helps sustain plants and animals.

A major role of the atmosphere is to provide oxygen gas needed for respiration. This activity will help you understand that role.

YOUR TURN

Breath Composition and Glucose "Burning"

1. Consider Figure 1 (page 378), which compares the composition of gases we inhale with that of gases we exhale.

   a. Summarize the changes in air's composition that result from its entering and then leaving the lungs.

   b. How do you account for these changes?

   c. Assuming an average of 14 breaths each minute, how many breaths do you take in a 24-hour day?

   d. What factors could make your answer to Question 1c change? Why?

   e. Assume you inhale 500 mL of air in one breath.

      (1) How many liters of air do you inhale each minute?

      (2) Each day?

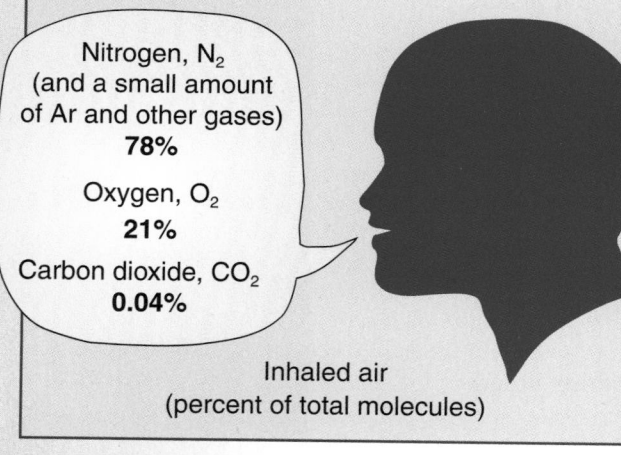

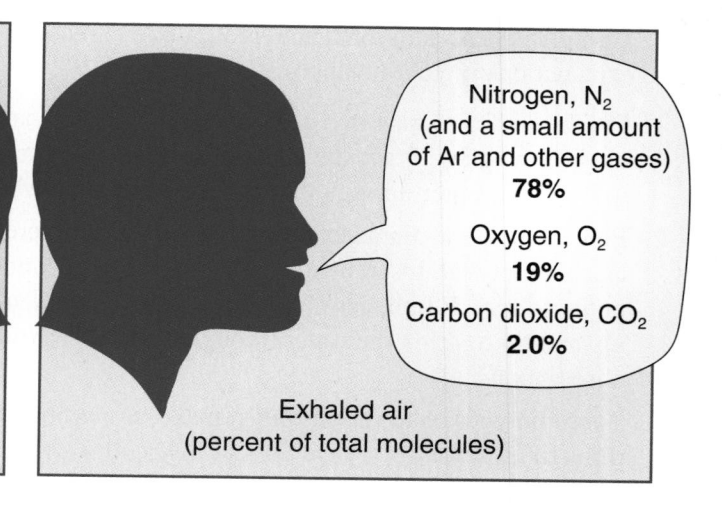

Figure 1 The composition of inhaled and exhaled air.

▶ *Refer to page 258 to review limiting reactants.*

▶ *Fat burning is also a source of body energy. Metabolizing one gram of fat produces about 38 kJ of energy.*

2. According to Figure 1, your lungs extract only a portion of the oxygen gas available in inhaled air. What do you think determines the amount of oxygen gas used?

3. Write the chemical equation for the overall biochemical conversion in your body of 1 mol of glucose ($C_6H_{12}O_6$) reacting with oxygen gas (O_2) to produce carbon dioxide (CO_2) and water (H_2O). This equation represents the overall changes involved in a series of glucose-"burning" reactions in the body.

  a. If the same number of oxygen gas and glucose *molecules* were available, which would be the limiting reactant? Why?

  b. If the same *masses* of oxygen gas and glucose were available, which would be the limiting reactant? Why?

  c. Assume that your body extracts about 20 mol of oxygen gas from air daily. Given that information and the "glucose burning" equation you wrote, how many moles of glucose can your body burn each day?

  d. A mole of glucose, $C_6H_{12}O_6$, has a mass of 180 g. How many grams of glucose could your body burn daily?

  e. Burning a gram of glucose produces about 17 kJ of energy. Assuming that all of your body's energy comes from glucose, how much energy does your body generate daily?

4. a. What substance produced in your body's "burning" of glucose is not shown in Figure 1?

  b. How could you verify that this substance is present in your exhaled breath?

Plant life needs a continuous supply of carbon dioxide—a waste product of animal respiration. Through photosynthesis in green plants (discussed in the Food unit), carbon dioxide combines with water, ultimately forming more glucose and oxygen gas. Thus, photosynthesis and respiration are complementary processes—reactants for one become products for the other.

The Earth's atmosphere cannot restore and cleanse itself if natural systems are burdened by excessive pollution, whether the pollution source is natural or results from human activities. Important environmental problems can arise if human activities which we could control overwhelm natural recycling and cleansing systems. We will investigate some of these issues later in this unit. But first, more fundamental questions: What *is* the atmosphere? What are some of its key physical properties?

 # PART A: SUMMARY QUESTIONS

1. Air is both similar to and different from other resources you have studied, such as water, minerals, and petroleum. Describe
   a. two similarities.
   b. two differences.

2. Defend or refute each of these statements about gas behavior. Base your position on gas properties you have observed in Part A.
   a. *An empty open bottle is not really empty.*
   b. *Atmospheric pressure acts only in a downward direction.*
   c. *Gases naturally "mix" by moving from regions of lower concentration to regions of higher concentration.*
   d. *All colorless gases have similar physical and chemical properties.*

3. From a chemical point of view, why do people need to take deeper breaths when they hike at high altitudes than when they hike similar trails at lower altitudes?

4. A well-trained athlete takes about 30 breaths each minute during vigorous exercise (such as sprinting), inhaling about 4 L of air with each breath. Assume that under these conditions the athlete extracts about 0.15 mol O_2 from the air each minute.
   a. If all the extracted oxygen gas is used in glucose-burning, how many grams of glucose will be "burned" by the athlete each minute? (*Hint:* Refer to the *Your Turn* on page 377.)
   b. How much energy (in kilojoules, kJ) does this athlete expend each minute? (Refer to the same *Your Turn* for helpful information.)
   c. As you may recall from the Food unit, a Calorie (kilocalorie) equals about 4.2 kJ. How many Calories does this athlete expend each minute in vigorous exercise?

Hikers at high altitudes must breathe deeper than those at low altitudes do.

B INVESTIGATING THE ATMOSPHERE

Human-made air-quality problems began with the discovery and use of fire. Fires for heating, cooking, and metalworking created air pollution problems in ancient cities. In 61 A.D., the philosopher Seneca described Rome's "heavy air" and "the stink of the smoky chimneys thereof." The Industrial Revolution, beginning in the early 1800s, added more air pollution around large towns. Meanwhile, chemists and physicists were increasing their studies of the atmosphere's properties and behavior.

In the following sections you will extend your general understanding of atmospheric gases, based in part on demonstrations you have observed (page 376). This knowledge will help you predict specific gas behaviors. It will also clarify how solar energy interacts with atmospheric gases, creating weather and climate.

We will begin by preparing and investigating some atmospheric gases.

B.1 LABORATORY ACTIVITY: THE ATMOSPHERE

GETTING READY

Air, as noted earlier (Figure 1), is a mixture composed mostly of nitrogen gas (N_2) and oxygen gas (O_2), together with much smaller amounts of carbon dioxide and other gases. Each of air's components has distinct physical and chemical properties.

A high concentration of carbon dioxide helps to produce larger, healthier, and faster-growing plants in a greenhouse.

Liquid nitrogen is used for freezing beef.

Oxygen gas extracted from air is used in a blowtorch.

In this activity your class will generate two atmospheric gases—oxygen and carbon dioxide—and examine some of their chemical properties. In addition, your teacher may either provide samples of nitrogen gas for testing or demonstrate some properties of nitrogen and of oxygen. Oxygen gas will be generated by this reaction:

$$2\ \underset{\substack{\text{hydrogen}\\\text{peroxide}}}{H_2O_2(aq)} \xrightarrow{\text{FeCl}_3} 2\ \underset{\text{water}}{H_2O(l)} + \underset{\text{oxygen}}{O_2(g)}$$

► *Iron(III) chloride, FeCl₃, is a catalyst in this reaction.*

Carbon dioxide will be produced by adding antacid tablets to water. These tablets contain a mixture of sodium bicarbonate ($NaHCO_3$) and potassium bicarbonate ($KHCO_3$) together with citric acid (an organic acid present in lemons and oranges). When dissolved in water, bicarbonate ions (HCO_3^-) react with hydrogen ions (H^+) from the citric acid to produce carbon dioxide gas:

$$H^+(aq) + HCO_3^-(aq) \rightarrow H_2O(l) + CO_2(g)$$

Your teacher will demonstrate the behavior of oxygen gas and of carbon dioxide in the presence of burning magnesium and hot steel wool. You will investigate burning wood in oxygen and in carbon dioxide. The key question: Do oxygen and carbon dioxide support the combustion of these materials?

Next, you will observe whether these gases react with limewater—a solution of calcium hydroxide, $Ca(OH)_2$. Finally, the acid–base properties of oxygen gas and carbon dioxide will be investigated.

Acidic substances produce H^+ ions when dissolved in water; basic substances produce OH^-. In the Water unit, page 55, you learned that the pH scale can be used to express the acidity or basicity of a solution.

► *Acids, bases, and pH are discussed further on pages 441–443 of this unit.*

Most pH indicators change color quite sharply at one point on the scale. Universal indicator contains a mixture of pH-sensitive compounds; each changes color at a different pH.

If a tank of compressed nitrogen gas is available, your class will also investigate similar properties for nitrogen gas.

Read the procedures below. Set up a data table in your notebook to record observations on the behavior of these gases.

PROCEDURE

Part 1: Gas Preparation

Label nine 19 × 150 mm test tubes accordingly: three "Sample 1," three "Sample 2," and three "Sample 3."

Your teacher will assign you to a group.

NITROGEN GAS

Follow either Option A or B as directed by your teacher.

Option A: Your teacher will carry out a chemical reaction that produces nitrogen gas and will demonstrate the Part 2 nitrogen gas tests. Record all observations in your data table.

Option B: If this option is followed, the nitrogen gas will be dispensed (under your teacher's supervision) from a compressed nitrogen gas cylinder. Fill three test tubes labeled "Sample 1," "Sample 2," and "Sample 3" with nitrogen gas, as follows:

1. Fill the three test tubes with water. Place them in a test-tube holder as your teacher directs.

2. The gas-collecting trough should be filled with enough tap water to cover the bottom to a 3-cm depth.

3. Place the end of the tubing from the nitrogen gas cylinder under water in the trough but not into a test tube.

4. Allow the nitrogen gas to bubble into the trough for 10–15 seconds.

5. Hold a finger tightly over the opening and invert the test tube in the water tray. Direct the stream of nitrogen gas into the inverted test tube mouth. When gas fills the tube, stopper the tube mouth while it is still inverted and under water. Then remove the test tube from the tray. Collect two more test tubes of nitrogen gas.

6. Complete the gas tests outlined in Part 2.

CARBON DIOXIDE GAS

1. Omitting the burner, set up the apparatus illustrated in Figure 2, page 383.

2. Fill three test tubes labeled "Sample 1," "Sample 2," and "Sample 3" with water. Place them in a test-tube holder as your teacher directs. Fill the gas-collecting trough with enough tap water to cover the bottom to a 3-cm depth.

3. Add 125 mL of water to the 500-mL flask. Drop one antacid tablet into the water. (Break the tablet to fit through the flask neck, if necessary.)

4. Quickly place the stopper with glass and rubber tubing into the flask. Place the other end of the tubing under water in the trough, but not into a test tube.

5. Allow the carbon dioxide gas to bubble into the trough for about ten seconds. Then begin to collect carbon dioxide gas samples in the test tubes.

6. Hold a finger tightly over the opening and invert the test tube in the water tray. Direct the stream of gas into the inverted test tube mouth. When gas fills the tube, stopper the tube mouth while it is still inverted and under water. Then remove the test tube from the tray. Collect two more test tubes of carbon dioxide gas.

7. If carbon dioxide gas evolution slows down before you are finished, add another antacid tablet to the generating flask.

8. Complete the gas tests outlined in Part 2.

OXYGEN GAS

1. Set up the apparatus for generating oxygen gas, as illustrated in Figure 2.

2. Fill the three remaining test tubes with water. Fill the gas-collecting trough with enough tap water to cover the bottom to a 3-cm depth.

3. Weigh one gram of iron(III) chloride ($FeCl_3$) on a clean paper square. Then carefully pour the iron(III) chloride sample into a 500-mL flask.

4. Pour 100 mL of 3% hydrogen peroxide (H_2O_2) into the flask.

5. Quickly place the stopper with attached glass and rubber tubing in the flask. Place the other end of the tubing under water in the trough, but not into a test tube. (*Note:* If the reaction is too slow to generate a steady stream of oxygen gas bubbles, heat the mixture *gently*—do not allow it to boil. Continue to heat gently as you move to Step 6.)

6. When the reaction is proceeding smoothly, allow gas to bubble into the water for one minute. Then begin to collect oxygen gas samples in the test tubes.

7. Place a finger tightly over the opening and invert the test tube in the water tray. Direct the stream of gas into the inverted test tube mouth. When gas fills the tube, stopper the tube mouth while it is still inverted under water. Then remove the test tube from the tray. Collect two more test tubes of oxygen gas.

If the reaction stops before you have collected enough gas, remove the stopper from the generating flask, turn off the burner, and allow the flask to cool for five minutes. Then add 20 mL of hydrogen peroxide solution to the reaction flask.

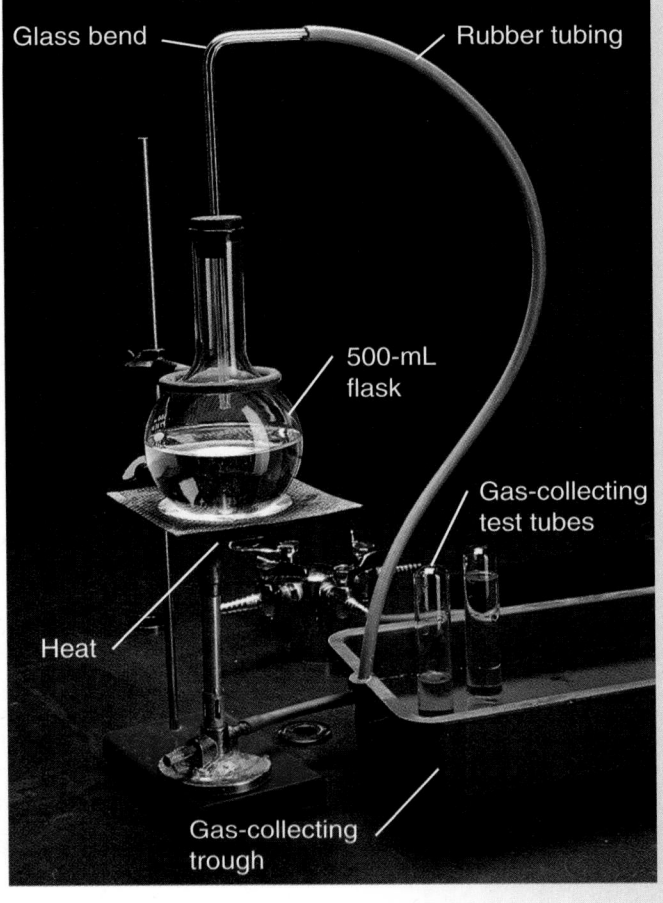

Glass bend — Rubber tubing

500-mL flask

Gas-collecting test tubes

Heat

Gas-collecting trough

Figure 2 Gas-generating setup.

CAUTION

Avoid inhaling any $FeCl_3$ dust.

8. When enough gas has been collected, remove the stopper from the generating flask. Then turn off the burner.

9. Complete the gas tests outlined in Part 2.

Part 2: Gas Tests

Your teacher will demonstrate magnesium and steel wool tests with the gases. Record your observations of these demonstrations in your data table.

Each student group will perform the following tests on its assigned gas, and share results with the class. Record observations for *all* gas tests in your data table.

1. Use a test-tube holder to hold the test tube containing the gas sample. Hold a wood splint with tongs and light the splint in a burner flame. Blow out the splint flame; then blow on the embers until they glow. Remove the stopper from Sample 1; quickly plunge the *glowing* splint into the test tube. Record your observations. Turn off the burner.

4. Add about 2 mL of limewater to Sample 2. Stopper the tube and shake it carefully. Note any changes in the liquid's appearance.

5. Add two drops of universal indicator solution to Sample 3. Stopper the tube and shake carefully. Match the resulting color with the indicator chart. Report the estimated pH to the class.

6. Wash your hands thoroughly before leaving the laboratory.

QUESTIONS

1. Why was it important to allow some gas to bubble through the water before you collected a sample?

2. Which gas appears to be
   a. most reactive?
   b. least reactive?

3. a. Which gas appears to support both rusting and combustion?
   b. Roughly what percent of air molecules does this gas represent? (See Figure 1, page 378.)
   c. If the atmosphere contained a *higher* concentration of this gas, what might be some consequences?

4. Carbon dioxide is used in some types of fire extinguishers. Explain why.

5. Figure 1 suggests that human lungs expel carbon dioxide. Describe two ways you could verify that your exhaled breath contains carbon dioxide gas.

What gases—in addition to nitrogen, oxygen, and carbon dioxide—are present in the atmosphere? Using a variety of sampling and measuring techniques, scientists have pieced together a detailed picture of the atmosphere's chemical composition. This is the subject of the next section.

B.2 A CLOSER LOOK AT THE ATMOSPHERE

Most of the atmosphere's mass and all of its weather are located within 10 to 12 km of the Earth's surface. This region, called the **troposphere** (Greek *tropos,* turn + *sphaira,* ball), is the part of the atmosphere in which we live. Let's examine the troposphere further.

Continuous gas mixing occurs in the troposphere, leading to a reasonably uniform chemical composition around the world (Table 1). Air trapped in glacial ice has about the same chemical makeup, suggesting that there has been little change in air throughout human history.

In addition to gases listed in Table 1, air samples can contain up to 5% water vapor, although in most locations the range for water vapor is 1–3%. Other gases naturally present in concentrations below 0.0001% (1 ppm)

► *The amount of water vapor in air determines the humidity.*

The only component of the atmosphere visible from space is condensed water vapor in the form of clouds.

Table 1	Tropospheric Air	
Substance	**Formula**	**Percent of Gas Molecules**
Major components		
Nitrogen	N_2	78.08
Oxygen	O_2	20.95
Minor components		
Argon	Ar	0.93
Carbon dioxide	CO_2	0.033
Trace components		
Neon	Ne	0.0018
Ammonia	NH_3	0.0010
Helium	He	0.0005
Methane	CH_4	0.0002
Krypton	Kr	0.00011
Others	see text	<0.0001 each

include hydrogen (H_2), xenon (Xe), ozone (O_3), nitrogen oxides (NO and NO_2), carbon monoxide (CO), and sulfur dioxide (SO_2). Human activity can alter the concentrations of carbon dioxide and some of air's trace components. Such activity can also add new substances that may lower air quality, as we will soon see.

When air is cooled under pressure, it condenses to a liquid that boils at about -185 °C. Liquid air, like petroleum, can be distilled to obtain its components in pure form. Distilling liquid air produces pure oxygen, nitrogen, and argon.

▶ The process of liquid air distillation is basically the same as petroleum distillation.

Imagine that we seal gaseous oxygen, nitrogen, and argon in three very flexible balloons. One balloon contains one mole of oxygen, the second a mole of nitrogen, and the third a mole of argon.

Here are four observations we could make regarding these samples:

I. Each of the three balloons has the same volume, assuming external air pressure and temperature are the same for each.

II. All three balloons shrink to half this volume when outside air pressure is doubled and the temperature is held constant.

III. All three balloons expand to the same extent when the gas temperature is increased by 10 °C and the external air pressure is held constant.

IV. All three balloons shrink to half their original volume when half the gas is removed and neither the external air pressure nor the temperature is changed.

A balloon filled with one mole of *any* gas is observed to expand and contract in exactly these ways, provided temperature does not drop too low or external pressure rise too high. In other words, all gases share similar expansion and contraction properties.

Now it's your turn.

YOUR TURN

Avogadro's Law

The four observations described above suggest some important conclusions about gases. One of these is this: ***Equal volumes of gases at the same temperature and pressure contain the same number of molecules.*** This idea was first proposed by an Italian chemistry teacher, Amedeo Avogadro, in the early 1800s. It is commonly known as ***Avogadro's law.***

If you look carefully, you can see the law at work in each of the four observations. First note that each balloon starts with the same number of molecules—a mole of them, or 6.02×10^{23} molecules.

In Observation I, all three balloons have the same volume at the same temperature and pressure—just as Avogadro suggested—because they contain the same number of molecules.

In Observation II, the external pressure is doubled and the gas volume becomes half of what it was. Despite these changes, all three balloons retain the same temperature and contain the same number of molecules—just as Avogadro's law specifies.

▶ Avogadro's full name was Lorenzo Romano Amedeo Carlo Avogadro di Quarequa e di Cerreto.

Explain how Avogadro's law is illustrated

a. in Observation III.

b. in Observation IV.

An important consequence of Avogadro's law is that all gases have the same molar volume if they are at the same temperature and pressure. The *molar volume* is the volume occupied by one mole of a substance. Scientists refer to 0 °C and one atmosphere (1 atm) pressure as *standard temperature and pressure,* or *STP. At STP, the molar volume of any gas is 22.4 L.* There is no corresponding simple relationship between moles of various solids or liquids and their volumes.

The fact that all gases have the same molar volume under the same conditions simplifies our thinking about reactions involving gases. For example, consider these equations:

$$N_2(g) + O_2(g) \rightarrow 2\ NO(g)$$

$$2\ H_2(g) + O_2(g) \rightarrow 2\ H_2O(g)$$

$$3\ H_2(g) + N_2(g) \rightarrow 2\ NH_3(g)$$

You have learned that the coefficients in such equations indicate the relative numbers of molecules or moles of reactants and products. Using Avogadro's law (equal numbers of moles occupy equal volumes), you can also interpret the coefficients in terms of gas volumes:

$$1\ \text{volume } N_2(g) + 1\ \text{volume } O_2(g) \rightarrow 2\ \text{volumes } NO(g)$$

$$2\ \text{volumes } H_2(g) + 1\ \text{volume } O_2(g) \rightarrow 2\ \text{volumes } H_2O(g)$$

$$3\ \text{volumes } H_2(g) + 1\ \text{volume } N_2(g) \rightarrow 2\ \text{volumes } NH_3(g)$$

The actual volumes could be any particular values, such as 1 L, 10 L, or 100 L, but the coefficients all change accordingly. For example, in the third equation, we could combine 100 L of N_2 and 300 L of H_2, and expect to produce 200 L of NH_3 if the conversion is complete. These volume relationships allow chemists to monitor gaseous reactions more easily by measuring the gas volumes involved.

Complete the following activity to check your understanding of key concepts discussed in this section.

▶ At STP, 22.4 L of gas, *about the volume of three basketballs, contains one mole of molecules.*

YOUR TURN

Molar Volume and Reactions of Gases

1. What volume would be occupied at STP by 3 mol of $CO_2(g)$?

2. In a certain gaseous reaction, 2 mol of NO reacts with 1 mol of O_2. How many liters of O_2 would react with 4 L of NO gas?

3. Toxic carbon monoxide (CO) gas is produced when fossil fuels such as petroleum burn without enough oxygen gas. The CO eventually converts into CO_2 in the atmosphere. Automobile catalytic converters are designed to speed up this conversion:

Carbon monoxide gas + Oxygen gas → Carbon dioxide gas

a. Write the equation for this conversion.
b. How many moles of oxygen gas would be needed to convert 50 mol of carbon monoxide to carbon dioxide?
c. How many liters of oxygen gas would be needed to react with 1120 L of carbon monoxide? (Assume both gases are at the same temperature and pressure.)

Now that you have explored the composition and properties of gases at the bottom of the sea of air in which we live, let's continue our study of the atmosphere by moving up to higher levels.

B.3 YOU DECIDE: ATMOSPHERIC ALTITUDE

If you were to dive beneath the ocean's surface, you would encounter increasing pressures and decreasing temperatures with increasing depth. Creatures deep in the ocean must adapt to conditions quite unlike those near the surface.

Likewise, the atmospheric environment at sea level is quite different from that at higher altitudes. From early explorations in hot-air and lighter-than-air balloons to current research about the atmosphere and beyond, scientists have gained a good understanding of conditions in all parts of the atmosphere.

Imagine yourself in a craft designed to fly upward from the Earth's surface to the farthest regions of our atmosphere. You have set the imaginary craft's instruments to record the altitude, air temperature, and air pressure. You can take one-liter gas samples at any time. Automated instruments will give you reports on each sample's mass, number of molecules, and composition.

During your ascent, the composition of the air remains essentially constant at 78% N_2, 21% O_2, 1% Ar, plus traces of other gases. At 12 km you notice you are above the clouds. Even the tallest mountains are far below. The sky is light blue; the sun shines brilliantly. The craft is now above the region where most commercial aircraft fly and above where weather occurs.

Above 12 km, the air composition is about the same as at lower altitudes, except that your instruments detect more ozone (O_3). You also notice the air is quite calm, with little of the turbulence you encountered below 12 km.

What limits the height to which these balloons can rise?

▶ A 12-km height equals about 7.5 miles or 39,000 feet.

Air samples taken at 50–85 km contain relatively few atoms or molecules. Those present are ions such as O_2^+ and NO^+. Above 200 km, your radar detects various communication satellites in orbit.

Your aircraft now returns to the Earth's surface. It's time to analyze the data you collected.

PLOTTING THE DATA

Table 2 summarizes the data recorded during the flight. The last two columns provide comparisons of equal-volume (1-L) samples of air at different altitudes.

Table 2		Atmospheric Data		
Altitude (km)	Temp. (°C)	Pressure (mmHg)	Mass (g) of 1-L Sample	Total Molecules in 1-L Sample
0	20	760	1.20	250×10^{20}
5	−12	407	0.73	150×10^{20}
10	−45	218	0.41	90×10^{20}
12	−60	170	0.37	77×10^{20}
20	−53	62	0.13	27×10^{20}
30	−38	18	0.035	7×10^{20}
40	−18	5.1	0.009	2×10^{20}
50	2	1.5	0.003	0.5×10^{20}
60	−26	0.42	0.0007	0.2×10^{20}
80	−87	0.03	0.00007	0.02×10^{20}

▶ *The unit mmHg— millimeters of mercury—is a common unit of pressure. See page 392.*

Prepare two graphs. On one, plot temperature versus altitude; on the other, pressure versus altitude. Arrange your axes so that each graph nearly fills a sheet of graph paper. The *y*-axis scale (altitude) for each graph should range from 0 to 100 km. The *x*-axis scale (temperature) on the first graph should range from −100 °C to +40 °C. The *x*-axis scale (pressure) in the second graph should extend from 0 to 780 mmHg.

Plot the data summarized in Table 2. Connect the points with a smooth line. (Note that the line may be straight or curved.) Use these graphs and knowledge gained from your flight to answer the questions below.

QUESTIONS

1. Compare the ways air temperature and air pressure change with increasing altitude.

   a. Which follows a more regular pattern?

   b. Try to explain this behavior.

2. Would you expect air pressure to rise or fall if you traveled from sea level to

   a. Pike's Peak (4270 m above sea level)?

b. Death Valley (85 m below sea level)?

c. Explain your answers.

3. Here's an observation you can make for yourself: When two rubber plungers ("plumber's helpers") are pressed together, they are quite difficult to separate.

a. Why?

b. Would it be easier or harder to separate them on top of a mountain?

c. Why?

4. a. Suppose you took one-liter samples of air at several altitudes. How would the following change?

(1) mass of the air sample

(2) number of molecules in the air sample

b. If you were to plot those two values (mass versus number of molecules) in a new graph, what would be the appearance of the plotted line?

c. Why?

5. Scientists often characterize the atmosphere as having four layers: *troposphere* (nearest Earth), *stratosphere, mesosphere, and thermosphere* (outermost layer).

a. Which flight data support the idea of such a "layered" atmosphere?

b. Mark the graphs with vertical lines to indicate where you think the boundaries between layers might be.

In previous sections we have used the term "air pressure." In the next section we will clarify what scientists mean by pressure.

B.4 AIR PRESSURE

In everyday language, we speak of being pressured—meaning that we feel too busy, or feel forced to behave in certain ways. The greater the pressure, the more "boxed in" we feel. To scientists, pressure also refers to force and space, but in quite different ways.

Pressure represents the force applied to one unit of surface area:

$$\text{Pressure} = \frac{\text{Force}}{\text{Area}}$$

▶ *This force (or weight) is due to gravitational attraction. In outer space, the object would be weightless.*

In the modernized metric system, force is expressed in a unit called the *newton* (N). To visualize a newton, imagine holding a personal-sized bar of soap (with a mass slightly greater than 100 g) in your hand. The bar of soap would exert a force (or "push") downward on your hand of one newton (1 N). Expressed another way, at Earth's surface, a 1-kg object exerts a downward force of about 9.8 N.

In the modernized metric system, if force is expressed in newtons (N) and area is expressed in units of square meters (m²), the resulting pressure unit is the pascal (Pa).

$$\frac{\text{Force in newtons (N)}}{\text{Area in square meters (m}^2\text{)}} = \text{Pressure in pascals (Pa)}$$

A pressure of one pascal is a relatively small value—it's roughly the pressure exerted on a slice of bread by a thin layer of butter.

To get a clearer idea of the notion of pressure, imagine a building brick that is 0.19 m long, 0.093 m wide, and 0.055 m high. Its mass is about 1.8 kg. At the Earth's surface, this brick would exert a force of 18 N.

Although the brick's mass remains constant, the pressure it exerts—for example, on a floor—depends upon which side of the brick faces downward (see Figure 3). Here is the calculation of the brick's pressure (in pascals) in each position:

Position A: $\dfrac{18 \text{ N}}{(0.093 \text{ m} \times 0.055 \text{ m})} = \dfrac{3.5 \times 10^3 \text{ N}}{\text{m}^2} = 3.5 \times 10^3 \text{ Pa}$

Position B: $\dfrac{18 \text{ N}}{(0.19 \text{ m} \times 0.093 \text{ m})} = \dfrac{1.0 \times 10^3 \text{ N}}{\text{m}^2} = 1.0 \times 10^3 \text{ Pa}$

Note that in Position A, the force of the brick is distributed over 0.0051 m², (0.093 m × 0.055 m). In Position B, the same force is spread over a much larger area—0.018 m², (0.19 m × 0.093 m). Because the area is about four times greater in Position B, the pressure in Position B is only about one-fourth as great as that in Position A.

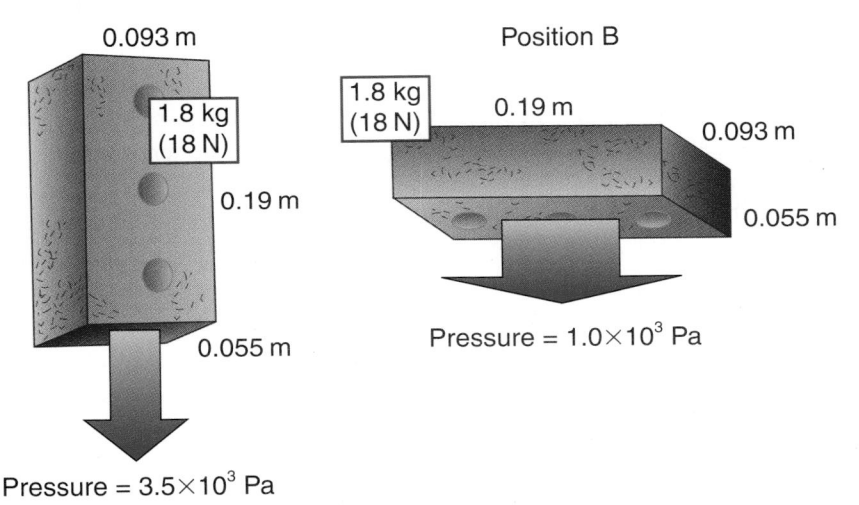

Figure 3 The pressure exerted by the brick depends on which side is down. For the same brick, the pressure is greatest for the side with the smaller area, in this case, Position A, the 0.093 m × 0.055 m side.

Let's consider some practical applications of a scientific view of pressure.

CHEMQUANDARY

PRESSURE PUZZLES

1. At various times, stiletto-heeled shoes are popular. Unfortunately, during these times, floors become scarred with tiny dents. When jogging shoes or shoes with broad heels are worn, no such problem develops. Why? (See Figure 4, page 392).

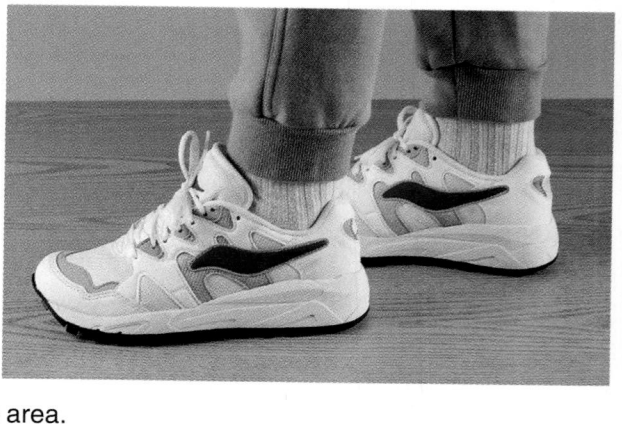

Figure 4 The pressure a shoe heel exerts depends upon its area.

2. A person who accidentally walks onto thin ice on a lake can lower the risk of "falling through" by lying down on the ice or standing on a large piece of plywood. Why?

3. Beach or desert vehicles that need to move easily over loose sand are often equipped with special tires. Would you expect these tires to be wide or thin? Why?

How do building bricks and shoes relate to the atmosphere? As this unit's opening demonstrations showed, the atmosphere exerts a force on every object in it. On a typical day at sea level, air exerts a force of about 100,000 N on each square meter of your body—a resulting pressure of about 100 kilopascals (kPa). This pressure is equal to one atmosphere (atm).

But there is yet another way to express pressure values. Pressure readings can also be given in **mmHg (millimeters of mercury)** or even inches of mercury. A weather report may include "the barometric pressure is 30 inches of mercury." Apparently, air pressure can be expressed as the height of a mercury column. Why? A simple experiment will clarify matters.

Fill a graduated cylinder (or soft drink bottle) with water. Cover it with your hand and invert it in a container of water. Remove your hand. What happens to the water level inside the cylinder? What force supports the weight of the column of water in the cylinder?

Now imagine repeating the experiment with a taller cylinder, and again with an even taller cylinder. If the cylinder were made taller and taller, at a certain height water would no longer fill the cylinder entirely when it is inverted in a container of water. Why?

This experiment was first performed in the mid-1600s. Researchers discovered that one atmosphere of air pressure could support a column of water only as tall as 10.3 m (33.9 ft). If the experiment is tried with even taller cylinders, the water still reaches only to the 10.3-m (10.3×10^3 mm) level. The space in the cylinder above the liquid is a partial vacuum.

Obviously, a barometer based on water would be extremely awkward to handle—it would be much too long. Scientists replaced the water with mercury, a liquid that is 13.6 times more dense than water. The resulting mercury barometer (see Figure 5, page 393) is shorter than the water barometer by a factor of 13.6. Thus, at one atmosphere of pressure, the mercury column has a height of 760 mm.

▶ *10.3 m =*
10.3×10^3 mm;
10.3×10^3 mm/13.6 =
760 mm.

In the following sections we will study the effects of pressure and temperature changes on gas volume. This knowledge will help us understand how gases behave in our bodies, or trapped below ground, or in the atmosphere. We will consider the effect of changes in pressure first.

B.5 BOYLE'S LAW: PUTTING ON THE SQUEEZE

Unlike the volume of a solid or a liquid, the volume of a gas sample can easily be changed. Applying pressure to a sample of gas decreases its volume—it becomes "compressed." By compressing a gas, you can store a greater mass of the gas in a given container. Tanks of compressed gas are quite common: Propane gas tanks are used in campers, mobile homes, and houses in isolated areas; welders use tanks of oxygen and acetylene gas; and hospitals supply oxygen gas from tanks to patients with breathing problems.

Tanks of gas are often used in chemistry laboratories. That observation leads to a puzzling *ChemQuandary*.

CHEMQUANDARY

A VOLUME DISCOUNT?

A chemistry teacher notices a large difference in the price of hydrogen gas sold by two companies.

Company A offers a 1-L cylinder of hydrogen gas for $28.
Company B offers a 1-L cylinder of the same gas for $50.

The teacher discovers that Company B offers a better bargain. How can that be?

How does the volume of a gas sample change when its pressure is increased? You already know the answer, from Observation II of the balloons on page 386. When the pressure on the balloons was doubled, their volumes decreased by one half.

This volume change illustrates a relationship common to all gases. Suppose we work at a constant temperature with 12 L of gas at a pressure of 2 atm. We observe that doubling the pressure to 4 atm cuts the volume to 6 L. If we double the pressure once again, to 8 atm, the volume decreases to 3 L.

Let's summarize these data and complete a simple calculation:

Pressure (atm)	Volume (L)	Pressure × Volume (atm·L)
2	12	24
4	6	24
8	3	24

For a given sample of gas at constant temperature, the product of its pressure and volume always has the same value. Such a relationship holds for any change in either the volume or pressure *at constant temperature*. If the volume of that gas sample is decreased from 6 L to 4 L, the pressure will increase to 6 atm (4 L × 6 atm = 24 atm·L, as before).

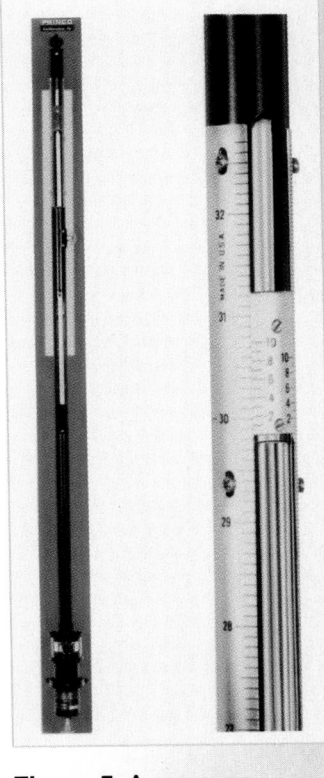

Figure 5 A mercury barometer. At sea level the atmosphere will support a column of mercury 760 mm high. The pressure unit "atmosphere" is thus related to pressure in millimeters of mercury. 1 atm = 760 mmHg.

This pressure–volume relationship is called **Boyle's law,** after Robert Boyle, the seventeenth-century English scientist who first proposed it. One way to state Boyle's law is this: *At a constant temperature, the product of the pressure and the volume of a gas sample is a constant value.*

The general mathematical statement of Boyle's law is

$$P_1V_1 = P_2V_2$$

where P_1 and V_1 are the original pressure and volume of a gas, and P_2 and V_2 are the new pressure and volume at the same temperature. Plotting a series of pressure and volume values for any gas sample gives a curve similar to that in Figure 6. Boyle's law allows us to predict changes in the pressure or volume of a gas sample at constant temperature whenever any three of the values P_1, V_1, P_2, or V_2 are known.

Consider this example:

A certain steel gas tank in a chemistry laboratory has a volume of 1.0 L. It contains gas at a pressure of 56 atm. What volume would the gas from such a tank occupy at 1.0 atm at the same temperature?

We can use Boyle's law to solve the problem this way:

1. Identify the starting and final conditions:

$$P_1 = 56 \text{ atm} \qquad P_2 = 1.0 \text{ atm}$$
$$V_1 = 1.0 \text{ L} \qquad V_2 = ? \text{ L}$$

2. Rearrange the equation $P_1V_1 = P_2V_2$ to solve for the unknown, in this case, V_2, the volume after the pressure decrease:

$$V_2 = \frac{P_1V_1}{P_2}$$

3. Substitute the values into the equation and solve for V_2:

$$V_2 = \frac{56 \text{ atm} \times 1.0 \text{ L}}{1.0 \text{ atm}} = 56 \text{ L}$$

► *Boyle's Law: P↑,V↓; P↓,V↑.*

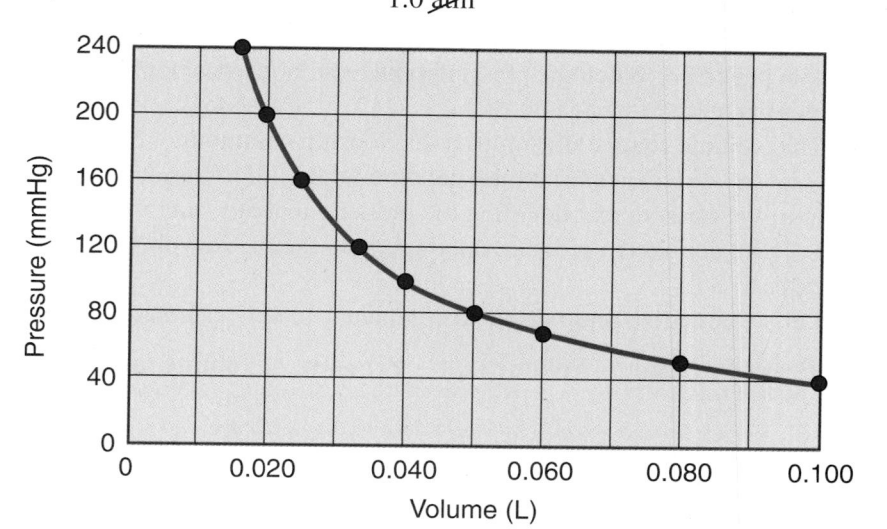

Figure 6 Boyle's law: The volume of a gas, maintained at constant temperature, is inversely proportional to its pressure. Therefore the product of P × V is constant. A plot of pressure vs. volume for any gas sample at constant temperature will be similar to this one.

A reasoning method can be used to solve the same problem: If the pressure *decreases* from 56 atm to 1.0 atm, the volume will *increase* by a proportional amount. Therefore, to find the new, larger volume, the known volume must be multiplied by a pressure ratio larger than one:

$$1.0 \text{ L} \times \frac{56 \text{ atm}}{1.0 \text{ atm}} = 56 \text{ L}$$

If the gas pressure increases, then the volume must decrease. (This assumes the temperature remains constant.) In that case the pressure ratio used must be less than one. Similar reasoning applies to problems in which the initial and final volume are known and the final pressure is to be found.

Work these problems, which use Boyle's law.

YOUR TURN

P–V Relationships

1. Explain each statement:

   a. *Even when they have an ample supply of oxygen gas, airplane passengers become uncomfortable if the cabin loses its pressure.*

   b. *Carbonated soft drinks "pop" when the container is opened.*

   c. *Tennis balls are sold in pressurized containers.*

   d. *When you climb a mountain or ride up in a tall building in an elevator, your ears may "pop."*

2. You can buy helium gas in small aerosol cans to inflate party balloons. Assume that according to the container label, the can contains about 7100 mL (0.25 cubic feet) of helium at 1 atm pressure. If the can's volume is 492 mL, what is the pressure of the gas inside the can (at the same temperature)?

3. Suppose that a tornado passes near your high school. The air pressure inside and outside your classroom (volume = 430 m^3) is 760 mmHg before the storm. At the peak of the storm, pressure outside the classroom drops to 596 mmHg. To what volume would the air in the room have to change to equalize the pressure difference between the inside and the outside? (Assume no change in air temperature.)

4. In the United States, automobile tire pressure is usually measured in non-metric units of pounds per square inch (lb/in.^{2})—often (and incorrectly) expressed simply as "pounds" of air pressure. The tire gauge actually reports the *difference* in pressure between air inside the tire and the atmospheric pressure. Thus, if a tire gauge reads 30 lb/in.^{2} on a day when atmospheric pressure is 14 lb/in.^{2}, the *total* tire pressure is 44 lb/in.^{2}.

   a. What volume of atmospheric air (at 14 lb/in.^{2}) would be needed to fill a 40.0-L tire to a tire gauge reading of 30 lb/in.^{2}?

   b. Why does pumping up an automobile tire by hand take such a long time?

Dramatic changes in air pressure are associated with the passage of a tornado.

In all these problems, we assumed that gas temperature remained constant. How does temperature affect gas volume? The answer to this question is important, because many reactions in laboratories, in chemical processing, and in food preparation are done at roughly the same atmospheric pressure. Did you ever wonder why cakes and bread rise as they bake? Read on and find out why.

B.6 LABORATORY ACTIVITY: T–V RELATIONSHIPS

GETTING READY

Most matter expands when heated and contracts when cooled. As you know, gas samples expand and shrink to a much greater extent than either solids or liquids.

In this activity you will investigate how the temperature of a gas sample influences its volume—assuming pressure remains unchanged. To do this, you will heat a thin glass tube containing a trapped air sample, and then record changes in air volume as the sample cools.

PROCEDURE

1. Fasten a capillary tube to the lower end of a thermometer, using two rubber bands (Figure 7). The open end of the tube should be placed closest to the thermometer bulb and 5–7 mm from the bulb's tip.

2. Immerse the tube and thermometer in a hot oil bath that has been prepared by your teacher. Be sure the entire capillary tube is immersed in oil. Wait for your tube and thermometer to reach the temperature of the oil (approximately 130 °C). Record the temperature of the bath.

3. After your tube and thermometer have reached constant temperature, lift them until only about one quarter of the capillary tube is still in the oil bath. Pause here for about three seconds to allow some oil to rise into the tube. Then quickly carry the tube and thermometer (on a paper towel, to avoid dripping) back to your desk.

4. Lay the tube and thermometer on a paper towel on the desk. Make a reference line on the paper at the sealed end of the melting-point tube. Also mark the upper end of the oil plug (Figure 8, page 397). Alongside this mark write the temperature at which the air column had this length.

5. As the temperature of the gas sample drops, make at least six marks representing the length of the air column trapped above the oil plug at various temperatures; write the corresponding temperature next to each mark. Allow enough time so that the temperature drops by 80 to 100 degrees. Because the tube has a uniform diameter, length serves as a relative measure of the gas volume.

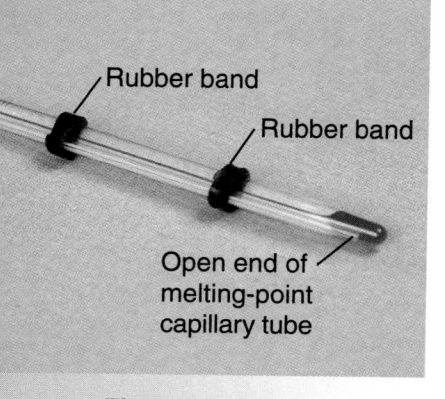

Figure 7 Apparatus for studying how gas volume changes with temperature.

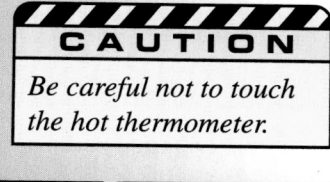

Be careful not to touch the hot thermometer.

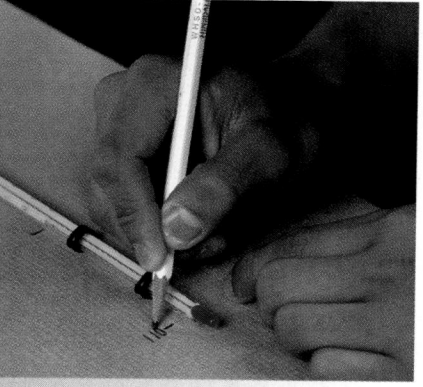

6. When the thermometer shows a steady temperature (near room temperature), make a final observation of length and temperature. Discard the tube and the rubber bands according to your teacher's instructions. Wipe the thermometer dry.

7. Measure and record (in centimeters) the marked lengths of the gas sample.

8. Wash your hands thoroughly before leaving the laboratory.

9. Prepare a sheet of graph paper for plotting your data. The vertical scale (length) should range from 0 to 10 cm; the horizontal scale (temperature) should include values from −350 °C to 150 °C. Label your axes and arrange the scales so the graph fills nearly the entire sheet. Plot the temperature–length data. Draw the best straight line through the plotted points. Using a dashed line, extend this straight graph line so that it intersects the x axis. Turn in one copy of your data; keep another copy to use in answering the following questions.

QUESTIONS

1. At what temperature does your extended graph line intersect the x axis?

2. What would be the volume of your gas sample at this temperature?

3. Why is this volume only theoretical?

4. Now renumber the temperature scale on your graph, assigning the value zero to the temperature at which the graph intersects the x axis. The new scale expresses temperature in kelvins (K)—the kelvin temperature scale. One kelvin is the same size as one degree Celsius. However, unlike zero degrees Celsius, zero kelvin is the lowest temperature theoretically possible—*absolute zero.*

5. Based on your graph, what temperature in kelvins would correspond to 0 °C, the freezing point of water? What temperature would correspond to 100 °C, the boiling point of water at one atmosphere pressure?

Your plotted data indicate that there seems to be a relationship between gas volume and temperature (at constant pressure). But what is that relationship?

In the 1780s, French chemists (and hot-air balloonists) Jacques Charles and Joseph Gay-Lussac studied the changes in gas volume caused by temperature changes at constant pressure. Data for oxygen and nitrogen gas are shown in Figure 9 (page 398). The plots for different gases and different sample sizes end at different points. However, if all graph lines are extended to the x axis (an extrapolation), the lines meet at the same temperature. Lord Kelvin (William Thomson, an English scientist) used the work of Charles and Gay-Lussac to establish a simple mathematical temperature–volume relationship for gases. He based his relationship on his own new temperature scale.

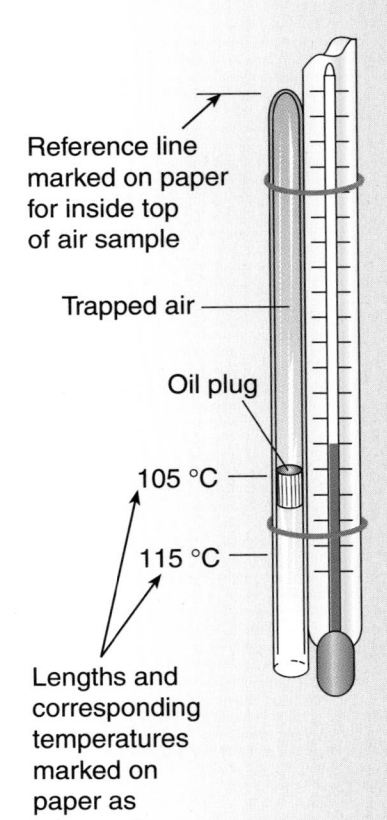

Reference line marked on paper for inside top of air sample

Trapped air

Oil plug

105 °C

115 °C

Lengths and corresponding temperatures marked on paper as tube cools

Figure 8 Marking the length and temperature of the air sample. The air trapped in this tube has cooled from 115 °C to 105 °C and shrunk as shown.

▶ *Temperatures on the kelvin temperature scale are written as 10 K, for example, not 10 °K.*

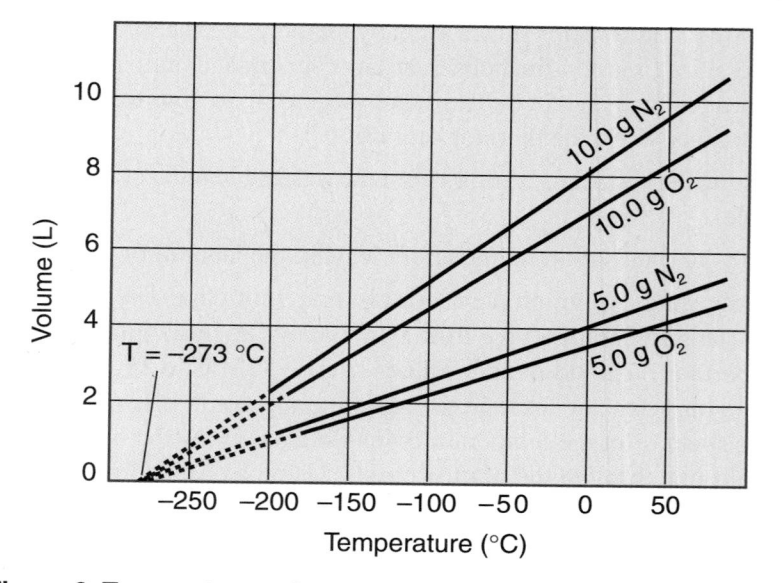

Figure 9 Temperature–volume measurements of various gas samples at 1 atm pressure. Extrapolation has been made for temperatures below liquefaction (O_2 is -183 °C, N_2 is -196 °C).

▶ *Kelvin temperature = °C + 273.*

    Doubling the *kelvin* temperature of a gas doubles its volume. Halving its *kelvin* temperature causes the gas volume to decrease by half, and so on. Such relationships are summarized in ***Charles' law: At constant pressure, the volume of a given gas sample divided by its kelvin temperature is a constant value.*** Or, expressing it another way,

$$\frac{V_1}{T_1} = \frac{V_2}{T_2}$$

CHEMQUANDARY

BEHAVIOR OF GASES

Apply Charles' law in answering these questions:

1. Why does bread rise when baked?

2. What would happen to the volume of gas in a balloon originally at 20 °C if you took it outdoors to a temperature of -37 °C?

3. Why must an anesthesiologist take into account that anesthetic gases during surgery are used both at room temperature (18 °C) and also at the patient's body temperature, 37 °C?

4. If you can install only one thermostat in a two-story house, should it be placed on the first or second floor? Explain your answer.

    Try the following gas behavior problems using any approach you prefer. In each problem identify the variable (pressure or volume) that is assumed to be constant.

T–V–P Relationships

1. a. If the kelvin temperature of a gas sample in a steel tank increases to three times its original value, what will happen to the pressure of the gas? (*Hint:* "Steel tank" is a good indication that gas volume will remain constant.)

    b. Give an example of such a situation.

2. a. If a sample of gas is cooled at a constant pressure until it shrinks to one-fourth its initial volume, what change in its kelvin temperature must have occurred?

    b. Give an example of such a situation.

3. Explain why car owners in severe northern climates often add more air to their tires in winter and release air from them in summer.

4. a. When the volume of a gas sample is reported, the pressure and the temperature must also be given. Why?

    b. This is not necessary for liquids and solids. Why?

5. Use gas laws to explain each situation:

    a. High, anvil-shaped clouds form below the level where warm, moist air and cold air masses meet.

    b. A weather balloon steadily expands as it rises.

6. Why shouldn't aerosol cans be put into a fire?

Why do all common gases at normal atmospheric conditions behave according to Boyle's law and Charles' law? An explanation will be developed in the next section.

B.7 IDEALLY, GASES BEHAVE SIMPLY

By the early 1800s, scientists had discovered the gas laws you have just explored. Balloon flights also provided valuable information on the atmosphere's composition and structure. However, the reasons for such consistent and similar behavior among gases were not yet understood.

However, bit by bit, scientists pieced together a comprehensive and satisfactory explanation of gas behavior—the kinetic molecular theory of gases.

To understand the theory, you must first understand the concepts of kinetic energy and of molecular motion. *Kinetic energy,* the energy possessed by a moving object, is sometimes called "energy of motion." Kinetic energy depends on the mass of the object and on its velocity.

► *Scientific laws describe the behavior (the "what") of nature, but do not provide explanations. Scientific theories and models address "how" aspects of our natural world.*

Traveling at the same velocity, a more massive object has greater kinetic energy than does a less massive object. For example, even though they are both moving at 60 mph, an 18-wheel truck has much greater kinetic energy than a small car.

If we consider two objects having equal masses, the one traveling faster has greater kinetic energy due to its greater velocity. This helps explain, for example, the difference of damage to an automobile when it hits a tree at 10 mph or at 50 mph.

The *kinetic molecular theory* of gases is based on several postulates:

▶ *A postulate is an accepted statement used as the basis of further reasoning and study.*

- Gases consist of tiny particles (atoms or molecules) having negligible size compared with the great distances separating them.

- Gas molecules are in *constant, random motion.* They often collide with each other and with the walls of their container. Gas pressure is the result of molecular collisions with the container's walls.

- Molecular collisions are elastic. This means that although individual molecules may gain or lose kinetic energy, there is no net (or overall) gain or loss of kinetic energy from these collisions.

- At a given temperature, molecules in a gas sample have a range of kinetic energies. However, the average kinetic energy of the molecules is constant and depends only on the kelvin temperature of the sample. Therefore, molecules of different gases at the *same* temperature, have the *same* average kinetic energy. As temperature increases, so do the average velocities and kinetic energies of the molecules.

An analogy might be helpful in understanding the molecular motions presented by the kinetic molecular theory.

YOUR TURN

Kinetic Molecular Theory: An Analogy

Imagine that a large group of dancers on an enclosed dance floor represent gas molecules bouncing around in a container. The dancers move back and forth across the floor, but not up to or down from the ceiling.

1. Identify which of the four variables—volume, temperature, pressure, and number of molecules—is most like the following:
   a. The number of dancers
   b. The size of the room
   c. The tempo of the music
   d. The number and force of collisions among the dancers

2. Which law—Boyle's, Charles', Avogadro's, or others—best applies to each of the following situations?
   a. The tempo of the music and the number of dancers remain the same, but the size of the dance floor decreases.
   b. The size of the dance floor and number of dancers remain the same, but the tempo of the music increases.

c. The size of the dance floor and the tempo of the music are kept the same, but the number of dancers increases.

Motion of people and tempo of the music are related.

Other analogies for gases at the molecular level may include a room swarming with tiny gnats, or full of bouncing superballs. Although helpful in giving us "mental pictures" of molecular behavior, such analogies, including dancers, fail to represent these characteristics:

- The extremely small size of the gas molecules relative to the total volume of a gas sample. For example, at room temperature less than one thousandth of the volume of a carbon dioxide gas sample is actually occupied by molecules. The remaining space is empty.
- The extremely high velocities of gas molecules. At room temperature, for example, nitrogen gas molecules travel—on average—nearly 1700 km/h (1100 mph).
- The extremely high frequency of collisions.

Because of these differences, the actual behavior of gases can be precisely represented only through mathematical expressions.

Complete the following exercises to check your understanding of the kinetic molecular theory of gases.

YOUR TURN

Gas Molecules in Motion

You will use the kinetic molecular theory to explain five common observations regarding gas behavior. Where appropriate, you should consider such factors as the kinetic energies, spaces between molecules, and molecular collisions. A sketch of molecules in motion may also be useful for some items.

Here is a worked-out example of how the kinetic molecular theory can be applied in explaining a common observation:

Observation: Increasing the volume of a gas sample causes its pressure to decrease if the temperature remains constant.

Explanation: Increasing the volume of a sample of gas molecules gives the molecules more room to move around. So, molecules must travel farther (on average) before they collide with (and bounce off) the container walls; there are fewer molecular collisions with the walls in a given time. Because pressure is caused by such molecular collisions, the gas pressure will decrease.

Now it's your turn. Use the kinetic molecular theory to explain each of these observations:

1. Decreasing the volume of a gas sample causes the gas pressure to increase, if the temperature is held constant.
2. At constant volume, the pressure of a gas changes when the temperature changes.
3. The atmosphere exerts pressure on our bodies, yet this pressure does not crush us.
4. Filled balloons eventually leak even when they are tightly sealed.
5. Helium-filled balloons leak faster than those inflated with air.

Gases that behave as kinetic molecular theory predicts are called ***ideal gases.*** At very high pressures or very low temperatures, gases do not behave ideally. The gas laws we have considered do not accurately describe gas behavior under such conditions. At low temperatures, molecules move more slowly. The weak attractions between them may become so important that the gas will condense into a liquid, such as when air is liquefied. At high enough pressures—if the temperature is not too high—the gas molecules are forced so close together that these same forces of attraction may again cause a gas to condense. However, under the usual conditions encountered in our atmosphere, most gas behavior approximates that of an ideal gas. Such gas behavior is well explained by the kinetic molecular theory.

Now that you understand how chemists account for some properties of gases, you are ready to explore how the gases in our atmosphere interact with solar energy to create weather and climate.

? PART B: SUMMARY QUESTIONS

1. Oxygen gas is essential to life as we know it. Our atmosphere is approximately 21% oxygen gas. Would a higher percent of atmospheric oxygen be desirable? Explain.

2. Explain the chemical change summarized below, using
   a. molecular pictures.
   b. a chemical equation.

1 L H_2	+	1 L Cl_2	→	2 L HCl
hydrogen gas		chlorine gas		hydrogen chloride gas

3. The diagrams below represent a cross-section of a steel tank containing only helium gas. The dots represent helium atoms.

   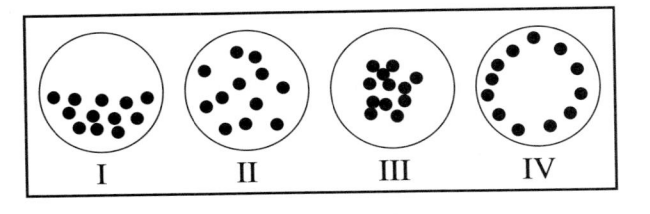

   | I | II | III | IV |

   a. Which diagram correctly illustrates the distribution of helium atoms at 20 °C and 3 atm? Explain your choice.
   b. Which diagram correctly illustrates the distribution of helium atoms when the temperature of the tank is lowered to −20 °C at 3 atm? The boiling point of helium is −269 °C. Explain your choice.
   c. Which diagram correctly illustrates the distribution of helium atoms at 20 °C if the pressure is reduced to 1.5 atm? Explain your choice.

4. Explain why a suction-cup hook can support the weight of a small object.

5. Explain why living creatures are not normally found above the troposphere.

6. Solve these problems. As part of your answer, identify which gas variable is assumed constant:
   a. A small quantity of the inert gas argon (Ar) is added to light bulbs to reduce the vaporization of tungsten atoms from the filament. What volume of argon at 760-mmHg pressure is needed to fill a 0.210-L light bulb at 1.30-mmHg pressure?
   b. A tank at 300 K contains 0.285 L of gas at 1.92 atm pressure. The tank is capable of withstanding a maximum pressure of 5.76 atm. Assuming that doubling the kelvin temperature causes the internal pressure to double,
      (1) at what temperature will the tank burst?
      (2) would it explode in a fire at a temperature of 1275 K?
   c. A 2.50-L container is filled at 298 K with butane gas, a fuel used in barbecue grills. At a constant pressure, the temperature is lowered to 280 K. What is the volume of the butane at this temperature?

7. Convection, the process whereby warm air rises and cold air falls, is important to the natural circulation and cleansing of the troposphere. Explain in molecular terms why convection occurs.

EXTENDING YOUR KNOWLEDGE
(OPTIONAL)

- In what ways is an "ocean of gases" analogy useful in thinking about the atmosphere? How does the atmosphere differ from the hydrosphere? (You may wish to compare a dive into the ocean with an ascent into the atmosphere.)

- In terms of pressure exerted at various depths, how is our atmosphere similar to the oceans?

- Boyle's law is illustrated each time you breathe in or breathe out. Explain.

- The gas behavior described by Boyle's law is a matter of life and death to scuba divers. On the surface of the water, the diver's lungs, tank, and body are at atmospheric pressure. However, under water a diver's body is under the combined pressure of atmosphere and water.

  a. When divers are 10.3 m below the surface of the ocean, how much pressure do their bodies experience?

  b. Why are pressurized tanks needed when diving to depths greater than 10 meters?

  c. What would happen to the volume of the tank if it were not strong enough to withstand such pressure?

  d. Why is it necessary to exhale and rise slowly when ascending from the ocean depths?

  e. How do the problems of a diver compare with those of a pilot climbing to a high altitude in an unpressurized plane?

- If air at 25 °C has a density of 1.28 g/L, and your classroom has a volume of 2×10^5 L

  a. What is the mass of air in your classroom?

  b. If the temperature increased, how would this affect the mass of air in the room (i) if the room is completely sealed; and (ii) if the room is not completely sealed?

- How would you convert a given temperature in degrees Celsius to its corresponding kelvin temperature?

- The kelvin temperature scale has played an important role in theoretical and applied chemistry.

  a. Find out how close scientists have come to zero kelvin (0 K) in the laboratory.

  b. Research from the field of *cryogenics* (low-temperature chemistry and physics) has many possible applications. Do research in the library to learn about activities in this field.

C ATMOSPHERE AND CLIMATE

Imagine a place where mid-day sunshine makes a rock hot enough to fry an egg, where nights are cold enough to freeze carbon dioxide gas to "dry" ice, and where the sun's ultraviolet rays can burn exposed skin in minutes. That place is Earth's moon. These extreme conditions are due to the absence of a lunar atmosphere.

The combination of the sun's radiant energy and the Earth's atmosphere helps to maintain a hospitable climate for life on this planet. The atmosphere delivers the oxygen gas we breathe and makes exhaled carbon dioxide available to plants for photosynthesis. It also serves as a sink for airborne wastes from people, industry, and technology—a role that is causing increasing concern. And, the atmosphere protects our skin from the sun's ultraviolet rays.

To understand the basis for our hospitable climate, it's important to know how solar energy interacts with our atmosphere. The sun warms the Earth's surface. The Earth's warm surface, in turn, warms the air above it. Because warm air expands, its density decreases, and so the warm air

Mongolian sheepherders utilizing the sun's energy.

rises. Colder, more dense air falls. These movements of warm and cold air masses help create continual air currents that drive the world's weather. Let's look more closely at the sun's energy. What makes it so useful?

C.1 THE SUNSHINE STORY

► *Nuclear fusion is discussed on page 353.*

► *For further background on electromagnetic radiation, see the Nuclear unit, page 299.*

► *Wavelength ↓, energy ↑.*

► *Such photon–electron interactions are taking place right now in double bonds of molecules in your eyes, making it possible for you to see this page.*

The enormous quantity of energy produced by the sun is a result of the fusion of hydrogen nuclei into helium. Most of this energy escapes from the sun as electromagnetic radiation. About 9% of solar radiation is in the ***ultraviolet (UV)*** region of the electromagnetic spectrum, 46% is in the ***visible*** region, and 45% is in the ***infrared (IR)*** region. The complete solar spectrum is shown in Figure 10.

Electromagnetic radiation is composed of photons, each possessing a characteristic quantity of energy. Photon energy determines the effect of radiation on matter. ***The shorter the wavelength, the higher the energy of the photon.***

Infrared radiation causes molecules to vibrate faster. As we noted in connection with the kinetic molecular theory (page 400), this raises the temperature.

Photons of visible light have more energy than infrared radiation photons. Visible light can energize electrons in some chemical bonds, with each photon delivering its energy to the electrons in just one chemical bond. Visible light also interacts with the electrons in chlorophyll molecules in green plants, providing the energy needed for the reactions involved in photosynthesis.

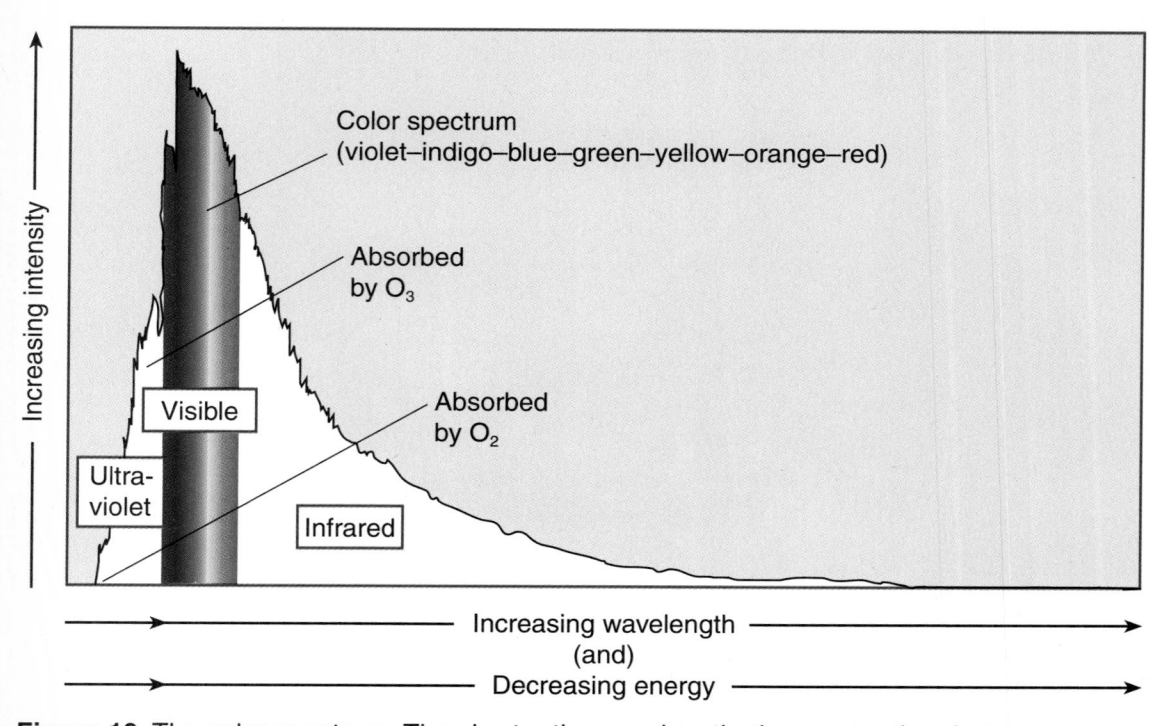

Figure 10 The solar spectrum. The shorter the wavelength, the greater the photon energy of the radiation. Intensity, plotted on the *y* axis, is a measure of the quantity of radiation at a given wavelength.

Ultraviolet photons have even greater energy than visible radiation. In fact, UV photons have sufficient energy to break single covalent bonds. As a result, chemical changes can take place in materials exposed to ultraviolet radiation—including damage to tissues of living organisms. This explains why you are cautioned to wear sunscreen when exposed to sunlight for extended periods (See Personal Chemistry and Choices unit, page 491).

As solar radiation passes through the atmosphere, it interacts with molecules and other particles. Once it reaches Earth's surface it is absorbed or used in other ways. To better appreciate the effects of solar radiation, let's briefly consider some factors that affect average global temperature.

▶ *Energies: IR < Visible < UV*

▶ *Wavelengths: IR: 10^3– 10^4 nm (1 nm = 10^{-9} m); visible: about 380–780 nm; UV: 180–380 nm.*

C.2 EARTH'S ENERGY BALANCE

The mild average temperature (15 °C, or 59 °F) at the Earth's surface is determined partly by the balance between the inward flow of energy from the sun and the outward flow of energy into space. However, certain properties of the Earth determine how much thermal energy our planet can hold near its surface—where we live—and how much energy it radiates back into space.

Figure 11 shows the fate of solar radiation as it enters our atmosphere. Some incoming solar radiation never reaches the Earth's surface, but is reflected directly back into space by clouds and particles in the atmosphere. A small amount of solar radiation is also reflected when it strikes such materials as snow, sand, or concrete at the Earth's surface.

About one fourth of all incoming solar energy is used to power the hydrologic cycle—the continuous cycling of water into and out of the atmosphere by evaporation and condensation.

▶ *Visible light reflected in this way allows the Earth's illuminated surface to be seen from space.*

▶ *For more information, see the Water unit (page 14).*

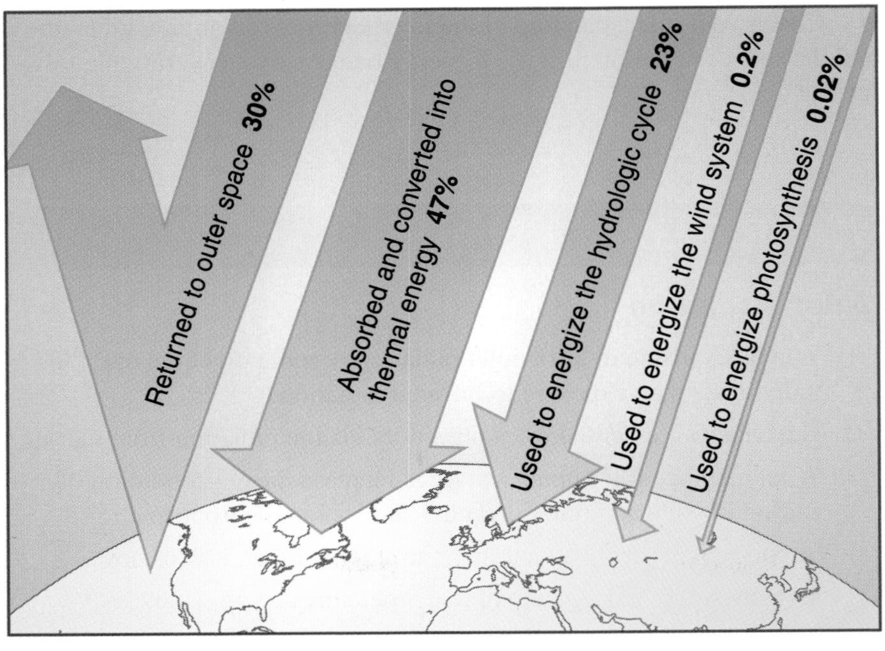

Figure 11 The fate of incoming solar radiation.

Almost half of solar energy is absorbed by the Earth, warming the atmosphere, oceans, and continents. All objects above absolute zero radiate energy, the quantity of energy radiated being directly related to the object's temperature. Thus, the Earth's surface reradiates most of the absorbed radiation, but not at its original wavelength. Instead, the radiation is re-emitted at longer wavelengths—in the infrared region of the spectrum. This reradiated energy plays an extremely important role in the world's energy balance. These lower-energy infrared photons are absorbed by molecules in the air more easily than the original solar radiation, thus warming the atmosphere.

Carbon dioxide (CO_2) and water (H_2O) are good absorbers of such infrared radiation, as are methane (CH_4) and halogenated hydrocarbons such as CF_3Cl, a CFC (CFCs are discussed in more detail on page 418). Clouds (concentrated droplets of water or ice) also absorb infrared radiation. Energy absorbed by these molecules in the atmosphere is once again radiated toward the Earth. Energy can pass back and forth between the Earth's surface and molecules in the atmosphere many times before it finally escapes into outer space.

This trapping and returning of infrared radiation by carbon dioxide, water, and other atmospheric substances is known as the *greenhouse effect,* because it resembles the way heat is held in a greenhouse on a sunny day. Such atmospheric substances are known as *greenhouse gases.*

Without water and carbon dioxide molecules in the atmosphere to absorb and reradiate thermal energy to Earth, our whole planet would reach thermal balance at a bone-chilling $-25\,°C$ ($-13\,°F$), quite close to the average temperature on Mars. At the other extreme is the planet Venus, an example of a runaway greenhouse effect. Its atmosphere is composed mainly of thick carbon dioxide clouds that prevent the escape of most infrared radiation. This causes the average temperature on Venus to be much higher ($450\,°C$) than on Earth ($15\,°C$).

Check your understanding of the interaction of radiation with matter and the role of radiation in Earth's energy balance by answering the following questions.

► *Gaseous substances that are good infrared absorbers—such as carbon dioxide, water vapor, methane, and CFCs—are called greenhouse gases.*

YOUR TURN

Solar Radiation

1. Why is exposure to ultraviolet radiation of more potential harm to individuals than exposure to infrared radiation?

2. Describe two essential roles played by visible radiation from the sun.

3. What may occur if human activities increase the concentration of carbon dioxide and other greenhouse gases in the troposphere?

4. Suppose our planet had a thinner atmosphere than it does now.

   a. How would average daytime temperatures be affected? Why?

   b. How would average nighttime temperatures be affected? Why?

Climate is influenced not only by the interaction of solar radiation with the atmosphere, but also by other factors. These factors include the Earth's rotation (causing day and night and influencing wind patterns), its revolution around the sun (causing seasons), its tilt on its axis (causing the uneven distribution of solar radiation and influencing wind patterns), and differing thermal properties of materials on its surface. In the next section we will examine the last of these factors.

CHEMQUANDARY

GRAB ANOTHER BLANKET!

Why does the outdoor temperature drop faster on a clear night than on a cloudy night?

C.3 AT THE EARTH'S SURFACE

If you have visited or lived in the South or Southwest, you have probably noticed that many cars (and their interiors) are light colored. A property of materials called *reflectivity* helps keep these vehicles cooler than vehicles with darker surfaces. When light photons strike a surface, some are absorbed—increasing the surface temperature—and some are reflected. The reflected radiation does not contribute to raising the object's temperature.

Light-colored surfaces reflect more radiation and therefore they remain cooler than do dark-colored surfaces. Clean snow, for example, reflects almost 95% of solar radiation, while forests are not very reflective. Similarly, variations in the reflectivity of materials at the Earth's surface help determine its surface temperature. On a hot day it is much more comfortable to walk barefoot across a plowed field than across an asphalt parking lot. The plowed field reflects almost 30% of the sun's rays, while the asphalt reflects very little.

Each material at the Earth's surface has a characteristic reflectivity and heat capacity, which together determine how fast the material warms up. *Heat capacity* is the quantity of thermal energy (heat) needed to raise the temperature of a given mass of material by 1 °C. In effect, heat capacity is a measure of a material's storage capacity for thermal energy. The lower a material's heat capacity, the more its temperature increases for a given quantity of added energy. The higher the heat capacity, the smaller the temperature increase for a given quantity of added energy. Thus, materials with higher heat capacities can store more thermal energy.

Water has several properties that create its unique role in the world's climate. One such property is its very high heat capacity (4.2 J/g °C), higher than that of any other common material. Because of this property, bodies of water are slow to heat up or cool down, and can store large quantities of thermal energy. By contrast, land surfaces have much lower heat capacities, cool off much more rapidly, and reach lower temperatures at night.

Oceans, lakes, and rivers, therefore, have a moderating effect on temperature. For example, on a warm day breezes blow from the ocean toward the land, because the temperature over the land is higher.

► *Heat capacity is often reported in units of joules per gram per degree Celsius, J/(g °C).*

The warmer, less dense air rises, allowing cooler and more dense air from the ocean to move in. Then, in the evening, the land cools off more rapidly. The breezes shift direction, blowing from the now-cooler land toward the ocean. In general, temperatures fluctuate less near oceans or large lakes than they do in regions far from large bodies of water.

YOUR TURN

Thermal Properties of Materials

1. Explain why each of the following occurs on a hot day in the sun.

   a. A white convertible's fender is cooler to the touch than its deeply colored maroon seats.

   b. An asphalt sidewalk is warmer to the touch than a concrete sidewalk.

2. Would you expect the average temperature to be lower during a winter with large quantities of snow or in one with very little snow? Why?

3. Why is water—heat capacity 4.2 J/(g °C)—a better fluid for a hot pack than ethyl alcohol—heat capacity 2.6 J/(g °C)?

4. Beach sand feels hotter than grass on a hot day. Which is more responsible for this observation—heat capacity or reflectivity?

5. Why does the temperature drop more on a clear, cool night than on a cloudy, cool night?

6. Which of the following two medium-sized cities, at the same latitude and longitude, would you expect to be hotter in the summer? Why?

   a. A city with many asphalt roads and concrete buildings.

   b. The same city, but located near a large body of water.

Can the thermal balance of our planet be upset? Is human activity affecting climate? We'll examine these questions in the next section.

C.4 CHANGES ON THE EARTH'S SURFACE

More than 100 years ago it was suggested that rapidly increasing the burning of fossil fuels might release enough carbon dioxide into the atmosphere to change the Earth's climate. You have learned (page 408) that water vapor, carbon dioxide, and other atmospheric greenhouse gases act as one-way screens. They let in sunlight, and then limit the escape of reradiated infrared photons. This produces the greenhouse effect that keeps our planet at a comfortable average temperature.

The greenhouse effect will remain stable as long as carbon dioxide and other greenhouse gases in the atmosphere remain at their normal levels, and no significant additional amounts of greenhouse gases are added. Both the hydrologic cycle (see page 14) and the carbon cycle (Figure 12) maintain constant concentrations of atmospheric water and carbon dioxide. However, human activity must also be taken into consideration.

The atmosphere contains about 12 trillion metric tons of water vapor, a quantity so large that it might seem impossible that human activity could significantly affect it. However, if global temperatures increase, more of the Earth's slightly warmer water will evaporate, increasing the atmospheric concentration of water vapor. As an important greenhouse gas, this increased water vapor would cause an even greater increase in global temperatures due to absorption and release of infrared radiation. A "spiraling-up" effect could occur—warmer temperatures producing more water vapor, producing warmer temperatures, and so on.

We know that human activity has increased atmospheric carbon dioxide level by about 30% since 1800. We increase CO_2 levels in several ways: When limestone—a form of calcium carbonate, $CaCO_3$—is converted to calcium oxide (CaO) for making concrete, some CO_2 is released. Clearing forests removes vegetation that would ordinarily consume CO_2 through photosynthesis. Cuttings and scrap timber are then burned,

Figure 12 The carbon cycle. How does this figure illustrate the cycling of carbon throughout the biosphere?

Deforestation contributes to increases in atmospheric CO_2.

releasing CO_2 to the atmosphere. Most significantly of all, burning fossil fuels releases CO_2 into the air, as these equations clearly illustrate:

$$\text{Burning coal: } C(s) + O_2(g) \rightarrow CO_2(g)$$

$$\text{Burning natural gas: } CH_4(g) + 2\,O_2(g) \rightarrow CO_2(g) + 2\,H_2O(g)$$

$$\text{Burning gasoline: } 2\,C_8H_{18}(g) + 25\,O_2(g) \rightarrow 16\,CO_2(g) + 18\,H_2O(g)$$

If more CO_2 is added to the atmosphere than can be removed by natural processes, its concentration will increase significantly. Eventually, if sufficient carbon dioxide is added, the atmosphere could retain enough additional infrared radiation to raise our world's surface temperature.

Has this warming effect already begun? Has the world's average temperature increased? Eight of the warmest years (on average) in the last 100 years have occurred since 1980 (Figure 13). An international panel of scientists has predicted that under a business-as-usual scenario (in which no steps are taken to control release of excess greenhouse gases), the global mean temperatures will rise about 0.3 °C each decade for the next century. This is a faster rate of warming than any since the last ice age ended, 10,000 years ago.

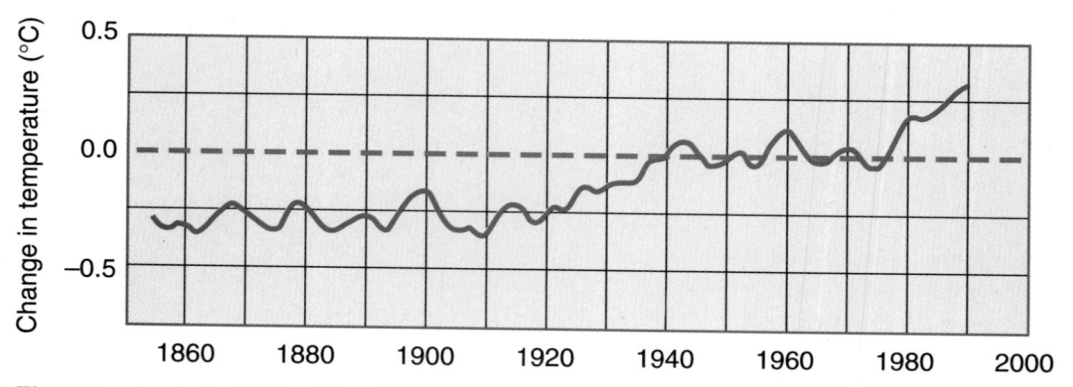

Figure 13 Global warming. The average daily temperature in January 1940 has been used as the zero point for comparison.

CHEMISTRY AT WORK

Keeping Cities Green Is a Nonstop Job

Imagine having a job that lets you do things you enjoy, while improving your neighborhood and other places around your hometown—and meeting interesting, dedicated neighbors as well.

Brent Schmidt believes he has found his "dream job": working as an arboriculturist for Seattle City Light, the municipal power company of Seattle, Washington. To help keep his city green, Brent uses his knowledge of botany, biology, and soil chemistry.

Brent works with communities to determine the best types of trees to plant near power lines, then helps coordinate volunteer participation. For example, he recently coordinated the planting of 600 trees by volunteers in a local community. The project had many details that Brent needed to oversee, such as the type and size of the trees to be planted, soil types, watering requirements, tools and equipment needed, and zoning ordinances.

In the past, utility companies were responsible for maintaining power lines, providing power to customers, and little else. If a tree grew into the area of the power lines, the power company sent a crew out to trim it back, or even cut it down as a safety precaution.

Today, Seattle takes a different approach. Instead of waiting for problems to happen, the company works closely with the community to plan ways for trees and power lines to coexist.

Each year, City Light plants as many as 2,500 trees that will coexist with power lines in Seattle's public areas. As a result, the city enjoys beautiful, beneficial greenery, and the company reduces the amount of tree trimming and removal needed to keep power lines unobstructed.

Try to imagine a world without trees, wilderness areas, or national forests.

- What would cities be like if there were no trees? Write a story from the point of view of a person living in a large, mostly barren city.
- What aspects of daily life would be different?
- What would the air be like?
- Are there things you take for granted now that you would no longer be able to do or enjoy?

413

The 1995 report of the United Nations Intergovernmental Panel on Climate Change states that Earth's average global air temperature has increased by 0.3 °C to 0.6 °C over the past 100 years. The 2,000-page report was written by 500 scientists and reviewed by 500 more climate experts. Almost all climate specialists agree that Earth's average temperature will increase 1.0 to 3.5 °C over the next century. This increase is likely because, sometime between 2050 and 2100, the level of atmospheric carbon dioxide is projected to be double the current value of approximately 360 ppm. Few climate experts expect the warming to be less than 1.0 °C, and even fewer predict a temperature decline during that time.

What are the projected effects of global warming? Based on the business-as-usual scenario, the oceans are predicted to rise about 5 cm each decade over the next century, with approximately a 20-cm increase by 2030. Such changes could bring widespread flooding in major coastal cities throughout the world. Regional climate changes would be expected, including reduced summer precipitation and soil moisture in North America. The greatest temperature increases are predicted above 40° north latitude, where fossil fuel burning and seasonal changes in plant growth are greatest. This could have a major impact on important food-growing areas. Some regions could lose arable land, while the best agricultural regions could shift locations.

Although increasing carbon dioxide would have a warming influence on the globe, no one can yet predict exactly what the climate will do. The climate is affected by many factors, interacting in ways not yet completely understood. Human settlements warm the Earth by lowering its reflectivity—altering it with cities and farms that replace forests and plains. Automobiles and air pollution affect local temperatures. Smog particles can both warm and cool the climate. On top of all this, the world's climate runs in cycles of alternating ice ages and warm periods. We may just be in the warm part of such a cycle.

C.5 YOU DECIDE: TRENDS IN CO_2 LEVELS

The CO_2 level data given in Table 3 show average measurements taken at the Mauna Loa Observatory in Hawaii.

PART 1

1. Plot a graph of the data in Table 3. Prepare the horizontal axis to include the years 1900 to 2050, and the vertical axis to represent CO_2 levels from 280 ppm to 600 ppm. Plot the appropriate points, and draw a smooth curve representing the trend of the plotted points. (A smooth curve shows general trends, and so do not draw a straight line or attempt to connect every point.)

2. Assuming the trend in your smooth curve will continue, extend your curve with a dashed line from the year 1994 to 2050. This extrapolation is a prediction for the future, based on past trends.

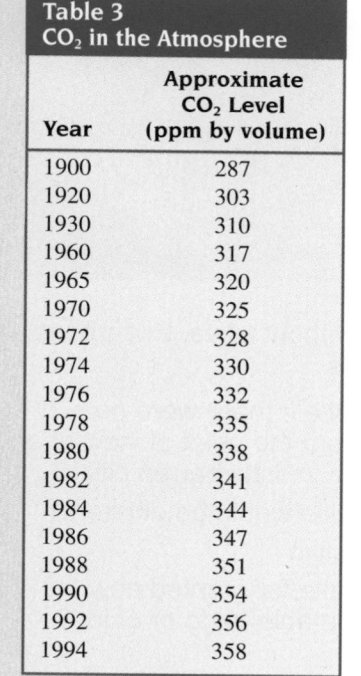

Table 3 CO_2 in the Atmosphere	
Year	Approximate CO_2 Level (ppm by volume)
1900	287
1920	303
1930	310
1960	317
1965	320
1970	325
1972	328
1974	330
1976	332
1978	335
1980	338
1982	341
1984	344
1986	347
1988	351
1990	354
1992	356
1994	358

▶ A "curve" on a graph may appear either curved or straight.

PART 2

You will now make some predictions using the graph you have just completed. You will also evaluate these predictions.

▶ *Keep in mind that such extrapolations are always tentative. Completely unforeseen factors may arise in the future.*

1. What does your graph indicate about the general change in CO_2 levels since 1900?

2. Based on your extrapolation, predict CO_2 levels for

   a. the current year.

   b. the year 2010.

   c. the year 2050.

3. Does your graph predict a doubling of the 1900 CO_2 level?

4. Which predictions from Question 2 are the most likely to be accurate? Why?

5. Describe factors that might cause your extrapolations to be incorrect.

6. What assumption must you make when extrapolating from known data?

C.6 LABORATORY ACTIVITY: CO₂ LEVELS

GETTING READY

The air we usually breathe has a very low CO_2 level. However, the CO_2 concentration in a small space can be substantially increased by burning coal or petroleum, decomposing organic matter, or accumulating a crowd of people or animals.

In this activity you will estimate and compare the amount of CO_2 in several air samples. To do this, the air will be bubbled through water that contains an indicator, bromthymol blue. Carbon dioxide reacts with water to form carbonic acid:

$$CO_2(g) + H_2O(l) \rightarrow H_2CO_3(aq)$$

As the concentration of carbonic acid in solution increases, bromthymol blue's color changes from blue to green to yellow.

PROCEDURE

Part 1: CO₂ in Normal Air

1. Pour 125 mL of distilled or deionized water into a filter flask and add 10 drops of bromthymol blue. The solution should be blue. If not, add a drop of 0.5 M NaOH and gently swirl the flask.

2. Pour 10 mL of the solution prepared in Step 1 into a test tube labeled "control." Put this *control* aside; you will use it for comparison.

3. Assemble the apparatus illustrated in Figure 14 (page 416), without the candle.

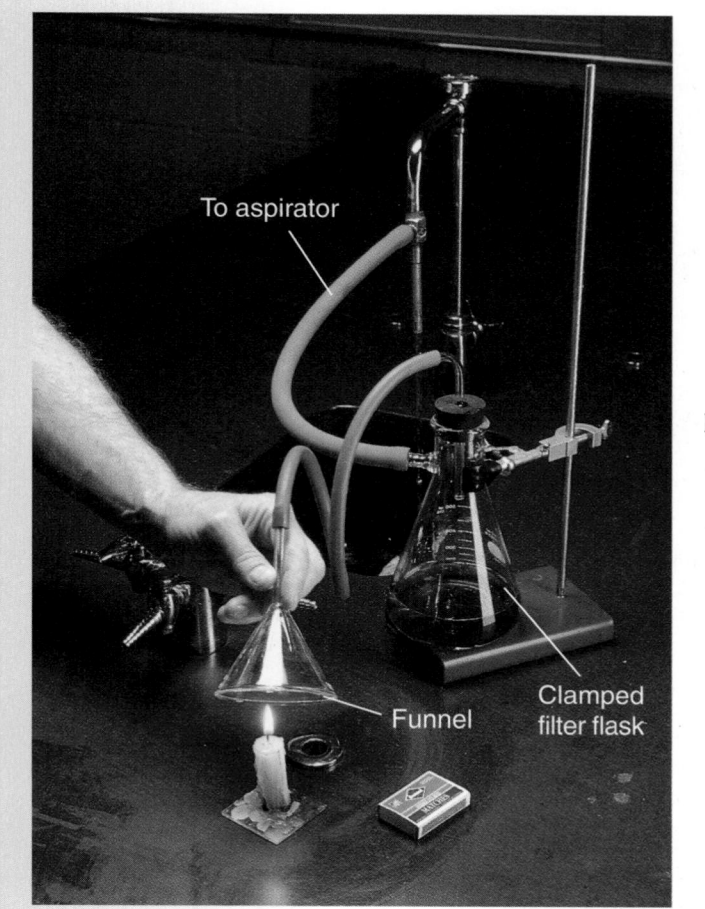

Figure 14 Apparatus for collecting air. Part 1 is done without the candle.

4. Turn on the water tap so that the aspirator pulls air through the flask. Note the position of the faucet handle so you can run the aspirator at this same flow rate each time.

5. Let the aspirator run for five minutes. Turn off the water. Remove the stopper from the flask.

6. Pour 10 mL of the used indicator solution from the flask into a second test tube labeled "normal air." Compare the color of this sample with the control. Save the "normal air" test tube.

Part 2: CO_2 from Combustion

1. Empty the filter flask and rinse it thoroughly with distilled or deionized water. Label a clean test tube "CO_2 combustion."

2. Fill the flask with unused indicator solution according to Step 1 in Part 1. Reassemble the apparatus as in Step 3 in Part 1.

3. Light a candle and position it and the funnel so that the tip of the flame is level with the funnel base.

4. Turn on the water tap to the position you observed in Part 1. Note the starting time. Run the aspirator until the indicator solution turns yellow. Record the time this takes in minutes. Turn off the water.

5. Pour 10 mL of the solution into a clean test tube labeled "CO_2 combustion." Compare its color with that of the "normal air" and the "control" solutions. Record your observations.

Part 3: CO_2 in Exhaled Air

1. Pour 125 mL of distilled or deionized water into an Erlenmeyer flask and add 10 drops of bromthymol blue. As before, add a drop of 0.5 M NaOH if the color is not blue.

2. Note the time, then exhale into the solution through a clean straw, until the indicator color changes to yellow. Record the time this takes.

3. Pour 10 mL of the solution into a clean test tube labeled "CO_2 breath." Compare its color with that of your other three solutions. Record your observations.

4. Wash your hands thoroughly before leaving the laboratory.

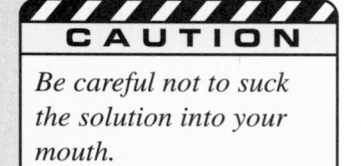

CAUTION

Be careful not to suck the solution into your mouth.

QUESTIONS

1. Compare the colors of the solutions from each test. Which sample contained the most CO_2?

2. Compare the times and color changes using normal air and air from above the candle. Explain the differences.

3. Which contained more CO_2, air in the absence of the burning candle or exhaled air?

4. What effect would each of the following have on the time needed for the indicator solution to change color?

   a. Many plants growing in the room near the indicator solution.

   b. Fifty more people entering and remaining in the room.

   c. Better ventilation in the room.

   d. Half the people in the room lighting up cigarettes.

C.7 YOU DECIDE: REVERSING THE TREND

Scientists understand the influence of CO_2 and other greenhouse gases on world climate better than they did even a decade ago. The use of sophisticated computer modeling has enhanced their research. The new understanding confirms the pattern of a global warming trend. Authorities in many nations feel that the potential risks of global warming justify major action now.

Answer these questions. Be prepared to discuss your opinions in class.

1. Describe possible actions the world community and individuals might take to halt the increase in atmospheric CO_2 from human sources.

2. a. Do you think the public would seriously consider taking action to halt such a global warming trend?

   b. If so, which action in your answer to Question 1 do you believe would gain strongest support?

C.8 OFF IN THE OZONE

Although a small dose of ultraviolet radiation is necessary for health, too much is dangerous. In fact, if all ultraviolet radiation in sunlight reached the Earth's surface, all life would be seriously damaged. Ultraviolet photons, as we noted earlier, have enough energy to break covalent bonds. Resulting chemical changes can cause sunburn and cancer in humans and damage to many biological systems (Personal Chemistry and Choices unit, page 489).

Fortunately, the Earth has an ultraviolet shield high in the stratosphere. The shield consists of a layer of gaseous ozone, O_3, that absorbs ultraviolet radiation. However, our stratospheric "ozone shield" is a fragile system with a remarkably small concentration of ozone. If all its ozone molecules were located on the earth's surface at atmospheric pressure, they would form a layer only 3 mm thick (about the thickness of a hardback book cover).

Here's how Earth's vital ozone shield works. As sunlight penetrates the stratosphere, very high energy ultraviolet photons react with oxygen gas molecules, splitting them into individual oxygen atoms.

$$O_2 + \text{very high energy UV photon} \rightarrow O + O$$

Individual oxygen atoms are very reactive. They immediately react further, most combining with oxygen molecules to produce ozone. A third molecule (typically N_2 or O_2, represented by M in the equation below) carries away excess energy from the reaction but is unchanged.

$$O_2 + O + M \rightarrow O_3 + M$$

Each ozone molecule formed can absorb a medium-energy ultraviolet photon in the stratosphere. Decomposition of the ozone results, producing an oxygen molecule and an oxygen atom:

$$O_3 + \text{medium-energy UV photon} \rightarrow O_2 + O$$

These products can then carry on the cycle by repeating the reaction already described, producing more ozone:

$$O_2 + O + M \rightarrow O_3 + M$$

The ozone layer is vitally important to supporting and protecting life on Earth. One indication of its importance is what would happen if the layer were even modestly reduced. A National Research Council study suggests that for each 1% decrease in stratospheric ozone, a 2–5% increase in cases of skin cancer will occur. Such ozone depletion could also lower yields of some food crops, sharply increase sunburns and eye cataracts, and damage some aquatic plant species.

Human activities seem already to have endangered this ozone layer. The major culprit has been identified as chlorine atoms from chlorinated hydrocarbon molecules, also called ***chlorofluorocarbons (CFCs),*** such as the Freons®—CCl_3F (Freon-11) and CCl_2F_2 (Freon-12). Chlorofluorocarbons are synthetic substances that have been used as propellants in aerosol cans, cooling fluids in air conditioners and refrigerators, and cleaning solvents for computer chips. Developed in the 1930s, the nontoxic and nonflammable CFCs quickly replaced toxic ammonia or sulfur dioxide as coolants in refrigerators and air conditioners. For almost 50 years, CFCs seemed to be the ideal substitute.

In 1974, two chemists—Sherwood Rowland and Mario Molina—proposed that chlorofluorocarbons (CFCs) posed a threat to the world's stratospheric ozone layer. Since then, their research and that of others have uncovered the true magnitude of the threat. Rowland and Molina received the 1995 Nobel Prize in Chemistry in recognition of their pioneering research in stratospheric ozone depletion.

A satellite launched in 1978 carried instruments to monitor changes in worldwide stratospheric ozone concentrations. By 1986, data sent from the satellite showed the formation of a seasonal "ozone hole" in the stratosphere above Antarctica (Figure 15, page 419). The "hole" itself is natural; some thinning of the ozone layer occurs there every September and October (spring in the Southern Hemisphere). However, the satellite data indicated a significant annual ozone depletion of 1–2.5%, well beyond the normal depletion. In fact, in September and October of 1993,

This man is recovering freon from an automobile cooling system.

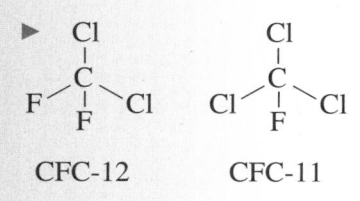

CFC-12   CFC-11

▶ *Methane and other hydrocarbons, as well as nitrogen monoxide, can also deplete stratospheric ozone.*

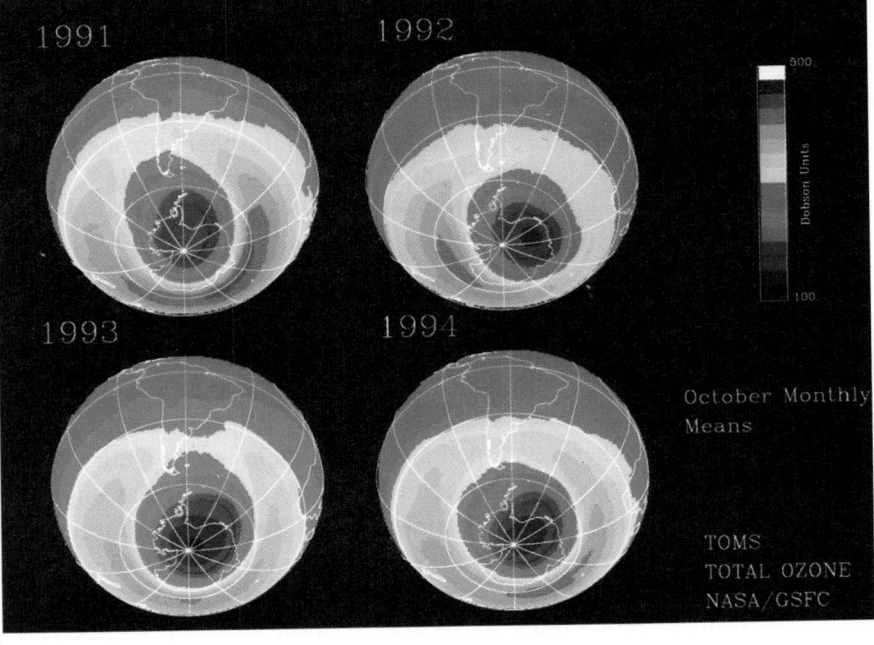

Figure 15 Average October concentrations of ozone in the stratosphere over Antarctica, 1991 to 1994. Reds and greens indicate high ozone concentration; blues and purples indicate low ozone concentration. The lower the Dobson units, the lower the ozone concentration.

nearly 70% of the stratospheric ozone disappeared over Antarctica, creating an ozone "hole" as large as North America. There also is recent evidence that stratospheric ozone is being depleted over Arctic regions as well, though not to the extent as that over Antarctica.

Several lines of evidence gathered since the mid-1980s have converged, clearly linking the Antarctic ozone depletion to chlorine atoms from CFCs, rather than to chlorine atoms from natural sources. The CFCs are highly stable molecules in the troposphere. However, in the stratosphere at 30 km, high-energy ultraviolet photons split chlorine atoms from CFCs by breaking the C–Cl bond. The freed chlorine atoms are very reactive and can participate in a series of reactions that destroy ozone by converting it to O_2:

▶ *Breaking a C–Cl bond in CFC-11 by a high-energy ultraviolet photon.*

$$\cdot Cl + O_3 \rightarrow \cdot ClO + O_2$$

$$\cdot ClO + O \rightarrow \cdot Cl + O_2$$

The sum of the last two reactions is

$$O + O_3 \rightarrow 2\,O_2$$

Notice that the chlorine atom used in the first reaction is regenerated by the second reaction through the formation and decomposition of chlorine monoxide (ClO). Thus, each chlorine atom can participate in not just one,

> ▶ *It is estimated that a single chlorine atom can destroy as many as 100,000 ozone molecules.*

but a great number of these ozone-destroying reactions. The net effect is depletion of ozone molecules (O_3) by their conversion into oxygen molecules (O_2), triggered by free chlorine atoms released from CFC molecules (see Figure 16).

Concerned about the growing threat to the stratospheric ozone layer, representatives of the United States and 55 other countries met in Montreal in 1987 and signed a significant and historic agreement, *The Montreal Protocol on Substances That Deplete the Ozone Layer*. This initial treaty, with its 1990 and 1992 amendments, established a timetable for phasing out *all* CFC production in developed nations by 1996.

The Montreal Protocol, now signed by more than 140 countries, seems to be working. In February 1992, President George Bush ordered CFC production in the United States to stop by 1996. Worldwide production of CFCs in other developed nations is declining as well. For example, 1992 CFC production in industrialized countries was only 57% as much as the approximately one billion kilograms produced in 1986. The amended treaty also includes a multimillion dollar fund ($510 million for 1994–96) to assist developing nations in phasing out their CFC production by the year 2010 by helping to build industries that do not rely on CFCs. This is necessary because CFC use by developing countries increased from 88,000 tons to 138,000 tons from 1986 to 1992.

Because of the Montreal Protocol restrictions, significant chemical research has been done to develop CFC substitutes that will not harm the ozone layer. If you drive an air-conditioned car manufactured after 1993, it likely uses a CFC substitute as the air conditioning coolant.

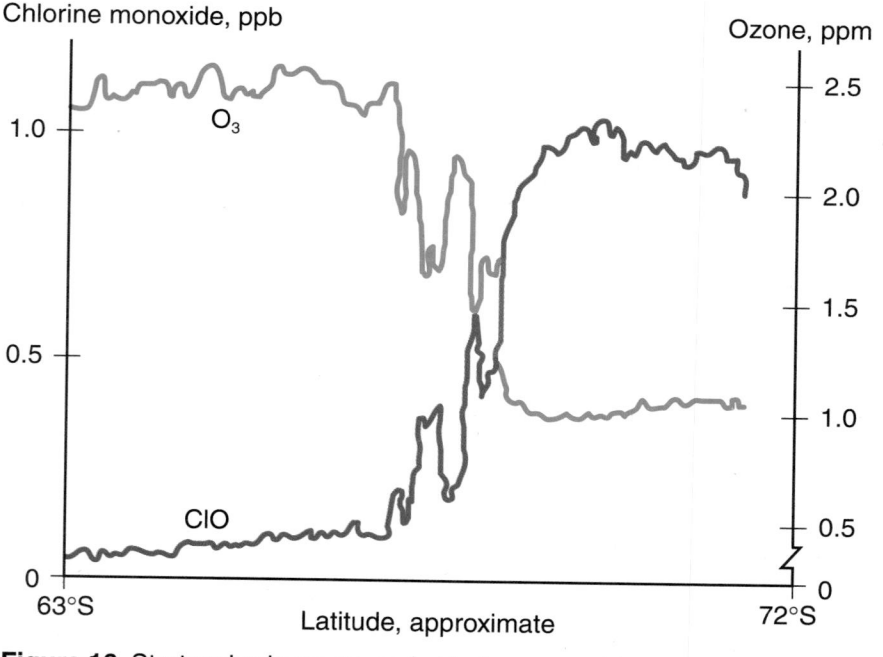

Figure 16 Stratospheric ozone and chlorine monoxide concentration over Antarctica.

Chemists have synthesized these substitutes by modifying CFCs' molecular structures. The current substitutes, known as **hydrochlorofluorocarbons (HCFCs)**, replace some of the offending chlorine atoms with atoms of hydrogen and fluorine. The hydrogen and fluorine atoms, unlike chlorine atoms in CFCs, allow HCFCs to decompose in the troposphere. For example, CHF_2Cl has only about 5% of the ozone-depleting capacity of CFC-12.

It turns out that HCFCs are not the perfect substitute for CFCs. Although some decomposition of them occurs in the troposphere, a significant portion of undecomposed HCFCs rise into the stratosphere. Also, HCFCs, like CFCs, are greenhouse gases, and there is some concern that they may contribute to global warming. Because of these potential problems, HCFCs are considered as only transitional replacements for CFCs. The 1992 amendments to the Montreal Protocol call for HCFC production to be phased out gradually by 2030.

Chemists continue research to develop even better CFC substitutes. That research has produced a group of synthetic compounds known as **hydrofluorocarbons, HFCs.** These molecules, such as CH_2FCF_3, contain just carbon, fluorine, and hydrogen atoms. The HFCs contain no ozone-depleting chlorine atoms, and the hydrogen atoms promote decomposition in the lower atmosphere.

The chemistry of the atmosphere is quite complex and difficult to study. We have seen that human activity has the potential to create large-scale effects that may modify the world's average temperature through altering the normal atmospheric concentrations of substances such as carbon dioxide. Human activities using CFCs also have increased our potential exposure to ultraviolet radiation by thinning the stratospheric ozone layer. Additional evidence regarding human effects on the atmosphere will be explored in the following sections.

▶ *A HFC, CH_2FCF_3*

$$\begin{array}{cc} H & F \\ | & | \\ H-C-C-F \\ | & | \\ F & F \end{array}$$

? PART C: SUMMARY QUESTIONS

1. a. Compare visible, infrared, and ultraviolet radiation in terms of the relative energies of their photons.

   b. Cite one useful role each of these types of radiation plays in a life process or a process important to life.

2. a. Explain why our atmosphere can be regarded as a one-way screen.

   b. How does this make the Earth more hospitable to life?

3. List and briefly describe two functions of the stratosphere.

4. From a scientific viewpoint, why do many desert dwellers wear white or light-colored clothing?

5. a. Compare ocean water and beach sand in terms of how quickly they heat up in the sun and how quickly they cool at night.

   b. What property of these two materials accounts for these differences?

A Lean, Green, Dry-cleaning Machine

Dr. **Sid Chao** is the president of Hughes Environmental Systems in El Segundo, California. We talked to Sid about a special project his company is working on: a new, environmentally friendly way to dry clean clothes using carbon dioxide.

Q. Can you provide a little background on your company?

A. A few years ago, we started doing business as a subsidiary of Hughes Aircraft. Our mission is to find new uses and customers for products and technologies that Hughes has developed for the defense industry. Our specialty is developing products that help protect or preserve the environment—for example, our experimental, liquid CO_2 dry-cleaning process. Some people refer to such processes as examples of "green chemistry." To us, they're primarily good products that we believe will sell.

Q. How did you happen to focus on dry-cleaning?

A. We knew there was a market demand for better, more cost-effective dry-cleaning methods. And we saw an opportunity to introduce a process that was better for the environment.

Our parent company had already developed a product that uses liquid CO_2 at ultrahigh pressures to clean the surfaces of metals, glasses, and materials used in defense equipment, while leaving behind certain desirable substances. We theorized that the same basic idea might be the basis of a good alternative dry-cleaning solvent, if we could "scale it down" to work in a typical dry-cleaning shop. We developed a liquid CO_2 system that operates at much lower pressures and at temperatures that a typical dry-cleaning shop can handle easily and cost effectively.

Q. How is it better than regular dry cleaning?

A. For one thing, it's much better for the environment. Typical dry-cleaning processes are responsible for adding significant levels of halogenated hydrocarbons into the atmosphere. Our process reuses CO_2 which has already been produced, so we're not adding to existing levels. We are also designing a process for recovering the used solvent and "cleaning" it so that it can be used again and again.

Q. Will we be seeing these systems soon at our own dry cleaners?

A. Right now we're building machines for field testing. Using the results of this testing, we're planning to develop a process that is easy to operate, and costs no more to buy or use than what's already on the market. We still have to fine-tune the marketing infrastructure for the product—that is, the manufacturing, delivery and service systems, and other details. But in the next few years, we hope to have our product available on the market.

Photograph courtesy of Hughes Environmental Systems Inc.

6. Describe how atmospheric concentrations of CO_2 and water help to maintain moderate temperatures at the Earth's surface.

7. List the two major ways humans increase the amount of CO_2 in the atmosphere.

8. What changes in atmospheric composition could result in
   a. an increase in the average surface temperature of the Earth?
   b. a decrease in the average surface temperature of the Earth?

9. Describe an important role of the world's ozone shield.

10. Explain the role of CFCs in stratospheric ozone depletion.

11. What is the relationship between the Montreal Protocol and stratospheric ozone depletion?

12. Explain how substituting HCFCs for CFCs could reduce stratospheric ozone depletion.

EXTENDING YOUR KNOWLEDGE (OPTIONAL)

- The presence of carbon dioxide in the atmosphere is only one part of the global carbon cycle. Investigate the current understanding of the carbon cycle, with particular attention to the role of oceans as a major "carbon sink."

- What effect might global warming have on ocean temperatures, and indirectly on the concentration of greenhouse gases in the atmosphere?

- Scientists use ice samples from the Antarctic to estimate what the carbon dioxide concentrations in the atmosphere were many thousands of years ago. Prepare a report on how the samples reveal this information.

- The phaseout of CFC production has significant economic implications. Explore the various ways in which citizens have been affected economically by the halt in U.S. CFC production.

D HUMAN IMPACT ON AIR WE BREATHE

The United States at night showing lighted metropolitan areas.

Dirty air is so common in the United States that weather reports for some major cities now include the levels of certain pollutants in the air. Depending on where you live, cars, power plants, or industries may be the major polluters. However, air pollution is not only an outdoor problem. Indoor air can be polluted by cigarette smoke, fumes from heating sources, and by gases that spontaneously evaporate from certain products, such as from polymeric materials in furniture, rugs, and building materials. In Section E we will take a closer look at indoor air pollution.

Air pollution smells bad, looks ugly, and blocks the view of stars at night. But beyond being unpleasant, air pollution causes billions of dollars of damage every year. It corrodes buildings and machines. It stunts the growth of agricultural crops and weakens livestock. It causes or aggravates diseases such as bronchitis, asthma, emphysema, and lung cancer, adding to the world's hospital bills. Clean air has become a battle cry for an increasing number of individuals, organizations, municipalities, and nations.

D.1 TO EXIST IS TO POLLUTE

Though many pollutants enter the air as the result of natural processes, in most cases natural pollution occurs over such a wide area that we do not notice it. Furthermore, the environment may dilute or transform such substances before they accumulate to harmful levels.

By contrast, air pollution from human activities is usually generated within more localized areas, such as in the vicinity of effluent pipes, smoke stacks, or in urban regions. When the quantity of an air pollutant overwhelms the ability of nature to dispose of it or disperse it, air pollution becomes a serious problem. Many large cities are prone to high concentrations of pollutants. If air pollutants generated by large cities were evenly spread over the entire nation, they would be much less noticeable and would have substantially lowered concentrations. Unfortunately, such reduction in concentrations does not occur.

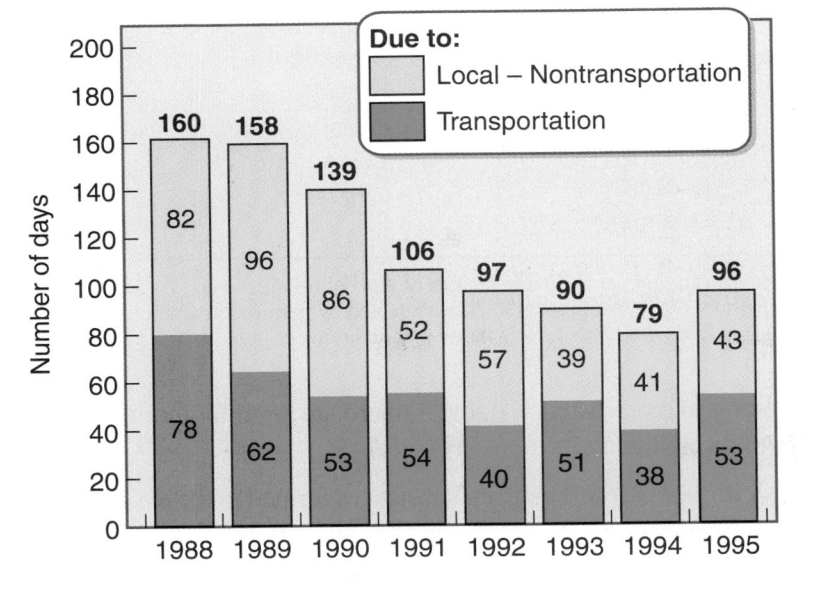

In San Diego, exhaust emissions from slow-moving traffic on crowded freeways are a major source of the photochemical smog shown here.

Number of Days San Diego Exceeded California Clean Air Standard

Due to:
- Local – Nontransportation
- Transportation

Year	1988	1989	1990	1991	1992	1993	1994	1995
Total	160	158	139	106	97	90	79	96
Local – Nontransportation	82	96	86	52	57	39	41	43
Transportation	78	62	53	54	40	51	38	53

Number of days

Table 4 lists the quantities of major air pollutants emitted worldwide from human and natural sources. These substances are all *primary air pollutants*—that is, they enter the atmosphere in the chemical form listed. For example, methane (CH_4, the simplest hydrocarbon) that leaks into the atmosphere is a by-product of fossil fuel use and a component of natural gas. Methane is also produced in prodigious quantities by anaerobic bacteria and termites as they break down organic matter. It is also important to recognize from Table 4 that *natural sources produce more of almost every major air pollutant than what is generated by human sources.*

Table 4	Worldwide Emission of Air Pollutants (10^{12} g/yr or 10^6 metric tons/yr)			
Pollutant	Human Source	Quantity	Natural Source	Quantity
CO_2	Combustion of wood and fossil fuels	22,000	Decay; release from oceans, forest fires, and respiration	1,000,000
CO	Incomplete combustion	700	Forest fires; photochemical reactions	2,100
SO_2	Combustion of coal and oil; smelting of ores	212	Volcanoes and decay	20
CH_4	Combustion; natural gas leakage	160	Anaerobic decay and termites	1,050
NO_x	High-temperature combustion	75	Lightning; soil bacteria	180
NMHC	Incomplete combustion	40	Biological processes	20,000
NH_3	Sewage treatment	6	Anaerobic bacterial decay	260
H_2S	Petroleum refining; sewage treatment	3	Volcanoes; anaerobic decay	84

Source: Adapted from Stern et al. Fundamentals of Air Pollution, 2nd ed.: Academic Press, Inc.: Orlando, FL, 1984; pp. 30–31. Table adapted by permission of Elmer Robinson, Mauna Loa Observatory. Data based on conditions prior to 1980.

► NO_x = nitrogen oxides

► NMHC = nonmethane hydrocarbons

In addition to those listed in Table 4, there are several other important categories of air pollutants:

• *Secondary air pollutants.* These are substances formed in the atmosphere by chemical reactions between primary air pollutants and/or natural components of air. For example, atmospheric sulfur dioxide (SO_2) and oxygen react to form sulfur trioxide (SO_3)—the two oxides are always present together. Further reactions with water in the atmosphere convert these sulfur oxides to sulfates (SO_4^{2-}) or sulfuric acid (H_2SO_4), a secondary pollutant partly responsible for acid rain (discussed in Section D.9, page 439).

- *Particulates.* This major category of pollutants includes all solid particles that enter the air either from human activities (power plants, waste burning, road building, and mining, for example) or from natural processes (such as forest fires, wind erosion, and volcanic eruptions). Particulates include visible emissions from smoke stacks or automobile tail pipes.

- *Synthetic substances.* Some pollutants, such as the CFCs, are produced only by human activity; otherwise, they would not be present at all. A case in point is the fluorine released into the stratosphere when high-energy UV photons break C–F bonds in CFCs. The presence of stratospheric fluorine is taken as further confirming evidence of the role CFCs play in stratospheric ozone depletion because no natural sources of stratospheric fluorine are known.

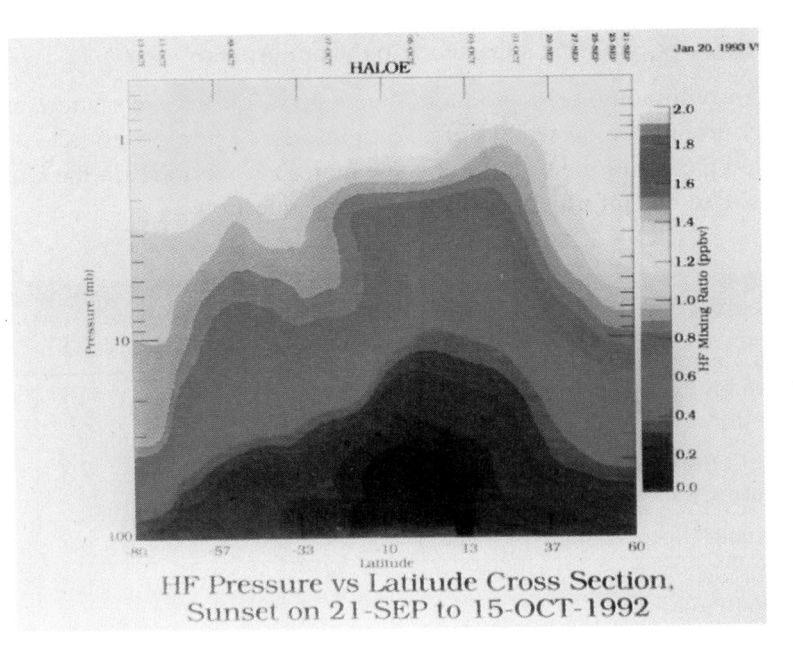

HF Pressure vs Latitude Cross Section.
Sunset on 21-SEP to 15-OCT-1992

This color image depicts global hydrogen fluoride amounts as measured by a satellite. The color scale at the side of the figure shows the mixing ratios of hydrogen fluoride in parts per billion (ppb) by volume.

▶ *1 ppb equals 1 molecule of contaminant in 1 billion air molecules. This is the equivalent of 1 pinch of salt in 10 tons of potato chips.*

D.2 YOU DECIDE: WHAT ARE THE MAJOR AIR POLLUTANTS?

Use Table 4 (page 426) to answer these questions.

1. a. What pattern was used to organize the data?
   b. For what reason might this pattern have been chosen?

c. Should the relative quantities of pollutants from human sources determine which pollutants should be reduced or controlled?

d. If not, what other factors should be considered?

2. Natural contributions of these potentially polluting substances, except one, greatly exceed contributions from human activities.

a. Does this imply that human contributions of these substances can be ignored? Explain your answer.

b. For which substance does the contribution from human activity exceed that from natural sources?

c. What does this suggest about modern society?

3. Refer to Table 1 (page 385). Which pollutants listed in Table 4 are also found naturally in the atmosphere at concentrations of 0.0001% or more?

4. What is the major source of human contributions to air pollutants listed in Table 4?

5. Considering the large quantities of potential air pollutants from natural sources, why might it pose a problem to add other, non-naturally occurring substances to the atmosphere?

Air pollution is a by-product of manufacturing, transportation, and energy production, as well as of natural processes. Table 5 gives a detailed picture of the sources of some major air pollutants in the United States. Use this information to answer these questions.

► TSP = Total suspended particulates
SO_x = Sulfur oxides (SO_2 and SO_3)
NO_x = Nitrogen oxides (NO and NO_2)
HC = Volatile organic compounds (methane and other hydrocarbons)
CO = Carbon monoxide

Table 5	U.S. Pollution, 1994 (in 10^6 metric tons/yr)					
Source	TSP	SO_X	NO_X	HC	CO	Total
Transportation (petroleum burning)	1.4	0.9	8.4	6.0	40.7	57.4
Fuel burning for space heating and electricity	1.8	16.4	10.3	2.3	7.2	38.0
Industrial processes	2.5	3.1	0.6	8.3	4.7	19.2
Solid waste disposal and miscellaneous	1.3	0.0	0.2	3.0	8.8	13.3
Totals	7.0	20.4	19.5	19.6	61.4	127.9

1. a. Overall, is industry the main source of air pollution?

b. If not, what is the main source?

2. For which pollutants is one-third or more contributed

a. by industry?

b. by transportation?

c. by burning fuel for space heating and electricity (usually referred to as "stationary fuel burning")?

How much pollution do *you* produce? If you drive, you use a major pollution source—the automobile. Automobiles contribute about half the total mass of air pollutants. When we spend time in heated or cooled buildings, use electricity, or buy food and other products, air has been polluted for our benefit. As a comic strip character once observed, "We have met the enemy, and he is us."

There are times when meteorological conditions combine with air pollutants to form a potentially hazardous condition called **smog,** a combination of smoke and fog. Smog can kill. In 1952 a deadly smog over London, England, lasted five days and killed nearly 4,000 people. Four years earlier in the coal-mining town of Donora, Pennsylvania, smoggy air killed 20 people and hospitalized hundreds of others. Similar episodes, although less deadly, have occurred in other cities.

D.3 SMOG: HAZARDOUS TO YOUR HEALTH

Because the pollutants in smog endanger health, their levels in the air are of major public interest; many weather forecasters report an air quality index along with humidity and temperature. The U.S. Environmental Protection Agency has devised an index based on concentrations of pollutants that are major contributors to smog over cities. You can see in Table 6 (page 430) that the combined health effects of these pollutants can become quite serious.

The composition of smog depends on the type of industrial activity and power generation in an area, on climate, and on geography. Above large cities containing many coal- and oil-burning industries, power-generating stations, and homes, the principal components of smog are sulfur oxides and particulates. Other areas, particularly those with plenty of sunshine, may have a smog problem caused by nitrogen oxides rather than sulfur oxides. This photochemical smog will be described in Section D.6.

Coal and petroleum both contain varying quantities of sulfur, from which sulfur oxides form during combustion. One successful approach to improving air quality sets limits on the quantity of sulfur allowed in the fuels burned.

Particulates from burning fossil fuels consist of unburned carbon or solid hydrocarbon fragments and trace minerals. Some particles contain toxic compounds of metals such as cadmium, chromium, lead, and mercury. Fatality rates in severe smogs have been higher than predicted from known hazards of sulfur oxides or particulates alone. According to some researchers, this increase may be due to **synergistic interactions** (in which the combined effect of the two pollutants is greater than the sum of the separate effects alone).

Before discussing smog resulting from automobile emissions, we will survey what can be done to decrease air pollution, and what industry is doing to clean up its smoke.

| Table 6 | U.S. Pollutant Standards Index |

Air Pollutant Levels (micrograms per cubic meter)						
Air Quality Index Value	Air Quality Description	Total Suspended Particulate Matter (24 Hours)	Sulfur Dioxide (8 Hours)	Carbon Monoxide (1 Hour)	Ozone (1 Hour)	Nitrogen Dioxide (1 Hour)	Effects and Suggested Actions
500	Hazardous	1,000	2,620	57,000	1,200	3,750	Normal activity impossible. All should remain indoors with windows and doors closed. Fatal for some in high-risk group.*
400	Hazardous	875	2,100	46,000	1,000	3,000	High-risk group should stay quietly indoors. Others should avoid outdoor activity.
300	Very unhealthful	625	1,600	34,000	800	2,260	High-risk group has more symptoms, and should stay indoors and reduce physical activity. All persons notice lung irritation.
200	Unhealthful	375	800	17,000	400	1,130	Those with lung or heart disease should reduce physical exertion. Healthy persons notice irritations.
100	Moderate	260	365	10,000	235	Not reported	Some damage to materials and plants. Human health not affected unless level continues for many days.
50	Good	75	80	5,000	118	Not reported	No significant effects.

*High-risk group includes elderly people and those with heart or lung diseases.

D.4 POLLUTION CONTROL AND PREVENTION

There are several basic ways to limit and prevent air pollution:

- Energy technologies that cause air pollution can be *replaced* with technologies that don't require combustion, such as solar power, wind power, and nuclear power.

- Pollution from combustion can be *reduced* by energy conservation measures, such as getting more energy from what we burn, and therefore burning less.

- Pollution-causing substances can be *removed* from fuel before burning. For example, most sulfur can be removed from coal.

- The combustion process can be *modified* so that fuel is more completely oxidized.

- Pollutants can be *trapped* after combustion.

All pollution reduction options cost money. When deciding upon a pollution prevention or control measure, the decision makers must answer two key questions: What will the prevention or control cost? What benefits will it offer?

Power plants and smelters generate more than half of the particulate matter emitted in the United States. However, notable progress has been made in cleaning up air pollution from these sources and others over the past decade. Since 1988, there has been a significant decrease in the amount of most air pollutants in the United States (Table 7).

Table 7 Changes in Average Concentrations of Air Pollutants in the United States, 1975–1991			
Pollutant	**1975**	**1991**	**Change**
Sulfur dioxide	0.0132 ppm	0.0075 ppm	43% decrease
Nitrogen oxides	0.021ppm	0.021 ppm	1% decrease
Carbon monoxide	10 ppm	6 ppm	40% decrease
Ozone	0.147 ppm	0.115 ppm	22% decrease
Particulates	63 $\mu g/m^3$	47 $\mu g/m^3$	22% decrease
Lead	0.68 $\mu g/m^3$	0.048 $\mu g/m^3$	93% decrease

There has also been a reduction in toxic materials released into the atmosphere from U.S. chemical manufacturing plants (Figure 17). The data come from the Toxics Release Inventory (TRI), an EPA annual report required by law.

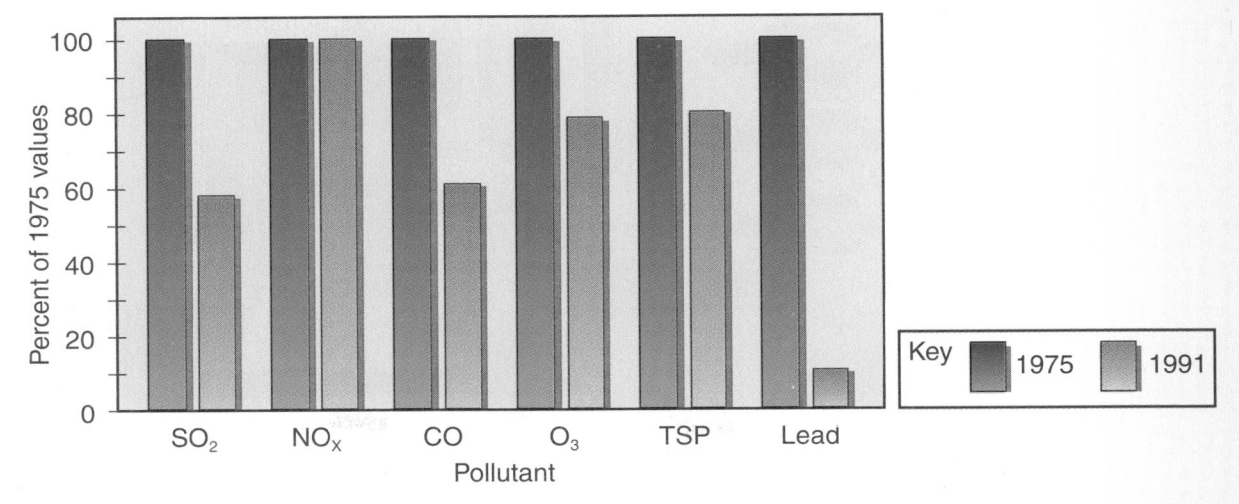

Figure 17 U.S. EPA Toxics Release Inventory.

Manufacturers have used large-scale techniques to reduce particulates coming from industrial plants. Several cost-effective methods, such as those described below, are used for controlling particulates and other emissions.

Electrostatic precipitation. This is currently the most important technique for controlling pollution by particulates. Combustion waste products are passed through a strong electrical field, where they become charged. The charged particles collect on plates of opposite charge. This technique removes up to 99% of particulates, leaving only particles smaller than one tenth of a micrometer (0.1 μm, where 1 μm = 10^{-6} m in diameter). Dust and pollen collectors installed in home ventilation systems are often based on this technique.

Mechanical filtering. This works much like a vacuum cleaner. Combustion waste products pass through a cleaning room (bag house) where huge filters trap up to 99% of the particles.

Cyclone collection. Combustion waste products pass rapidly through a tight circular spiral chamber. Particles thrown outward to the chamber's walls drop to the base of the chamber where they are removed. This technique removes 50–90% of the larger, visible particles, but relatively few of the more harmful particles (those smaller than 1 μm).

Scrubbing. This method controls particles and sulfur oxides accompanying them.

In ***fluidized bed combustion,*** air blows pulverized coal and powdered limestone (calcium carbonate, $CaCO_3$) into a combustion chamber where the heat from the burning coal decomposes the limestone (left side of Figure 18).

$$CaCO_3(s) + Heat \rightarrow CaO(s) + CO_2(g)$$

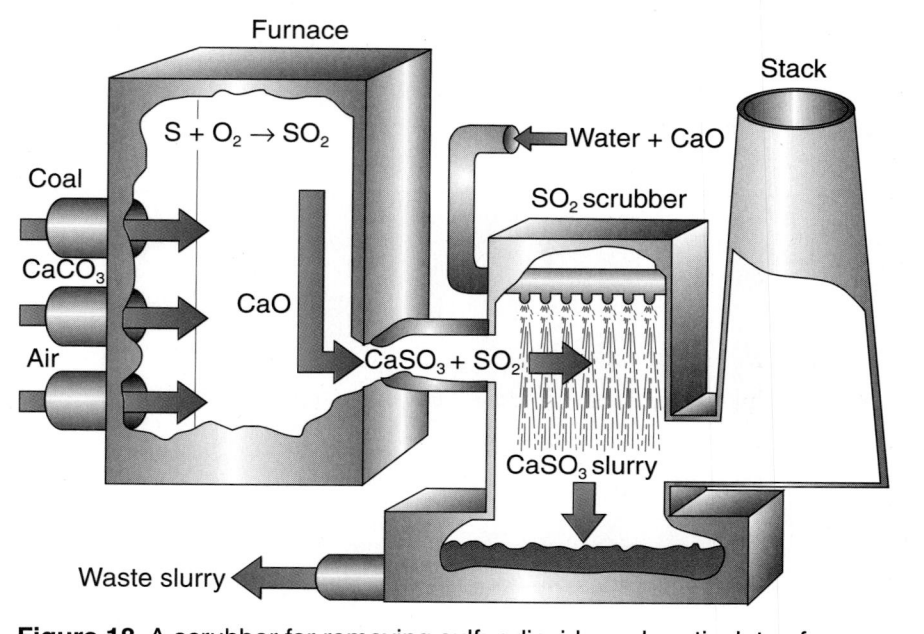

Figure 18 A scrubber for removing sulfur dioxide and particulates from products of industrial combustion processes. Dry scrubbing occurs in the furnace, and wet scrubbing in the SO_2 scrubber.

The lime (CaO) then reacts with sulfur oxides to form calcium sulfite ($CaSO_3$) and calcium sulfate ($CaSO_4$):

$$CaO(s) + SO_2(g) \rightarrow CaSO_3(s)$$

$$CaO(s) + SO_3(g) \rightarrow CaSO_4(s)$$

These products are washed away as a slurry (a mixture of solids and water).

In an example of wet scrubbing (Figure 18), sulfur dioxide gas is removed by using an aqueous solution of calcium hydroxide, $Ca(OH)_2$(aq). The sulfur dioxide reacts to form solid calcium sulfite, $CaSO_3$:

$$SO_2(g) + Ca(OH)_2(aq) \rightarrow CaSO_3(s) + H_2O(l)$$

Scrubbers that can remove up to 95% of sulfur oxides are required for all new U.S. coal-burning plants. Unfortunately, their use adds significantly to the cost of electrical power.

D.5 DEMONSTRATION: CLEANSING AIR

Your teacher will demonstrate two pollution control methods.

Part 1: The Electrostatic Precipitator
1. Observe what happens to the smoke. Record your observations.
2. Observe the chemical reaction that takes place on the copper rod. Record your observations.

Part 2: Wet Scrubbing
1. Observe the color of the liquid and pH paper in each flask as the reaction proceeds. Record your observations.

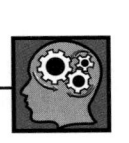

Wet scrubbing.

QUESTIONS
1. Write an equation for the reaction between HCl and NH_3.
2. What information did the universal indicator provide about each flask in Part 2?
3. What information did the pH paper at the neck of each flask provide?
4. What was the overall effect of the scrubbing, as shown by the indicators?
5. List the two ways in which the quality of the air in the reaction vessel was changed by wet scrubbing.
6. a. What advantages do precipitators have over wet scrubbers?
   b. What are their disadvantages?

CHEMQUANDARY

STEPS TOWARD CLEAN AIR

Explain why (1) treating smoke before it is released from power plants is an important goal, but why (2) it may also mislead the general public about how to obtain clean air.

D.6 PHOTOCHEMICAL SMOG

The ill effects of pollution from automobiles were first noted in the Los Angeles area in the 1940s. A brownish haze that irritated the eyes, nose, and throat and damaged crops appeared in the air. Researchers were puzzled for some time because Los Angeles has no significant industrial or heating activities. However, the city has an abundance of automobiles and sunshine, and has mountains on three sides. Although Los Angeles is smog-prone primarily because of its valley location, there is more to the story, because smog there was much worse than seemed reasonable.

Normally, air at the Earth's surface is warmed by solar radiation and by reradiation from surface materials. This warmer, less dense air rises, carrying pollutants with it. Cooler, less polluted air then moves in. In a ***temperature inversion,*** a cool air mass is trapped beneath a less dense warm air mass, often in a valley or over a city (Figure 19). In Los Angeles, the combination of sunny weather and mountains produces temperature inversions about 320 days each year. During a temperature inversion, pollutants cannot escape and their concentration may rise to dangerous levels.

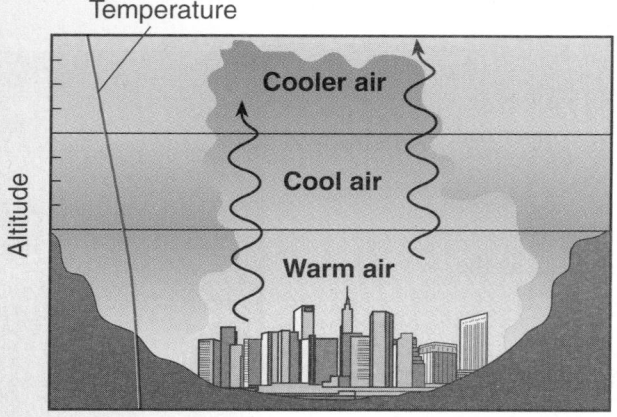

(a) Normal conditions

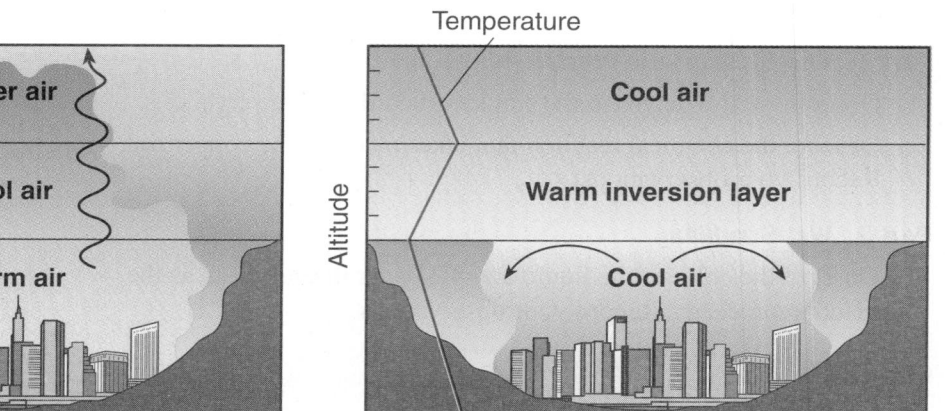

(b) Temperature inversion

Figure 19 A temperature inversion.

From *Chemistry: An Environmental Perspective* by P. Buell and J. Girard, © 1994. Adapted by permission of Prentice-Hall, Inc., Upper Saddle River, NJ.

As sometimes happens in science, a serendipitous discovery answered a piece of the puzzle of smog. In 1952, chemist Arie J. Haagen-Smit was attempting to isolate the ingredient in pineapples that is responsible for their odor. Working on a smoggy day, he detected a greater concentration of ozone (O_3) in his experiment than is normally found in clean, tropospheric air. He shifted his research to identify the ozone's source. Within a year, he published a ground-breaking paper, *"The Chemistry and Physics of Los Angeles Smog,"* describing the importance of sunlight in smog chemistry, and coined the term ***photochemical smog.*** Its key ingredients are sunlight (the *photo* in photochemical) and fossil fuels burned in cars and trucks. As you will soon learn, photochemical smog is different from, but potentially as serious as, the London-type smog discussed in Section D.3.

For our purposes, the equation below represents the key ingredients and products of photochemical smog: Hydrocarbons, carbon monoxide, and nitrogen oxides from car and truck exhausts are irradiated by sunlight in the presence of oxygen. The irradiation and reaction form a potentially dangerous mixture of products. The products include other nitrogen oxides, ozone, and irritating organic compounds, plus carbon dioxide and water vapor.

$$\text{Auto exhaust} + \text{Sunlight} + O_2 + CO + NO_x \rightarrow$$
$$\text{(hydrocarbons)}$$

$$O_3 + NO_x + \text{Organic compounds} + CO_2 + H_2O$$
$$\text{(oxidants and irritants)}$$

Nitrogen oxides are essential ingredients of such smog. At the high temperature and pressure of automotive combustion (**several hundred** °C and about 10 atm), nitrogen gas and oxygen gas react in the air in the car's engine to produce the pollutant, colorless nitrogen monoxide (NO).

$$N_2(g) + O_2(g) + \text{Energy} \rightarrow 2\ NO(g)$$

In the atmosphere, nitrogen monoxide in the car's exhaust is oxidized to orange-brown nitrogen dioxide (NO_2), visible in polluted urban atmospheres.

$$2\ NO(g) + O_2(g) \rightarrow 2\ NO_2(g)$$

Carbon monoxide is also present in automobile exhaust.

The photochemical smog cycle begins as photons from sunlight dissociate NO_2 into NO and oxygen (O) atoms. The atomic oxygen then reacts with oxygen molecules (O_2) in the troposphere to produce ozone in the same way that it does in the stratosphere.

$$NO_2(g) + \text{Sunlight} \rightarrow NO(g) + O(g)$$

$$O_2(g) + O(g) \rightarrow O_3(g)$$

We now have accounted for two of the harmful and unpleasant ingredients of photochemical smog. Nitrogen dioxide has a pungent, irritating odor. At relatively low concentrations (0.5 ppm) nitrogen dioxide can inhibit plant growth, and cause respiratory distress in humans at 3–5 ppm for one hour. Ozone is a very powerful oxidant. At concentrations as low as 0.1 ppm, ozone can crack rubber, corrode metals, and damage plant and animal tissues.

The highly reactive ozone undergoes a complex series of reactions with hydrocarbons—principally volatile organic compounds (VOCs)—the third essential ingredient of photochemical smog. The hydrocarbons escape from gasoline tanks or are emitted during incomplete combustion of gasoline. The products of these reactions cause burning eyes, are harmful to individuals with respiratory or heart disease, and can injure plants and damage materials such as rubber and paint.

The concentrations of these three essential components of smog change during a 24-hour period. In the following section, you will complete an activity that considers some factors that influence their concentration changes.

▶ *Any reaction initiated by light is a photochemical reaction.*

▶ *NO and NO_2 are referred to as NO_x ("nocks").*

▶ *The ozone produced in photochemical smog largely remains in the troposphere.*

▶ *The organic compounds in photochemical smog can cause damage to the eyes and lungs.*

D.7 YOU DECIDE:
AUTOS AND SMOG

Use data in Figure 20 to answer these questions.

1. a. Between what hours do the concentrations of nitrogen oxides and hydrocarbons peak?

   b. Account for this fact in terms of automobile traffic patterns.

2. Give two reasons why a given pollutant may decrease in concentration over several hours.

3. The concentration maximum for NO_2 occurs at the same time as the concentration minimum of NO. Explain this phenomenon.

4. Although ozone is necessary in the stratosphere to protect us from ultraviolet light, on the surface of the Earth it is a major component of photochemical smog.

   a. Determine from Figure 20 which chemicals, or species, are at minimum concentrations when O_3 is at maximum concentration.

   b. What does this suggest about the production of O_3 in polluted tropospheric air?

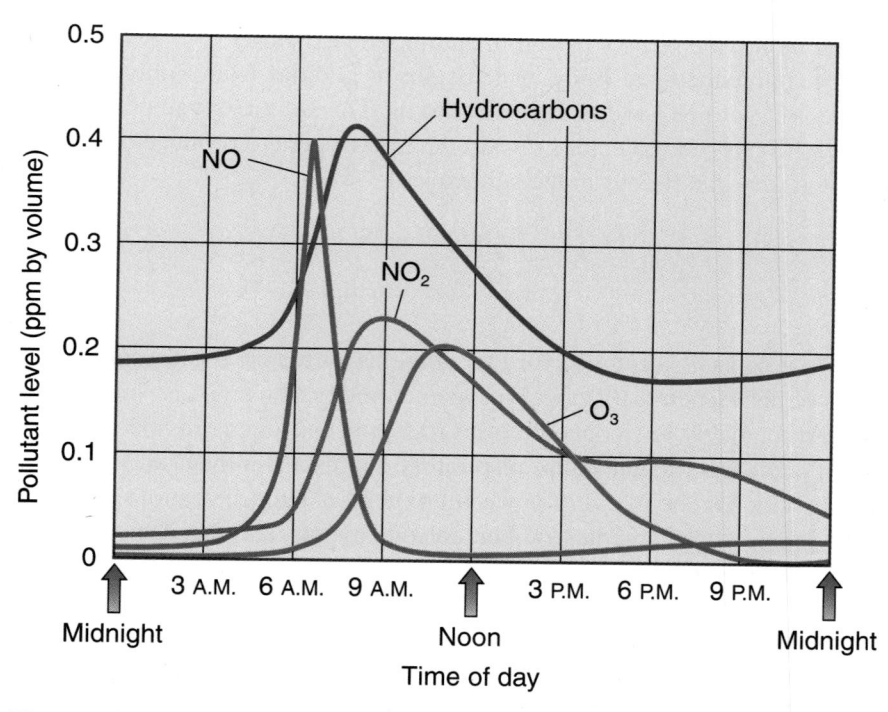

Figure 20 Photochemical smog formation.

Smog is being produced faster than the atmosphere can dispose of it. However, the nation has made considerable progress in smog control. Many cities have cleaner air than they did 30 years ago. The following section explores the control of pollution from automobiles.

D.8 CONTROLLING AUTOMOBILE EMISSIONS

The Clean Air Act of 1970 authorized the Environmental Protection Agency to set emissions standards for new automobiles. Maximum limits were set for hydrocarbon, nitrogen oxide, and carbon monoxide emissions.

These levels were to be achieved in gradual steps by 1975. Improvements in automobile engines were made between 1970 and 1975 by modifying the air–fuel ratio, adjusting the spark timing, adding carbon canisters to absorb gasoline that would normally evaporate before combustion, and installing exhaust gas recirculation systems.

However, additional measures were required to decrease emissions enough to meet the standards completely. The result was development of the *catalytic converter,* a reaction chamber attached as part of the exhaust system. In the catalytic converter, exhaust gases and outside air pass over several catalysts that help convert nitrogen oxides to nitrogen gas, and hydrocarbons to carbon dioxide and water. The carburetor air–fuel ratio is set to produce exhaust gases with relatively high concentrations of carbon monoxide and hydrogen. These gases enter the first half of the catalytic converter where nitrogen oxides are reduced. For example,

$$2\,NO(g) + 2\,CO(g) \xrightarrow{\text{Catalyst}} N_2(g) + 2\,CO_2(g)$$

$$2\,NO(g) + 2\,H_2(g) \xrightarrow{\text{Catalyst}} N_2(g) + 2\,H_2O(g)$$

The second half of the converter then further oxidizes carbon monoxide and hydrocarbons to carbon dioxide and water.

What exactly is a catalyst? You have encountered catalysts several times now in other connections. Enzymes that aid in digesting food and in other body functions are organic catalysts. The iron(III) chloride added in generating oxygen gas in the laboratory (page 383) was an inorganic catalyst. In every case, whether inorganic or organic, the catalyst increases the rate of a chemical reaction that, without the catalyst, would proceed too slowly to be useful. Although the catalyst participates in the reaction, it is not considered a reactant, because the catalyst is unchanged when the reaction is over.

How can a catalyst speed up a reaction and escape unchanged? Reactions can occur only if molecules collide with sufficient energy and at the correct orientation to disrupt bonds. The minimum energy required for such effective collisions is called the *activation energy.* You can think of the activation energy as an energy barrier that stands between the reactants and products (Figure 21, page 438). Reactants must have enough energy to get over the barrier before reaction can occur. The higher the barrier, the fewer the molecules that have the energy to mount it, and the slower the reaction proceeds.

Emissions testing has become an essential component of tuning an automobile.

▶ *A catalyst is regenerated by the end of the reaction.*

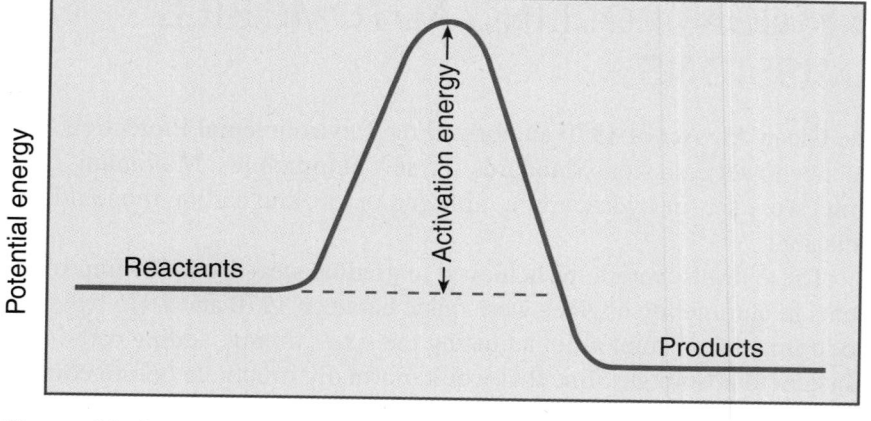

Figure 21 Activation energy diagram.

A catalyst increases the rate of a chemical reaction by providing a different reaction pathway, one with a lower activation energy. In effect, the catalyst lowers the barrier. The result is that more molecules have sufficient energy to react and form products within a given period of time. In automotive catalytic converters, a few grams of platinum, palladium, or rhodium act as catalysts. For this use, the metal samples are embedded in pellets of a solid carrier such as aluminum oxide, Al_2O_3.

The Clean Air Act Amendments of 1990 have set new tailpipe emission standards for U.S. automobiles manufactured after 1992. Some of these are summarized in Table 8, along with comparable values in the earlier Clean Air Act. The 1990 Amendments specify that at least 40% of 1994 model year autos, 80% of 1995 model year autos, and 100% of 1996 and later model year autos must meet the new criteria. These new national tailpipe emission levels set what are believed to be achievable goals, which can help significantly to reduce photochemical smog.

Table 8	National Tailpipe Emission Standards	
Year	**Hydrocarbons (g/mile)**	**Nitrogen oxides (g/mile)**
1975	1.5 (0.9)	3.1 (2.0)
1980	0.41 (0.41)	2.0 (1.0)
1985	0.41 (0.41)	1.0 (0.4)
1990	0.41 (0.41)	1.0 (0.4)
1995	0.25	0.4
2003	0.125 if necessary*	0.2 if necessary*

*if 12 of 27 seriously polluted cities remain smoggy; California standards in parentheses.

CHEMQUANDARY

CONTROLLING AIR POLLUTION

From a practical viewpoint, why might controlling air pollution from more than 180 million automobiles, buses, and trucks in the U.S. be more difficult than controlling air pollution from our power plants and industries?

D.9 ACID RAIN

Acid rain was first reported in Scandinavia. Then, it appeared in the north-eastern United States. Although New England, Germany, and Scandinavia are among the regions hardest hit by acid rain, it is widespread in the industrialized world. Even the Grand Canyon has been damaged by acid rain due to air pollution from coal-fired power plants miles away.

The surfaces of limestone and concrete buildings, and of marble statues, crumble. Fish no longer populate some major lakes. In some areas, crops grow more slowly and forests begin to die. Problems attributed to acid rain are complex and may involve factors in addition to acid rain. But, what does the term "acid rain" mean?

Naturally occurring substances, chiefly carbon dioxide, have always caused rainwater to be slightly acidic. Normally, rainwater's pH is about 5.6 due to the reaction of carbon dioxide and water in the air to form carbonic acid :

$$CO_2(g) + H_2O(l) \rightarrow H_2CO_3(aq)$$

> ▶ *Sometimes the more general term "acid precipitation" is used to include acidic fog, sleet, and snow, as well as rain.*

Rain with a pH below 5.6 is **"acid rain."** Oxides of sulfur and nitrogen emitted from fossil-fuel-burning power plants, various industries, and automobiles also react with rainwater, to form acids that have lowered the pH of rainwater to 4–4.5 in the northeastern United States. (A pH of 4–4.5 is about the acidity of orange juice.) The key reactions in forming acid rain are

$$H_2O(l) \ + \ SO_2(g) \ \rightarrow \ H_2SO_3(aq)$$
$$\text{sulfurous acid}$$

$$H_2O(l) \ + \ SO_3(g) \ \rightarrow \ H_2SO_4(aq)$$
$$\text{sulfuric acid}$$

$$H_2O(l) \ + \ 2\,NO_2(g) \ \rightarrow \ HNO_3(aq) \ + \ HNO_2(aq)$$
$$\text{nitric acid} \qquad \text{nitrous acid}$$

> ▶ *The lower the pH, the more acidic the solution.*

Occasionally the levels of these oxides in air produce enough acid to lower the pH to 3. (A pH of 3 is about the pH of vinegar.)

This more acidic rain lowers the pH of lakes, killing fish eggs and other aquatic life; some species are more sensitive than others. Statues and monuments (such as the Parthenon in Greece) that have stood uneroded for centuries recently began corroding because of acid rain and sulfates. The acid attacks calcium carbonate in limestone, marble, and concrete:

$$H_2SO_4(aq) + CaCO_3(s) \rightarrow CaSO_4(s) + H_2O(l) + CO_2(g)$$

Calcium sulfate is more water-soluble than calcium carbonate. Therefore, the stonework decays as the calcium sulfate is washed away to uncover fresh calcium carbonate that can react further with the acid rain.

Salts of sulfuric acid, which contain the sulfate ion (SO_4^{2-}), are also present in acid rain and atmospheric moisture. In the air the sulfate particles scatter light, causing haze. And they are potentially harmful to human health when they are deposited in sufficiently large amounts in the lungs.

Air pollutants are responsible for the partial disintegration of this sculpture.

Human Impact on Air We Breathe

Many different and interrelated reactions are involved in producing acid rain. Studies are continuing in the hope of gaining a more complete understanding of its causes. One puzzle is how sulfur dioxide is oxidized to sulfur trioxide. Oxygen gas dissolved in water very slowly oxidizes sulfur dioxide. The reaction may be accelerated in the atmosphere by sunlight or by metal ion catalysts such as iron, manganese, or vanadium in soot particles.

The control of acid rain is made difficult because air pollution knows no political boundaries. Acid rain, carried by air currents, often shows up hundreds of kilometers from the pollution sources. For example, much of the acid rain in Scandinavia is thought to come from industrial sections of Germany and the United Kingdom. The acid rain that falls on New England may come largely from the industrial Ohio Valley, in the Midwest.

The legislation of the U.S. Clean Air Amendments of 1990 is one attempt to reduce acid rain. These amendments state that the annual sulfur dioxide emissions in the lower forty-eight states and the District of Columbia must be lowered by 10 million tons from 1980 levels. Sulfur dioxide emissions decreased by nearly 3 million metric tons from 1980 to 1990. In addition, nitrogen oxide emissions must be lowered by 2 million tons annually from 1980 levels. Almost 7 million metric tons less of nitrogen oxides were emitted in 1991 than in 1980.

To achieve these reductions, the EPA issues permits to major electric utility plants specifying the maximum amount of sulfur dioxide each plant can release each year. As a bonus for reducing emissions, plants with low emissions can auction or sell their excess permits to plants that produce more.

Have you made any personal observations of the effects of acid rain? In the following laboratory activity you will have the opportunity to observe first-hand the action of solutions with acidity approximately equal to that of acid rain.

D.10 LABORATORY ACTIVITY: MAKING ACID RAIN

GETTING READY

In this activity you will create a mixture similar to that in acid rain by generating and dissolving a gas in water. You will observe the effects of the dissolved gas's acid rain chemistry on plant material (represented by an apple skin), on an active metal (Mg), and on marble ($CaCO_3$).

PROCEDURE

1. Peel two pieces of skin from an apple and place them on a paper towel. One piece will be tested in step 7; the second piece will be used as a control for comparison in that step.

2. Weigh out 1 g of sodium sulfite (Na_2SO_3). Put it into the bottom of a 1 pint plastic bag that has a "sealing strip."

3. Carefully fill a Beral pipet with 6 M hydrochloric acid.

4. Put this Beral pipet into the bag containing the sodium sulfite. Lay the bag on the laboratory bench and, *without pressing on the pipet*, carefully smooth the bag to remove most of the air from it. Then, seal the bag by closing the sealing strip.

5. Keeping the bag sealed, hold the bag by one of its top corners. Slowly squeeze the Beral pipet so that the hydrochloric acid drips on the solid sodium sulfite.

6. Use a graduated cylinder to measure 10 mL of distilled water. Carefully open one corner of the bag by partially opening the sealing strip, and add the distilled water to the bag. Quickly reseal the bag. Gently swirl its contents about 30 seconds.

7. Carefully open one corner of the bag. Use a Beral pipet to remove a small amount of the solution from the bag. *Reseal the bag.* Add 4–5 drops of the solution to one of the pieces of apple skin, and let it remain there for about 30 seconds. Do not put any of the solution on the second apple skin. Observe the two pieces after the 30 seconds and record your observations.

8. Using a clean stirring rod, place a drop of distilled water on a piece of red litmus paper, on a piece of blue litmus paper, and on a piece of pH paper. Record your observations.

9. Carefully open one corner of the bag. Using a clean Beral pipet, transfer two pipetsful of the solution from the plastic bag to a test tube. *Reseal the bag.* Use a clean stirring rod to test the solution with each color of litmus paper and with pH paper. Record your observations. Acid rain has a pH of 4–4.5. Is your solution more acidic or less acidic than this?

10. Drop a 1-cm length of magnesium ribbon into the liquid in the test tube. Observe it for at least three minutes before recording your observations.

11. Add two small marble chips to the solution remaining in the plastic bag. *Reseal the bag.* Observe the marble chips for at least three minutes. Record your observations.

12. Wash your hands thoroughly before leaving the laboratory.

QUESTIONS

1. Give the name and formula of the gas formed in the bag by the reaction of hydrochloric acid and sodium sulfite.

2. What effect did this dissolved gas have on the acidity of the distilled water?

3. If a liquid similar to the solution in the plastic bag coated a marble statue or steel girders supporting a bridge, what effect might it have?

D.11 pH AND ACIDITY

Water and all its solutions contain both hydrogen (H^+) and hydroxide (OH^-) ions:

- In *pure water* and *neutral solutions*, the concentrations of these ions are equal, but very small.

▶ *The concept of pH was introduced in the Water unit, Section C.6.*

- In *acidic solutions*, the hydrogen ion concentration is higher than the hydroxide concentration. In very acidic solutions, the hydrogen ion concentration is much higher.

- In *basic solutions,* the hydroxide ion concentration is higher than the hydrogen ion concentration. In very basic solutions, the hydroxide ion concentration is much higher.

- The pH concept is built on the relationship between hydrogen ions and hydroxide ions in water. The term pH stands for "power of hydrogen ion"—where "power" refers to a mathematical power (exponent) of 10. For example, a solution with 0.001 mol H^+ (1×10^{-3} mol H^+) per liter has a pH of 3.

A pH of 7 in a water solution at 25 °C represents a neutral solution— one in which $H^+(aq)$ and $OH^-(aq)$ each have a concentration of 1×10^{-7} mol per liter, or 1×10^{-7} M. The pH of pure water is 7 (Figure 22).

▶ *A solution concentration expressed as moles of solute per liter of solution is called its* **molarity** *or* **molar concentration,** *symbolized M.*

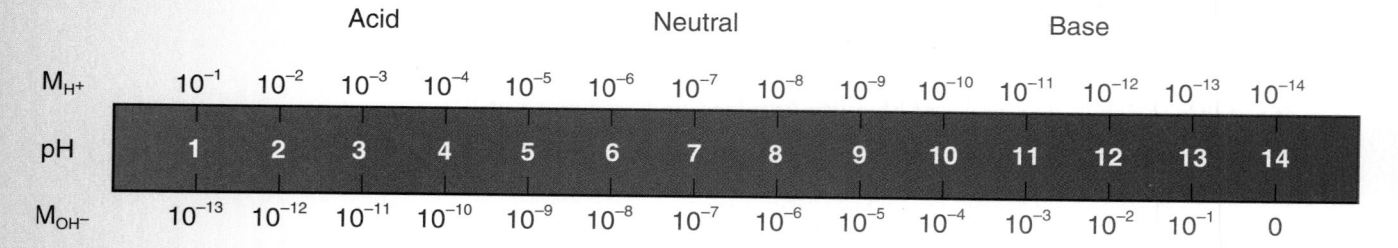

Figure 22 The relationships among pH, molarity of H^+, and molarity of OH^-.

▶ *As pH ↓, H^+ concentration ↑.*

▶ $\dfrac{H^+ \text{ conc} ↑ \qquad pH ↓}{\Delta}$

▶ *As pH ↑, OH^- concentration ↑.*

Values of pH lower than 7 represent acidic solutions. *The lower the pH value, the greater the solution's acidity.* In an *acidic* solution, the concentration of H^+ ion is *greater* than that of OH^- ion. An acidic solution at pH 1 contains 1×10^{-1} mol (0.1 mol) H^+ in each liter of solution—its H^+ concentration is 0.1 M. This is ten times more acidic than a solution at pH 2, where the hydrogen ion concentration is 1×10^{-2} mol per liter, or 0.01 M.

By contrast, *values above pH 7 represent basic solutions. In a basic solution, the OH^- ion concentration is greater than that of H^+ ion.* A solution with a pH of 14 contains 1×10^{-14} mol H^+ and 1 mol OH^- per liter—1×10^{-14} M for H^+ and 1 M for OH^-. That is considerably more basic than a solution at pH 8—with concentrations of 1×10^{-8} M for H^+ and 1×10^{-6} M for OH^-. The higher the pH above 7, the more basic is the solution.

Each step down the pH scale increases the acidity by a factor of 10. Thus, lemon juice, pH 2, is 10 times more acidic than a soft drink at pH 3 which, in turn, is 10,000 times more acidic than pure water at pH 7 (four steps away on the pH scale).

pH

1. Following are listed some common aqueous solutions with their typical pH values. Classify each as acidic, basic, or neutral, and arrange them in order of increasing hydrogen ion concentration.

   a. Stomach fluid, pH = 1

   b. A solution of baking soda, pH = 9

   c. A cola drink, pH = 3

   d. A solution of household lye, pH = 13

   e. Milk, drinking water, pH = 6

   f. Sugar dissolved in pure water, pH = 7

   g. Household ammonia, pH = 11

2. How many times more acidic is a cola drink than milk?

? PART D: SUMMARY QUESTIONS

1. Identify the major types of air pollutants.

2. In what sense is "pollution-free" combustion an impossibility?

3. Identify the major components of smog and their sources.

4. Define the term "synergism" and explain its relevance to air pollution.

5. In which region of the United States is acid rain most prevalent? Why?

6. a. Name the major chemicals in acid rain.

   b. What are their sources?

7. How might efforts to control air pollution result in other kinds of pollution?

8. Pollution has sometimes been defined as "a resource out of place." Name a substance that is a resource in one part of the atmosphere, a pollutant in another part.

9. Write an ionic equation for the reaction of NO_2 with rainwater.

10. Which of these compounds are acids? Which are bases? Which are neutral?

    a. NaOH

    b. HNO_3

    c. CH_4

    d. $C_{12}H_{22}O_{11}$ (table sugar)

    e. H_2SO_3

11. Which of these solutions has the lowest pH? the highest pH?

    a. lemon juice (10^{-2} M H^+ concentration)

    b. stomach fluid (10^{-3} M H^+ concentration)

    c. drain cleaner (NaOH) (10^{-1} M OH^- concentration)

12. Identify two ways that photochemical smog differs from London-style smog.

13. In what two ways do automobiles contribute to photochemical smog production?

- Carbon monoxide can interfere with the body's O_2 transport and exchange system. Investigate the health effects of CO and its relationship to traffic accidents.

- Investigate the advantages and disadvantages of various alternatives to the standard internal combustion engine. Options include the electric engine, gas turbine, Rankine engine, stratified charge engines, Wankel engine, Stirling engine, diesel engine, and expanded mass transportation systems.

- Make an acid–base indicator at home from red cabbage juice and test its properties using common household materials. (Consult your teacher for instructions.)

- The Toxics Release Inventory (TRI) is available from the EPA as a book or on computer disks. If your school has a copy, you can use the TRI to check on how well your state or metropolitan area has done in reducing air pollution.

This unit has focused on atmospheric gases formed by nature and human activities. Both sources can create air pollution, an unwanted situation that can endanger our quality of life and of human life itself. We have also described methods that decrease and control pollutant emissions.

Now we turn to consider air quality from a different perspective—indoor air.

Scientists at the Department of Energy's Argonne National Laboratory are studying the effects of sulfur dioxide and ozone on crops.

E PUTTING IT ALL TOGETHER: INDOOR AIR QUALITY

Although we generally can choose what we eat or drink, we can't usually choose the air we breathe. Each of us daily breathes 10,000 to 20,000 L of air. For most of us, 85 to 90 percent of that air is indoor air. Thus, indoor air quality is important because of how much time we spend indoors at home, school, and work.

Indoor air quality refers to the properties of indoor air that affect the health and well-being of those exposed to it. Concerns about indoor air quality are likely as old as humankind. Our early ancestors had to deal with the poor quality of the air in their caves and crude dwellings. Fires for cooking and heating filled their living space with smoke and combustion products to much higher levels than we would accept in our dwellings today. Much later, miners used canaries as a crude air quality detector. If the canaries in the mine became ill or died, miners knew it was unsafe to continue working in that space.

Indoor air is a complex mixture, but it has only two sources—outside air that enters buildings, and air coming from within the buildings. What are some of the main sources of air pollution within a home, an apartment, or an office? What are the most common pollutants? What measures can be taken to reduce indoor air pollution?

▶ *More than 20 million people work in offices in the United States.*

E.1 INDOOR AIR POLLUTANTS

As many as 900 different contaminants have been detected in indoor air at the parts per billion level (ppb) or higher. They include the following:

- More than 350 volatile organic compounds (VOCs). The most common VOCs are formaldehyde (H_2CO), benzene (C_6H_6), toluene (C_7H_8), chloroform ($CHCl_3$), acetone (C_2H_6O), and styrene (C_8H_8). These can come from cleaning solvents; dry-cleaned clothes; plastics; perfumes and hair sprays; paints; carpets and furnishings; plywood; and other products.
- Biological contaminants—molds, fungi, bacteria, viruses, pollen, and insects. Because high humidity promotes growth of many biological contaminants, the relative humidity in buildings should not exceed 50 percent.
- Combustion products such as carbon monoxide, nitrogen dioxide, sulfur dioxide, tobacco smoke components, and polyaromatic hydrocarbons (PAHs) such as benzo[*a*]pyrene (See Personal Chemistry and Choices unit, page 458).
- Contaminants generated by office equipment and supplies such as ozone from computer terminals and photocopiers, ethanol and methanol from duplicating machines, and acetone from typewriter correction fluid.

▶ *For most of the contaminants, a 1 ppb level is well below the concentration needed to produce low air quality.*

- Environmental tobacco smoke (ETS). It is estimated that almost 470,000 tons of tobacco are burned indoors each year generating both mainstream and sidestream smoke (See Personal Chemistry and Choices unit, page 457).

Table 9 illustrates the classification, examples, and sources of some common indoor air pollutants.

Table 9	Some Indoor Pollutants and Their Sources	
Phase of Matter	**Source**	**Pollutant**
Solid/particulate	Plants	Molds, mildew, bacteria, viruses
	Pets	Pet dander, dust, flea feces
	Floor tile	Asbestos
Liquids/gases	Clothes	Dry-cleaning fluid (1,1,2 trichlorotrifluoroethane), moth balls (p-dichlorobenzene)
	Furniture	Formaldehyde
	Particle board	Formaldehyde
	Carpet	Styrene
	Paint and paint thinners	Methanol, methylene chloride
	Glues and solvents	Xylene, acetone, toluene
	Cigarette smoke	Formaldehyde, carbon monoxide, benzo[a]pyrene
	Faulty furnaces and/or kerosene space heaters	Carbon monoxide
	Rocks under house	Radon from decay of uranium
	Water from showers	Chloroform
	Electric arcing	Ozone

In the next exercise we will consider sources of air pollution within a home. Note that "homes" include apartments, which can have many of the same indoor air quality problems as houses because they have the same pollution sources.

YOUR TURN

Air in the Home

Pollutants can come from the building's structural materials and furnishings, from activities within the building, and from the occupants of the building. Figure 23, page 447 illustrates several areas of a house. Each area has at least one potential source of an indoor air pollutant.

On a separate sheet of paper, list the numbers from Figure 23 identifying locations.

1. Give at least one potential source of indoor air pollution for each location.

2. What locations are most likely to be the greatest sources of pollutants?

3. Name pollutants that are likely to originate in the
   a. kitchen.
   c. ground under the house.
   b. garage.
   d. family room or den.
4. Because outdoor air enters buildings, some indoor air pollutants are the same as outdoor ones. Identify three such pollutants.
5. Some people feel that having some houseplants in homes, shops, and offices helps to remove significant quantities of air pollutants. Check out this assumption by finding evidence to support or contradict it.

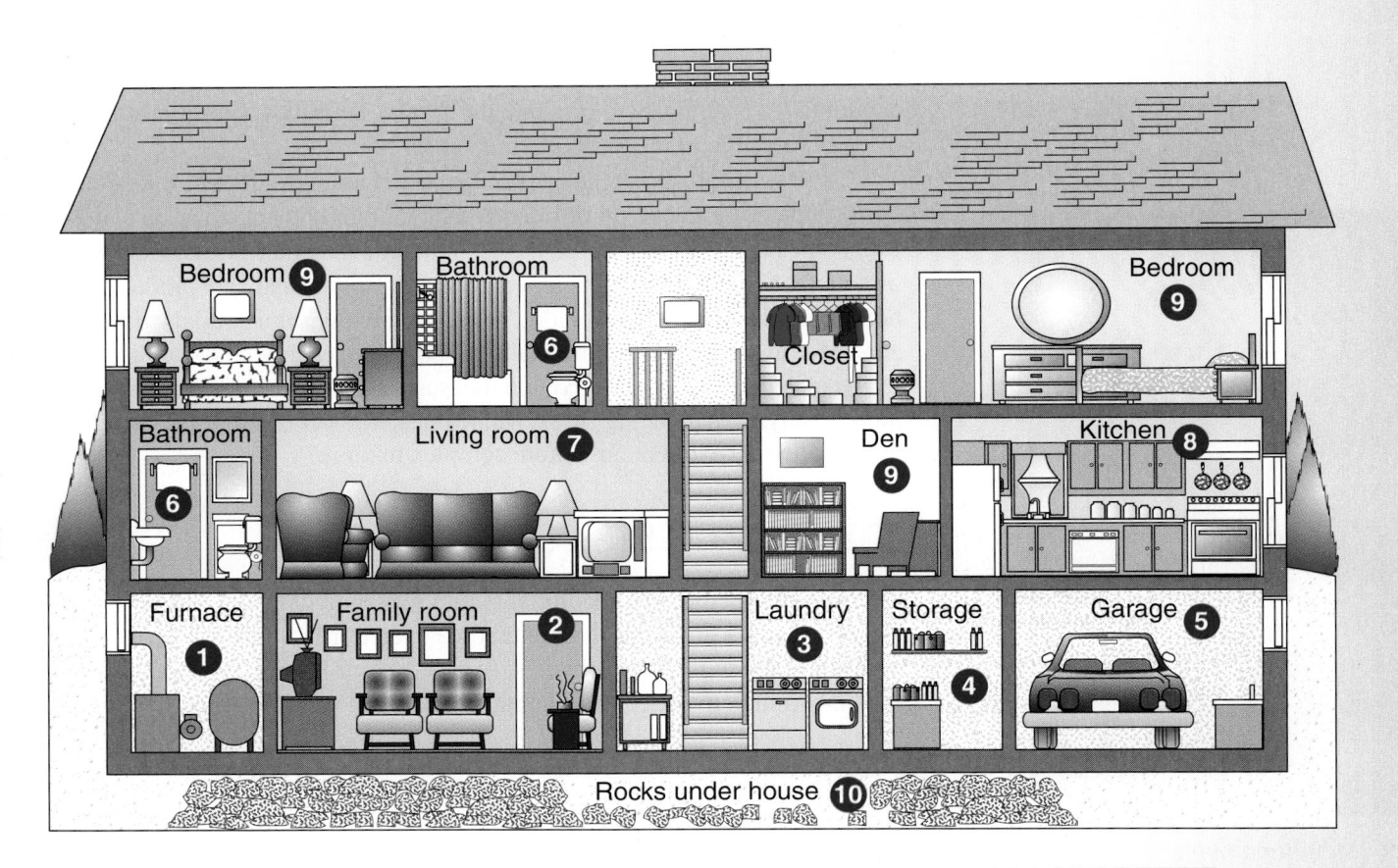

Room	Source	Pollutant
Bedrooms	Kerosene heater	Nitrogen oxides
Bathroom	Water from shower	Chloroform
Closet	Clothes	Dry cleaning fluid
Living room	Furniture	Formaldehyde
Den, bedrooms	Carpet	Styrene
Rocks under house	Uranium	Radon
Furnace	Faulty operation	Carbon monoxide
Storage room	Insect control	Pesticides
	Paint cans	Methylene chloride
Garage	Oil and gas spills	Hydrocarbons
Family room	Tobacco smoke	Carbon monoxide

Figure 23 Sources of indoor air pollution in a house.

We need to recognize that the mere presence of a pollutant does not mean poor indoor air quality. Rather, it is only when the pollutant exceeds a certain minimum concentration that air quality becomes a concern. Air quality standards have been federally established for outdoor air, but no such standards currently exist for indoor air. A major obstacle in trying to create indoor air quality standards is the enormous variability of indoor air among the more than 80 million buildings in the United States. However, research continues into the concentration level and length of exposure to various indoor pollutants needed to create specific health problems.

The concentration of pollutants in indoor air is controlled by a number of factors. These factors include: (1) the *air exchange rate*—the rates of outdoor air moving inside and indoor air flowing outside; (2) the rate at which pollutants are generated indoors; and (3) reaction rates among the indoor air components. If the air exchange rate is low, pollutants can accumulate, potentially to problematic levels. Tightly sealed homes are more susceptible to this problem than other homes.

Several factors have changed the kinds and amounts of indoor air pollutants. Wide-scale use of synthetic materials for building construction and furnishings first occurred in the 1950s. Synthetic materials release new types of pollutants, sometimes in significant quantities. Personal care products, household cleaners, and pesticides, some containing new materials, have appeared on the market. Additionally, responses to energy crises of the 1970s included energy conservation measures such as sealing buildings more tightly, thus reducing air exchange rates. The net result of these factors has been to make indoor air a more complex mixture.

E.2 A CLOSER LOOK

There are many indoor air pollutants that we could consider further. However, space and time require that we keep our list fairly short. Therefore, we will look more closely at only four indoor air pollutants—radon, formaldehyde, environmental tobacco smoke, and asbestos.

RADON

Radon is a radioactive, colorless, odorless, and tasteless gas formed by the nuclear decay of U-238 in ground under and around buildings (see also the Nuclear unit, page 327). Because it is a gas, radon and its decay products can be inhaled. Studies indicate that lung cancer can result from extended exposure to high levels of radon. The EPA contends that radon causes approximately 14,000 deaths annually in the United States. However, controversy surrounds indoor radon and its health effects. The EPA estimates of lung cancer deaths due to radon and the minimum acceptable radon concentration in a dwelling are especially controversial.

Epidemiologists do not agree on the minimum radon level required for significant health problems to develop or for remedial action to be taken, called the "action level." That level is expressed as picocuries per liter of air (pCi/L). A *picocurie* is a unit of radioactivity equal to about 4 nuclear disintegrations per minute. Thus, 4 pCi/L is about 16 nuclear disintegrations per minute in a liter of air. Currently, the action level is

Madame Marie Curie

▶ *A picocurie is a subunit of a larger unit, the Curie, named to honor Marie Sklodowska Curie. Madame Curie was a Polish scientist who won two Nobel prizes, one in chemistry and one in physics. 1 Curie = 3.7×10^{10} radioactive disintegrations per second, dps.*

4 picocuries per liter of air (4 pCi/L) in the United States; 4 pCi/L for new construction and 10 pCi/L for existing buildings in Sweden; 10 pCi/L in England; and 20 pCi/L in Canada. However, epidemiologists agree that a radon level higher than 20 pCi/L represents a significant health risk.

Figure 24 indicates variation in radon concentration by location in the United States, based on EPA estimates. The average radon value in U.S. homes is 1.3 pCi/L. An action level of 4 pCi/L would affect an estimated 6 million homes in the United States (about 6% of the houses) at a cost of at least $50 billion to homeowners and landlords. If the United States used the Canadian standard of 20 pCi/L, as some experts suggest, only about 50,000 homes, rather than 6 million homes, would need remediation.

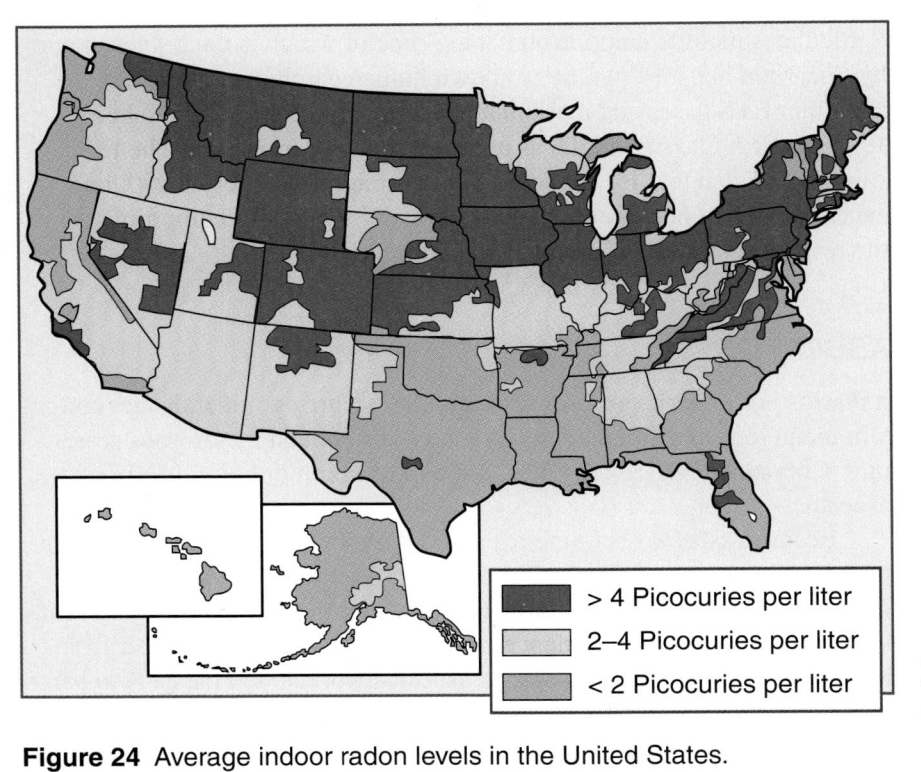

> 4 Picocuries per liter

2–4 Picocuries per liter

< 2 Picocuries per liter

Figure 24 Average indoor radon levels in the United States.

FORMALDEHYDE

Formaldehyde, H_2CO, is a pungent-smelling gas found outdoors as well as indoors. However, its outdoor concentration is usually about 10 ppb in urban areas compared to indoor concentrations that can be 10 to 100 times greater than those outdoors. In fact, formaldehyde is one of the most common indoor VOCs, coming from carpets, cigarettes, cabinets, and insulation. Phenol–formaldehyde and urea–formaldehyde resins are commonly used bonding agents in plywood, chipboard (particle board), paneling, and in urea–formaldehyde foam insulation (UFFI). Higher-than-normal formaldehyde levels (up to 10 ppm) are often found in mobile homes because particle board and UFFI are both commonly used in mobile home construction. Because of releasing formaldehyde, UFFI was banned in 1982, but court action reinstated its use in 1983.

Formaldehyde

Phenol

Urea

Putting It All Together: Indoor Air Quality

Formaldehyde irritates the eyes, nose, and throat. Studies done to determine the possible carcinogenic (cancer-causing) effects of formaldehyde among people who are routinely exposed to it at work have been contradictory. No direct cause-and-effect relationship has yet been established between formaldehyde and increased risk of cancer.

ENVIRONMENTAL TOBACCO SMOKE (ETS)

ETS, also called *secondhand smoke,* is a complex mixture of gases and particulates that come from the tips of burning cigarettes and the mainstream smoke exhaled by smokers (Personal Chemistry and Choices, page 457). Among the gases are carbon monoxide, nitrogen oxides, formaldehyde, VOCs, and polonium, a highly radioactive element. Particulates include nicotine and "tars," one of which is the polyaromatic hydrocarbon benzo[a]pyrene, a known human carcinogen. Given these and other ETS gases and particulates, it may not be surprising that in 1993 the EPA classified ETS as a known human carcinogen. The EPA study concluded that ETS is responsible for approximately 3,000 lung cancer deaths of nonsmoking adults annually while also compromising the respiratory health of infants and children.

ASBESTOS

Asbestos is a generic term for six naturally occurring minerals that contain metal ions and silicon and oxygen atoms bonded in a fibrous structure. Chrysotile, $Mg_3Si_2O_5(OH)_4$, is the most commonly used form of asbestos.

Because asbestos is flame resistant, it has been used for nearly a century as a thermal insulator for pipes and walls, and in automobile brake linings. Its fibers act as a binding agent in floor tiles, pipe wrap, and roofing materials. However, evidence of high incidence of lung cancer from inhalation of asbestos fibers among asbestos workers led the EPA to ban the manufacturing, importing, and processing of most asbestos products by 1997. Also, smokers have an increased risk of lung cancer from asbestos because the asbestos fibers act *synergistically* with cigarette smoke; the toxic effect is greater than for asbestos fibers or cigarette smoke alone.

Although asbestos removal has been mandated in some states, its removal remains controversial. A number of studies have shown only very low asbestos fiber concentrations in indoor air under normal building use. The results of other studies question whether chrysotile use should even be banned, or whether existing asbestos should be removed from buildings. Some experts feel that asbestos should be left intact unless there is specific evidence that it has been damaged and its fibers have become airborne. Others feel that it is difficult to predict when asbestos will become damaged, and it should be removed before that can happen. One thing is certain—unless stringent precautions are taken, removing asbestos insulation significantly raises airborne asbestos concentrations. Therefore, only trained personnel should remove asbestos.

E.3 EVALUATING A POLLUTANT

You will be assigned to a group that will consider a particular room and a specific indoor air pollutant found in that room. The pollutant will be radon, formaldehyde, ETS, or asbestos. Your group will gather sufficient information about the assigned pollutant to answer the following:

- How could the presence of the pollutant be determined? You will not actually conduct the test, but you are to develop an experimental strategy to determine the presence (or absence) of the pollutant.

- Develop a strategy or method to determine how much of the pollutant is present.

- Suppose that the pollutant concentration is determined to be higher than the safe level. Describe actions that can be taken to lower the pollutant concentration to an acceptable level if

    a. the pollutant concentration is just slightly above the safe level.

    b. the pollutant concentration is 5 to 10 times greater than the safe level.

- Would the concentration of the pollutant differ in a different room? If so, how might it differ and why?

- Each of the pollutants assigned is controversial. Do further research in the library to identify the following for your pollutant:

    a. the nature of the controversy

    b. the latest evidence presented by each side of the controversy

    c. the major point to be clarified or demonstrated

    d. data needed to resolve the controversy

You will share your group's information with groups that studied the same pollutant and report your findings to the entire class.

    a. What similarities are there among the four pollutants?

    b. What differences are there among the four pollutants?

E.4 LOOKING BACK

This unit explored some ways matter and energy interact in our atmosphere—interactions which involve a variety of important chemical reactions on a grand scale. You also considered key technological issues affecting society, issues surrounding the ways we use our air as a source of reactants and as a depository for reaction products from natural and human chemical activities.

The "sea of air" that surrounds us is vital to the world's climate, economy, and people. Personal decisions as well as those made at all levels of government will establish the future quality of this invaluable resource. Although clean air is invisible, the quality of air—whether indoors or outdoors—has a continuing visible impact on the health of every inhabitant of this planet. We hope that what you have learned in this unit has helped you think more carefully about this essential natural resource and to act accordingly in the future to preserve it.

Personal Chemistry and Choices

- ► What everyday decisions affect our level of personal well-being?

- ► How do such choices make a difference to our well-being?

- ► In what ways can risks be assessed and applied to health-related choices?

- ► How can we balance the risks and benefits of health-related choices?

- ► How is chemistry related to medicines, personal-care products, and drugs?

- ► What chemistry is involved in our body's structures and processes?

INTRODUCTION

There is some level of risk in everything we do; driving, exercising (or not exercising), taking medications—even eating—present possible risks. Some activities or products have sufficient risk that they carry warning labels.

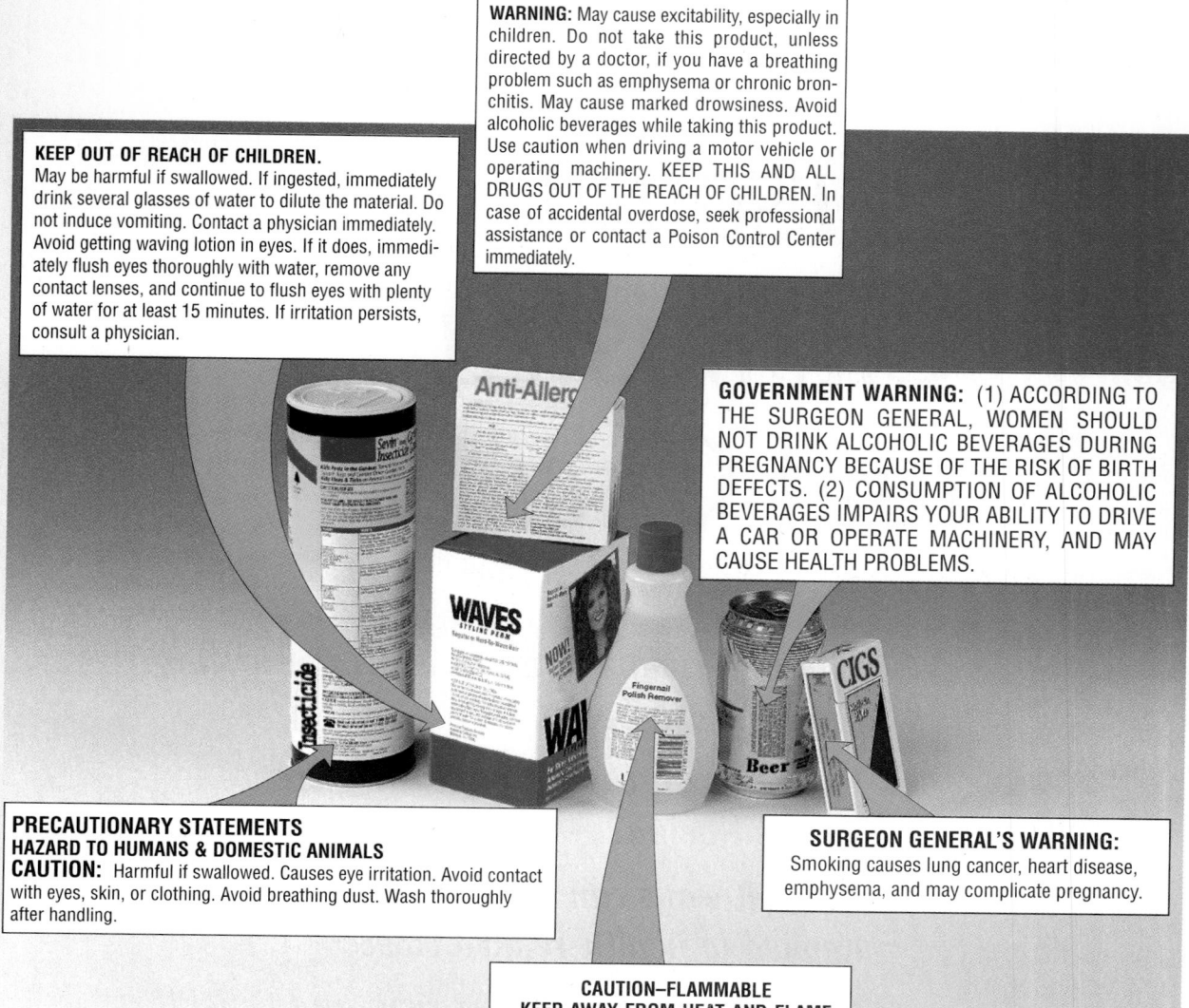

WARNING: May cause excitability, especially in children. Do not take this product, unless directed by a doctor, if you have a breathing problem such as emphysema or chronic bronchitis. May cause marked drowsiness. Avoid alcoholic beverages while taking this product. Use caution when driving a motor vehicle or operating machinery. KEEP THIS AND ALL DRUGS OUT OF THE REACH OF CHILDREN. In case of accidental overdose, seek professional assistance or contact a Poison Control Center immediately.

KEEP OUT OF REACH OF CHILDREN.
May be harmful if swallowed. If ingested, immediately drink several glasses of water to dilute the material. Do not induce vomiting. Contact a physician immediately. Avoid getting waving lotion in eyes. If it does, immediately flush eyes thoroughly with water, remove any contact lenses, and continue to flush eyes with plenty of water for at least 15 minutes. If irritation persists, consult a physician.

GOVERNMENT WARNING: (1) ACCORDING TO THE SURGEON GENERAL, WOMEN SHOULD NOT DRINK ALCOHOLIC BEVERAGES DURING PREGNANCY BECAUSE OF THE RISK OF BIRTH DEFECTS. (2) CONSUMPTION OF ALCOHOLIC BEVERAGES IMPAIRS YOUR ABILITY TO DRIVE A CAR OR OPERATE MACHINERY, AND MAY CAUSE HEALTH PROBLEMS.

PRECAUTIONARY STATEMENTS
HAZARD TO HUMANS & DOMESTIC ANIMALS
CAUTION: Harmful if swallowed. Causes eye irritation. Avoid contact with eyes, skin, or clothing. Avoid breathing dust. Wash thoroughly after handling.

SURGEON GENERAL'S WARNING:
Smoking causes lung cancer, heart disease, emphysema, and may complicate pregnancy.

CAUTION–FLAMMABLE
KEEP AWAY FROM HEAT AND FLAME
Keep away from small children. Harmful if taken internally. In case of accidental ingestion, consult a physician and contact a Poison Control Center. Harmful to synthetic fabrics, wood finishes, and plastics.

A PERSONAL CHOICES AND RISK ASSESSMENT

Many of the choices we make—to do a certain activity, or to use a product—have significant impact on our present and future well-being. Each product just mentioned obviously carries a level of risk. Otherwise, a warning would not be needed. The caution or warning implies a risk to our body's chemistry associated with a chemical component or components in the product. A chemical agent in the product might interfere with the body's delicate biochemical balance that controls our well-being.

Normally, we are born with beneficial chemical interactions operating within our bodies. These chemical systems make possible such basic activities as producing energy, growing and maintaining cells and tissues, and processing sensory data (vision, hearing, and so on). Such interactions are beyond our direct control. However, others are affected by our choices, and those choices may be made unwisely. For example, many U.S. citizens die much earlier than necessary because they have made unwise health choices. According to the U.S. Centers for Disease Control and Prevention, more than half of early deaths could be prevented by changes in personal behavior. Our attitudes and actions directly control many potential risks.

Generally, warning labels are developed by state and federal regulatory agencies after extensive research to assess the level of risk of a product or activity. Paying attention to such warnings is just common sense, part of the need to assess risks to make informed choices for our well-being. One such choice is whether to smoke or not, a major decision that will influence your health. The following activity will give you a chance to consider the risks and benefits connected with that choice.

YOU DECIDE: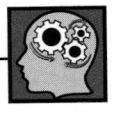
RISK ASSESSMENT AND SMOKING

One way of doing a preliminary risk assessment is to develop a grid of responses related to an activity and its risks (sometimes called burdens) and benefits. In this exercise, the activity is smoking or not smoking, an action directly within your personal control.

On a separate sheet of paper, copy the following grid. Working individually or in small groups as assigned by your teacher, complete the blocks in the grid. Analyze all the positive and negative factors that might influence your answers. Be prepared to discuss your results with the class.

	Smoking	Not smoking
Risks		
Benefits		

We now turn to a closer look at cigarette smoking. Smoking releases chemicals that affect our internal chemistry, and we will examine the results. In this way, we will look at a personal choice—to smoke or not to smoke—and at some of the chemical consequences of the choice.

A.1 CIGARETTE SMOKING

Probably some people you know smoke cigarettes. Approximately 20% of U.S. high school seniors smoke cigarettes daily. Smoking is considered risky because of medical problems it can cause for smokers and to those around them.

Yet, there are those who find cigarette smoking a pleasurable, social activity. The tobacco industry claims that cigarette smoking does no harm. Through its advertisements, the industry responds to anti-smoking groups by emphasizing that smoking is a personal choice, one allowed as part of individual freedom. The industry also points out that tobacco helps the economy in every area where it is grown as a "cash crop."

In 1964, about 40% of the U.S. population smoked, but the current level is about 25%, amounting to about 46 million adult Americans who still smoke cigarettes. Each of them smokes an average of 10,000 cigarettes (500 packs) annually. Total sales of cigarettes in the United States peaked in 1981 and have gradually declined since then. On the other hand, smoking has increased among U.S. adolescent females by about 55% since 1968.

What are some chemical components associated with cigarette smoke? How do they affect our well-being?

Decision making is a part of everyone's life.

A.2 CHEMICAL COMPONENTS OF CIGARETTE SMOKE

The glowing tip of a cigarette reaches a temperature of 850 °C. This temperature is high enough to set off chemical reactions that produce more than 4,000 substances with each puff. About 400 of these substances, including carbon monoxide and nicotine, are toxic. Forty are carcinogenic (cancer-causing). Thus, the 50,000 puffs taken annually by a one-pack-a-day smoker create a significant chemical assault on the mouth, lungs, bloodstream, and heart of the smoker.

Some of the compounds found in cigarette smoke are listed in Table 1. They are found both in **mainstream smoke,** the smoke inhaled by the smoker, and in **sidestream smoke,** which goes into the air around the smoker. Sidestream smoke, also called "secondhand" or "passive" smoke, is inhaled by nonsmokers as well. Data from the Centers for Disease Control and Prevention indicate that each year about 3,000 nonsmokers die from cancer and over 30,000 die from heart disease due to secondhand smoke exposure. People who sit in nonsmoking sections of restaurants and other public places reduce their exposure to secondhand smoke, but do not entirely escape it.

Table 1 Some Compounds in Cigarette Smoke in Addition to Carbon Monoxide and Nicotine	
Compound	**Formula**
Acetone	C_3H_6O
Ammonia	NH_3
Aniline	C_6H_7N
Benzene	C_6H_6
Benzo[a]pyrene	$C_{20}H_{12}$
Formaldehyde	H_2CO
Hydrogen cyanide	HCN
Methanol	CH_3OH
Naphthalene	$C_{10}H_8$
Nitric oxide	NO
Pyridine	C_5H_5N

YOUR TURN

People tend to use the word "toxic" rather loosely. However, the term "toxic" has very specific legal and medical definitions applied to individual materials. To evaluate whether a compound is toxic or can cause cancer, we need to consider how much of the material enters the body and what happens to it chemically within the body. We also must know how exposure to the compound can occur. This includes knowing how and where the compound is used.

Curiously, a number of compounds classified as toxic or carcinogenic are used to make nontoxic consumer products. These compounds, and others like them, are reactants or intermediates in the manufacturing of industrial and consumer products. Carefully controlled manufacturing conditions minimize the exposure of workers to the potentially toxic or carcinogenic materials.

Use the information in reference materials suggested by your teacher to answer the following questions about the compounds in Table 1.

1. Identify which compounds are considered to be toxic to humans. Also identify the level of toxicity for each substance.

2. Which compounds are considered carcinogenic to humans? At what levels?

A.3 SOME OTHER COMPOUNDS IN CIGARETTE SMOKE

The carbon monoxide in cigarette smoke reduces the blood's ability to carry oxygen gas to the cells by about 8%. Carbon monoxide gas binds with the iron ions in hemoglobin 200 times more strongly than oxygen gas does. Thus, carbon monoxide binding prevents hemoglobin from picking up and carrying oxygen. The body responds to this reduction in oxygen-carrying capacity by increasing the number of red blood cells. This thickens the blood, adding stress on the heart as it pumps.

Carried in the blood of a pregnant woman, carbon monoxide seems to have a detrimental effect on the fetus. No one knows precisely what happens, but there is no doubt that the fetus is deprived of necessary oxygen. It is known that women who smoke during pregnancy have babies weighing an average of 200 g less than babies born to nonsmoking women. About 2,500 infant deaths yearly in the United States are caused by mothers who smoke.

Tars from smoke accumulate in the lungs. In a response that may clear the tars, the enzyme elastase is activated. While catalyzing the chemical breakdown of tars, the enzyme also seems to catalyze chemical attack on the lung's membranes. This action scars the lungs and reduces their ability to transfer oxygen gas to the blood. Emphysema often results.

Numerous compounds in tobacco and smoke can cause cancer. Among these are nitrosamines and polycyclic aromatic hydrocarbons (PAHs), examples of which are given in Figure 1. Because of the carcinogenic compounds and other highly reactive materials in cigarette smoke, regular smokers run a high risk of getting lung cancer—10 to 20 times greater than a nonsmoker—and have four times the risk of esophageal cancer. In addition to the risk of cancer, smokers also face increased risk of strokes and heart attacks; many others die from emphysema and other chronic breathing problems.

About 40% of those who try smoking become regular users, especially if they begin smoking before age 20. About 90% of these young smokers become addicted. Those who are addicted smoke at least 10 cigarettes daily

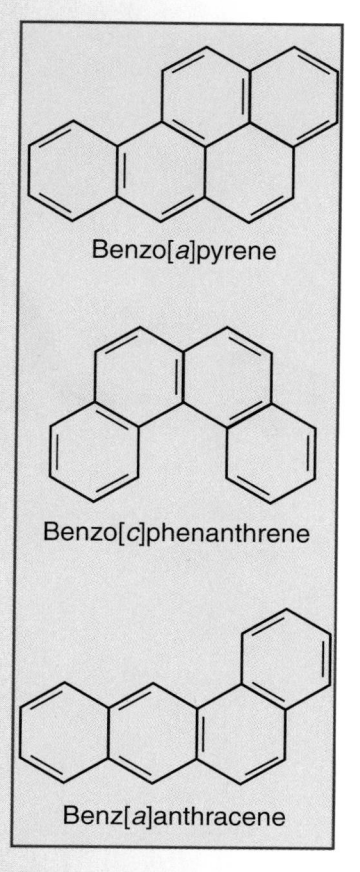

Benzo[a]pyrene

Benzo[c]phenanthrene

Benz[a]anthracene

Figure 1 Some carcinogens in cigarette smoke. These are all polycyclic aromatic compounds.

and have trouble with not smoking. Lack of knowledge is not the problem—90% of cigarette smokers indicate that they know smoking is harmful, and 30% annually try to quit smoking. Unfortunately, only 3% of them are successful in quitting.

A smoker's health can be greatly improved by quitting. In fact, after quitting, the risks of smoking-caused health problems decline steadily for about 15 years, until the risk of death becomes nearly as low as for a person who never smoked.

Recently, skin patches have been developed for people who want to reduce or stop their smoking (Figure 2). At the beginning of treatment, the patch contains a day's supply of nicotine. (A "day's supply" is the number of cigarettes usually smoked daily multiplied by 9 mg, the amount of nicotine in one typical cigarette.) Over a period of time, a smoker switches to patches that have decreasing amounts of nicotine. The release of nicotine from the patch into the body must be carefully controlled so that the patient does not get too much at one time, because too much nicotine can be fatal. (A lethal dose is only 60 mg.) Therefore, nicotine patches are available only by prescription. The nicotine patch can be modified to limit the amount of nicotine that enters the patient's body to about 20 mg per day.

▶ *The Surgeon General reports that "smokeless tobacco" products are not a safe alternative to cigarettes. Their use can cause addiction, inflamed and receding gums, and mouth cancer. Smokeless tobacco containers carry warnings such as: "WARNING: This product may cause gum disease and tooth loss."*

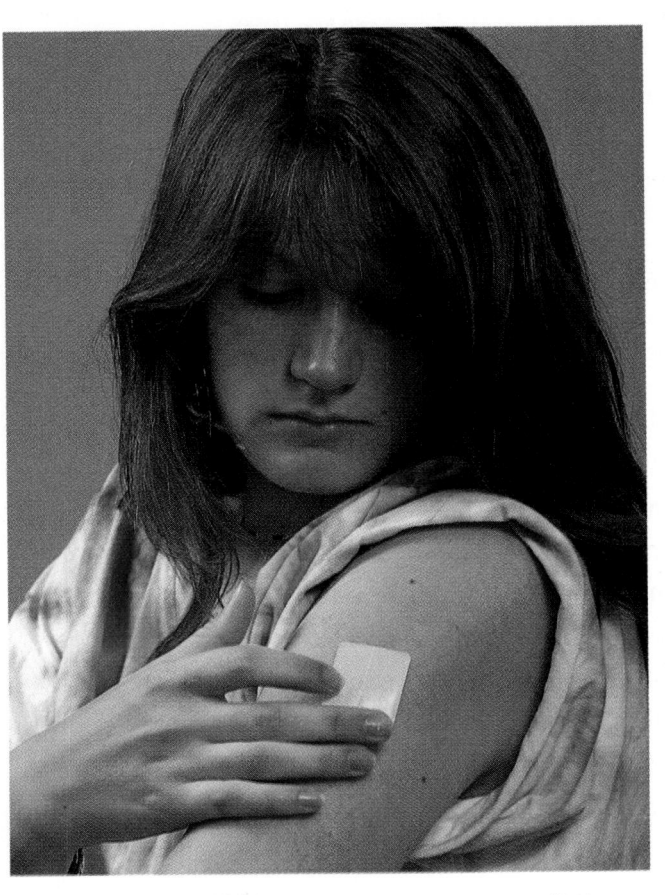

Figure 2 The nicotine "patch" releases controlled amounts of nicotine to help smokers reduce their cigarette smoking.

A.4 MAKING JUDGMENTS ABOUT RISK

Informed decision making is important in personal health-care.

Scientists often identify and investigate events that routinely occur together, either at the same time or in a predictable sequence. Such research often involves factors or events related to personal and public health. Situations or incidents that regularly happen together are said to be *correlated.* When correlations are found, one scientific goal is to determine whether or how the two events are related.

Sometimes advocates of a given health-related position may imply that a high correlation means one event caused the other. In fact, other events may have caused both. For example, when people start wearing overcoats, home heating bills generally increase. But overcoats do not cause heating bills to rise. The onset of cold weather directly causes both the appearance of overcoats and increased heating bills.

Correlation may also be pure coincidence. For many years, skirt hem heights and stock market prices increased and decreased together. Do you think there was an actual connection between these two trends?

YOUR TURN

Events: Related or Not?

Examine the pairs of events listed below and express the relationship between them by choosing which phrase applies in each case: "correlates with," "directly causes," or "is not related to." (You may find that more than one phrase applies.)

For example, the events "spring" and "hay fever" are correlated for some people. After some allergy tests, a cause–effect relationship might be established. A given sufferer might find out that pollen in spring directly causes the hay fever symptoms. Thus, even though spring and hay fever are correlated, the direct cause of the hay fever—at least in this example—would be pollen.

1. Number of outdoor picnics, number of thunderstorms
2. Average temperature, season of the year
3. Number of ocean swimmers, tide height
4. Classes attended, grade in chemistry
5. Availability of guns, number of homicides
6. Average family income, amount of nutritious food per family
7. Water temperature, amount of dissolved oxygen
8. Length of hydrocarbon chain, boiling point
9. Alcoholism, stress
10. Correctly finishing chemistry homework, solving chemistry problems in class

11. Pleasant personality, number of dates

12. Smoking, lung cancer

13. Glucose metabolism, energy production

14. Stock prices, American Football Conference Super Bowl victories

A.5 STUDYING HUMAN DISEASE

As you discovered from the previous *Your Turn,* actual cause-and-effect relationships may be hard to prove among pairs of events that initially might seem correlated. Finding actual correlations in research depends partly on what is being studied, and on how difficult it is to control important variables. In chemical research in a laboratory, it is relatively easy to control most variables. In medical research, it is never easy to do so, partly because it is unacceptable to expose human subjects to potentially harmful treatments.

For that reason, medical scientists may begin a study by treating animals such as mice or rabbits, which are fairly closely related to humans. Such experiments can produce useful results, but cannot definitely predict how humans will respond to the same treatments. Human body chemistry is different in some ways from that of other animals. In addition, some conditions can't be duplicated with animals.

Animal studies are supplemented with *epidemiological studies.* In such studies, scientists study specific populations of people in search of factors that cause or prevent disease. For example, they study victims of heart attacks, gathering information about their living habits, in an effort to determine what makes some people likely to have heart attacks. Or they study groups that have a much lower than average incidence of heart attacks to identify what may protect these individuals.

A problem arises, however, in that a single epidemiological study may reveal correlations, but no evidence of a cause-and-effect relationship. At times, such findings may even mislead researchers. For example, one study concluded that coffee drinking causes heart attacks. But the researchers had ignored smoking among coffee drinkers as a possible cause of heart disease. It would have been more useful if the results for coffee-drinking smokers had been compared with those for coffee-drinking nonsmokers.

Despite these difficulties, important and useful information continues to be generated by health-related research and epidemiological studies. Such data and findings can help guide informed risk analysis and decision making, ranging from a variety of personal choices to health-related societal policies.

Your personal health and well-being are, in large measure, maintained by chemical systems that were already in operation when you were born. In the next section we will explore some key features of these interrelated systems of "internal chemistry."

Preventing Illness, Easing Suffering, Giving Hope

Dr. **Jeff Olliffe** is a family practice physician in Seattle, Washington, specializing in treating patients who have the Human Immunodeficiency Virus (HIV).

People who are concerned that they have been exposed to HIV arrange an office visit to consult with Jeff. Together, Jeff and the patient assess the patient's risk of HIV infection. If warranted, Jeff draws a blood sample and performs a test known as HIV ELISA (short for Enzyme-Linked Immuno-Sorbent Assay), that detects those antibodies the body makes when one is exposed to the virus.

Having the HIV virus does not mean that a patient is dying. People can be HIV positive and be otherwise perfectly healthy. However, being positive can eventually lead to developing Acquired Immune Deficiency Syndrome, or AIDS. The most common ways people become exposed to HIV are through the intimate exchange of bodily fluids by having unprotected sex or using contaminated needles to inject drugs.

If a patient tests negative for HIV antibodies, Jeff advises him or her about how to stay that way. If a patient tests positive for HIV, then Jeff will order additional tests to confirm the ELISA findings. If the tests confirm the presence of HIV, Jeff counsels his patient on how to stay healthy, and how to avoid transmitting the virus to others.

For his patients with AIDS, Jeff treats the symptoms, provides emotional support, and prescribes medicines to prevent opportunistic infections. "Although there is not yet a cure for AIDS," he notes, "a patient's quality of life— physically and psychologically—can be improved with life-prolonging treatments."

Jeff takes part in "compassionate release medicine" studies to help test the effectiveness of certain experimental drugs in attacking and

controlling HIV. Jeff prescribes these drugs in various concentrations, depending on his patient's symptoms and other criteria. Then he monitors and records any possible side effects of the drugs. Periodically, he reports results to appropriate governmental medical agencies.

Jeff routinely uses biochemistry and pharmacology in his work, and relies on his knowledge of chemistry to understand the physiological and chemical interactions of the body with medicines. However, he also emphasizes the human or social side of his work. "There is a true art," Jeff explains, "in dealing with people who are facing crises."

What Else Is Being Done?

Doctors and scientists around the world are trying to develop a cure for AIDS. Read about some of the new approaches to treatment that were announced over the last year. How do they differ from previously available approaches? How are they being tested?

Photograph courtesy of Jeff Olliffe, M.D.

? PART A: SUMMARY QUESTIONS

1. Carbon monoxide in cigarette smoke can combine with oxygen to form carbon dioxide. Write a balanced chemical equation for the burning of carbon monoxide to form carbon dioxide.

2. Nicotine is one of many compounds in cigarette tobacco. The chemical formula for nicotine is $C_{10}H_{14}N_2$.

   a. When a cigarette burns, nicotine reacts with oxygen to produce carbon monoxide, nitric oxide (NO), and water. Write a balanced chemical equation for this reaction.

   b. Calculate the molar mass of nicotine.

   c. A typical cigarette contains 9 mg of nicotine. Calculate how many moles of nicotine are in a cigarette.

   d. Although 9 mg of nicotine might seem to be a very small mass, it is a measurable amount. However, even 9 mg of nicotine contains a very large number of nicotine molecules. Calculate how many nicotine molecules there are in 9 mg of nicotine. To give you an idea of how large that number of nicotine molecules is, it would take all of the nearly 6 billion people on earth counting one nicotine molecule per second for 24 hours a day almost 170 years to complete the task.

3. In your own words, explain to a classmate or a household member how a nicotine patch works to reduce the number of cigarettes a smoker has in a day.

4. a. Name three major components of cigarette tobacco smoke.

   b. Describe the effects of each on the body.

B YOUR BODY'S INTERNAL CHEMISTRY

B.1 BALANCE AND ORDER: KEYS TO LIFE

Even when you dive into icy water or jog on a hot day, your body temperature rarely deviates more than two degrees from 37 °C. If it does, you are probably ill. Likewise, your blood pH remains virtually constant, as do concentrations of many substances in your blood and cells. Your body would quickly fail if its chemistry became unbalanced.

Maintenance of balance in body systems is *homeostasis*. Failure of homeostasis leads to illness, and—if not corrected—eventually leads to death. The best way to stay healthy is to help your body chemistry stay in balance. For example, eating a balanced diet provides your body with the building-block molecules and ions it needs to make new tissues. Exercise is also important. It keeps your cardiovascular system (heart, lungs, veins, and arteries) fit, so that blood circulates to all your cells.

Body chemistry's most impressive and important feature is its controlled complexity. Every minute your cells make hundreds of different compounds, and do so more quickly and accurately than any chemical factory. These compounds take part in the multitude of reactions occurring simultaneously within each cell.

To function effectively, cells must be supplied with needed chemical substances, their waste products must be removed promptly, and a proper temperature and pH must be maintained. Blood serves as the conduit for supplies and wastes. Also, blood protects cells from harmful changes in physical and chemical conditions, such as those caused by invading bacteria.

Under normal conditions, your body and its external environment both contain well-balanced bacterial ecosystems. Although harmful bacteria are always around us, they generally fail to cause disease. In large part that is because they must compete with harmless bacteria that limit the number of harmful microbes. Sickness often results if the harmless bacteria are suppressed and the disease-causing bacteria have a population explosion.

Harmful bacteria proliferate in a dirty environment. Before municipal sewage treatment began, drinking water supplies were continually threatened by armies of bacteria. Water supplies were often contaminated. In considering data about the control of infectious diseases, we often overlook the fact that improved municipal sewage treatment has greatly enhanced public health.

Despite modern medicine and sanitation, occasionally we become sick. Fortunately, the body is very good at regaining its balance. Given a healthful environment, proper care, and a good diet, most people recover from most diseases by themselves. But in some cases the body needs assistance, and chemists have developed many drugs that promote recovery and relieve pain. Aspirin and the host of modern antibiotics and anti-depressants are products of chemists' fertile imaginations and laboratories.

However, body chemistry is so delicately balanced that a medicine added to help one system regain equilibrium may throw other systems out of balance. This is what doctors mean by *side effects.* For instance, certain useful drugs may harm the kidneys or livers of a small percentage of patients. Sometimes we need to decide which creates greater problems—the disease or the cure. In the following sections, we will explore the body's composition and chemistry, and investigate some built-in mechanisms that promote homeostasis.

B.2 ELEMENTS IN THE HUMAN BODY

Table 2 lists twenty-six elements making up your body, along with the quantities normally present in a healthy adult. These elements are found in biomolecules and as ions in body fluids. They must be present in appropriate amounts in order for homeostasis to be maintained.

Table 2 Elemental Composition of a 70-kg Human Adult	
Elements	**Percent (%) of Body Mass**
Hydrogen, oxygen, carbon, nitrogen, calcium	98
Phosphorus, chlorine, potassium, sulfur, magnesium	2
Iron, zinc	0.01
Copper, tin, manganese, iodine, bromine, fluorine, molybdenum, arsenic, cobalt, chromium, lithium, nickel, cadmium, selenium	Each less than 0.001

Oxygen, carbon, hydrogen, nitrogen, and calcium are the most abundant. Together, they make up about 98% of body mass. Sixty-three percent of all atoms in the body are hydrogen, 25% are oxygen, 10% are carbon, and 1.4% are nitrogen. Most of these atoms are present in organic compounds such as the carbohydrates, proteins, and fats described in the Food unit. Calcium—1.5% of the body's mass—is an important structural element in teeth and bones, and is involved in the passage of nerve impulses, muscle contraction, and blood clotting.

You will recall from the Food unit (page 275) that elements such as iodine, selenium, copper, and fluorine are essential to health, even though they are normally present as less than 10 parts per million (ppm) of body mass. Table 3 (page 466) describes functions of some trace elements in the body.

Table 3	Minerals	
Mineral	**Source**	**Deficiency Condition**
Macrominerals		
Calcium (Ca)	Canned fish, milk, dairy products	Rickets in children; osteomalacia and osteoporosis in adults
Chlorine (Cl)	Meats, salt-processed foods, table salt	Convulsions in infants
Magnesium (Mg)	Seafoods, cereal grains, nuts, dark green vegetables, cocoa	Heart failure due to spasms
Phosphorus (P)	Animal proteins	Poor bone maintenance
Potassium (K)	Orange juice, bananas, dried fruits, potatoes	Poor nerve function, irregular heartbeat, sudden death during fasting
Sodium (Na)	Meats, salt-processed foods, table salt	Headache, weakness, thirst, poor memory, appetite loss
Sulfur (S)	Proteins	None
Trace minerals		
Chromium (Cr)	Liver, animal and plant tissue	Loss of insulin efficiency with age
Cobalt (Co)	Liver, animal proteins	Anemia
Copper (Cu)	Liver, kidney, egg yolk, whole grains	Anemia and low white blood cell count
Fluorine (F)	Seafoods, fluoridated drinking water	Dental decay
Iodine (I)	Seafoods, iodized salts	Goiter
Iron (Fe)	Liver, meats, green leafy vegetables, whole grains	Anemia; tiredness and apathy
Manganese (Mn)	Liver, kidney, wheat grains, legumes, nuts, tea	Weight loss, dermatitis
Molybdenum (Mo)	Liver, kidney, whole grains, legumes, leafy vegetables	None
Nickel (Ni)	Seafoods, grains, seeds, beans, vegetables	Cirrhosis of liver, kidney failure, stress
Selenium (Se)	Liver, organ meats, grains, vegetables	Kashan disease (a heart disease found in China)
Zinc (Zn)	Liver, shellfish, meats, wheat germ, legumes	Anemia, stunted growth

YOUR TURN

Elements in the Body

1. Select five elements found in the body. Using Tables 2 and 3, and what you know about these elements, answer these questions:

   a. Where in the body (or in what type of biomolecule) is each element found?

   b. What role does each element play in maintaining health?

2. Is it correct to say that elements highly abundant in the body—such as calcium and phosphorus—are more essential than trace elements such as iron and iodine? Why? (*Hint:* Consider the notion of a limiting reactant.)

3. a. Why might a change in diet be more likely to cause a deficiency of trace elements than of abundant elements?
   b. Why are cases of overdose-poisoning more common among trace elements? (*Hint:* Consider percent change and chemical balance.)

To understand how cellular chemistry contributes to health and well-being, and what you can do to assist your chemistry, we will now investigate some of the basic principles involved.

B.3 CELLULAR CHEMISTRY

How quickly can you pull back your hand from a hot object? How fast can you move your foot from the accelerator to brake a car to a stop? Cellular chemical reactions must take place extremely rapidly to allow you to do these things.

Cells are the scene of action in your body. Cellular chemistry is fast, efficient, and precise. You have learned that food is "burned" to supply the body's continuous energy needs. This "burning" occurs in the cell, where energy is released and used with great speed and efficiency. The body's building needs are also met with incredible efficiency within cells. Each healthy cell makes hundreds of different substances—each at exactly the right moment and in precisely the amount required. In addition, the cell generates its own energy and chemical machinery to do so.

What's the secret behind this impressive performance? It lies with enzymes, which are present in all cells. **Enzymes** are biological catalysts. Like all catalysts, they speed up reactions without undergoing any lasting change themselves. The speed of an enzyme-catalyzed reaction is hard to comprehend. In one second, a single enzyme molecule in your blood can catalyze the decomposition of 600,000 H_2CO_3 molecules:

$$H_2CO_3(aq) \xrightarrow{enzyme} H_2O(l) + CO_2(g)$$

In that same second, 600,000 molecules of carbon dioxide are liberated to move into your lungs as you exhale. Also in one second, an enzyme in saliva can free 18,000 glucose molecules from starch:

$$-glucose-glucose-glucose + H_2O(l) \xrightarrow{enzyme} -glucose-glucose + glucose$$
(portion of starch molecule)   (remaining portion of starch molecule)

Before exploring further how enzymes do their jobs, you will conduct the following laboratory activity. Witness for yourself the speed of an enzyme-catalyzed reaction!

B.4 LABORATORY ACTIVITY: ENZYMES

GETTING READY

You will investigate how an enzyme affects the rate of decomposition of hydrogen peroxide to water and oxygen gas:

$$2 H_2O_2(aq) \rightarrow 2 H_2O(l) + O_2(g)$$

This reaction was used to generate oxygen gas in the Air unit, Laboratory Activity: The Atmosphere (page 380).

You will be assigned a particular material—a piece of apple, potato, yeast, or liver—to test as a catalyst. You will test both a fresh piece of the material, and one that has been boiled, to see whether either material catalyzes the decomposition of hydrogen peroxide.

Read the procedures below. Prepare a data table with appropriately labeled columns for your data.

PROCEDURE

1. Obtain two pieces of your assigned material, one that is fresh and one that has been boiled.

2. Label two 16 mm × 124 mm test tubes, one "fresh," the other "boiled."

3. Add the fresh piece to the "fresh" test tube. Add the boiled piece to the other test tube.

4. Add 5 mL of hydrogen peroxide, H_2O_2, to each test tube. Observe whether any gas is evolved, a sign of decomposition.

5. Discard and dispose of the materials and solutions as directed by your teacher.

6. Wash your hands thoroughly before you leave the laboratory.

QUESTIONS

1. What can you conclude from the observations of the test with the fresh catalyst?

2. Explain your observations of the results when the boiled material was used.

3. Compare your experimental data with other class members who used:
   a. the same catalyst material.
   b. a different catalyst material.

4. Why does commercial hydrogen peroxide contain preservatives?

5. There is an old belief that when hydrogen peroxide is added to a cut, foaming shows that infection is present. What is the real reason for the foaming?

Let's find out why enzymes are so effective.

B.5 HOW ENZYMES WORK

Enzymes speed up reactions by making it easier for molecules to react. Both enzymes themselves and many of the molecules they help to react are very large. If specific sites or functional groups on the molecules are to react with each other, the molecules must approach so these functional groups are close to each other.

In general, enzymes function in this manner:

- Reactant molecules—known as *substrates*—and enzymes are brought together. The substrate molecules fit into the enzyme at an *active site* where the key functional groups are properly positioned (Figure 3).
- The enzyme then interacts with the substrate molecule(s), weakening key bonds and making the reaction more energetically favorable. (Our discussion of automobile catalytic converters in the Air unit, page 437, described how catalysts function—the reaction is speeded up because the activation energy barrier is lowered.)
- The reaction occurs. The products then depart from the enzyme surface, freeing the enzyme to interact with other substrate molecules.

Some enzymes require the presence of coenzymes, such as vitamins. For example, B vitamins are coenzymes in the release of energy from food molecules. Figure 4 illustrates how some coenzymes function.

Enzymes are as selective as they are fast. They complete only certain reactions, even though substrates for many other reactions are also available. How does the cell "know" what to do? Why do your cells *oxidize* glucose to CO_2 and water, instead of *reducing* it to CH_4 and water? Why do stomach cells digest protein rather than starch or fat? The answer to all these questions is the same: A given enzyme can catalyze only one specific class or type of reaction. Thus, a cell can only complete reactions for which it has enzymes. Thanks to the specificity of enzymes, cells have precise control over which reactions take place.

When someone drinks an alcoholic beverage, the ethanol (ethyl alcohol) is initially acted on in the stomach by the enzyme alcohol dehydrogenase (ADH). Dehydrogenases are enzymes that facilitate the removal of hydrogen atoms from a substrate. ADH catalyzes the conversion of ethanol (C_2H_5OH) to acetaldehyde (C_2H_4O). It is believed that acetaldehyde in high concentration in the brain and other tissues is responsible for

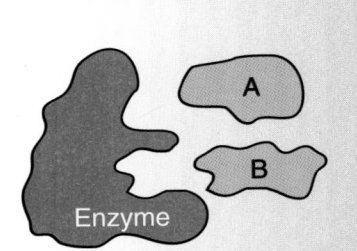

Without the coenzyme, substances A and B cannot be led to the active site of the enzyme and hence cannot react with each other.

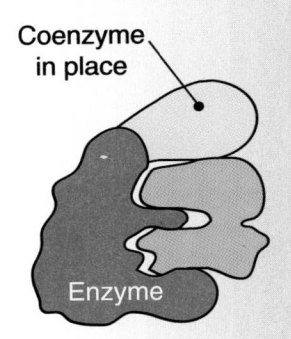

With the coenzyme, substances A and B are oriented at the active site of the enzyme. Here they can react with each other.

Figure 4 Diagram of a vitamin serving as a coenzyme.

Substrate molecules   Product molecule

Active site   Active site   Active site
Enzyme   Enzyme–substrate   Enzyme
                        complex

Figure 3 Model of the interactions of enzyme and substrate molecules.

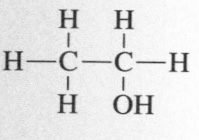

Ethanol

Acetaldehyde

the effects felt with a "hangover," or with continued alcohol abuse. Alcohol that is unconverted in the stomach enters the bloodstream and ultimately goes to the liver, where ADH converts it to acetaldehyde.

B.6 HOW ENERGY IS RELEASED AND STORED

Body chemistry is not only fast and selective, but also extremely energy-efficient. Here's why: Some cellular reactions release energy; others require energy. Just as money is often stored in a bank until needed, reserves of body energy are stored in carbohydrates and fats. Between its release from these energy-rich molecules and its use in the cell, energy is stored briefly in biomolecules called ATP (adenosine triphosphate). Think of it this way—between obtaining cash and spending it, you store the money in your pocket.

ENERGY FROM GLUCOSE

The energy in food may be used quickly after you eat carbohydrates. More often, it is stored in fats that can be broken down to glucose molecules later. In either case, cells needing energy begin with glucose.

The body's primary energy-release reaction is the oxidation of glucose, which we can describe by this overall equation for each mole of glucose:

$$C_6H_{12}O_6(aq) + 6\,O_2(g) \rightarrow 6\,CO_2(g) + 6\,H_2O(l) + 2.87 \times 10^3\,kJ$$

This equation summarizes a sequence of more than 20 chemical reactions, catalyzed by more than 20 enzymes. Molecule by molecule, glucose is broken down in this sequence in cells throughout your body. Energy contained in glucose is set free bit by bit in individual reactions and immediately is placed in short-term ATP storage.

ATP provides the energy needed for human activity.

In this energy-storage reaction, ADP (adenosine diphosphate) adds a phosphate group to form ATP and water:

$$\text{Energy} + \text{ADP(aq)} + \text{HPO}_4^{2-}\text{(aq)} \rightarrow \text{H}_2\text{O(l)} + \text{ATP(aq)}$$

(31 kJ/mol ADP)   hydrogen
phosphate ion

Oxidation of each mole of glucose produces enough energy to add 38 mol of ATP to the cell's short-term energy storage.

Thus, the quantity of energy stored in each ATP molecule is conveniently small and can be used as needed to power individual steps in cellular reactions. Some steps in reactions require less energy than is stored in a single ATP molecule, while others must use the energy from several ATP molecules.

ATP*   ADP*

* The box represents two C–N ring structures plus a sugar unit.

ENERGY FROM ATP

Your body stores and later releases energy from at least 6.0×10^{25} molecules (100 mol) of ATP daily. The structural formulas for ATP and ADP depict ionic forms for the molecules—the forms of ATP and ADP found in body fluids.

The extraordinary efficiency of energy release, storage, and subsequent release is due to enzyme speed and specificity.

YOUR TURN

Enzymes and Energy in Action

Consider this sample problem: Table 2 on page 244 indicated that a 55-kg (120-lb) individual burns 4.2 Cal/min while walking.

a. If this energy is obtained from ATP, and each mole of ATP yields 7.3 kcal (31 kJ), how many moles of ATP provide the energy needed by a 55-kg person to walk for an hour? (1 Cal = 1 kcal)

b. If each mole of glucose burned produces 38 mol of ATP, how many moles of glucose, $C_6H_{12}O_6$, would be needed to provide this energy? How many grams of glucose is that?

First, we find the moles of ATP needed for one hour of walking:

$$4.2 \text{ Cal/min} \times 60 \text{ min/1 h} = 252 \text{ Cal/h} = 252 \text{ kcal/h}$$

$$252 \text{ kcal/h} \times 1 \text{ mol ATP/7.3 kcal} = 35 \text{ mol ATP/h}$$

Because 38 mol ATP can be obtained from one mole of glucose, our need for 35 mol ATP can be met with less than a mole of glucose:

35 mol ATP × 1 mol glucose/38 mol ATP = 0.92 mol glucose

How many grams is that? We first find that a mole of glucose ($C_6H_{12}O_6$) has a mass of 180 g (its molar mass). Because we need less than one mole of glucose, its mass must be less than 180 g:

0.92 mol glucose × 180 g glucose/1 mol glucose = 166 g glucose

Now try these items:

1. Assume your body produces approximately 100 mol of ATP daily.
   a. How many moles of glucose must be consumed to produce this?
   b. How many grams of glucose is this?

2. One minute of muscle activity requires about 0.001 mol of ATP for each gram of muscle. How many moles of glucose must be oxidized for 454 g (one pound) of muscle to dribble a basketball for one minute?

CHEMQUANDARY

CHEW ON THIS PROBLEM!

Amylase, an enzyme present in saliva, catalyzes the release of glucose from starch. Under optimum conditions of temperature and pH, one molecule of amylase can generate as many as 18,000 molecules of glucose each second.

1. See if you can detect a change in the taste of a soda cracker (containing starch) as you chew it for a minute or more before swallowing it.
2. Describe any taste change you detect.
3. Explain what caused the change.

Besides controlling the release and storage of energy, enzymes catalyze reactions that break down large food molecules into building blocks your cells can use. You will be invited to investigate this role of enzymes in the next activity.

B.7 LABORATORY ACTIVITY: AMYLASE TESTS

GETTING READY

In this laboratory activity, the activity of the enzyme amylase will be detected by noting the products of the enzyme-catalyzed reaction.

An enzyme in saliva, amylase, breaks down starch molecules into individual glucose units. The glucose will be detected by Benedict's reagent as a yellow-to-orange precipitate. The amount of precipitate increases with the concentration of glucose.

The activity of amylase will be observed and compared at two temperatures (room temperature and refrigerator temperature) and at several pH values. You will be assigned to a group that will examine either the activity of amylase at room temperature or amylase that has been refrigerated.

Read the procedures below. Prepare a data table with appropriately labeled columns for data collected by each group member.

DAY 1: PREPARING THE SAMPLES

1. Label each of five test tubes (near the top) with the temperature of your assigned test, and with a pH value of 2, 4, 7, 8, or 10.

2. Using pH solutions provided by your teacher, place 5 mL of the pH 2, 4, 7, 8, and 10 solutions in the appropriate tubes.

3. Add 2.5 mL of starch suspension to each tube.

4. Add 2.5 mL of 0.5% amylase solution to each tube. Continue with Steps 5–8 below.

5. Insert a stopper in each tube and shake the tube well.

6. Leave the tubes that are to remain at room temperature in the laboratory overnight as directed.

7. Give your teacher the tubes to be refrigerated.

8. Wash your hands thoroughly before leaving the laboratory.

DAY 2: EVALUATING THE RESULTS

1. Prepare a hot-water bath by adding about 100 mL of tap water to a 250-mL beaker. Add a boiling chip. Heat the beaker on a ring stand above a Bunsen burner.

2. Place 5 mL of Benedict's reagent in each tube. Replace the stoppers—being careful not to mix them up—and shake well.

3. Make sure the tubes are clearly labeled. *Remove the stoppers* and place the tubes in the water bath.

4. Heat the water bath until the solution in at least one tube has turned yellow or orange. Continue heating for 2–3 minutes more.

5. Use a test tube clamp to remove the tubes from the water bath. Arrange them in order of increasing pH.

6. Observe and record the color of the contents of each tube.

7. Share your data with other members of your group.

8. Wash your hands thoroughly before leaving the laboratory.

QUESTIONS

1. At which temperature was each enzyme most effective? At which pH?

2. Make a general statement about the effect of temperature on an enzyme-catalyzed reaction.

The body's proteins—especially enzymes—are very sensitive to changes in temperature, acidity, or concentrations of foreign materials (such as heavy metal ions) in their environment. Yet acids and bases are continually added to and removed from cells and body fluids. How can the body handle these materials without adversely affecting vital body chemistry? Some background on acids and bases will provide the answer in the next part of this unit.

? PART B: SUMMARY QUESTIONS

1. What is meant by homeostasis? How is it related to the normal functioning of your body?

2. Identify some factors that contribute to homeostasis in a healthy human body.

3. Name the five most common elements in the human body, in order of decreasing abundance.

4. Explain to a classmate the form in which each element is found in the body:
   a. chlorine
   b. sodium
   c. iron
   d. magnesium
   e. zinc

5. Most of your body's atoms are carbon, hydrogen, and oxygen, combined into molecules. For which two major classes of nutrients are these atoms mainly used?

6. a. List three characteristics of chemical reactions in cells.
   b. Show how each is important in everyday activities.

7. a. What class of cellular compounds is primarily responsible for the characteristics in Question 6?
   b. Describe how these compounds work.

8. How is energy released from the oxidation of food materials stored in the body?

9. The production of one gram of protein requires about 17 kJ of energy. The molar mass of the protein albumin is 69,000 g/mol. How many ATP molecules must react to energize the formation of one mole (69,000 g) of albumin?

Eating right is one way to keep the body in balance.

ACIDS, BASES, AND BODY CHEMISTRY

C.1 STRUCTURE DETERMINES FUNCTION

In 1883 the Swedish chemist Svante Arrhenius defined an acid as any substance that, when dissolved in water, generates hydrogen ions (H^+). He defined a base as a substance that generates hydroxide ion (OH^-) in water solutions. For example, hydrogen chloride gas, $HCl(g)$, dissolved in water produces hydrochloric acid, $HCl(aq)$, the primary acid in gastric juice (stomach fluids). The hydrogen ion, $H^+(aq)$, makes a solution acidic and reactive enough to behave like a typical acid.

ACIDS AND BASES

Most common acids generally are molecular compounds containing covalently bonded hydrogen atoms. At least one hydrogen atom in each acid molecule can dissociate from the acid molecule in water. The molecular acid dissociates (breaks apart) in water by releasing a hydrogen ion (H^+), leaving behind an electron. The remainder of the original acid molecule—with the hydrogen's electron attached—becomes an anion. As shown in the equations below for three common acids, the *net result is that the molecular acid dissociates in water into a hydrogen ion, H$^+$(aq), and an aqueous negative ion.*

$$\text{Hydrochloric acid: } HCl(aq) \rightarrow H^+(aq) + Cl^-(aq)$$

$$\text{Nitric acid: } HNO_3(aq) \rightarrow H^+(aq) + NO_3{}^-(aq)$$

$$\text{Sulfuric acid: } H_2SO_4(aq) \rightarrow H^+(aq) + HSO_4{}^-(aq)$$

The acidity of a solution is indicated by its pH (see page 442). Acidic solutions have a pH value less than 7; *the lower the pH, the more acidic the solution.*

Many common bases, unlike acids, are ionic compounds. They already contain ions, generally metal ions and hydroxide ions, OH^-. When such a base dissolves in water, the ions separate and disperse uniformly in the solution. The hydroxide ions (OH^-) in solution give basic solutions their characteristic properties. Here are equations for the dissolving of some bases in water:

▶ *In acidic solutions, the hydrogen ion is bonded to water. Such a combination, H(H$_2$O)$^+$, or more commonly H$_3$O$^+$, is called a* **hydronium ion.**
$$H^+ + H_2O \rightarrow H_3O^+$$

Acids, Bases, and Body Chemistry

$$NaOH(s) \xrightarrow{\text{water}} Na^+(aq) + OH^-(aq)$$
sodium hydroxide

$$KOH(s) \xrightarrow{\text{water}} K^+(aq) + OH^-(aq)$$
potassium hydroxide

$$Ba(OH)_2(s) \xrightarrow{\text{water}} Ba^{2+}(aq) + 2\,OH^-(aq)$$
barium hydroxide

Basic solutions have a pH value greater than 7; *the higher the pH, the more basic the solution.*

NEUTRALIZATION

Acidic solutions are characterized by $H^+(aq)$, and $OH^-(aq)$ is characteristic of basic solutions. When equal amounts of each of these ions react, water is produced, which is neither acidic or basic. Water is a neutral compound (pH = 7.0). Consequently, this reaction is known as ***neutralization*** and can be represented by the equation:

$$H^+(aq) + OH^-(aq) \rightarrow H_2O(l)$$

For the neutralization of hydrochloric acid, HCl(aq), by sodium hydroxide, NaOH(aq), the *total* ionic equation is:

$$H^+(aq) + Cl^-(aq) + Na^+(aq) + OH^-(aq) \rightarrow H_2O(l) + Na^+(aq) + Cl^-(aq)$$

Note that Na^+ and Cl^- ions appear on both sides of the equation; they undergo no net change. Because they do not take part in the reaction, they are called ***spectator ions.*** Omitting spectator ions from the total ionic equation leaves a *net* ionic equation, which shows only the reaction that takes place. The net ionic equation for this neutralization reaction is:

$$H^+(aq) + OH^-(aq) \rightarrow H_2O(l)$$

You can see why the acidic and basic properties of the reactants disappear. During neutralization, water, a neutral substance, forms.

Why are some acids and bases harmful to skin, while others—including vinegar, sour milk, and citric acid—are safe for humans? In other words, what determines the strength of a given acid or base? We will explore this question in the next section.

C.2 STRENGTHS OF ACIDS AND BASES

Acids are classified as strong or weak according to the extent they produce hydrogen ions, $H^+(aq)$, when the acid dissolves in water. Similarly, bases are classified as strong or weak according to the extent they produce hydroxide ions, $OH^-(aq)$, in water. When a strong acid dissolves in water, almost every acid molecule dissociates into a hydrogen ion and an anion; scarcely any of the *undissociated* molecular acid remains. In other words, *strong acids are strongly (highly) dissociated in water.* For example, nitric acid, such as found in some types of acid rain, is a strong acid.

Formation of an aqueous nitric acid solution can be represented by this equation:

$$HNO_3(aq) \rightarrow H^+(aq) + NO_3^-(aq)$$

Virtually 100% of the original HNO_3 molecules form hydrogen ions and nitrate ions.

In a weak acid, only a few acid molecules dissociate into hydrogen ions and anions; most of the acid remains *undissociated* in solution as acid molecules. *Weak acids are only weakly (slightly) dissociated.* Acetic acid, $HC_2H_3O_2$, is a weak acid. The equation for formation of an acetic acid solution is written:

$$HC_2H_3O_2(aq) \rightleftharpoons H^+(aq) + C_2H_3O_2^-(aq)$$

A solution of acetic acid contains many more $HC_2H_3O_2$ molecules than $H^+(aq)$ and $C_2H_3O_2^-(aq)$ ions. Note the double arrow written in the equation. One meaning of a double arrow is that the reactants are not completely consumed—the solution contains both reactants and products.

Some acids present in acid rain are weak, such as sulfurous acid (H_2SO_3), nitrous acid (HNO_2), and carbonic acid (H_2CO_3). Others are strong, such as sulfuric acid (H_2SO_4) and nitric acid (HNO_3).

Ionic bases, such as sodium hydroxide (NaOH) and potassium hydroxide (KOH), are strong and highly caustic bases. Their solutions contain only cations and OH^- ions in relatively high concentration. Sometimes, however, the OH^- ion concentration in solution is limited by the low solubility of the base. Magnesium hydroxide, $Mg(OH)_2$, sometimes used as an antacid ingredient, is such a base.

A weak base common in the environment is ammonia, which forms OH^- ions in solution by this reaction:

$$NH_3(g) + H_2O(l) \rightleftharpoons NH_4^+(aq) + OH^-(aq)$$

Now it's your turn.

YOUR TURN

Acids and Bases

1. Hydrochloric acid, HCl, is a strong acid found in gastric juice (stomach fluid).

   a. Write an equation for the formation of ions when this acid is dissolved in water.

   b. Name each ion.

2. Here are two acids found in acid rain. For each, give the name and write an equation for formation of its ions in aqueous solution.

   a. H_2SO_4 (a strong acid): One of its ionization products is responsible for damaging buildings and monuments.

b. H_2CO_3 (a weak acid): Decomposition of this acid gives the fizz to carbonated soft drinks.

3. Each base whose formula is given below has important commercial and industrial applications. Give the name of each and write an equation showing the ions formed in aqueous solution.

   a. $Mg(OH)_2$. This is the active ingredient in milk of magnesia, an antacid.

   b. $Al(OH)_3$. This compound is used to bind dyes to fabrics.

4. A person takes a dose of the antacid $Mg(OH)_2$ to relieve "excess stomach acid," $HCl(aq)$.

   a. Write the total ionic equation for the reaction between $HCl(aq)$ and $Mg(OH)_2(aq)$.

   b. Write the net ionic equation for this reaction.

Under normal conditions, the pH of your blood rarely rises above 7.45 or drops below 7.35. In fact, if the blood's pH varies very much beyond these values, sickness and even death can occur. The normal constancy of blood's pH is quite remarkable when you consider that large quantities of acid enter the blood daily from the foods you eat and from products of their digestion. What chemical mechanisms does your body use for keeping blood pH so constant? Read on.

C.3 ACIDS, BASES, AND BUFFERS IN THE BODY

Much of the acid that invades the human bloodstream comes from the carbon dioxide that cells give off as they oxidize glucose. Carbon dioxide reacts with water in the blood to produce carbonic acid, H_2CO_3, which then forms H^+ and HCO_3^-.

$$CO_2(g) + H_2O(l) \rightleftharpoons H_2CO_3(aq) \rightleftharpoons H^+(aq) + HCO_3^-(aq)$$

The amount of carbon dioxide dissolved in the blood determines the concentration of carbonic acid.

Your body produces 10–20 mol of carbonic acid daily. Other acids, including phosphoric acid (H_3PO_4) and lactic acid ($CH_3CHOHCOOH$) are produced by the digestion of foods. Considering this continuous influx of acids, how does the body maintain a fairly constant pH?

It uses chemical buffers. A **buffer solution** is one that can neutralize the addition of acid (H^+) or base (OH^-) to it. Thus, a buffer contains two components: an acid to neutralize the addition of OH^-, and a base to neutralize H^+ from the addition of an acid.

How does such a buffer solution maintain pH? Here's how: When base (OH^-) is added to a buffer solution, the acid in the buffer provides H^+ ions to neutralize the base, thus preventing a large change in pH. Or, if some acid (H^+) is added to this buffer solution, the base in the buffer neutralizes the added H^+, again preventing a large change in pH.

The blood's primary buffer system is made up of carbonic acid (H_2CO_3) and sodium bicarbonate, $NaHCO_3$. The addition of acid or base to this system results in these reactions:

$$H^+(aq) \quad + \quad HCO_3^-(aq) \quad \rightarrow \quad H_2CO_3(aq)$$
from bicarbonate carbonic acid
added from $NaHCO_3$
acid in the buffer

$$OH^-(aq) \quad + \quad H_2CO_3(aq) \quad \rightarrow \quad HCO_3^-(aq) \quad + \quad H_2O(l)$$
from carbonic bicarbonate ion water
added acid in
base buffer

The reaction products themselves then become part of the buffer system that keeps the pH constant.

Other buffer systems also help maintain body pH balance. One involves the oxygen-carrying protein hemoglobin, a weak acid. Another is the phosphate buffer system composed of $H_2PO_4^-$ and HPO_4^{2-}.

Let's investigate buffer action in the laboratory. You will have the opportunity to witness the unique behavior of buffers, and build background that will help explain your body's ability to maintain "friendly" pH levels in its systems.

C.4 LABORATORY ACTIVITY: BUFFERS

GETTING READY

In this activity, you will prepare and test a bicarbonate buffer. The results of adding acid and base to water will be compared with adding acid and base to the buffered solution.

Prepare a data table like that on page 480.

1. Place a 24-well wellplate on a white sheet of paper. The white paper will help you more easily see the color changes during the experiment. Add 20 drops of distilled water to one well of the wellplate. Add one drop of universal indicator to that well and record the color of the resulting solution in the data table.

2. Add 20 drops of 0.1 M sodium bicarbonate ($NaHCO_3$) solution to another well. Cut off the stem of a clean Beral pipet so that you have a miniature straw. Place the tip of the straw into the $NaHCO_3$ solution. Blow gently through the straw for 2–3 minutes, keeping the tip of the straw in the solution.

3. Remove the straw from the solution, which is now the buffer solution. Add one drop of universal indicator to the buffer solution and record the resulting color.

Preparing the bicarbonate buffer.

4. Add 5 drops of 0.1 M NaOH to the distilled water well used in Step 1. Record the color produced.

5. Counting the number of drops, add 0.1 M NaOH dropwise to the well containing the buffer solution and universal indicator until its color matches the color in the well from Step 4. Record the number of drops required.

6. Repeat Steps 1, 2, and 3 using clean, dry wells.

7. Add 5 drops of 0.1 M HCl to the new well containing distilled water and universal indicator. Record the color produced.

8. Counting the drops, add 0.1 M HCl dropwise to the well containing the buffer solution and universal indicator until its color matches the color in the well from Step 7. Record the number of drops required.

9. Save the solutions until the data table information is completed.

10. Dispose of the solutions as instructed by your teacher.

11. Wash your hands thoroughly before leaving the laboratory.

Data Table				
Step number	Starting material	Added	Color	Approximate pH
1	Distilled water	Universal indicator		
3	Buffer solution	Universal indicator		
4	Distilled water + universal indicator	0.1 M NaOH		
5	Buffer solution + universal indicator	0.1 M NaOH		
7	Distilled water + universal indicator	0.1 M HCl		
8	Bufffer solution + universal indicator	0.1 M HCl		

QUESTIONS

1. a. Did your bicarbonate buffer solution perform as a buffer?

   b. What observations support this answer?

2. a. Compare the number of drops of base (NaOH) needed to reach the same pH when added to water and to the bicarbonate buffer.

   b. Do the same for the HCl.

3. The bicarbonate buffer contains bicarbonate ions (HCO_3^-). Write an equation showing how bicarbonate ions would prevent the pH of the blood from rising if a small quantity of base (OH^-) were added.

C.5 BALANCING pH

Even in many abnormal situations, body pH balance is maintained by the combined action of blood buffers, breathing rate, and the kidneys.

Changes in breathing rate affect the concentration of dissolved carbon dioxide, which, we have seen, is a major source of acid in the blood.

Added in large enough concentrations, any acid or base can overwhelm a buffer. The kidneys help reduce this potential risk by another mechanism for maintaining the body's pH balance on a minute-by-minute basis. Kidneys can excrete excess acidic or basic substances. In addition, the kidneys provide a backup buffer system for abnormal conditions. If the blood is too acidic, the kidneys produce the weak bases HCO_3^- (aq) and HPO_4^{2-} (aq) to neutralize the excess acid. If the blood is too basic, the kidneys neutralize the excess base with the weak acids H_2CO_3(aq) and $H_2PO_4^-$ (aq). The reaction products are then eliminated in urine.

If blood pH drops and stays below 7.35, a condition known as *acidosis* exists. Mild acidosis can occur if the lungs temporarily fail to expel CO_2 efficiently, causing a buildup of carbonic acid in the blood. The central nervous system is affected by acidosis, in mild cases causing mental confusion and disorientation. In extreme cases shock, coma, and even death can result.

When the pH of blood rises and stays above 7.45, *alkalosis* develops. Temporary alkalosis can occur after severe vomiting in which large amounts of hydrochloric acid are lost from the stomach. Chronic alkalosis can produce weak, irregular breathing and muscle contractions. Severe alkalosis can lead to convulsions and death.

Apply what you have learned about pH balance in the body by completing this *Your Turn:*

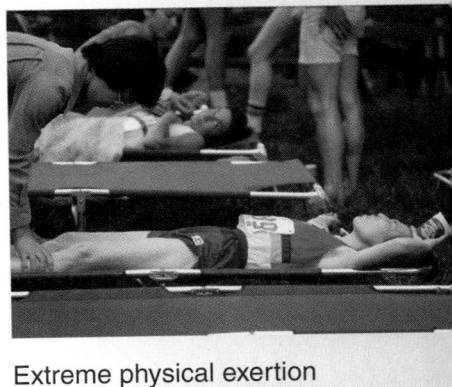

Extreme physical exertion can cause body functions to become unbalanced temporarily.

YOUR TURN

Conditions That Affect pH Balance

1. Hyperventilation is a condition of forced, heavy breathing that typically occurs when an individual is frightened or near physical exhaustion.

    a. How would hyperventilation affect the concentration of CO_2 in the blood?

    b. How would this affect the blood's pH? (Fortunately, the body has an automatic response—fainting—that stops hyperventilation before serious damage is done.)

    c. One first-aid treatment to get blood back to the proper pH is to have the hyperventilating individual breathe into a paper bag for a short time. Comment on how this treatment works chemically to re-establish the blood's pH.

2. If you held your breath, how would this affect

    a. the concentration of CO_2 in the blood?

    b. the pH of the blood?

3. Strenuous muscle activity, such as long-distance running, can cause lactic acid to accumulate in the blood, resulting in a wobbly feeling in leg muscles. How would the accumulation of lactic acid affect blood pH?

4. Cardiac arrest occurs when the heart stops and blood circulation is halted, but cellular reactions continue. When this occurs, doctors often inject sodium bicarbonate solution, $NaHCO_3$(aq), directly into the heart muscle, even before restarting the heart.

   a. What effect would cardiac arrest have on blood pH? Why?

   b. How would the injection counteract this effect?

5. What condition of acid–base imbalance might be caused by a large overdose of aspirin (acetylsalicylic acid)?

We have explored the chemical composition of the human body and some mechanisms within it. It is time to explore some chemistry at the body's surface, where we have more control over what happens.

? PART C: SUMMARY QUESTIONS

1. What is the chemical difference between
   a. an acid and a base?
   b. a strong acid and a weak acid?
   c. a strong base and a weak base?

2. You are given a sample of a compound unknown to you and told that it contains hydrogen. Can you conclude that the compound is an acid? Explain.

3. Describe the composition and function of a buffer system.

4. Name four conditions that can overload the body's buffer systems. In each case describe the chemistry of the overload and its effect on blood pH.

5. A solution of lactic acid (HLac) and sodium lactate (NaLac) is sometimes used in hospitals as a buffer. Write a chemical equation to represent how this buffer would compensate for:

   a the addition of acid to it.

   b. the addition of base to it.

EXTENDING YOUR KNOWLEDGE (OPTIONAL)

Typically, a buffer contains relatively high concentrations of a weak acid and a salt of that acid. The anion of the salt acts as a base by accepting H^+ ions.

In a library find information about the Bronsted–Lowry acid–base theory. Then, apply this information to the following questions.

   a. Identify the Bronsted–Lowry acid or base in each pair of compounds. Classify the acid or base as a strong or weak acid or base.

   (1) KCl and HCl
   (2) NaOH and H_2O
   (3) $NaNO_3$ and HNO_3
   (4) $NaC_2H_3O_2$ and $HC_2H_3O_2$

   b. Which one of the four pairs would make the best buffer system? Why?

D CHEMISTRY AT THE BODY'S SURFACE

Keeping clean is important, not just for the sake of appearance, but for good health. Considerable chemistry is involved in personal cleanliness, and that chemistry is—at least in large part—under your personal control. Understanding this important aspect of your body's "external chemistry" will help you make better decisions concerning your well-being.

First, let's consider the skin's surface and substances that can keep it clean without damaging it.

D.1 KEEPING CLEAN WITH CHEMISTRY

Some glands in human skin produce oils that keep skin flexible. Chapped lips show how uncomfortable a shortage of these oils can be. On the other hand, our skin must be washed fairly often to prevent the build-up of excess oil.

Sweat is another gland-produced skin secretion. We sweat to keep our body temperatures at a relatively constant level during hot weather or during heavy exercise. Sweat—mostly water—is about 1% sodium chloride. If sweat were just salty water, there would be little need to bathe with soap. However, areas of the body where sweat glands are concentrated also secrete the oils that produce body odors. These odors quickly become offensive as bacteria convert the oils to foul-smelling compounds.

So, keeping clean involves both removing dirt and grease we pick up from outside sources, and removing sweat and excess oil from the skin before much bacterial action occurs. Mainly, though, skin cleaning involves removing oils and greases.

A variety of cleaning agents remove oil and dirt by dissolving them. For example, a water solution of ammonia, $NH_3(aq)$, reacts with grease, making ammonia an effective cleaner for the kitchen, but not one recommended for skin.

Water alone will not dissolve oil or grease, because water is polar and oil and grease are nonpolar substances. Therefore, bathing or showering without soap fails to clean your skin thoroughly; soap is needed, as well as water.

To understand soap's cleaning ability, it is useful to review the molecular basis of solubility. The polarity of water molecules gives rise to hydrogen bonding among water molecules. *Hydrogen bonding* is an attraction between a polar (partially positive) hydrogen atom and a strongly electron-attracting atom like nitrogen or oxygen having an unshared electron pair. The hydrogen atom can be in one molecule and the nitrogen or oxygen in

Keeping clean may be important, but it is not always possible.

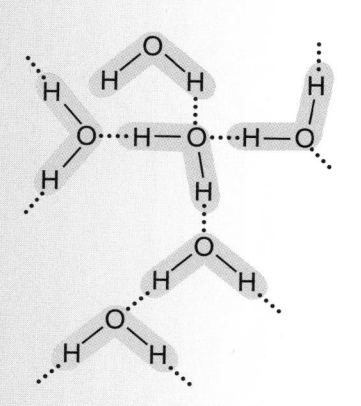

Figure 5 Hydrogen bonding in water.

$$CH_3 - \overset{\displaystyle OH}{\underset{\displaystyle H}{\overset{|}{\underset{|}{C}}}} - CH_3$$

Isopropyl alcohol

another molecule. Very large molecules, like proteins, can have the H and N or O atoms in the same molecule, and then hydrogen bonding also occurs within the molecule or between protein molecules.

The partially positive hydrogen atoms of one water molecule are attracted to the unshared electron pairs on the partially negative oxygen atoms of nearby water molecules (Figure 5). The dots in Figure 5 represent the hydrogen bonds between water molecules.

Many substances dissolve in water because they are polar. Rubbing alcohol (isopropyl alcohol) is an example. Its molecules have a polar —OH group that hydrogen bonds with water, causing the alcohol to dissolve in water. When an ionic compound like salt (NaCl) dissolves in water, the Na^+ and Cl^- ions in salt are attracted to the polar water molecules. The partially negative oxygen atom in water molecules align with the positive sodium ions; the partially positive hydrogen atoms in water are attracted to the negative chloride ions. The attraction between water and the Na^+ and Cl^- ions is enough to pull them out of solid NaCl, causing them to dissolve.

Hydrogen bonds between water molecules are much weaker than the H–O covalent bonds that hold hydrogen and oxygen atoms together within water molecules. But hydrogen bonds are strong enough to pull water molecules into clusters, causing water to "bead" on a flat surface. By contrast, nonpolar oil molecules have no hydrogen bonds and are less attracted to each other. Nonpolar substances, on the other hand, do not dissolve in water. When water and nonpolar substances are mixed, hydrogen bonds hold the water molecules together so tightly that the nonpolar molecules are forced to form a separate layer. Oils and greases are nonpolar and insoluble in water because they contain large numbers of C–H bonds without any, or with only a few, oxygen atoms in their molecules. The nonpolar molecules have a greater attraction for each other than for polar molecules such as water.

Earlier you explored the meaning of "like dissolves like" and found that, in general, polar or ionic substances dissolve in water, and nonpolar substances dissolve in nonpolar solvents. Compatible attractive forces are the key to solubility. As you will see, some large molecules can be soluble in polar solvents at one end and in nonpolar solvents at the other end.

Complete the following activity to review the connection between polarity and solubility.

YOUR TURN

Polarity and Solubility

1. Indicate whether each substance below is nonpolar, polar, or ionic. Then predict whether it is soluble in water or in nonpolar solvents. Base your reasoning on bond polarity, charge, molecular shape, and the like-dissolves-like rule.

a. NaOH, sodium hydroxide (a base used in making soaps)

b. Diethyl ether (an industrial solvent)

$$H-\overset{\displaystyle H}{\underset{\displaystyle H}{\overset{|}{\underset{|}{C}}}}-\overset{\displaystyle H}{\underset{\displaystyle H}{\overset{|}{\underset{|}{C}}}}-O-\overset{\displaystyle H}{\underset{\displaystyle H}{\overset{|}{\underset{|}{C}}}}-\overset{\displaystyle H}{\underset{\displaystyle H}{\overset{|}{\underset{|}{C}}}}-H$$

c. CH_2OHCH_2OH, ethylene glycol (antifreeze)

$$HO-\overset{\displaystyle H}{\underset{\displaystyle H}{\overset{|}{\underset{|}{C}}}}-\overset{\displaystyle H}{\underset{\displaystyle H}{\overset{|}{\underset{|}{C}}}}-OH$$

d. KNO_3, potassium nitrate (a fertilizer)

2. Ethanol (CH_3CH_2OH) is a water-soluble organic molecule. Its solubility in water is due to its small size and its polar —OH group, which allow for hydrogen bonding with water. Cholesterol ($C_{27}H_{46}O$), an animal fat associated with heart disease, is not water-soluble even though, like ethanol, it contains an —OH group.

a. Why isn't cholesterol water-soluble?

b. Would you expect cholesterol to dissolve in nonpolar solvents? Why?

3. Consider the chemical structures of a fatty oil (a triglyceride) and glucose:

$$CH_2-O-\overset{\displaystyle O}{\overset{\|}{C}}-CH_2-CH_2-CH_2-CH_2-CH_2-CH_2-CH_2-CH_3$$
$$CH_2-O-\overset{\displaystyle O}{\overset{\|}{C}}-CH_2-CH_2-CH_2-CH_2-CH_2-CH_2-CH_2-CH_3$$
$$CH_2-O-\overset{\displaystyle O}{\overset{\|}{C}}-CH_2-CH_2-CH_2-CH_2-CH_2-CH_2-CH_2-CH_3$$

A triglyceride

Glucose

a. Which will dissolve in water? Why?

b. Which will dissolve in gasoline? Why?

c. How do the two molecules differ?

4. Sodium lauryl sulfate, $Na^+CH_3(CH_2)_{11}OSO_3^-$, is a synthetic detergent. Do you think this compound is soluble in water? In oil? Explain your answer in terms of the rule "like dissolves like."

Soap works because it is strongly attracted to both oil and water.

Skin cleansers, soaps, and detergents have molecules with special ends that affect their solubility. At one end, these molecules are polar; at the other end, nonpolar. For example, the long hydrocarbon chain in the soap sodium stearate (Figure 6) dissolves in oil, but not in water. However, the oxygen atom at the other end of the structure has a negative charge, which is strongly attracted to polar water molecules.

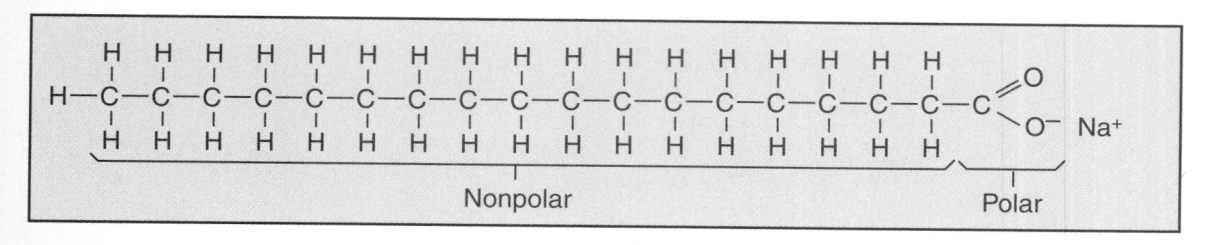

Figure 6 Sodium stearate, a soap. Commercial soaps are mixtures of structures such as these.

Thus, sodium stearate and other soaps and detergents are strongly attracted to both water and oil. One end (the polar part) of the molecule dissolves in water; the other end (the nonpolar portion) dissolves in the oil, causing the oil and water to mix. The long, skinny molecules cluster around an oil droplet, with their oil-soluble ends pointing inward into the oil, and their charged ends pointing out into the water (Figure 7, page 487). This enables wash water containing soap or detergent to remove oil and grease. The oily droplets enter the wash solution and are rinsed away.

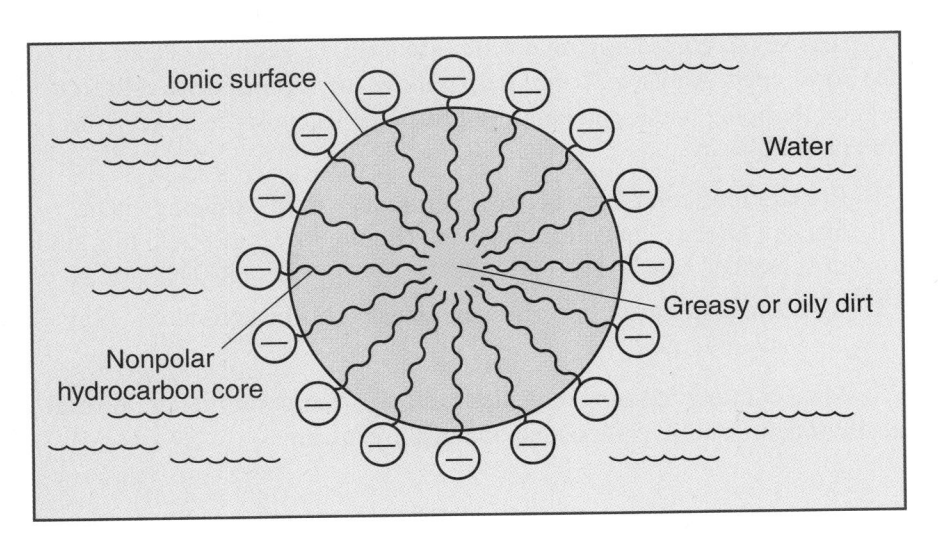

Figure 7 Spherical cluster of soap molecules in water. This is how soaps and detergents remove grease and oil from soiled surfaces.

Today, most people buy commercial soaps and detergents. In previous centuries, however, people often made their own soap from lye (sodium hydroxide, NaOH) and animal fat (glyceryl tripalmitate, for example). Glyceryl tripalmitate reacts with lye (sodium hydroxide) to form the soap sodium palmitate:

$$CH_3-(CH_2)_{14}-\overset{\overset{O}{\|}}{C}-O-\overset{\overset{H}{|}}{C}-H$$
$$CH_3-(CH_2)_{14}-\overset{\overset{O}{\|}}{C}-O-\overset{|}{C}-H \ + \ 3 \ NaOH\longrightarrow$$
$$CH_3-(CH_2)_{14}-\overset{\overset{O}{\|}}{C}-O-\underset{\underset{H}{|}}{\overset{|}{C}}-H$$

glyceryl tripalmitate

$$3 \ Na^+ \ {}^-O-\overset{\overset{O}{\|}}{C}-(CH_2)_{14}-CH_3 \ + \ Glycerol$$

sodium palmitate

Because lye soap contains unreacted lye, it can damage skin.

CHEMQUANDARY

BENZOYL PEROXIDE AND ACNE

Many adolescents produce excess skin oil (sebum). Dead cells can clump together and close the pores in oil glands. Bacteria that consume skin oil can then cause the small infections seen as acne pimples.

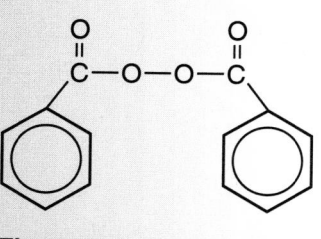

Figure 8 Benzoyl peroxide structure. This substance is used in controlling acne.

The key to controlling mild acne is to keep the skin very clean and the pores open. Abrasive lotions help remove the layer of dead cells, dirt, and oil. Keeping hands and hair away from the face also helps keep dirt and bacteria away.

1. Preparations containing benzoyl peroxide, a mild oxidizing agent, also are effective in controlling acne (Figure 8). How would this kind of substance help?

2. In severe acne cases, doctors sometimes prescribe antibiotics. Why are they effective?

Most commercial acne remedies offer only temporary relief. Severe cases of acne require professional medical treatment.

D.2 SKIN: IT'S GOT YOU COVERED

Skin represents your body's contact with the outside world as well as your barrier against it. It protects you from mechanical damage and from harmful bacteria. Sweating helps keep your body temperature relatively constant. The skin's sense of touch provides important information about the physical environment. And, it even plays a role in communicating human emotions.

Human skin is about 3 mm thick and has two major layers (Figure 9). The inner layer, the *dermis,* contains protein fibers that give the skin strength and flexibility. Sweat and oil glands are located in the dermis, with tubes (ducts) leading to pores on the skin surface.

The cells of the dermis continually divide, pushing the outer layer of cells outward to form the *epidermis,* the skin's outer layer. Oil secreted by glands in the dermis forms a protective film on the surface, keeping epidermal cells soft and preventing excessive water loss. This oily layer is slightly acidic, thereby protecting protein fibers whose structures can be altered by alkaline conditions.

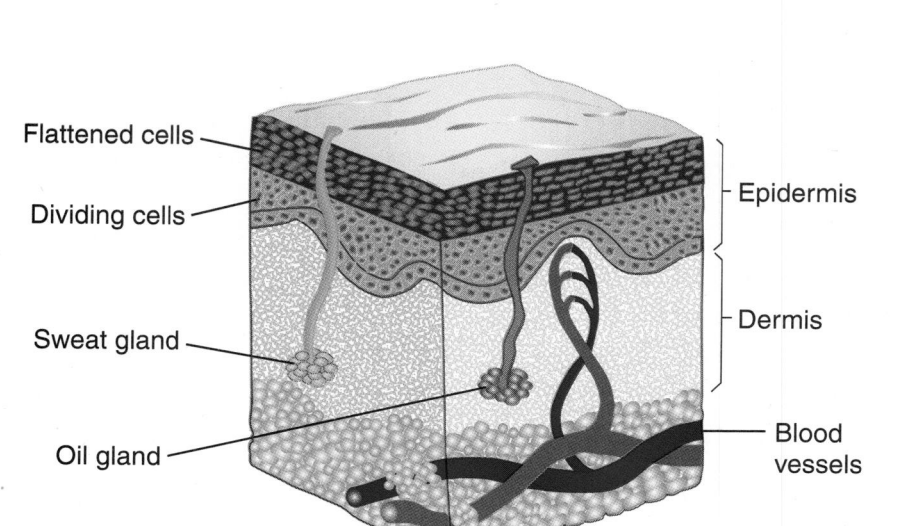

Figure 9 Human skin.

Dead cells at the skin's surface continually flake off; usually the flakes are too small to be visible. Dandruff is composed of dead epidermal cells that stick together, falling off in relatively large pieces. Dandruff has nothing to do with germs, although some people waste considerable time and money trying to "cure" dandruff with germ-killing shampoos. The best dandruff shampoos contain substances that prevent dead epidermal cells from clustering.

Maintaining healthy skin is a fairly simple matter. Eat wisely. Keep clean, but don't overdo it. Too much bathing will remove too much of your skin's oils, drying out and possibly damaging your skin. Most of all, avoid spending too much time in the sun.

D.3 PROTECTING YOUR SKIN FROM THE SUN

The sun's radiation provides energy and light that we need and welcome, but its ultraviolet component can seriously damage exposed skin. The skin-darkening material that forms during a suntan is *melanin,* the same black pigment responsible for black hair and naturally dark skin. When an ultraviolet photon of suitable energy strikes a melanin-producing cell, it activates an enzyme that triggers the oxidation of the amino acid tyrosine. Less-energetic ultraviolet photons cause the modified tyrosine to form a polymer. The final melanin structure is a tree-like branching of chains of modified tyrosine.

SKIN DAMAGE

The most immediate effect of excessive exposure to sunlight is sunburn. Even minimal sunburn injures cells and temporarily swells the dermal blood vessels. It may also increase the rate of cell division. A bad sunburn kills many cells. The skin blisters and the dermal cells divide more rapidly, thickening the epidermis.

Ultimately, tanning—and particularly sunburning—can lead to wrinkled skin, even to skin cancer. Melanoma skin cancers are believed to be linked to excessive exposure to the sun. These cancers are rare, but in very many cases are fatal, killing about 10,000 yearly in the United States. Scientists have conclusively linked the less-dangerous nonmelanoma cancers to overexposure to sunlight. Such skin cancers have increased dramatically over the past decade in the United States, annually accounting for nearly 600,000 new cases. Although nonmelanoma skin cancers are rarely fatal, they are to be avoided.

Ultraviolet light and cancer are linked because of changes in cell molecules. Ultraviolet (UV) radiation that reaches the Earth's surface from the sun has sufficient energy to energize outer electrons in molecules and even break or rearrange some types of covalent bonds. These breaks are most threatening and destructive when they occur in DNA, the genetic-instruction molecules. Although cells have mechanisms for repairing many of these breaks, if the breaks go unrepaired they can cause mutations in DNA. This kind of damage is greatest in rapidly dividing tissue, since the

▶ *This UV radiation is in the 280–400 nm wavelength range (1 nm = 10^{-9} m)*

mutations are quickly passed on to daughter cells. Skin cells, for instance, are replaced monthly. Mutations caused by ultraviolet exposure are partly responsible for age spots, wrinkling, and general aging of skin. Eventually such mutations can lead to cancer. As you may remember from the Air unit, increased numbers of skin cancer might arise from a further thinning of the stratospheric ozone layer that shields us from damaging ultraviolet radiation.

Dark-skinned individuals are better protected than light-skinned people from the damaging effects of solar radiation. Melanin in their epidermis absorbs much of the sun's ultraviolet radiation, dispersing the energy and preventing it from damaging living, dividing cells in layers below the epidermis. But getting a tan isn't very protective if you are naturally light-skinned.

Your susceptibility to the sun's UV radiation is measured by a unit dermatologists call a *minimal erythemal dose (MED)*. An MED is defined in terms of the number of hours it takes sunlight in a given time and place to turn skin barely pink. A low MED means a higher susceptibility to sun exposure. Conversely, if you have a high MED, it would take you a longer time to redden under the same exposure conditions. For example, an individual with a low MED will redden in 0.33 hours (20 min) at noon on a clear day in midsummer, whereas a person with an average MED will need 0.6 hours (35–40 min) for the same effect to develop. Persons with a very high MED can wait about 1.2 hours for the same results.

Location is another factor in determining MEDs. An individual with an average MED who can take 0.6 hours of sun at 40–44° north latitude (the latitude of New York City, Northern California, or Chicago) could stand only 0.33 hours in Miami, and even less in the tropics near the equator.

To increase the public's awareness of the danger caused by overexposure to the sun's radiation, the National Weather Service issues a daily Ultraviolet Index that appears nationally in many newspapers and on television weather broadcasts (Figure 10, page 491). The Index values are based on how much time it takes for skin damage to occur as shown in Table 4.

Table 4	UV Index	
Index	Minutes for some damage to occur in light-skinned people who "never tan"	Minutes for some damage to occur in dark-skinned people who "never tan"
0–2 Minimal	30 minutes	More than 120 minutes
3–4 Low	15–20 minutes	75–90 minutes
5–6 Moderate	10–12 minutes	50–60 minutes
7–9 High	7–8.5 minutes	33–40 minutes
10–15 Very High	Less than 4–6 minutes	20–30 minutes

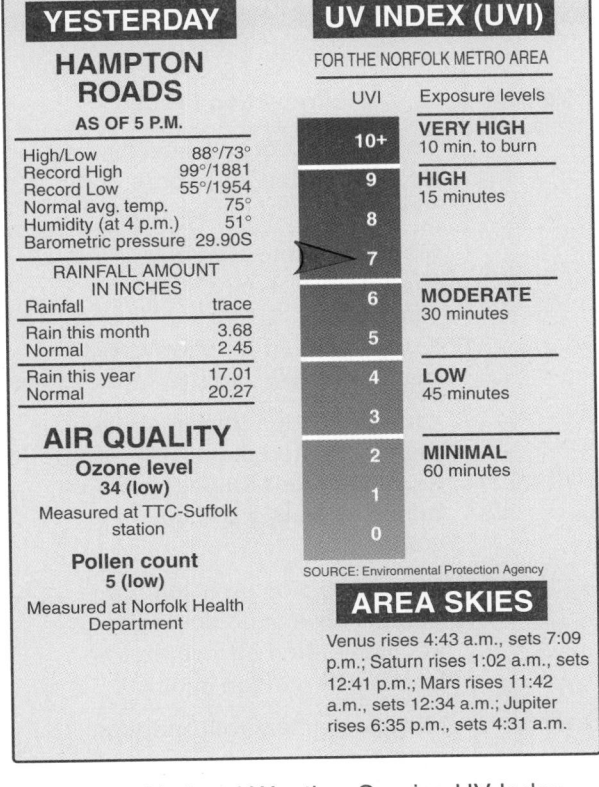

Figure 10 National Weather Service UV Index.

The sun's ultraviolet radiation is made up of UVA and UVB. UVB is the more energetic radiation, which causes sunburn. The somewhat less energetic UVA is still harmful. UVA is suspected in the skin aging and premature wrinkling caused by overexposure.

SUNSCREENS

Sunscreens having a high enough *SPF (sunscreen protection factor)* can reduce possible damage from harmful exposure to the sun. They generally contain compounds that absorb UV radiation. Table 5 (page 492) describes the level of protection offered by various SPF ratings that range from 2 to 15 or more.

A 15 SPF generally is adequate for most people. Although sunscreens with SPFs higher than 15 are commercially available, they do not screen an *appreciably* greater amount of UV radiation. For example, an SPF 15 filters about 93% of UV radiation compared with about 97% for SPF 30.

Which sunscreen is right for you? You can quickly estimate it by multiplying your MED by the SPF number to get the number of hours of outdoor protection the sunscreen will give you. (This assumes that the sunscreen is not washed or rubbed off during that time.) Let's say that you have an average MED (0.6 h) and use a 15 SPF sunscreen. This would give you 0.6 h × 15 = 9 hours of peak time protection at 40–44° north latitude. A low-MED person would only get 4–5 hours using the same sunscreen (0.33 h × 15 = 4.5 h).

Table 5	Sunscreen Protection Factor (SPF)	
Rating of Sun Protection Product	**SPF Values**	**Protection Level**
Minimal	2–4	Least protection; permits tanning. Recommended for people who rarely burn, but tan profusely.
Moderate	4–6	Moderate protection from sunburning; permits some suntanning. Recommended for people who burn minimally and tan well.
Extra	6–8	Extra protection from sunburn; permits limited tanning. Recommended for people who burn moderately and tan gradually.
Maximal	8–15	High protection from sunburn; permits little or no tanning. Recommended for people who burn easily and tan minimally.
Ultra	15 or more	Most protection from sunburn; permits no tanning. Recommended for people who burn easily and rarely tan.

Zinc oxide provides effective protection against solar radiation.

Some sunscreens provide protection in another way. Such preparations, sometimes labeled incorrectly as "chemical free," contain zinc oxide or titanium dioxide. These white pigments simply block or scatter UV light rather than absorb UV radiation.

You can explore the differences in sunscreen protection in the following activity.

D.4 LABORATORY ACTIVITY: SUNSCREENS

GETTING READY

In this activity you will observe the ultraviolet screening abilities of some commercial sunscreens having different SPFs (Table 5). Each product to be tested will be spread on a transparent plastic sheet that allows solar radiation to pass through; that is, the sheet itself does not absorb ultraviolet radiation. Mineral oil and zinc oxide will be used as controls in the activity. Mineral oil does not absorb sunlight and thus has an SPF of zero. Because zinc oxide completely blocks sunlight, it has an infinitely high SPF value. The relative darkness on the sun-sensitive paper created by the different sunscreens will be compared with the effects of the mineral oil and zinc oxide. After testing four samples with known SPF values, you will determine the SPF value of an additional sample.

Read the procedure below and prepare a data table in your notebook for recording your observations.

PROCEDURE

1. Use a pen or a wax crayon to draw seven widely separated circles on an acetate sheet according to instructions given by your teacher. Label the first circle as MO (mineral oil) and the second as ZnO (zinc oxide). Label the next four circles with the SPF numbers of the four known samples. Label a seventh circle with the code number of your unknown. You should now have an acetate sheet ready for testing seven samples: the two controls; four samples with known SPFs; and an unknown whose SPF you will determine (Figure 11).

2. Place one drop of each substance to be tested on the appropriate circles on the sheet. Place a second plastic sheet over the first. Gently lay your textbook on the second plastic sheet to smooth the samples as evenly as possible between the two sheets.

3. We assume the day is sunny! (If it is not, your teacher will indicate an alternative method.) Take your prepared samples and a *covered* piece of sun-sensitive paper outside. Lay the sun-sensitive paper on a flat surface in the sun before removing the cover from the paper. Lay the two acetate sheets on top of the paper.

4. Keep the samples in the sun until the visible regions of the sun-sensitive paper have faded to a very light blue color. The length of exposure time depends on the sunniness of the day.

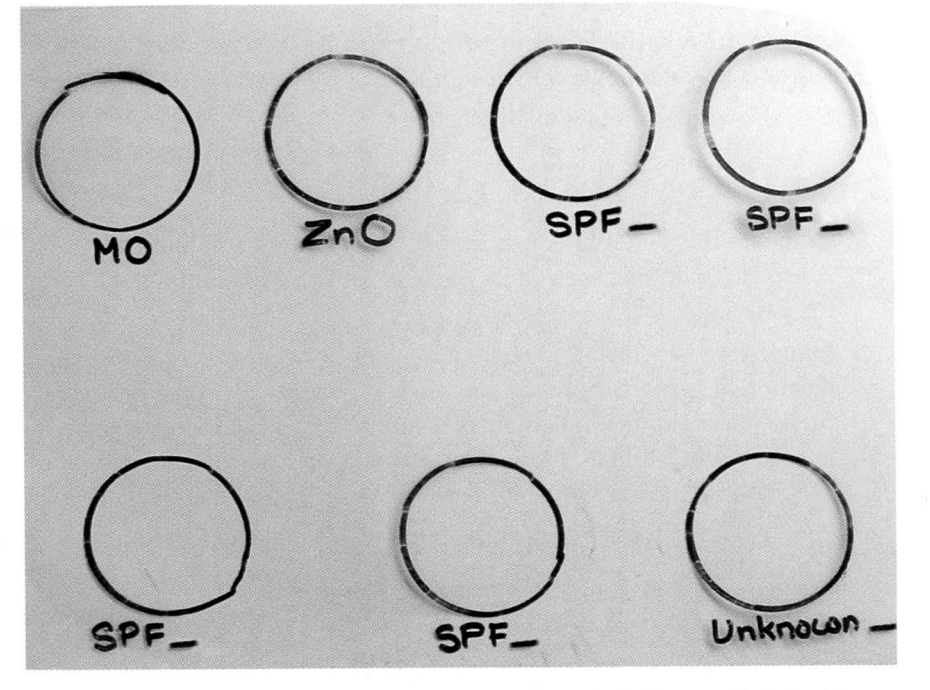

Figure 11 Acetate sheet marked with wax pencil circles and labeled for placement of mineral oil, zinc oxide, sunscreen lotions, and an unknown.

Students performing laboratory activity.

5. After the sun-sensitive paper has faded to the proper color, remove the acetate sheets from the paper. Quickly replace the cover on the sun-sensitive paper.

6. Back in the laboratory, place the sun-sensitive paper under cold water for 2–3 minutes to "develop" the images.

7. Record the relative darkening of the paper by the controls, the four sunscreens, and the unknown.

8. Dispose of the samples as instructed.

QUESTIONS

1. Compare the darkening of the paper allowed by the unknown sunscreen with that created by those of known SPF. Estimate the SPF for your unknown.

2. Suntanning is believed to be a major cause of skin cancer. However, most of us spend time in the sun anyway, trading off some probable future harm for the immediate pleasure of being outdoors, or of having a beautiful tan. Identify other trade-offs we make in everyday life that involve risks and benefits.

3. People with naturally dark skin have a lower incidence of skin cancer. Why?

4. Why is it inaccurate for any sunscreens to be designated as "chemical free?"

Ultraviolet radiation has its dangers, but a little ultraviolet light is actually necessary for health.

D.5 GETTING A "D" IN PHOTOCHEMISTRY

During the Industrial Revolution, many children were forced to work indoors for as long as 12 hours daily and had very little exposure to the sun. Many of these children developed *rickets,* a disease that caused their bones to become soft and easily deformed. Resting indoors only made the disease worse. Surviving victims frequently had bowed legs and other crippling disorders.

Doctors soon realized that children living outside cities rarely had rickets. City children having rickets were quickly cured if they were sent to the country. These clues eventually led to the discovery that sunlight and vitamin D prevented rickets. Vitamin D carries calcium ions (Ca^{2+}) from the digestive tract into the blood and deposits them in bones. Scientists learned that sunlight helps the body produce vitamin D from another compound, 7-dehydrocholesterol, which is present in all body cells (Figure 12).

Now, hardly anyone gets rickets. The quantity of sunlight needed to prevent calcium deficiencies varies with the season and latitude, and with the people themselves. In Boston during the summer, for example, a daily 15-minute exposure on the face alone is all that is needed to prevent rickets among Caucasians. In northern latitudes people of African descent suffer from vitamin D deficiency more often than Caucasians do. Can you explain this?

Some food additives also help prevent vitamin D deficiency. The body can convert these substances into vitamin D. For example, the ingredients listed on some milk cartons include the compound irradiated ergosterol, which has a chemical structure almost identical to that of 7-dehydrocholesterol. Today, doctors are concerned that the American diet may actually contain too much vitamin D. In excess it can cause unwanted calcium deposits, such as kidney and gall bladder stones.

CHEMQUANDARY

VITAMIN D SOLUBILITY

Look at the structure of vitamin D in Figure 12. In which body material, fat or blood, will this molecule tend to dissolve? Does this help explain why vitamin D "overdoses" might cause problems?

Figure 12 Sunlight produces vitamin D from a chemical relative of cholesterol. This is one example of how small differences in molecules may cause great differences in their chemical behavior.

D.6 OUR CROWNING GLORY

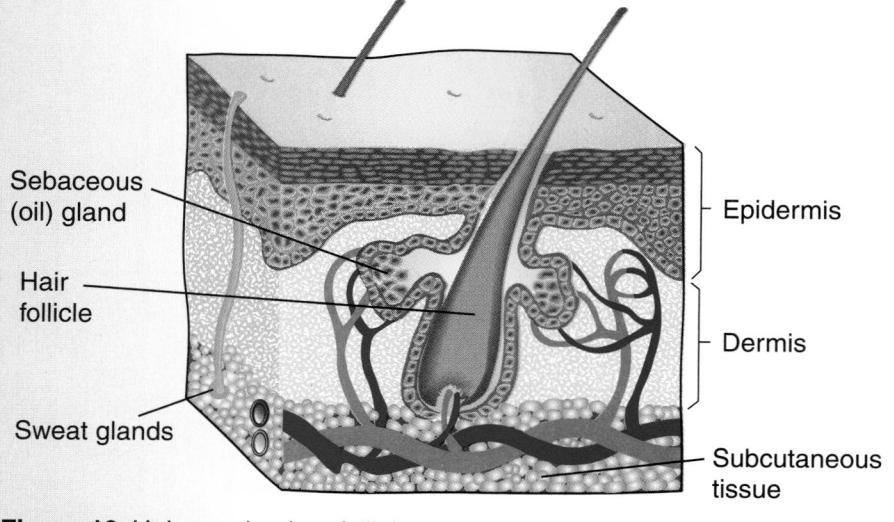

Among personal features in which people invest their time, money, and concern, hair ranks first for many individuals. It is—at various times—washed, waved, trimmed, bleached, curled or straightened, colored, cut, and combed.

Each hair grows out of a small, deep pocket in the scalp called a *follicle* (Figure 13). Although hair is nonliving fibrous protein, the growth of hair begins in living cells that are modified skin cells. These cells emerge from the bottom of the follicle, and move outward towards the surface of the scalp in the same way that dermal cells move outward to become epidermal cells. But before they reach the surface, they die, leaving behind only the amino acid chains they have added to the bottom of the hair.

The structure of hair is quite complex, as shown in Figure 13 and Figure 14. A protein chain, **alpha keratin,** is the basic structural unit. Three coiled chains of alpha keratin form a supercoiled, three-strand rope; eleven such three-stranded ropes make up a microfibril. Each hair is made of many bundles of microfibrils and is surrounded by a tough outer layer, the cuticle. The cuticle consists of several layers of flat cells that overlap like shingles on a roof.

Hair flexibility is made possible by protein chains in the cuticle layer separating from one another, shifting position slightly without breaking. This flexibility makes it possible to style hair, but it can also lead to hair damage. You'll test the condition of your own hair as part of the following laboratory activity.

Magnified human hair (1,000 ×).

Figure 13 Hair growing in a follicle.

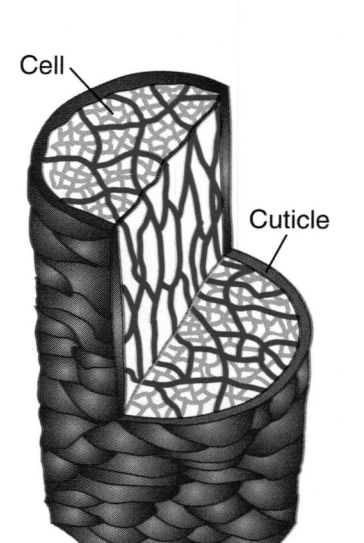

Figure 14 The microscopic structure of a hair.

D.7 LABORATORY ACTIVITY: CHEMISTRY OF HAIR

GETTING READY

Healthy hair has a well-developed structure. Scalp oils and the hair's natural acid coating protect this structure. The same kinds of oil glands that lubricate skin keep the scalp and hair from drying out. In addition, oil helps protect the scalp from growth of bacteria and fungi. However, many agents can remove the oils, destroy the acid coating, and in extreme cases even damage the hair structure itself. Because hair is nonliving, it has no way to repair itself.

To satisfy our hair styling needs—that is, to change the structure of our hair and make it curly or straight—we use some agents that may cause damage. In Part 1 you will test the effectiveness of various solutions in altering hair's structure. In Part 2 you will test some observable characteristics of your own hair to determine its health.

Read Parts 1 and 2 and prepare two data tables, one for each part.

PROCEDURE

Part 1: *Effects of Various Solutions on Hair*

For this part of the activity, you will be assigned to a four-student group. Each group member will be responsible for investigating the effect of one of these four treatments on a small hair sample:

Treatment A   pH 4, hydrochloric acid, HCl(aq)
Treatment B   pH 8, sodium hydroxide, NaOH(aq)
Treatment C   permanent wave solution
Treatment D   (1) permanent wave solution and
              (2) permanent wave neutralizer

1. Obtain a bundle of at least 20 hairs. Using a rubber band, securely fasten the hair to the end of a wooden splint as shown in Figure 15.

2. If you are completing Treatment A, B, or C, place 2 mL of your solution in a labeled test tube. If you are using Treatment D, place 2 mL of permanent wave solution in one test tube and 2 mL of permanent wave neutralizer in a second tube. Label both tubes.

3. For Treatment A, B, or C, place a splint with hair attached in the solution for 15 min. For Treatment D, place the splint in the permanent wave solution; allow it to remain for 10 min. Then rinse the hair under running water and place the splint in the test tube with the neutralizer for 5 min.

4. Begin work on Part 2 while the hair samples remain in the solutions as specified in Step 3.

5. After the specified times, remove the splints from the tubes and place them on paper toweling. Dab lightly with a folded paper towel to absorb excess moisture. Label the towel with the name of the solution in which the hair was immersed. Allow the hair to dry partially. (You can hasten drying with a light bulb or hair dryer.)

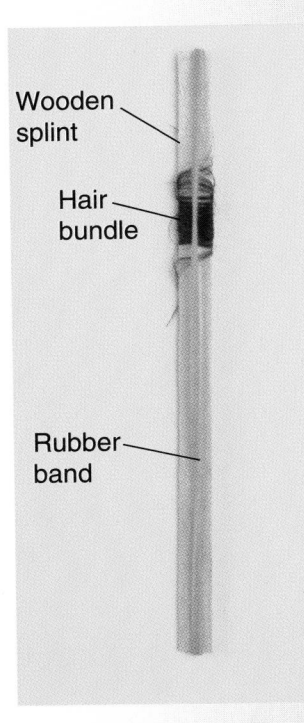

Figure 15 Hair bundle attached to wooden splint.

Wooden splint

Hair bundle

Rubber band

6. When the hair is fairly dry, remove the rubber band and carefully unwind the hair bundle from the stick. Do not disrupt the curl.

7. Compare the extent of curl in this hair sample with that observed in hair from the other three treatments. Rank the curliness of the four hair samples in 1-2-3-4 order, where 1 represents the curliest sample.

8. Remove about six hairs from your hair bundle. Grasp both ends of one hair and pull gently until it breaks. Repeat this test using a bundle of five hairs. Record your observations and share them with other members of your group.

9. Thoroughly moisten your bundle of hair under a gentle stream of tap water.

10. Squeeze out excess water from the hair by gently pressing as you pull the hair between two fingers. Do not squeeze or pull too hard.

11. Allow the hair to dry. (Again, you may hasten drying with a light bulb or hair dryer.)

12. Rank the extent of curl still present in each of the four hair samples.

13. Repeat the stretch test (Step 8) on one hair, and on a group of five hairs from your bundle. Record your observations and share with other members of the group.

Part 2: *Testing Your Hair*

1. Obtain a single 15-cm-long hair from your head by clipping it with scissors. If your hair has been colored, styled, permed, or otherwise treated, take your hair sample from just behind your ear (where it is least likely to have been affected by treatment.)

2. Hold one end of the hair sample between the thumb and index finger of one hand. Using your other hand, hold the hair between your thumbnail and index fingernail. Hold tightly as you slide your nails down the length of the hair, as though you were curling a gift ribbon.

3. Now hold the strand at both ends and gently pull. Stretch the strand for 15 seconds, then release and observe. If the curl is gone, your hair is structurally weak. The extent to which the curl returns indicates the relative strength of your hair structure. Record your observations.

STRETCH TEST

1. Obtain another strand of hair from your head. Hold the strand ends between the thumb and index finger of each hand; stretch the hair gently. Do not jerk it. Healthy hair will stretch up to 30% (at average humidity), somewhat like a rubber band. Unhealthy hair has little or no stretch. A seriously damaged hair will break easily. Record the results.

2. Obtain another strand of hair. Wet it with tap water and repeat Step 1 with this wet strand.

3. If time permits, repeat Steps 1 and 2 with a strand of hair donated by a classmate.

CUTICLE TEST

1. Hold a small bundle of your hair about 3 cm away from your scalp with your thumb and finger.

2. Grasp the bundle near the outer end with the thumb and finger of the other hand, and pull your hand along the bundle toward your head. Record whether the hair strands pile up in front of your finger and thumb, or lie flat.

3. Wash your hands thoroughly before leaving the laboratory.

QUESTIONS

1. It takes more force to stretch coarse hair than fine hair. Why?

2. Explain the results of the cuticle test.

3. Is your hair healthy or damaged?

4. Which solutions were the best curling agents prior to rewetting?

5. Which solutions helped retain the curl after wetting?

6. Which solutions made the hair most brittle?

7. What combination of conditions caused the most damage to hair?

8. Propose a chemical explanation for the difference in the curl produced by water and that produced by permanent wave solutions.

9. Is an acidic or basic solution more damaging to hair? Why?

10. Permanent wave solutions are basic. Compare the results observed from treatment with sodium hydroxide solution and permanent wave solution.

11. How might you protect your hair after swimming and before sunbathing?

D.8 HAIR STYLING AND CHEMICAL BONDING

As we have seen, hair is made of intertwined protein chains. The individual chains are held in place by hydrogen bonding, ionic bonding, and disulfide bridges.

Hydrogen bonding in protein chains is depicted in Figure 16. The hydrogen atom in such a bond is usually covalently bonded to a nitrogen atom. The hydrogen atom is strongly attracted to an unshared electron pair on an oxygen atom in a neighboring chain, or in a folded-back portion of its own chain.

All active proteins contain hundreds of hydrogen bonds relatively closely spaced along the protein chains. The protein chains are held in their ball-like or sheet-like structures by hydrogen bonds between and among protein chains. In many fibrous proteins, the chains do not remain fully extended; they are coiled, much like a "curly" telephone cord. Numerous hydrogen bonds keep the coils in place; others hold the chain to neighboring chains. Such protein coiling occurs in keratin, a fibrous hair protein, as illustrated in Figure 17 (page 500).

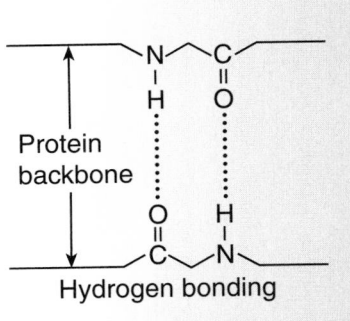

Figure 16 Hydrogen bonding between two chemical groups.

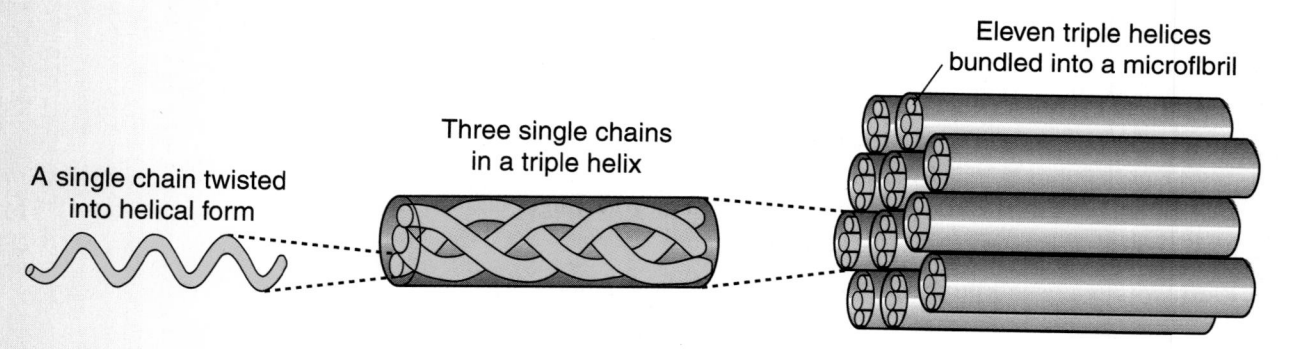

Figure 17 Hair keratin, a fibrous protein. The helical chains are themselves twisted into a triple helix, and these helices are bundled into a microfibril. Microfibrils form the core of each strand of hair.

Ionic bonding between protein chains (Figure 18) is an attraction between an ionic structure in one protein chain and an oppositely charged ionic structure in a neighboring chain or in a folded-back portion of the same chain. Such bonds are extremely important in hemoglobin, for example.

Nonpolar interactions (Figure 19) are relatively weak attractions between nonpolar chemical groups. They often occur in the interior of globular proteins. Although weak, many such interactions help keep the structure intact, largely by preventing polar water molecules from entering.

Many proteins contain a few covalent bonds that link together neighboring chains or folded-back portions of the same chain. Most often these are disulfide bonds, as illustrated in Figure 20. Disulfide bonds can form between two sulfur-containing regions between or within proteins. The amino acid cysteine contains such an —S—H unit. Two cysteines can react at their —S—H units, losing hydrogen and forming a disulfide bond, —S—S—:

$$\sim\sim\sim S\!-\!H \ + \ H\!-\!S\!\sim\sim \ \longrightarrow \ \sim\sim\sim S\!-\!S\!\sim\sim\sim \ + \ H_2$$

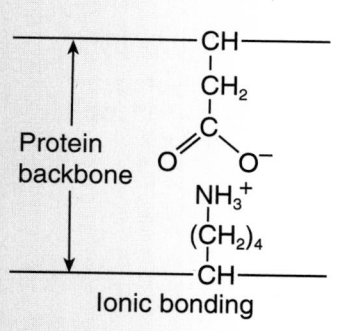

Figure 18 Ionic bonding involving two chemical groups.

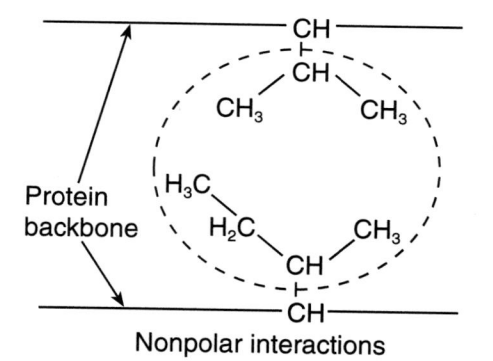

Nonpolar interactions

Figure 19 Interactions between two protein backbones. Nonpolar chemical groups are creating a region from which water molecules are excluded.

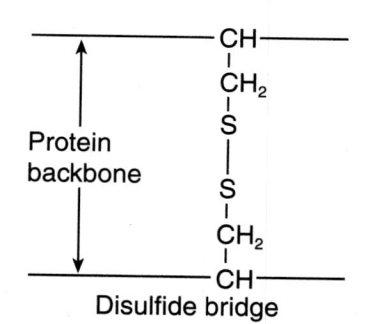

Disulfide bridge

Figure 20 Disulfide bond formed in cysteine units.

Disulfide bonds provide the key to understanding hair curling and straightening. Hair styling is a matter of breaking bonds between protein chains and forming new ones. If the protein chains separate but do not form new bonds, split ends can occur.

Your laboratory work with hair confirmed what you already knew: If you wet your hair, it becomes easier to style. Here's why: Wetting your hair allows water molecules to break hydrogen and ionic bonds between hair strands. Water forms new hydrogen bonds with the side chains. Thus separated, the side chains no longer hold the strands tightly in their natural position. As you comb your hair, the strands are free to slide into new positions. When the water evaporates, these side branches form new hydrogen or ionic bonds in their new positions, locking the chains in whatever position they happen to be.

Hair preparations that are too acidic can permanently weaken hair. Even after the hair dries, some excess hydrogen ions may remain bonded to the sidechains, preventing them from forming new ionic bonds.

Heat from blow-dryers and curling irons helps style hair quickly, because the higher temperatures speed up interactions between protein strands. But excessive heat can disperse hair oils, and even split protein chains.

The problem with the hair styling methods described so far is that they are temporary. Once hair is wet again, their effects are quickly erased. To get a waterproof hairstyle—a permanent—it is necessary to rearrange bonds that are not easily affected by water. These are the covalent bonds of the disulfide bond acting as a bridge between protein chains or between regions of a single protein chain.

The curliness of your hair depends on how disulfide bonds are joined between parallel protein chains. When a person gets a permanent, curls are created or removed in three steps, as shown in Figure 21. Here's the chemical recipe: First, break the disulfide links between protein chains. Next, use a form (curlers, rollers, etc.) to curl or uncurl the hair. Third, rejoin the disulfide links between protein chains in their new orientation.

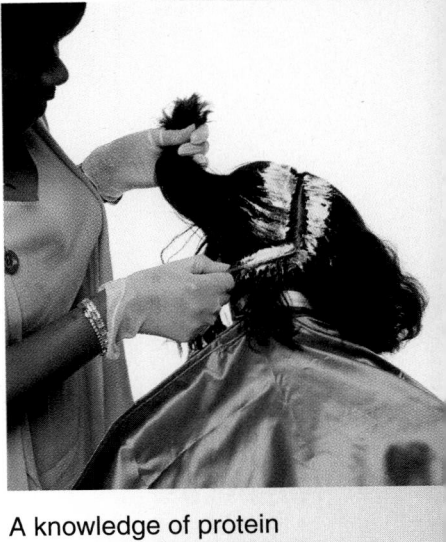

A knowledge of protein chemistry helps explain the effects of hair treatments.

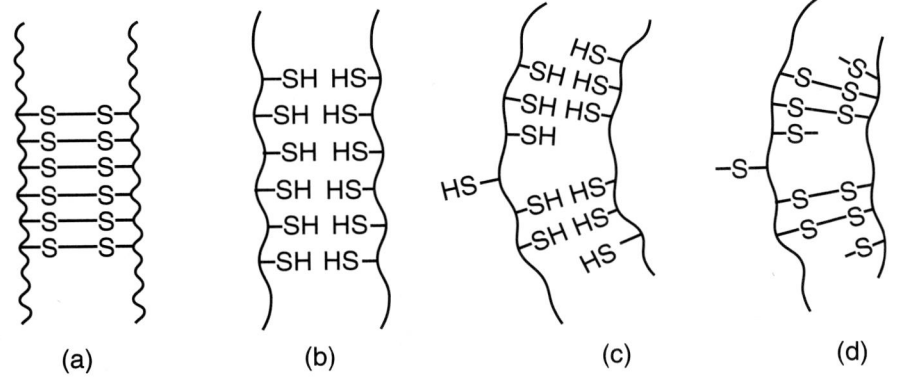

(a)   (b)   (c)   (d)

Figure 21 (a) Parallel protein chains before treatment, held in place by disulfide bonds. (b) A chemical reaction (reduction) breaks disulfide bonds and adds hydrogen atoms. Chains are now free to move relative to each other. (c) The hair is curled or straightened. (d) A chemical reaction (oxidation) reforms disulfide bonds between relocated chains.

As you have learned, chemistry and chemical reactions are responsible for much of what happens inside and outside your body. In the next part of this unit, you will learn how body chemistry can be altered by drugs and how it responds to toxic substances.

? PART D: SUMMARY QUESTIONS

1. a. What substances are removed during proper skin cleansing?

   b. Why must they be removed?

   c. What problems result from excessive skin cleansing?

2. Predict which of these materials are soluble in water and which must be washed away with soap or detergent.

   a. Salt (NaCl), a component of sweat

   b. Nondiet soft drink (essentially sugar water) spilled on your hands

   c. Grease left on your hands after eating french fries with your fingers

   d. Rubbing alcohol (70% 2-propanol, isopropyl alcohol, in water)

3. Explain the action of a soap or detergent in washing away grease.

4. What are the functions of your skin?

5. Describe the chemical process of tanning.

6. Briefly describe the link between tanning and skin cancer.

7. Describe how a sunscreen protects the skin.

8. a. Why does healthy hair stretch somewhat like a rubber band?

   b. Explain the difference in stretch between dry and wet hair.

9. Explain the different chemical effects of water and permanent wave solution on hair structure.

EXTENDING YOUR KNOWLEDGE (OPTIONAL)

• Why is dark skin an advantage for people living in tropical countries? Why is light skin a disadvantage?

• Investigate the causes and mechanism of formation for malignant melanoma, a serious form of cancer.

E CHEMICAL CONTROL: DRUGS AND TOXINS IN THE HUMAN BODY

As the science of synthetic organic chemistry has advanced, many medicines have been synthesized. These substances have helped alleviate much illness and suffering, making life for many much more enjoyable than it might have been. But rather than relieving suffering, some substances have contributed to it. Drug addiction is present in every segment of society.

What are drugs? How do they work? Are they totally foreign to the human system? What does the body do to protect itself from their negative effects?

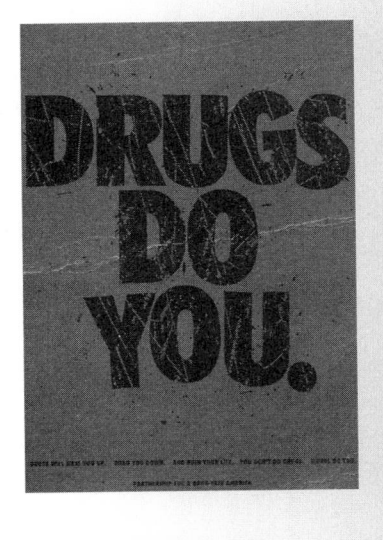

E.1 A GLIMPSE OF DRUG FUNCTION

Drugs alter the body's structure and function, including the brain's chemistry. In this sense, alcohol, aspirin, amphetamines, crack, caffeine, nicotine from cigarettes, and morphine, a powerful painkiller, are all drugs. However, drugs differ greatly in strength and effect. Some stimulate while others depress the brain. Some drugs relieve pain or stop infection. Still others make up for chemical deficiencies. All act at the most fundamental biochemical level, the molecular level. This action takes place most often within a specific area of the body, that is, certain cells or tissues.

By contrast, all *toxins* harm the body. Sodium cyanide (NaCN), carbon monoxide (CO), polychlorinated biphenyls (PCBs), polyaromatic

Most communities provide help and information regarding drug-related problems.

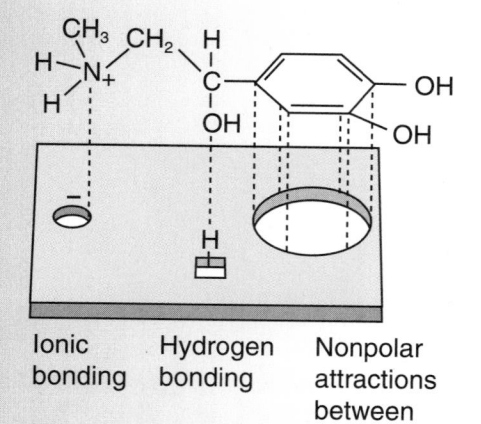

Ionic bonding   Hydrogen bonding   Nonpolar attractions between flat areas

Figure 22 An adrenaline molecule at a receptor.

The opium poppy is the source of morphine and other narcotic drugs. In medicine, these drugs are extremely valuable in helping control pain; on the street, they bring pain to addicts and their families.

hydrocarbons in tobacco smoke, and strychnine, a potent nerve poison, are all toxins. Chemical toxins, like drugs, function at the molecular level in the body. The difference between a drug and a toxin is often a matter of dose. Any substance, even water, can become toxic at too large a dose. An overdose of morphine can cause death by shutting down the respiratory system, but in very small doses, it is a useful anesthetic.

How does a drug function at its action site in the body? Drug specificity, like enzyme specificity, often depends upon molecular shape. Many drugs act on *receptors*—regions on proteins or cell membranes with just the shape and chemical properties needed to interact with the drug molecule and help it initiate the desired biological response.

As an example, consider adrenaline, a hormone produced naturally by the body. When you are suddenly frightened, the adrenal glands release adrenaline. This hormone circulates in the bloodstream, activating the heart and other organs. Settled at its receptor sites, adrenaline initiates a cascade of reactions that prepare the body in various ways for the physical activity of "fight or flight." (It increases heart rate, for example.) A model of an adrenaline molecule at its receptor is shown in Figure 22. Adrenaline is also used in many nasal drops and sprays used to relieve severe allergic symptoms.

To further illustrate drug action, let's look at some drugs that relieve intense pain, but at a price—they can become addicting. Classified as *narcotic analgesics,* these include morphine, meperidine, and methadone. Analgesics relieve pain by blocking nerve signals from the pain source on their journey to the brain. Blocking occurs when drugs of the right shape and composition interact with receptors on proteins in the membranes of key brain cells. The drug molecule distorts the shape of the membrane protein enough to block the pain signal.

The receptor sites for narcotic analgesics appear to have these features:

- a negative ion site that can bind an ammonium ion (NH_4^+) or its equivalent
- a flat surface to which a flat cyclic group of atoms (an aromatic group) can bind
- between these, a cavity into which a chain of atoms connecting the aromatic group and the ammonium ion may fit

All potent analgesics studied thus far either have the shape needed to allow bonding to these receptors or can adapt to them.

One might wonder why brain cells have receptors of just the right type for morphine-like drugs that are foreign to the body. It has been found that the brain's own natural painkillers, called *endorphins* and *enkephalins,* act at these same receptors.

Although the structural formulas of morphine and endorphins or enkephalins do not look much alike to the untrained eye (Figure 23, page 505), in fact several parts of the molecules are identical. These parts are believed to be a key to the action of these painkillers.

Morphine and other narcotic analgesics are used in cases of severe or enduring pain, but for common aches and pains we use another kind of analgesic such as aspirin. Unlike aspirin, which can be bought "over the counter," purchasing a narcotic analgesic requires a prescription.

Figure 23 Morphine and an enkephalin, both painkillers. The protein chain of the methionine enkephalin folds into a shape similar to that of morphine, a shape that fits the body's painkilling receptors.

E.2 YOU DECIDE:
PROS AND CONS OF ASPIRIN

The most widely used nonprescription painkillers belong to a chemical family called *salicylates.* Acetylsalicylic acid, commonly known as aspirin, is the most familiar. A versatile drug, aspirin can reduce fever and inflammation, and relieve pain.

Aspirin is an example of a drug made more useful by chemical modification. Originally, salicylic acid itself was given to reduce fever and pain, but it severely irritated the mouth and upper digestive tract. To reduce such irritation and help the molecule pass more rapidly into the blood and to appropriate target areas, chemists tried modifying the molecular structure of salicylic acid. Some modifications rendered it useless as a drug, and other alterations created drugs with unwanted side effects.

► *The term "aspirin" is a combination of "a" for acetyl and "spirin," the official name of the plant,* Spirea ulmaria, *from which salicylic acid was first extracted.*

Salicylic acid

Acetylsalicylic acid

However, replacing the H of the —OH group with an acetyl group, —CO—CH$_3$, produced acetylsalicylic acid, a drug with greatly diminished side effects that has significant effect on pain and fever. This drug we call aspirin has been a valued medication for nearly a century. Many other drugs have been tailored similarly to do specific jobs in the body.

Aspirin is a drug with highly beneficial uses. At low doses—usually one tablet daily—it can help prevent heart attacks. Normal doses of two or more tablets reduce pain and fever. Doctors may prescribe eight aspirin tablets or more daily for someone suffering from arthritis.

In most cases, aspirin is regarded as completely safe. However, it has some side effects. In some people, aspirin can induce bleeding ulcers and may promote fluid retention by the kidneys. And, because aspirin is an acidic compound, some people get an upset stomach when they take it. High, prolonged doses of aspirin may cause loss of balance and slight hearing loss. Several studies have suggested an association—not a cause–effect link, but a possible connection—between aspirin and Reye's syndrome. This rare, but sometimes fatal, disorder affects some children and teenagers recovering from influenza or chicken pox.

Imagine that you suffer from arthritis and your joints ache constantly, limiting your activity. Aspirin could relieve the pain.

1. Would you take the aspirin? In other words, would you prefer to live with the pain, or with the treatment's possible side effects?

2. If you decide to take aspirin, what might be done to counteract some of the long-term side effects?

3. Nonaspirin pain relievers such as ibuprofen (in Nuprin®, Advil®, Medipren®, and others) and acetaminophen (in Datril®, Tylenol®, and others) are also available without prescription.

   a. What information would you obtain to decide whether any of these medications would be preferable to aspirin for arthritis?

   b. Where might you get this information?

4. Why should you take any medications only when they are really needed?

Not all drugs act at specific receptors, as morphine and endorphins do. Alcohol appears to act on all or many neurons (nerve cells). It depresses transmission of nerve signals. This slows down functioning of the entire brain.

Below is a sampling of information obtained from reliable studies of alcohol. This information allows you to assess risks you and your classmates may face as you contemplate your future.

SUMMARIES OF ALCOHOL STUDIES

• Drivers who have been drinking are involved in over half of all highway deaths in the United States.

• Alcohol is directly or indirectly responsible for nearly 100,000 deaths yearly in the United States.

• Alcohol-related accidents are the number one cause of death among teenagers.

- Possible impairment due to alcohol abuse includes high blood pressure; diseases of the liver, pancreas, and intestines; blackout periods (long intervals during which there is no memory); severe vitamin deficiencies; cardiovascular damage; and neuroses.

- About 15 million Americans suffer from alcoholism or alcohol dependency. This is more than five times the total number of cocaine or crack users, and 30 times the number of all U.S. heroin addicts.

- Alcohol use in the United States extracts a cost of more than $15 billion in medical expenses and nearly $120 billion in lost worker productivity annually.

- The estimated life expectancy for heavy drinkers is 10–12 years shorter than that for the general public.

- Alcohol consumption during pregnancy can cause serious physical and mental deficiencies in infants, known as Fetal Alcohol Syndrome (FAS). Drinking during pregnancy is a major cause of mental retardation in the United States.

- Children of alcoholics are three to four times more likely to become alcoholics than other children are.

▶ *Mickey Mantle, a baseball Hall of Famer, received a liver transplant to replace his own liver, which was damaged by chronic alcohol abuse.*

E.3 YOU DECIDE:
EFFECTS OF ALCOHOL

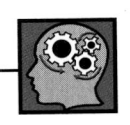

Concern over risks of excessive alcohol consumption and legal liabilities has led the federal government to require warning labels on alcoholic beverages, as noted at the beginning of this unit.

The effect of alcohol on a person's behavior varies with the amount the person consumed, the individual's weight, and time that has elapsed since the alcohol was consumed. Table 6 (page 508) presents data for estimating the percent alcohol in the blood and lists the effects of various blood alcohol levels on behavior. For this table, it is assumed that the drinks are consumed within 15 minutes. One "drink" is defined as a bottle of beer, a glass of wine, or a shot (1.5 oz; 45 mL) of whiskey or other 80-proof liquor. To find the change in the blood alcohol percent over time, subtract 0.015% for each hour.

Alcohol dehydrogenase is the enzyme that accelerates the breakdown of ethanol to acetaldehyde. Research has found lower alcohol dehydrogenase (ADH) levels in women's stomachs than in men's. Therefore, the amount of unconverted ethanol that reaches the bloodstream is higher in women than in men. The amount of ADH in the liver also depends on how much time has passed since a person has eaten before drinking. An empty stomach or fasting decreases ADH. Consequently, drinking when hungry causes the concentration of unconverted alcohol in the bloodstream to remain high as it reaches the brain. This leads to more rapid intoxication.

In most states, the legal blood alcohol limit for automobile drivers is 0.10% (0.10 g of alcohol in 100 mL of blood). Some states have decreased this to 0.08% (0.08 g alcohol in 100 mL of blood) or even less for underage drinkers.

Each is one "drink," even though the volumes differ.

Chemical Control: Drugs and Toxins in the Human Body

No. of Drinks	Blood Alcohol Level (Percent)* Body Weight (lbs)						
	100	125	150	175	200	225	250
1	0.03	0.03	0.02	0.02	0.01	0.01	0.01
2	0.06	0.05	0.04	0.04	0.03	0.03	0.03
3	0.10	0.08	0.06	0.06	0.05	0.04	0.04
4	0.13	0.10	0.09	0.07	0.06	0.06	0.05
5	0.16	0.13	0.11	0.09	0.08	0.07	0.06
6	0.19	0.16	0.13	0.11	0.10	0.09	0.08
7	0.22	0.18	0.15	0.13	0.11	0.10	0.09
8	0.26	0.21	0.17	0.15	0.13	0.11	0.10
9	0.29	0.24	0.19	0.17	0.14	0.13	0.12
10	0.33	0.26	0.22	0.18	0.16	0.14	0.13
11	0.36	0.29	0.24	0.20	0.18	0.16	0.14
12	0.39	0.31	0.26	0.22	0.19	0.17	0.16

Table 6 — Alcohol in the Blood

Effect on Behavior

Blood Alcohol Level (%)	Behavior
0.05	Lowered alertness; reduced coordination
0.10	Reaction time slowed 15–25%; visual sensitivity reduced up to 32%; headlight recovery 7–32 seconds longer
0.25	Severe disturbance to coordination; dizziness; staggering; senses impaired
0.35	Surgical anaesthesia; reduced body temperature
0.40	50% of people die of alcohol poisoning

*Data from MADD

However, driving with a blood alcohol level less than the legal limit does not guarantee safe motoring. A study of 13,000 drivers revealed that the probability of causing a traffic crash doubled at a blood alcohol level of just 0.06%. The National Highway Traffic Safety Administration classifies a vehicle crash as alcohol-related when alcohol can be detected at any level in the driver.

1. If a 125-lb person consumes two bottles of beer, predict the individual's blood alcohol level if it is

   a. measured immediately.

   b. measured after two hours.

2. List two kinds of behavior you might observe in an individual having a blood alcohol level of 0.15%.

3. If a 175-lb person has six drinks in rapid succession and drives an automobile one hour later, will the individual be considered legally intoxicated in most states?

4. If a 100-lb person consumes 10 drinks in rapid succession, will the individual be in any danger of death?

5. How long should a 200-lb person who has consumed three drinks wait to safely drive an automobile?

E.4 NOTES ON SOME ILLICIT DRUGS

Marijuana comes from the plant *Cannabis captiva.* Its principal active ingredient is tetrahydrocannabinol (THC). Effects of smoking marijuana include increased heart beat, time distortion, and difficulties with motor coordination. Loss of motor coordination is linked to abundant THC receptors in motor coordination control centers. The presence of THC on these receptors blocks normal motor control reactions. Marijuana use can cause euphoria (a "high") or a heightened sense of anxiety. Although users describe an impression of enhanced understanding, there is no evidence that marijuana use increases learning or insight.

Cocaine, like alcohol, acts on neurons, altering the transmission of nerve signals in several parts of the brain. It acts as an anesthetic by blocking the channels through which sodium ions move in and out of neurons. Without that movement, pain signals cannot be transmitted. Cocaine also interferes with the transmission of impulses between neurons. At the same time, it produces a "high," or euphoria. The biological basis for its euphoric effects is only partly understood. Cocaine can cause brain hemorrhages, dangerous increases in blood pressure, and even death.

Crack is a purified and considerably more addictive form of cocaine. Its addictiveness is due to the fact that crack is more potent than cocaine. Users often smoke crack in 10–15 minute "binges" over several hours, which can lead to addiction after the first use.

Ice is composed of rock-like crystals of methamphetamine, a central nervous system stimulant. When smoked, this highly addictive and dangerous drug gives the user the same general effects as crack cocaine. However, unlike crack, ice produces a high that lasts for hours rather than minutes. Illegal methamphetamine manufacture is a multibillion dollar industry.

LSD stands for *l*ysergic *a*cid *d*iethylamide, a compound found naturally in morning glories and in a fungus on rye grains. The LSD sold on the street is synthesized in illegal laboratories. LSD is a *hallucinogen,* a class of drugs that distort the user's perception of reality. LSD induces a "trip" that can last several hours in which the user experiences a severely altered state of reality. Laboratory studies indicate that LSD modifies nerve transmission to sensory centers in the brain. Other research has shown that LSD use leads to decreased mental capacity, and has a high potential with heavy use for creating serious mental psychoses. A particular danger of LSD, even used only once, is the potential of "flashbacks" that create "bad trips." Such flashbacks can occur months, even years, after the initial exposure to LSD.

Designer drugs are the products of laboratory-based chemical reactions. These reactions create new substances that, although structurally similar to legal drugs, produce the effects of drugs of abuse. Most designer drugs are narcotics or hallucinogens, going under such names as Ecstasy and China White. Although they are derived from legal substances, such designer drugs carry the same risks and dangers as illegal drugs.

Anabolic steroids have received public attention and notoriety due to their apparent ability to enhance athletic performance, mainly through stimulating growth of lean muscle mass. They are chemical variants of testosterone, the hormone that imparts masculine physical traits to individuals. Use of anabolic steroids carries risks of acne, kidney and liver damage, liver cancer, heart disease, and sterility. The use of anabolic steroids is formally outlawed for all athletes. Despite this, random checks of athletes continue to reveal cases of anabolic steroid abuse in international competitions, professional sports, and in college and high school athletics.

Considering all the damage that drugs and foreign substances can do within the body, you might wonder how we manage to survive. Fortunately, the body is not completely defenseless.

E.5 FOREIGN SUBSTANCES IN THE BODY

Your body has a number of defense mechanisms that help protect you from *limited* amounts of toxic substances and disease-causing organisms. For example, the membranes that cover body surfaces and line the digestive tract and lungs resist invasion by most disease-causing organisms. These membranes also protect you against many chemicals in the environment. Stomach acid (pH 1–2) helps destroy many microorganisms in the stomach before they can enter the bloodstream.

These external defenses are backed by internal defenses. For example, let's look at how the liver detoxifies the blood, and how the body deals with the presence of foreign proteins.

DETOXIFICATION IN THE LIVER

Some digestion of foods and other swallowed substances takes place in the stomach, but most occurs in the small intestine. Digested food, other small molecules, and some ions pass through the lining of the small intestine into the blood, which carries them to the liver. Undigested material and molecules or ions that do not pass through the intestinal wall proceed to the large intestine and are eliminated from the body. Some toxic substances that enter the stomach are removed from the body in this way.

The liver also separates useful from harmful or undesirable substances. Useful substances—including glucose and other simple carbohydrates, amino acids, and fatty acids—are released to the blood for general circulation to cells that need them. Some of these molecules are retained by the liver: glucose is stored as glycogen, the body's reserve carbohydrate; and amino acids are converted to proteins for use in the blood.

Some harmful or undesirable substances separated by the liver are eliminated from the body. Carried in bile from the liver to the intestines, they join other waste matter.

Other harmful substances undergo reactions in the liver that make them less toxic or more soluble in water. In this form they are more easily eliminated by the body. The conversion of ethanol to acetaldehyde by liver alcohol dehydrogenase was described in Section B.5. Another example of

CHEMISTRY *AT WORK*

Using Chemistry To Bring Criminals to Justice

For a moment, no one in the courtroom moved. Aware that everyone was staring at her, Susan could hear the sound of her own heart beating.

Then the prosecutor continued. "Mrs. Ragudo, please tell this court," he said, holding up the small plastic bag of white powder, "the results of your analysis of the substance labeled Exhibit A."

For many people, the thought of testifying in a criminal court case is enough to make them break out in a sweat. But for **Susan Ragudo,** it's just another part of her job as a Forensic Scientist employed by the Commonwealth of Virginia. Forensic science—the use of science in criminal investigations—can describe many types of inquiry. Susan's specialized area involves analyzing crime scene material for evidence of illegal substances.

To identify an unknown white substance, for instance, Susan begins with a color test, introducing the substance to a reagent. Depending on the resulting color change, Susan knows what the substance most likely is. For example, the drug heroin will turn purple, cocaine will turn blue, and so on. Next, Susan uses thin layer chromatography (TLC) to compare the substance to a known sample of the suspected drug. From the TLC, Susan is fairly certain what the substance is, and whether there is only one substance present. She finally tests her conclusion by using gas chromatography/mass spectroscopy. Then she returns the substance—and her analysis data—to the law enforcement officers.

Susan explains that in her field it's essential to know one's way around a laboratory. It's also important to have an analytical mind—to be able

to "zero in on" a substance logically, by conducting and interpreting tests. Susan adds that it also helps for her to be comfortable speaking in public, because she is sometimes required to testify in court about her findings.

Calling All Sleuths

Susan Ragudo uses chemistry to help determine whether substances are illegal drugs.

- In what other types of criminal investigation could chemistry be helpful?
- What types of tests could be used?
- Write a story about a crime in which everyone overlooks a small chemical clue—everyone but you, that is. Don't be bashful . . . make yourself the hero, and explain how you used a basic understanding of chemistry to unravel the crime . . . and convict the criminal!

Photograph by Robert J. Llano

chemical alteration in the liver is the conversion of benzene (a water-insoluble toxin) to catechol, which is water-soluble:

$$\text{benzene} + O_2 \xrightarrow{\text{enzyme}} \text{catechol}$$

benzene   catechol

Another typical reaction is the conversion of sulfite—a poison when present in sufficient amount—to sulfate:

$$2\ SO_3^{2-}(aq) + O_2(aq) \xrightarrow{\text{enzyme}} 2\ SO_4^{2-}(aq)$$

sulfite ion   sulfate ion

▶ *Molybdenum is a trace mineral essential to human nutrition. See the Food unit, page 275.*

The enzyme catalyzing this reaction contains molybdenum ions. Without tiny amounts of these ions in the liver, many foods would poison us. The liver also chemically alters and eliminates many hormones and drugs that accumulate in the blood.

The liver is the body's largest internal organ and is extremely important. In many ways it saves us from our bad eating, drinking, and perhaps even drug-taking habits. But it can take only so much abuse before damage occurs. The liver's ability to detoxify materials is limited. Excessive ingestion of harmful substances can place a heavy burden on it. For example, overdoses of alcohol, ibuprofen, or acetaminophen can cause liver damage. As a result, liver function may be diminished. This can cause problems in distribution of essential molecules—glucose and amino acids—and in synthesis of important proteins. Overburdening the liver also can result in the accumulation of harmful molecules in the body's fat reserves.

DEFENSE AGAINST FOREIGN PROTEIN

In most cases, foreign proteins enter the body as part of disease-causing agents such as viruses, bacteria, fungi, and parasites. Body chemistry depends so strongly on having exactly the right protein at the right spot at the right time that any invader protein is a signal to marshal a defense against the potential harm it might do. The body's defense strategy, carried out by its immune system, involves building a protein that surrounds part of the invader molecule. Once surrounded, the invader cannot react to cause harm.

Any foreign protein that sets this defense mechanism in motion is called an ***antigen;*** an ***antibody*** is the complementary substance created by the body to destroy the antigen's activity. Figure 24 illustrates the action of antibodies. Formation of an antigen–antibody complex precipitates the invader from suspension in the blood or other body fluids and allows it to be destroyed or otherwise removed by the body's waste disposal system.

Building antibodies with exactly the right complementary protein is no easy task. In your body, only certain kinds of white blood cells can do it. Once the blood has been exposed to a certain kind of bacteria or virus, some of its white cells can henceforth rapidly synthesize the antibodies needed to destroy the bacteria. A person then has acquired an immunity to the disease caused by that microorganism.

However, it is important to recognize that the structures of bacteria and other disease-causing organisms do not remain static. Through mutations

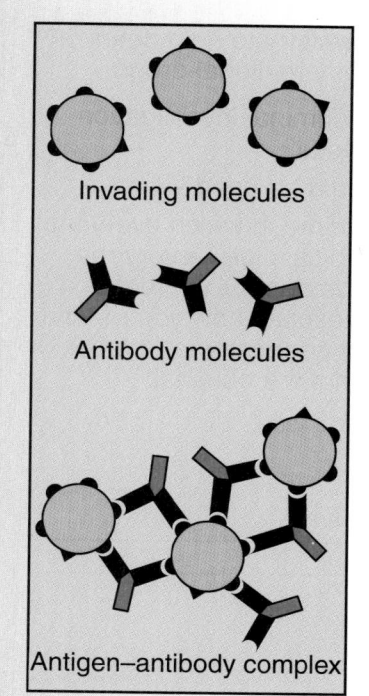

Invading molecules

Antibody molecules

Antigen–antibody complex

Figure 24 Antibodies precipitate invading molecules.

and variations, bacteria and other disease-causing organisms are modified to become less susceptible to drugs. Therefore, a drug that is successful now may not be useful in the future against a mutated version of the disease-causing agent. For this reason, teams of researchers—pharmaceutical chemists, biochemists, microbiologists, bacteriologists, and others—use computer modeling to better understand the molecular basis of a drug's action on a particular type of disease. Such understanding and modeling can lead to the more rapid development of new drugs to combat diseases.

Unfortunately, the body cannot protect itself against all disease-causing agents. The "common cold," caused by an influenza virus, is such a disease. Although we can take medication to help reduce the fever and other effects, no known drug can cure it directly. The flu symptoms last several days, depending on the strain of flu virus and its intensity, until the body is able to re-establish its normal chemistry.

Computer modeling of a trial drug.

E.6 DRUGS IN COMBINATION

The effect of some combinations of drugs is simply a sum of the desirable actions of each (e.g., aspirin for a fever and an antibiotic for an infection). In other cases, drugs that are beneficial separately produce mainly undesirable effects when used in combination. Still other drugs interfere with each other. For example, the antibiotic tetracycline kills bacteria by binding free iron and other important ions in your system. This makes these ions unavailable to bacteria that need them. So, if you are taking iron tablets to compensate for anemia (iron-poor blood) and your doctor prescribes tetracycline at the same time, you could lose the benefits of both. The tetracycline may bind to the iron from the tablets, putting both tetracycline and iron out of commission.

Some combinations of drugs are deadly. When a medication is prescribed for you, it is essential to tell the physician about any other drugs you are taking. Also, it is wise to understand the trouble you can get into by taking combinations of drugs on your own.

One of the deadliest combinations is alcohol and sedatives, which include tranquilizers, sleeping pills, and the addictive street drugs known as "downers." Sedatives slow brain activity by retarding communication between nerve cells. Combining alcohol with sedatives or certain other drugs is *synergistic*—the total effect is greater than that of either drug alone. Together, sedatives and alcohol can depress nerve activity so much that even involuntary functions, such as breathing, may stop. The lethal dose of a barbiturate (e.g., phenobarbital) is decreased by 50% when it is combined with alcohol. That is, it takes only half as much of the barbiturate to cause death under these circumstances.

Taking the tranquilizer diazepam (the active ingredient in Valium) with alcohol is also a potentially deadly combination. Even combining antihistamines and alcohol can cause excessive drowsiness that can result in accidents. The combination of narcotic analgesics and alcohol can lead to loss of control of body movements. In short, it is dangerous to combine alcohol with any other drug without recognizing the possible consequences.

1. a. How are narcotic analgesics, such as morphine, believed to work in the brain?

   b. How are these like the brain's own painkillers?

2. a. Describe the effect of alcohol on nerve cells.

   b. Compare that with the effect of cocaine on nerve cells.

3. Give two reasons why you should be sure your physician knows about all medications or drugs you are taking.

4. Should the fact that you are taking medication make any difference in your eating and drinking habits? Explain.

5. Describe at least four chemical processes that occur regularly in the liver.

6. List some ways the body deals with toxic substances.

7. Ethanol (ethyl alcohol) is a widely abused drug found in alcoholic beverages. As with many drugs, one-time use of alcohol can be such that the amount is just beyond the threshold amount needed to produce detectable physiological effects. Or, it can be so excessive that the one-time amount is lethal.

   a. Depending on body weight, the normal threshold dose for ethanol is about 50 mg of ethanol per liter of blood (0.050 g/L). The density of ethanol is 0.79 g/mL, and there are 1,000 mL in 1 L. Calculate the volume (mL) of ethanol needed to reach this threshold dose.

   b. The lethal dose for ethanol is 4,000 mg of ethanol per liter of blood (4.0 g/L). Calculate the volume (mL) of ethanol needed for this lethal dose.

   c. The greater the amount of ethanol consumed, the more severe its physiological effects (Table 6, page 508). Calculate the volume (mL) of ethanol that must be consumed to reach a blood-alcohol level of 0.12 g/100 mL, an amount that far exceeds the legal limit for driving in most states.

   In this regard, it is important to recognize that a simple equivalency exists among various types of alcoholic beverages. One 1.5-oz (45 mL) shot of 80 proof whiskey, 5 oz (150 mL) of wine, or one 12-oz (360 mL) beer contain *the same amount of pure ethanol*—0.5 oz (15 mL).

EXTENDING YOUR KNOWLEDGE (OPTIONAL)

• Some folk remedies for illnesses are effective. Chemical studies have shown that there is a molecular basis for their action. One example is the use of willow bark to treat pain. Look into the chemistry of folk remedies, and find why some are successful.

• Prepare a library report on some hazardous food–drug interactions.

F PUTTING IT ALL TOGETHER: ASSESSING RISKS

F.1 RISK ASSESSMENT

We previously examined the risks and benefits of smoking (page 456). In the Nuclear unit we analyzed the various risks and benefits associated with the application of nuclear technologies, including nuclear medicine and nuclear energy (page 359). We now turn to consider risk–benefit analysis as it applies to our general well-being. Identifying key underlying causes of premature death and assessing your own risk from these factors can guide personal decisions affecting your well-being and your chances of living your natural life span.

To assess your risk of having a heart attack, you can consider your family's medical history, your diet, your weight, and other personal habits such as smoking or exercise. Some of these factors are within your control by the choices you make; others, such as your genetic makeup, are beyond your direct control. To decrease the risks of a heart attack, you might decide to modify your diet, lose weight, and exercise more. However, there is no guarantee that these changes can prevent a heart attack. Though genetics may be the most important factor in not having a heart attack, it is still prudent to take proper care of yourself to guard against a heart attack, especially if your family medical history might predispose you to have a greater risk of such an attack.

You can control some health risks by your behavior. Eating properly, getting sufficient exercise, and avoiding substances known to cause health problems are obvious examples. You *can* decide whether or not you will smoke, drink alcoholic beverages, or take mind-altering drugs. As you know, the use and abuse of these substances is a grave concern for individuals, families, and society.

Generally, **risk** is considered as the overall potential for harm from an event associated with an activity or a technology. How risky something is relates to *both* the likelihood of the event *and* how severe its consequences can be. For example, the probability of dying in an automobile accident is calculated to be about 1 in 6,000, compared to a 1 in 4 million probability of dying in a commercial airplane crash. The severe consequence in either case is death. However, the probability of dying in an automobile accident is about four thousand times greater than in a commercial airplane crash. In spite of the overwhelming evidence that flying is safer than driving, many people who refuse to fly think nothing of driving a car.

The perception of risk is a concept that also differs among individuals. For example, good swimmers see little risk in going swimming. However, those who do not swim well might view swimming as a potentially high-risk activity. An important concept in evaluating risk is whether the actual risk is high or the activity or phenomena is *perceived* as very risky. When individuals feel that they have direct personal control over an activity, for example driving a car, they generally perceive the activity as less risky. Those activities in which they do not have direct personal control—for example, traveling in a commercial airplane—are perceived as more risky. Other factors entering the mixture of actual and perceived risk include:

- familiarity with the risk
- potential for disastrous, not slight, consequences
- perceived benefit compared to level of risk

Whether through federal policies or rules of the workplace, some people appear to seek the assurance of risk-free living, an unrealistic goal. Adam Wildavsky writing in the *American Scientist* expresses the irony of this pursuit in saying:

> "How extraordinary! The richest, longest lived, best protected, most resourceful civilization, with the highest degree of insight into its own technology, is on its way to becoming the most frightened.
> Is it our environment or ourselves that have changed? Would people like us have had this sort of concern in the past? . . . Today, there are risks from numerous small dams far exceeding those from nuclear reactors. Why is the one feared and not the other? Is it just that we are used to the old or are some of us looking differently at essentially the same sort of experience?"

What is your perception of risk? What activities do you perceive as riskier than others? Which do your classmates classify as the riskiest? In the following sections, you will assess the relative risk of a variety of activities and technologies.

F.2 RISK ASSESSMENT AND PERCEPTIONS

To do risk assessment on an activity or technology, the risk associated with it must first be identified. Once the risk is identified, then the level of risk can be assessed. The following *Your Turn* begins a series of three exercises associated with 16 common activities or technologies listed below.

YOUR TURN

What's the Risk?

In this activity you are asked to identify the risk associated with each of the following 16 activities or technologies.

Nuclear power	Motor vehicles	Police work	Motorcycles
Alcoholic beverages	Handguns	High school football	Surgery
Smoking	Hunting	Commercial aviation	X-rays
Swimming	Bicycling	Food preservatives	Pesticides

Your teacher will assign you to a group to work on this exercise. Each group will complete the following exercise.

1. Copy each of the 16 items on a separate sheet of paper.

2. On the paper list what your group considers to be the *major* risk or risks involved with each of the 16 activities or technologies.

3. Compare your group's list of risks with those of the other groups.

   a. Which of the same risks, if any, were identified by all the groups for a given item?

   b. For which risks were the lists of risks most different?

   c. Explain why the lists of risks might differ.

Now that you have identified the major risks associated with the 16 items, you will consider them again, this time from the standpoint of risk assessment. Why are some riskier than others? How can risk assessment be applied?

YOUR TURN

How Risky?

The potential for harm is determined by the probability that an event will occur, as well as the severity of its consequences. Thus, this *combination* of probability and severity of consequences establishes the level of risk associated with an event. Obviously, the riskiest events are those most likely to occur (high probability) *and* with very serious consequences—disabling injuries or death.

Such risk assessment can be made using a graph like that in Figure 25 (page 518). The *x* axis represents the estimated severity of the consequences—low, medium, or high—of a perceived activity or technology. The likelihood or probability of the event happening is indicated on the *y* axis, from low to high probability. Thus, a low-risk occurrence, one associated with events that happen infrequently and with low consequences, would be in the lowest left block (letter C).

You will use Figure 25 to classify the riskiness of the 16 technologies and activities listed in the previous *Your Turn*. Your teacher will assign you to a group to work on organizing the listed items in terms of risk. Each group will complete the following exercise.

1. Copy Figure 25 on a separate sheet of paper.

2. Assign each of the sixteen items to one of the blocks on your copy of the Figure. Some blocks might contain no entries; other blocks might contain more than one.

3. Which items are influenced by lifestyle?

4. Compare your Figure 25 with those of the other groups.

   a. Which items among all groups are classified as most risky?

   b. Which items are the lowest-ranked risks?

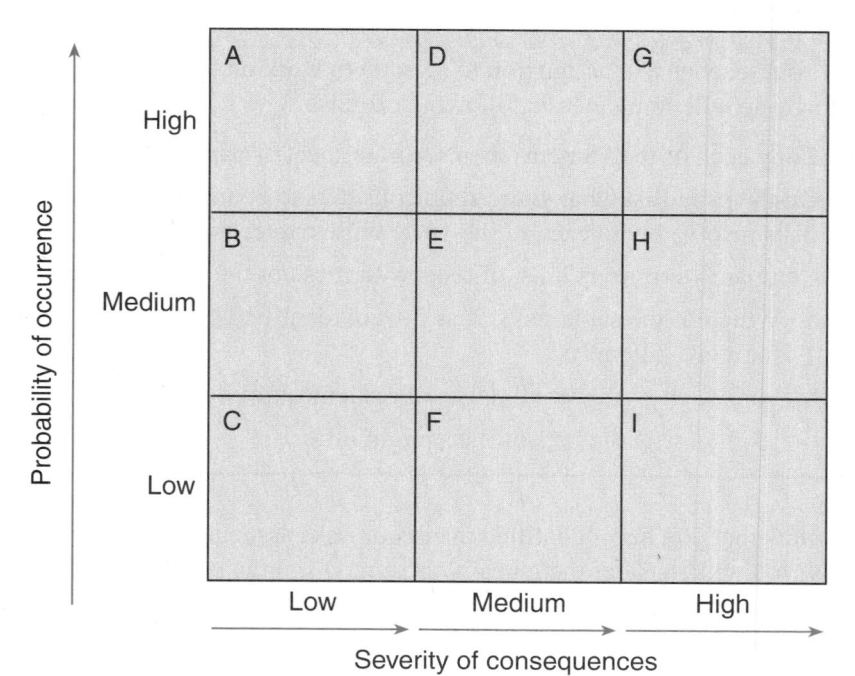

Figure 25 Risk assessment: probability and severity.

F.3 YOU DECIDE: LIVING WITH RISKS

We will now consider the sixteen items in a slightly different way. This time your group will assess and rank the sixteen items in terms of risk. Work within the group as follows.

1. Copy the list of sixteen items on a separate sheet of paper. There might be an activity or technology not listed that your group might decide to add to the list. You can add up to three new items. However, for each you add, omit one from the original list, so the total remains sixteen.

2. Using the list developed by the group, rank each of the 16 activities or technologies in the order of your group's estimate of their *decreasing* risk. The riskiest will be ranked first on *your* list; the least risky will be listed as 16th.

3. Compare your list with those of other groups.

   a. Which activities or technologies were added or deleted?

   b. Which activities or technologies among all groups are ranked most frequently among the top five (most risky)?

   c. Which are the five lowest ranked (least risky) activities or technologies?

   d. How might the rankings change depending on where the activities or technologies are located?

4. Which risks could vary based on your body chemistry?

5. Your teacher will provide you with a ranking done by experts in risk assessment of your 16 activities or technologies.

    a. In what ways do the experts' top five differ from yours?

    b. How do the experts' lowest five compare to yours?

    c. Which rankings by the experts surprised you? Why?

    d. How do the ideas of *actual risk* and *perceived risk* apply to your lists and those of the experts?

F.4 LOOKING BACK

This unit has given you a variety of perspectives regarding your body's continuous encounters with chemistry—both internally and externally. Some observers have noted that we live in a chemical world. Because everything (that is, every *thing*) is composed of chemical substances, such a remark cannot be challenged. However, this unit has emphasized that we are *more* than just inhabitants of a chemical world. Our bodies are also active parts of that world and are affected by the choices and the risks involved with living.

Some choices allow you to take control of your personal health and well-being. In this regard, certain decisions you face are clear-cut, while others involve considerable uncertainty. As this unit has illustrated, many of the health-related choices you encounter can be clarified and guided by a knowledge of the chemistry involved.

There is much you can do individually to control your health, life span, and well-being. Choose wisely!

The Chemical Industry: Promise and Challenge

▶ What are the major activities, products, and services of the chemical industry?

▶ What contributions has the chemical industry made to our way of life?

▶ How do contributions by the chemical industry compare to the potential risks?

▶ What should be the role of the chemical industry as a responsible social partner?

▶ What responsibilities does the public have in this partnership?

EKS or WYE May Spell JOBS

By Gary Franzen
Staff writer of *Riverwood News*

Riverwood now has what may be an exceptional opportunity. Two industrial firms are seeking permission from the town council to build a chemical manufacturing plant in the vacant Riverwood Corporation facility. The facility has been available for more than a year since the Riverwood Corporation, which was our town's major employer, went bankrupt. Only one company can occupy the site.

Both the EKS Nitrogen Products Company and the WYE Metals Corporation have expressed interest in establishing operations here. The EKS Company wants to manufacture ammonia (NH_3) to sell as a fertilizer and to use in producing other substances. EKS, with corporate headquarters in Delaware, is the nation's third largest producer of nitrogen-based chemicals. WYE would produce aluminum metal here for use by other companies across the country. The WYE Corporation is the top U.S. aluminum producer.

The prospect of jobs for local citizens comes as great news to most townspeople. Since the Riverwood Corporation closed, our community's unemployment rate has hovered near 10%. However, some citizens are also concerned that a chemical plant might reduce our quality of life.

The Riverwood Industrial Development Authority has participated in separate meetings with EKS and WYE corporate officials for several months. The companies are impressed with the advantages of locating a chemical manufacturing plant at Riverwood—our stable, well-educated workforce, access to abundant resources and electrical energy, a favorable tax base, excellent railroad and highway access, and an appealing lifestyle.

However, representatives of the Industrial Development Authority and of both companies realize that community acceptance will be important to the success of the venture, if a manufacturing company is approved to locate here. In the near future citywide discussions will be held, leading to a town council meeting to determine whether either company will be allowed to build a plant, and if so, which one to invite.

Mayor Edward Cisko welcomes the opportunity of such a plant providing a needed boost to our town's economy. He points out that following the recent town council recommendation to address high local unemployment, he formally invited both companies to consider locating a plant in Riverwood. Bill D. Moore,

see JOBS page 523

JOBS from page 522

President of the Riverwood Chamber of Commerce, has also expressed his enthusiasm for a chemical plant occupying the Riverwood Corporation building: "Such a plant would be a great boost to the area economy. I also have been concerned about the site becoming a public eyesore if allowed to remain vacant." EKS Public Affairs Director Jill Mulligan says the EKS plant would provide at least 200 new jobs in an environmentally safe setting. Nancy Belski, WYE Director of Public Information, notes that about 200 jobs would be created by the WYE Corporation, a firm she said has a long record of environmentally responsible manufacturing.

On the other hand, Riverwood Environmental League members question the wisdom of locating a chemical plant nearby. Spokesperson Aaron Fosa has expressed concerns regarding air quality, pos-sible chemical spills, and waste dis-posal problems. He suggests that jobs could be created by other industries, with fewer potential problems.

Reactions by citizens this writer interviewed were mixed. "I've got car payments and house payments to make," says Carson Cressey, an unemployed laboratory technician. "Our savings won't last much longer. I hope I can get a job at that chemical plant."

"Unemployment has really cut into my business," says Cynthia Shapiro, owner of the sporting goods store on Pine Street. "I had to lay off my assistant. A new employer in town would help get things going again."

"If it's going to smell like when my parents wash the windows, I don't want an ammonia plant," says ten-year-old Bobby Burns. "I hate the smell of ammonia! But, if it's the aluminum plant, maybe I can sell the aluminum cans for recycling."

Several important points are illustrated by this *Riverwood News* arti-cle. To reduce its unemployment, Riverwood needs new industry, but should it be a chemical industry? What is a chemical industry and what chemistry is involved in the products it manufactures? What are advan-tages and disadvantages of such a manufacturing plant for the community?

These questions and other issues will be addressed in this unit. The unit concludes with a town meeting where you will have opportunities to use chemical principles learned in this and previous units to debate whether Riverwood should invite one of these companies to establish a plant in this town.

The enormous pharmaceutical industry has been one result of applied chemistry.

▶ *The nation's chemical industry has maintained a positive trade balance for more than five decades.*

A THE CHEMICAL INDUSTRY

Throughout earlier *ChemCom* units you discovered that chemistry is concerned with the composition and properties of matter, changes in matter, and the energy involved in these changes. In this unit, you will explore the ways in which chemical industries apply basic chemical principles to produce material goods and services.

A.1 CHEMICAL PRODUCTS

The chemical industry's business is to change natural materials so as to make useful products for various purposes. New substances are also created as replacements for natural ones—for example, plastics as substitutes for wood and metals; synthetic fibers for cotton and wool.

The chemical industry is a worldwide multibillion-dollar enterprise that affects our daily lives through its products and economic impact. Yet, most people don't realize what happens in making a useful product. This creates an aura of mystery about how chemical industries operate, how products are manufactured, and what to expect from a chemical plant.

The chemical industry is an important partner in modern society, employing well over a million people worldwide. Over the past 80 years, it has grown through mergers of smaller companies and initiation of new large companies. During that time the manufacturing focus has enlarged from a rather limited range of basic products to over 60,000 different products. These hundreds of companies form the nation's third largest manufacturing industry; only the machinery and electrical equipment

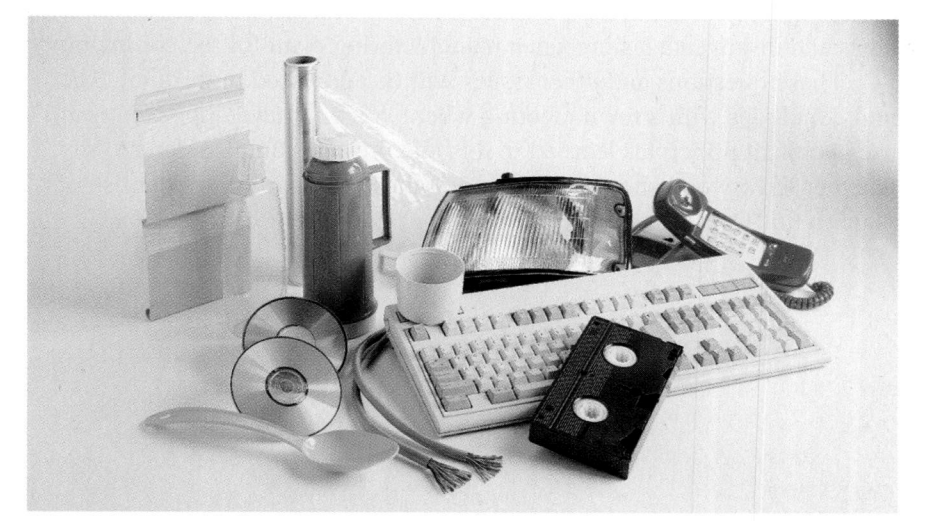

Plastics play many important roles in our daily lives.

industries are larger. Indeed, if the food and petroleum industries are included, the chemical industry is the world's largest. For example, over 50 billion pounds of plastic are manufactured in the United States annually.

Products of the chemical industry reach most of us indirectly, already used in other things. For instance, the automobile industry and home construction use enormous amounts of industrial chemicals. Paints and plastics are needed for automobile body parts such as bumpers, dashboard panels, upholstery and carpeting, and for the synthetic rubber used in tires and hoses. Home construction requires large amounts of plastics for flooring, insulation, siding, piping, and appliances; paints, metals, and air conditioning fluids are also used extensively.

The chemical industry's wide range of products can be classified into three general categories of materials, based on their identities and intended uses (Figure 1).

BASIC CHEMICALS
(25% of total output)

Acids
Bases
Salts
Common Organics

CHEMICAL PRODUCTS
Used in further manufacturing
(20% of total output)

Pigments
Plastics
Synthetic
Fibers

FINISHED PRODUCTS
Either for ultimate consumption or
materials in other industries
(>50% of total output)

Drugs
Detergents
Cosmetics
Paints
Fertilizers
Building
Materials

Figure 1 Products of the chemical industry.

Chemical Processing in Your Life

To sense how pervasive the products of chemical processing are in every-day affairs, try to list five items or materials around you that have not been produced or altered by the chemical industry. Start by considering everyday items—clothes, household items, means of transportation, books, writing instruments, sports and recreation equipment—whatever you routinely encounter.

Write answers to the following questions. Come to class prepared to discuss your answers.

1. a. Are any items on your list packaged in materials produced by the chemical industry?

   b. How important is the packaging material?

2. In what ways are items or materials on your list better or worse than manufactured or synthetic alternatives? Consider factors such as cost, availability, and quality.

3. If a product is "100% natural" does that mean it has not been processed by the chemical industry? Why? Support your answer with at least one example.

In Riverwood, the two companies being considered manufacture products made of nitrogen and aluminum. The EKS Company produces nitric acid and ammonia, which are both basic chemical products; sheet aluminum, made by the WYE Corporation, is an example of a finished chemical product. This unit offers you the opportunity to learn about the processes used by the EKS and the WYE companies. This understanding will help you later, when you participate in the debate about whether a chemical plant should be located at Riverwood, and if so, which one to invite.

A.2 WHAT DOES RIVERWOOD WANT?

Before learning that the EKS Company and the WYE Corporation were interested in locating in Riverwood, the town council had already established some general criteria to be met by any new Riverwood industry. The criteria are partly based on the Environmental Protection Agency's (EPA) Toxics Release Inventory (TRI). TRI is a state-by-state database of annually required reported releases of approximately 300 designated toxic materials into the air, water, or land by more than 20,000 manufacturers. TRI data are available on CD-ROM, computer disks, magnetic tapes, as well as on-line and in books. Thus, TRI is a major source of information directly accessible by citizens interested in data about releases of toxic materials in their local environment, such as the people who live in a town like Riverwood.

Here are the town council's criteria, developed in conjunction with citizen hearings. Any new industry must:

- be a safe place to work;

- be environmentally responsible in the manufacturing, storage, use, and transportation of its products and disposal of wastes;
- respond promptly to community concerns about its products and operations, and take any needed corrective measures, as part of the public's "right to know";
- develop, with the community, procedures necessary to handle plant emergencies that could endanger the community and its surroundings; and
- comply with appropriate federal, state, and local regulations and policies governing the manufacturing of its products.

Each company also intends to establish a Riverwood Community Advisory Panel. Such panels are increasingly more common as a form of public outreach by chemical manufacturers.

In addition, officials of the EKS Company and the WYE Corporation have pointed out that they are members of the Responsible Care initiative developed by the nearly 200 member companies and partners of the Chemical Manufacturers Association (CMA). Begun in 1988, Responsible Care is a program in which chemical manufacturing companies voluntarily agree to public scrutiny and evaluation, the only major U.S. industry to do so.

Companies such as EKS and WYE who participate in Responsible Care pledge to follow ten guiding principles:

- Recognize and respond to community concerns.
- Develop chemicals that are safe to make, transport, use, and dispose of.
- Make health, safety, and environmental protection priorities in planning products and processes.
- Report information on chemical-related health hazards promptly to officials, workers, and the public, and recommend protective measures.
- Advise customers on the safe use, transport, and disposal of chemicals.
- Operate plants in such a way as to protect the environment and the health and safety of workers and the public.
- Conduct and support research on health, safety, and environmental effects of products, processes, and waste generated.
- Resolve problems created by past handling and disposal of hazardous materials.
- Participate with government and others to create responsible laws and regulations to safeguard the community, workplace, and environment.
- Offer assistance to others who produce, handle, use, transport, and dispose of chemicals.

The *Your Turn* on page 528 will give you an opportunity to compare the Riverwood criteria and the Responsible Care guidelines.

To better understand these two companies, you will explore the chemistry behind the EKS Company's manufacturing of synthetic nitrogen products used for fertilizers and for explosives. Although fertilizers

Responsible Care ®
A Public Commitment

and explosives clearly have different uses, both are based on at least one common starting material—nitrogen gas. In Part B we will explore some nitrogen chemistry in order to provide some perspective on the EKS Company's products. Later, you will learn more about the WYE Metals Corporation and its products. As a result, you will be better able to help the town council decide whether to have either manufacturing facility in Riverwood.

YOUR TURN

Living with a Chemical Plant

The first of several town meetings to discuss the possibility of a chemical plant in Riverwood will be held soon. Representatives from both companies and town council members will attend, as well as interested local citizens.

1. List four questions related to the town council's criteria you would want answered by the EKS Company if you lived in Riverwood.

2. List four questions based on the criteria you would want the WYE Corporation to answer.

3. List four questions related to the Responsible Care guidelines you would want answered by the EKS Company.

4. List four questions related to the Responsible Care guidelines you would want answered by the WYE Corporation.

5. Consider your lists for Questions 1–4. In what ways are the questions
   a. similar?   b. different?

6. What criteria would you like added to the town council's original list? Why?

7. How would you expect citizens to serve on a community advisory panel? What would they do?

? PART A: SUMMARY QUESTIONS

1. Identify three principal issues associated with a chemical plant locating in Riverwood.

2. The EKS Nitrogen Products Company produces ammonia and nitric acid. Anhydrous ammonia, NH_3, is one form of ammonia used as a fertilizer. Another common form is ammonium nitrate, NH_4NO_3, formed by the reaction of ammonia with nitric acid, HNO_3. Write a balanced equation for this reaction.

3. The WYE Metals Corporation produces aluminum, a very familiar metal. Aluminum resists corrosion because it reacts with oxygen, O_2, to produce a protective film of relatively unreactive aluminum oxide, Al_2O_3. Write a balanced equation for this reaction.

4. The two companies have spoken of "an environmentally safe setting" and "environmentally responsible manufacturing." Describe what you think these phrases mean.

5. You already know that home construction and automobile manufacturing are two major users of manufactured chemicals. Name two other significant users of such chemicals.

CHEMISTRY AT WORK

Searching for Solutions in Research Chemistry

Some day when you reach for a pill to relieve an aching muscle or a headache, you might have **Todd Blumenkopf** to thank. Todd is a Research Chemist at Pfizer, Inc. Todd and his laboratory staff are working on medications that reduce the swelling or inflammation of joints and muscles.

Todd and his colleagues in the Chemical Research Division are working to conceive of and synthesize new compounds that target those enzymes and receptors in the human body that contribute to inflammation. Todd's team investigates the relationship between a chemical compound's biological activity and its structure. Then they devise ways to synthesize the compound in the laboratory. Later, each compound is tested by in vitro (in an artificial environment outside the body) and in vivo (in the living body of a plant or animal) methods to determine whether it has desirable results without unwanted side effects.

After a drug has passed the required laboratory testing, researchers collect additional data about the drug to file with the Food and Drug Administration (FDA). When trying to get a new drug approved, companies like Pfizer must prepare and submit a large quantity of documents, including laboratory and toxicology data.

As a youngster, Todd had a strong interest in science. His parents encouraged him to pursue a career in science, where they felt his disability might be met with less resistance than in other fields. During his undergraduate years, Todd decided to focus on research as a career. As he progressed in his studies, he chose a career in pharmaceutical research. After receiving his Ph.D. and completing postdoctoral work, Todd now works at Pfizer's pharmaceutical research facilities in Groton, Connecticut.

Todd also works on the American Chemical Society's Committee on Chemists with Disabilities, which sponsors the development of materials and other efforts to improve opportunities in the chemical sciences for individuals with disabilities.

Todd believes that qualities needed for success in chemistry research are curiosity, creativity, and motivation. He also stresses the importance of patience and perseverance, since the process of developing a new product from an initial idea can take many years.

B THE CHEMISTRY OF SOME NITROGEN-BASED PRODUCTS

Fertilizer may sound like an unexciting product, but the manufacture and sale of fertilizers is a worldwide multimillion-dollar business. It employs thousands of people and affects the lives of nearly everyone. The EKS Nitrogen Products Company is committed to producing high-quality fertilizer at a reasonable cost, using the best technology available. The company manufactures fertilizers that contain nitrogen, phosphorus, and potassium, the major nutrients needed by growing plants. The complete list of nutrients is slightly different for each fertilizer.

Farmers and home gardeners use fertilizers to supplement natural soil nutrients. If your family uses fertilizers on a lawn, a garden, or house plants, how do you know which fertilizer is best for the job? One way is to find out whether the fertilizer contains the proper ingredients. Complete fertilizers contain a mixture of three main ingredients—nitrogen, phosphorus, and potassium—in addition to inert filler material and trace ions. The photo below shows a label of such a commercial fertilizer.

A sequence of three numbers indicates the percentages of the three key ingredients—nitrogen, N; phosphorus, P (calculated as P_2O_5); and potassium, K (calculated as K_2O)—in alphabetical order. Thus, the 29-3-4 values shown on the label indicate 29% nitrogen, 3% P_2O_5, and 4% K_2O, respectively. The proportion of each varies according to crop needs. Most lawn grasses need nitrogen, and so a 20-10-10 fertilizer would be a good choice

▶ *Reporting P and K as P_2O_5 and K_2O originated during early research on plant fertilizers, when plants were burned and the resulting amounts of P_2O_5 and K_2O were weighed.*

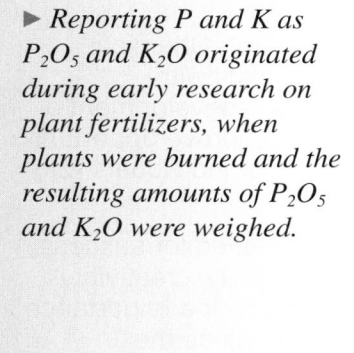

A bag of lawn fertilizer. The 29-3-4 formula refers to the percents of nitrogen, phosphorus (as P_2O_5), and potassium (as K_2O) present.

for them. When phosphorus is especially necessary, such as for fruits and vegetables, then a 10-30-10 mixture would be preferred over a balanced (10-10-10) composition.

Fertilizers usually contain small amounts of other important plant nutrients also, both cations and anions. The cations are likely to be potassium (K^+), magnesium (Mg^{2+}), and iron (Fe^{2+}); the anions—nitrate (NO_3^-), phosphate (PO_4^{3-}), and sulfate (SO_4^{2-}). Each cation and anion plays particular roles in plant development and growth. In the following laboratory activity you will confirm the presence (or absence) of various nutrients in a fertilizer solution.

B.1 LABORATORY ACTIVITY: FERTILIZER

GETTING READY

In this laboratory activity you will play the role of a technician in the Analytical Department of the EKS Company. You will be asked to test a fertilizer solution for six ions (three different anions and three different cations). So that you will know what a positive test looks like in each case, in Part 1 you will perform tests on known solutions of three anions and three cations used in many fertilizers. Then, in Part 2, by performing the same tests on the unknown fertilizer solution and comparing the results with those in Part 1, you should be able to identify which of the six different ions are present in the fertilizer solution.

Prepare the data table below in your notebook. The six "known" ions are listed across the top. Tests and reagents are listed in the left-hand column, keyed by procedure step numbers. The Xs in some cells indicate tests that are not needed.

Data Table—Ion Tests							
Ion Tests	**Known Cations**			**Known Anions**			**Unknown**
	Ammonium NH_4^+	**Iron(III) Fe^{3+}**	**Potassium K^+**	**Nitrate NO_3^-**	**Phosphate PO_4^{3-}**	**Sulfate SO_4^{2-}**	**# _____**
1. Color							
2a. NaOH							
2b. $BaCl_2$							
2c. $BaCl_2$ & HCl							
3. Brown ring test	×	×	×		×	×	
4. NaOH & litmus				×	×	×	
5c. Flame test	×	×		×	×	×	
5d. Flame test & cobalt glass	×	×		×	×	×	
6. KSCN test	×		×	×	×	×	

PROCEDURE

Part 1. Ion Tests

Put on your laboratory goggles before beginning work.

Prepare a water bath to use in Step 4. Put about 30 mL of distilled water into a 100-mL beaker. Place the beaker on a wire gauze on a ring stand and heat the water gently with a Bunsen burner flame. The water must be warm, but should not boil. Control the heat accordingly.

Use a toothpick when you are directed to mix a solution in a wellplate depression. To avoid contamination, a clean toothpick should be used each time.

COLOR TESTS

1. Obtain a Beral pipet set containing each of six known ions—nitrate (NO_3^-), phosphate (PO_4^{3-}), sulfate (SO_4^{2-}), ammonium (NH_4^+), potassium (K^+), and iron(III) (Fe^{3+}). Record the color of each solution in the data table.

NaOH and BaCl$_2$ TESTS

2. a. Label six of the depressions in a multiple-well wellplate with symbols for the known ion solutions. Then place two or three drops of each solution into the corresponding well. Test each sample solution individually by adding one or two drops of 3 M sodium hydroxide (NaOH) solution to it, using a new toothpick to mix each solution. Record your observations. If your wellplate still has at least six clean, unused depressions, go on to the next step without washing it. Otherwise, clean and rinse the wellplate.

   b. Repeat Procedure 2a, testing with 0.1 M barium chloride ($BaCl_2$) solution instead of NaOH. Record your observations. Do not clean the wellplate after these tests are completed.

   c. Add three drops of 6 M HCl to each of the six depressions containing $BaCl_2$ solution. Record your observations. Clean and rinse the wellplate.

BROWN-RING TEST for NITRATE

3. A mixture of Fe^{2+} and sulfuric acid produces a characteristic reaction indicating the presence of nitrate ion (NO_3^-): A brown ring forms at the interface of the two solutions.

   Follow this procedure to perform the brown-ring test:

   a. After putting a wellplate on a white sheet of paper, place three drops of nitrate ion solution in a clean depression.

   b. Carefully add two drops of concentrated sulfuric acid to the depression and gently mix the two solutions with a toothpick to form a single layer of solution.

   c. Working carefully and slowly, add two drops of iron(II) sulfate ($FeSO_4$) solution so the solution flows along the wall of the depression to form a second layer above the first one.

   d. Allow the wellplate to stand without mixing for two minutes.

e. Observe any change that occurs at the interface between the two liquid layers. Record your observations. A brown ring indicates the presence of nitrate ions in the original solution.

NaOH and LITMUS TESTING of CATIONS

4. a. Place four drops of each known cation test solution into a separate clean, small test tube.

    b. Moisten three pieces of red litmus paper with distilled water; place them on a watch glass.

    c. Add 10 drops of 3 M NaOH directly to the solution in one test tube. Do not allow any NaOH to contact the test tube lip or inner wall. Immediately stick a moistened red litmus paper strip onto the upper inside wall of the test tube. The strip must not contact the solution.

    d. Warm the test tube gently in the hot water bath for one minute. Note your observations after waiting about 30 seconds.

    e. Repeat Steps 4c and 4d for each remaining test tube.

FLAME TESTS

5. Various metal ions give off characteristic colors in a Bunsen burner flame. You will observe the flame test for potassium ions under two sets of conditions.

    a. Obtain two wood splints that have been soaked in a potassium ion solution.

    b. Set up and light a Bunsen burner. Adjust the flame to produce a light blue, steady inner cone, and a more luminous, pale blue outer cone.

    c. Insert a wood splint containing potassium ions into the outer cone of the flame. Note any change in flame color, the color's intensity, and the total time (seconds) the color lasts.

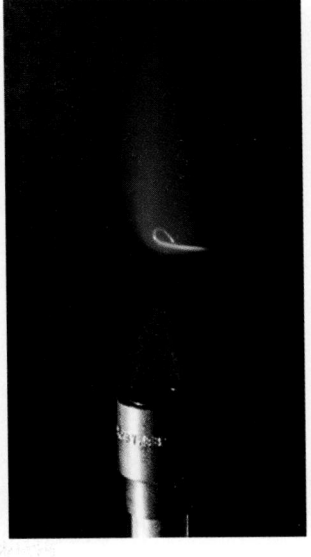

Flame tests: sodium (yellow), lithium (red), strontium (red), and calcium (orange).

d. Repeat the potassium ion flame test, this time observing the flame through a cobalt-blue glass (or a didymium glass). Again, note the color, intensity, and duration of the color. Your partner can hold the splint in the flame while you observe using the colored glass. Then change places. Record all observations in your data table.

KSCN TEST

In the Water unit (page 36) you learned that adding potassium thiocyanate (KSCN) to an aqueous solution containing Fe^{3+} ions produces a deep red color due to the formation of $FeSCN^{2+}$. This color confirms the presence of iron(III) in the solution.

6. a. Place three drops of the known Fe^{3+}-containing solution into a wellplate depression. Add one drop of 0.1 M KSCN solution to the depression. Record your observations.

   b. Clean and rinse the wellplate.

Part 2. Tests on Fertilizer Solution

1. Obtain a Beral pipet containing an unknown fertilizer solution. Record the unknown number of the solution in your data table. The solution will contain one of the anions and one of the cations tested in Part 1. Observe and record the unknown solution's color.

2. Conduct each test described in Part 1, but substitute the unknown fertilizer solution for the known solution. Record all observations. Repeat a test if you want to confirm your observations. *Note:* The test described below for potassium ions in the unknown fertilizer solution is modified from the test used in Part 1. Use the following flame test procedure to test for potassium ions in your unknown.

   a. Obtain a platinum or nichrome wire inserted into glass tubing or into a cork stopper.

   b. Set up and light a Bunsen burner. Adjust the flame to produce a light blue, steady inner cone, and a more luminous, pale blue outer cone.

   c. Place about 10 drops of 12 M HCl in a small test tube. Dip the wire into the HCl, and then insert the wire tip into the flame. Position the wire in the outer "luminous" part of the flame—not in the center cone. As the wire heats to a bright red, the flame may become colored. The color is due to metallic cations on the wire.

   d. Continue dipping the wire into the HCl and inserting the wire into the flame until there is little or no change in the flame's blue color when the wire is heated to redness. The wire is then clean.

   e. Place seven drops of the unknown solution into a clean depression in the wellplate. Dip the clean wire into the unknown fertilizer solution. Insert the wire into the flame as before. Note any change in flame color, the color's intensity, and the total time (seconds) the color lasts.

   f. Repeat the potassium ion flame test, this time observing the flame through a cobalt-blue glass (or a didymium glass). Again, note the color, intensity, and duration of the flame. Your partner can hold

▶ *The tubing or stopper serves as a handle.*

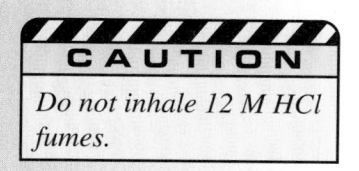

CAUTION

Do not inhale 12 M HCl fumes.

the wire in the flame while you observe using the colored glass. Then change places. Record all observations in your data table.

3. Wash your hands thoroughly when you finish the laboratory activity.

4. Compare your results from Part 2 with those obtained for the known solutions. Identify the anions and cations in your fertilizer solution by circling them in your data table.

QUESTIONS

1. Name and give the formulas of two compounds that could have supplied the ions present in your unknown fertilizer solution. For example, potassium chloride (KCl) supplies potassium ions to a solution; sodium nitrate ($NaNO_3$) furnishes nitrate ions to a solution.

2. Describe a test you could perform to decide whether a fertilizer sample contains phosphate ions.

3. If litmus paper indicates that a certain fertilizer solution is basic, which ion(s) studied in this activity could be present in the solution?

4. Could a candle be used as a replacement for a Bunsen burner in doing flame tests? Explain.

5. Does information gathered in this activity allow you to judge whether a given fertilizer is suitable for a particular use? Explain.

B.2 FERTILIZER'S CHEMICAL ROLES

Each year EKS manufactures more than 3 million tons of ammonia and more than 1.5 million tons of nitric acid. Most of this production is used to produce fertilizers sold to farmers and gardeners.

The agricultural use of industrially made fertilizers is a good example of materials substitution. Originally, manure served as the primary soil-enriching material for growing crops. Farmers add fertilizer to soil to increase the growth rate and yield of their crops. Recall the concept of a limiting reactant (Food unit, page 258). The purpose of all fertilizers is to add enough nutrients to soil so plants have adequate supplies of each one.

The raw materials used by growing crops are mainly carbon dioxide from the atmosphere, and water and nutrients from the soil. Water and

Lightning storms contribute to a series of reactions that "fix" atmospheric nitrogen as a dilute solution of nitric and nitrous acids, directly available for plant use.

Applying anhydrous ammonia fertilizer to the soil will help result in a healthy crop.

► *Ammonia can also be directly applied to soil as a nitrogen-rich fertilizer.*

nutrients such as nitrate (NO_3^-), phosphate (PO_4^{3-}), magnesium (Mg^{2+}), and potassium (K^+) ions are absorbed by plant roots from the soil. Phosphate becomes part of the energy-storage molecule ATP (adenosine triphosphate, page 471), the nucleic acids RNA and DNA, and other phosphate-containing compounds. Magnesium ions are a key component of chlorophyll, essential for photosynthesis. Potassium ions are found in the fluids and cells of most living things. Without adequate potassium ions, a plant's ability to convert carbohydrates from one form to another and to synthesize proteins would be diminished.

Nitrogen is critically important in plant growth. Plant cells are largely protein, and nitrogen makes up about 16% of the mass of protein molecules. Although molecular nitrogen gas (N_2) is abundant in the atmosphere, it is so unreactive that plants cannot use it directly. However, atmospheric nitrogen can be "fixed"—converted to compounds that plants can use. Lightning or combustion can "fix" atmospheric nitrogen by causing it to combine with other elements, especially hydrogen and oxygen, to form compounds used by plants. In addition, certain plants called legumes have nitrogen-fixing bacteria in their roots.

► *Conversion of N_2 to nitrogen compounds usable by plants is called* **nitrogen fixation.** *Clover and alfalfa are examples of legumes.*

Scientists are exploring biological methods, other than legumes, for making atmospheric nitrogen more available to plants. These include incorporating genes that produce nitrogen-fixing enzymes into microorganisms and even into higher plants. This would make it possible for any plants or their bacteria to produce their own nitrogen fertilizer, just as the bacteria on legumes do.

Ammonia (NH_3) and ammonium ions (NH_4^+), added to soil from decaying matter and other sources, are oxidized to nitrate ions (NO_3^-) by soil bacteria. Before building amino acids, plants first reduce the nitrate to nitrite ions (NO_2^-) and then to ammonia. They then use ammonia directly in amino acid synthesis. Unlike animals, higher plants can synthesize all the amino acids they need, starting with ammonia or nitrate ions.

When organic matter decays, much of the nitrogen released recycles among plants and animals, and some returns to the atmosphere. Thus, some nitrogen gas removed from the atmosphere through nitrogen fixation eventually cycles back to its origin. The nitrogen cycle, another global natural cycle like the water and carbon cycles (pages 14 and 411, respectively), is shown in Figure 2. It consists of these steps:

1. Atmospheric nitrogen (N_2) is converted to ammonia (NH_3) or ammonium ion (NH_4^+) by nitrogen-fixing bacteria that live in legume root nodules or in soil.

2. Ammonia and ammonium ions, in turn, are oxidized by various soil bacteria—first into nitrite ions (NO_2^-) and then into nitrate ions (NO_3^-).

3. Most nitrogen taken up by the roots of higher plants is in the form of nitrate.

4. The nitrogen then passes along the food chain to animals that feed on plants, and to animals that feed on other animals.

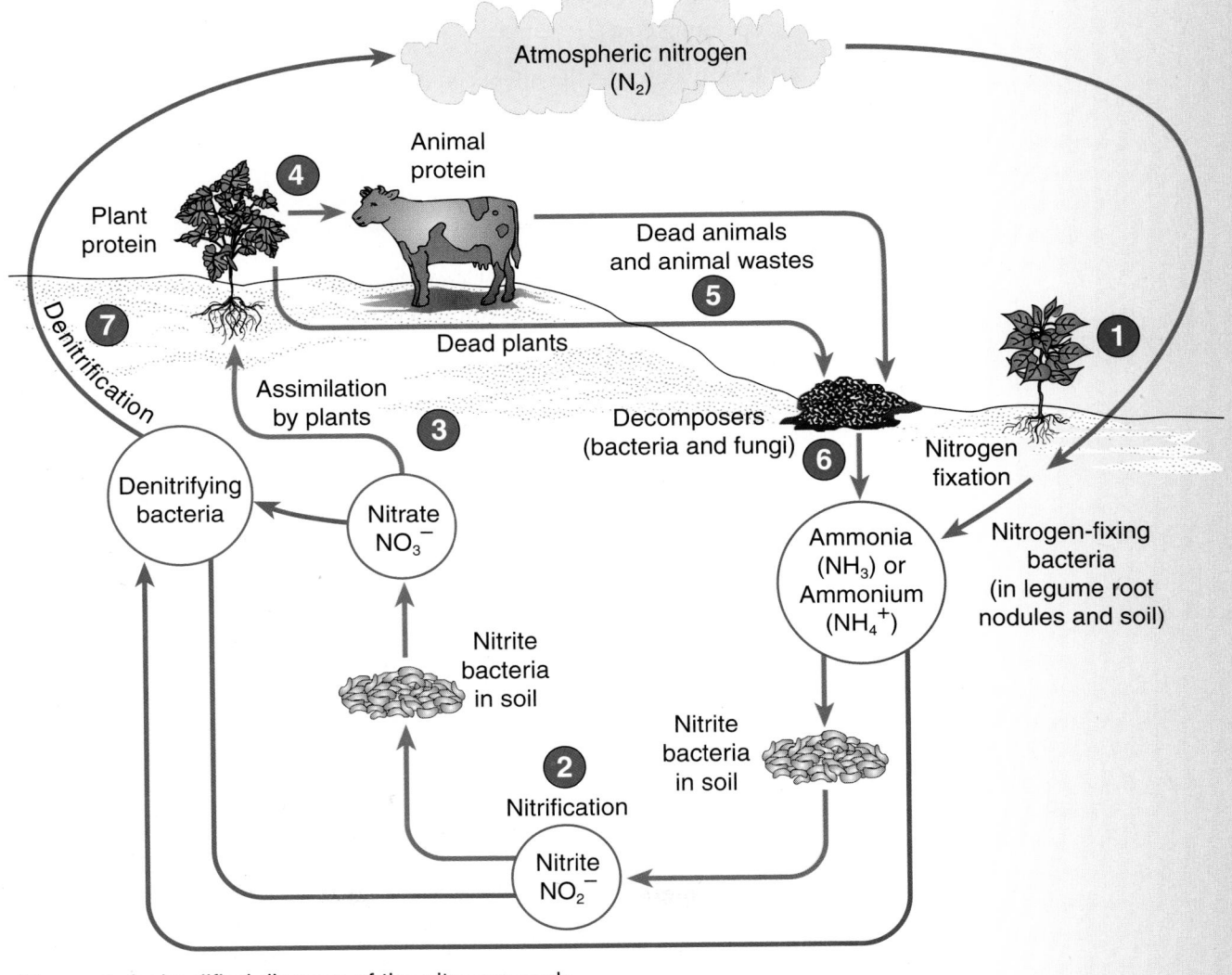

Figure 2 A simplified diagram of the nitrogen cycle.

5. When these plants and animals die, their proteins and other nitrogen-containing molecules are broken down chemically, mostly by bacteria and fungi.

6. Nitrogen not assimilated by bacteria and fungi is released as ammonia or ammonium ions. Thus, much nitrogen recycles through the living world without returning to the atmosphere.

7. Some nitrogen, however, is "lost" to the atmosphere after denitrifying bacteria convert ammonia, nitrite, and nitrate back to nitrogen gas.

YOUR TURN

Plant Nutrients

1. Some farmers alternate plantings of legumes with harvests of grain crops. Why?

2. Why is it beneficial to return nonharvested parts of crops to the soil?

3. How might research on new ways to fix nitrogen help lower farmers' operating costs?

4. Briefly describe the effects on a plant of having too few
   a. nitrate ions.
   b. phosphate ions.
   c. potassium ions.

In Laboratory Activity B.1 you identified some major ions present in fertilizer. In the next activity you will complete a quantitative study of one of these ions, phosphate (PO_4^{3-}), in a solution.

B.3 LABORATORY ACTIVITY: PHOSPHATES

GETTING READY

Most fertilizer packages list the percents (by mass) of the essential nutrients contained in the fertilizer. In this activity you will determine the mass and percent of phosphate ion in a fertilizer solution. The method you will use, called colorimetry, is based on the fact that the intensity of a solution's color is related to the concentration of the colored substance. So that you can use this method, you will employ a chemical reaction that converts colorless phosphate ions (PO_4^{3-}) to colored ions. To determine the percent of phosphate ion present, you will compare the intensity of color in the unknown solution with the color intensity of color standards, solutions with known phosphate concentrations. If the color intensity is the same in the sample as in the standard, then the phosphate concentrations are the same in the two solutions. If the color intensity of the sample is less than that of the standard, the sample has a lower phosphate concentration. To

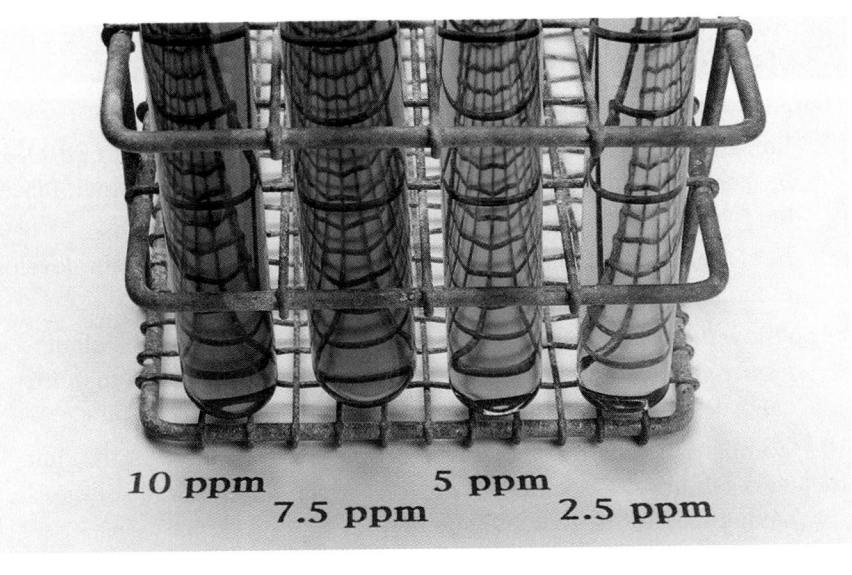

A set of color standards.

prepare the unknown fertilizer solution, you will dilute a solution of the fertilizer by a factor of 50. This dilutes it enough for comparison with the color standards.

PROCEDURE

1. Label five test tubes as follows: 10 ppm, 7.5 ppm, 5.0 ppm, 2.5 ppm, and X.

2. Complete these steps to prepare the unknown fertilizer solution:
   a. Place 0.50 g of fertilizer in a 400-mL beaker. Label the beaker "original."
   b. Add 250 mL of distilled water. Stir until the fertilizer is completely dissolved.
   c. Pour 5.0 mL of this solution into a clean, dry 400-mL beaker labeled "dilute." Discard the remaining 245 mL of original solution.
   d. Add 245 mL of distilled water to the 5.0-mL solution in the "dilute" beaker. Stir to mix.

3. Pour 20 mL of the diluted solution into the test tube labeled X. Discard the remaining diluted solution.

4. In the tube labeled 10 ppm, place 20 mL of standard 10 ppm phosphate ion solution provided by your teacher. Add solutions and water to the other three test tubes as listed below:

Water bath for heating samples.

Concentration (ppm)	Standard 10 ppm Phosphate Solution (mL)	Distilled Water (mL)
7.5	15.0	5.0
5.0	10.0	10.0
2.5	5.0	15.0

The Chemistry of Some Nitrogen-based Products

5. Add 2.0 mL of ammonium molybdate–sulfuric acid reagent to each of the four prepared standards and to the unknown.

6. Add a few crystals of ascorbic acid to each tube. Stir to dissolve.

7. Prepare a water bath by adding about 200 mL of tap water to a 400-mL beaker. Place the beaker on a ring stand above a Bunsen burner. Place the five test tubes in the water bath.

8. Heat the water bath containing the test tubes until a blue color develops in the 2.5-ppm solution. Turn off the burner.

9. Allow the test tubes to cool briefly. Then, using a test tube clamp, remove the test tubes from the water bath and place them in numerical order in a test tube rack.

10. Compare the color of the unknown solution with those of the standard solutions. Place the unknown between the standard solutions having the closest-matching colors.

11. Estimate the concentration (ppm) of the unknown solution from the known color standards. For example, if the unknown solution color falls between the 7.5-ppm and 5.0-ppm color standards, you might decide to call it 6 ppm, or 6 g PO_4^{3-} per 10^6 g solution. Record the estimated value.

12. Wash your hands thoroughly before leaving the laboratory.

CALCULATIONS

1. Calculate the mass (grams) of phosphate ion in the fertilizer using the following equation. In the blank, place the numerical value of the unknown solution concentration (in ppm) determined in Step 11.

$$\text{grams of } PO_4^{3-} = \frac{\underline{\hspace{1cm}} \text{g } PO_4^{3-}}{10^6 \text{ g solution}} \times 250 \text{ g solution} \times 50$$

The multiplication factor of 50 in the calculation takes into account the 50-fold dilution of the fertilizer solution. The phosphate concentration in the original fertilizer is 50 times greater than its value in the test sample. Record the calculated mass of phosphate ion.

2. Calculate the percent phosphate ion (by mass) in the fertilizer sample:

$$\% PO_4^{3-} = \frac{\text{Mass of phosphate ion (Step 11)}}{\text{Mass of fertilizer, 0.50 g}} \times 100\%$$

Record this value.

QUESTIONS

1. Name two household products or beverages for which you can estimate relative concentration just by observing color intensity.

2. Instruments called colorimeters are often used for determining solute concentration. They measure the quantity of light that passes through an unknown sample and compare it with the quantity of light that passes through a known standard solution. What are the advantages of a colorimeter over the human eye?

3. Explain this statement: The accuracy of colorimetric analysis depends on the care taken in preparing the standards.

4. How could a reaction that produces a precipitate be used to determine the concentration of an ion?

5. a. Why is it important for farmers to know the actual percent composition of fertilizers they use?

   b. What risks (or costs) are involved in applying more of a soil nutrient than is actually needed?

B.4 FIXING NITROGEN

In seeking ways to fix nitrogen artificially, scientists in 1780 first combined atmospheric nitrogen and oxygen by exposing them to an electric spark. However, the cost of electricity made this too expensive for any commercial use. A less expensive method, the Haber–Bosch process, replaced it. From 1912 to 1916, Fritz Haber and Karl Bosch in Germany developed the technique for making ammonia from hydrogen and nitrogen. Here is some of the chemistry behind the process used by EKS in its manufacturing facilities to produce ammonia.

Whenever atoms lose one or more of their electrons, the process is called oxidation. For example, the conversion of metallic sodium atoms into sodium ions (Na^+) is oxidation; electrically neutral sodium atoms are oxidized to +1 sodium ions. The opposite process, gaining electrons, is called reduction. For example, the formation of chloride ions, Cl^-, from electrically neutral chlorine atoms is reduction. Electrons can be transferred to or from atoms, molecules, or ions. As a result, elements and compounds can be oxidized or reduced, and products of oxidation–reduction reactions can be atoms, molecules, or ions.

The relative tendency of bonded atoms to attract electrons in compounds—that is, to become reduced—is called *electronegativity.* Numerical values have been assigned to this tendency. Nonmetals typically have higher electronegativities than metals. Electronegativity values for some common elements are shown in Figure 3, page 542.

Consider the key Haber–Bosch process reaction, depicted in the equation below with electron-dot formulas:

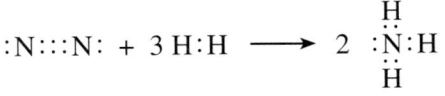

Note that each nitrogen atom originally shares six valence electrons with another nitrogen atom. Both nitrogen atoms have equal attraction for their shared electrons. As the reaction progresses, each nitrogen atom becomes covalently bonded to three hydrogen atoms. The bonded nitrogen and hydrogen atoms each share an electron pair—but they do not share equally. Nitrogen atoms are more electronegative than hydrogen atoms. Because nitrogen atoms have a greater attraction for the shared electrons than do hydrogen atoms, the nitrogen atom in each NH_3 molecule has acquired a greater share of hydrogen's electrons and been

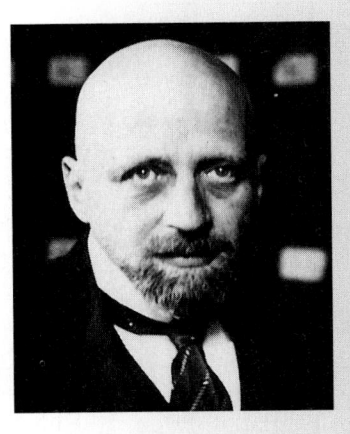

Fritz Haber (1868–1934) received a Nobel Prize in 1918 for the synthesis of ammonia from its elements.

▶ *For review of oxidation and reduction, see the Resources unit, page 142.*

Carl Bosch (1874–1940) received a Nobel Prize in 1931 for his invention and development of chemical high-pressure methods.

| | | | | | | H 2.1 | | | | | | | | | | |

												B 2.0	C 2.5	N 3.0	O 3.5	F 4.0
Li 1.0	Be 1.5											Al 1.5	Si 1.8	P 2.1	S 2.5	Cl 3.0
Na 0.9	Mg 1.2															
K 0.8	Ca 1.0	Sc 1.3	Ti 1.5	V 1.6	Cr 1.6	Mn 1.5	Fe 1.8	Co 1.9	Ni 1.9	Cu 1.9	Zn 1.6	Ga 1.6	Ge 1.8	As 2.0	Se 2.4	Br 2.8
Rb 0.8	Sr 1.0	Y 1.2	Zr 1.4	Nb 1.6	Mo 1.8	Tc 1.9	Ru 2.2	Rh 2.2	Pd 2.2	Ag 1.9	Cd 1.7	In 1.7	Sn 1.8	Sb 1.9	Te 2.1	I 2.5
Cs 0.7	Ba 0.9	La–Lu 1.0–1.2	Hf 1.3	Ta 1.5	W 1.7	Re 1.9	Os 2.2	Ir 2.2	Pt 2.2	Au 2.4	Hg 1.9	Tl 1.8	Pb 1.9	Bi 1.9	Po 2.0	At 2.2
Fr 0.7	Ra 0.9	Ac 1.1	Th 1.3	Pa 1.4	U 1.4	Np–No 1.4–1.3										

Decreasing electronegativity

Figure 3 Electronegativity values of selected elements.

▶ *Reduction is electron gain; oxidation is electron loss.*

reduced in its reaction with hydrogen. Consequently, each hydrogen atom has lost some of its share of electrons (has been oxidized) in the reaction.

The importance of the Haber–Bosch process is that it converts difficult-to-use nitrogen molecules from air into ammonia molecules, a form of "fixed" nitrogen (Figure 4). Once nitrogen is chemically combined with another element, it can be easily converted to other nitrogen-containing compounds. For example, under proper conditions, ammonia will react readily with oxygen gas to form nitrogen dioxide:

$$4\,NH_3 + 7\,O_2 \rightarrow 4\,NO_2 + 6\,H_2O$$

▶ *Oxidation–reduction reactions are sometimes called "redox" reactions.*

This is also an oxidation–reduction reaction. In forming NO_2, the nitrogen atom in ammonia has been oxidized—the nitrogen has lost its share of electrons. Why? Because oxygen is more electronegative, it attracts electrons more strongly than does hydrogen. So, each oxygen atom has been reduced—each oxygen has gained more control of electrons than it had in O_2.

Nitrogen (N_2) was reduced in the Haber–Bosch reaction. Because the nitrogen atom in NH_3 has a greater share of its electrons than it had in N_2, the nitrogen atom in NH_3 is said to have a negative oxidation state. But in

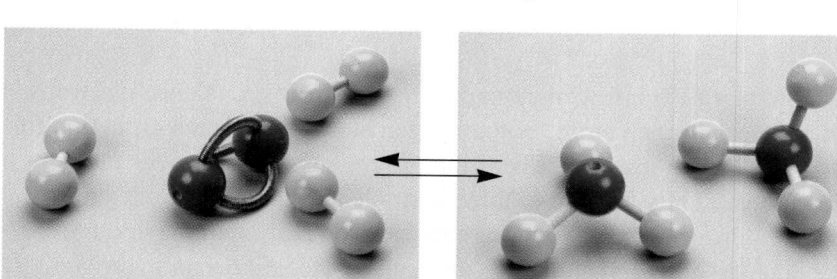

Figure 4 In the left photo, nitrogen molecules are shown in blue, and hydrogen molecules in yellow. They react to form ammonia molecules (right photo).

NO_2, the nitrogen atom has a lesser share of its electrons than it had in N_2 because oxygen attracts electrons more strongly than nitrogen. Consequently, the nitrogen atom in NO_2 has a positive oxidation state. In N_2 and O_2 (and atoms of all elements that are not combined chemically with any other element), each atom has a zero oxidation state.

YOUR TURN

Electronegativity and Oxidation State

Oxidation state is a convenient, although arbitrary, way to express the degree of oxidation or reduction of atoms in substances. Each atom in an element or compound can be assigned a numerical oxidation state. The higher (more positive) the oxidation state, the more the atom has been oxidized. The lower (less positive) the oxidation state, the more it has been reduced. In assigning oxidation states in binary compounds (compounds of two elements), atoms of lower electronegativity are assigned positive oxidation states—corresponding to loss of electrons (an oxidized state). Atoms of more electronegative elements are given negative oxidation state values—corresponding to a gain of electrons (a reduced state).

Using these guidelines we can determine, for example, which element in aluminum oxide (Al_2O_3) has the positive oxidation state and which the negative oxidation state. The electronegativity values in Figure 3 (page 543) indicate that aluminum has an electronegativity of 1.5 and oxygen has 3.5. This means that oxygen has greater electron-attracting ability than does aluminum. Therefore, in aluminum oxide, aluminum is assigned a positive oxidation state and oxygen a negative oxidation state.

The chemical combination of aluminum and oxygen

$$4\,Al + 3\,O_2 \rightarrow 2\,Al_2O_3$$

is an oxidation–reduction reaction in which aluminum metal becomes oxidized. This is because its oxidation state changes from zero (in the uncombined element) to a positive value. By contrast, oxygen gas becomes reduced—its oxidation state changes from zero to a negative value.

1. Consider these compounds. Using electronegativity values, decide which element in each compound is assigned a positive oxidation state and which is assigned a negative oxidation state.

    a. Sulfur trioxide, SO_3

    b. Hydrazine, N_2H_4

    c. Water, H_2O

    d. Hydrogen chloride, HCl

    e. Sodium chloride, NaCl

    f. Carbon monoxide, CO

    g. Iodine trifluoride, IF_3

    h. Manganese dioxide, MnO_2

2. a. Each of the following compounds is composed of a metallic and nonmetallic element. Decide which element in each compound has a positive oxidation state and which has a negative oxidation state.

▶ *All elements that are not combined with other elements are assigned zero oxidation state. Oxidation state is a useful concept in understanding redox reactions. "Oxidation state" is sometimes referred to as "oxidation number."*

▶ *The oxidation state of an* uncombined *element is zero. That is, when an element is not combined with any* other *element, its oxidation state is zero.*

▶ *During various reactions of the nitrogen cycle (page 537), nitrogen gas (N_2), in which nitrogen has an oxidation state of zero, is oxidized and reduced.*

(1) Lead(II) fluoride, PbF_2 (4) Nickel(II) oxide, NiO

(2) Sodium iodide, NaI   (5) Iron(III) chloride, $FeCl_3$

(3) Potassium oxide, K_2O (6) Lead(II) sulfide, PbS

b. Consider your answers to Question 2a. What conclusion can you draw about the oxidation states of metals and nonmetals in binary compounds?

3. In this oxidation–reduction reaction, which element is oxidized and which is reduced?

$$Ni + S \rightarrow NiS$$

4. Iron ions are part of a system essential to energy transfer in cells. In that system, Fe^{2+} ions are converted to Fe^{3+} ions. Is this oxidation or reduction?

The relative ease of conversion of one nitrogen compound to another is used in the chemical industry to make a large number of compounds from ammonia. Table 1 gives examples of the enormous impact ammonia production has on making available a wide variety of consumer products.

▶ *R = hydrocarbon group; M = metal ion.*

Table 1	Industrial Nitrogen Compounds	
Compound	Structure	Use
Ammonia, NH_3	NH_3	Fertilizer, explosives, fibers, plastics
Nitric Acid, HNO_3	$HO—NO_2$	Fertilizer, explosives, manufacturing
Hydrazine, N_2H_4	$H_2N—NH_2$	Rocket fuel, plastics
Amines, $R—NH_2$ (e.g., propylamine)	$CH_3CH_2CH_2NH_2$	Intermediate
Nitrates, MNO_3 (e.g., sodium nitrate)	$Na^+NO_3^-$	Fertilizer, explosives, food preservation
Nitrites, MNO_2 (e.g., sodium nitrite)	$Na^+NO_2^-$	Textile bleach, manufacturing, food preservation
Hydroxylamine, NH_2OH	H—N̈—OH with H below	Reducing agent
Phenylhydrazine, $C_6H_5N_2H_3$	⬡—$NHNH_2$	Dye intermediate
Urea, $(NH_2)_2CO$	H_2NCNH_2 with ‖ O below	Fertilizer, plastics

B.5 INDUSTRIAL NITROGEN FIXATION AT RIVERWOOD

Producing ammonia from nitrogen gas and hydrogen gas is a chemical challenge. A principal reason is that the reaction does not use up all the available nitrogen and hydrogen to produce ammonia; some nitrogen and hydrogen remain, even after ammonia is produced. When ammonia molecules form in the reactor, they immediately begin to decompose back to nitrogen and hydrogen molecules. This type of reaction—where products reform reactants while reactants form products—is known as a reversible reaction.

Chemists represent reversible reactions with double arrows linking the products and reactants.

$$N_2 + 3\,H_2 \rightleftharpoons 2\,NH_3$$

The reverse reaction (ammonia decomposition) occurs simultaneously with the forward reaction (ammonia production). This limits the amount of ammonia that can be obtained. A dynamic equilibrium is reached when the rate of product formation (in this case ammonia) equals the rate at which it decomposes. In a dynamic equilibrium, the reaction appears to stop, because the rates of forward and reverse changes exactly balance one another.

Although ammonia decomposes in the reactor due to the high temperature, it is stable once it cools to room temperature. One possible way to increase the yield of ammonia is to cool and remove the ammonia as soon as it forms. This shifts the overall reaction (indicated in the reversible equation above) to the right; additional ammonia continues to form. Although expensive, this procedure prevents the reaction from reaching equilibrium; ammonia is removed as it forms, thus preventing the reverse reaction (ammonia decomposition).

▶ *Modern ammonia plants give a 20%–30% conversion of reactants to ammonia, depending on the temperature and pressure used.*

▶ *All reversible reactions in closed containers reach equilibrium if conditions such as temperature remain constant.*

Worker inspecting machinery at an ammonia-processing plant.

Because ammonia is stable at lower temperatures, another yield-increasing plan might seem possible. Why not run the reactor at a low temperature, so the ammonia will not decompose? Unfortunately, this doesn't work. Simply mixing three volumes of hydrogen with one volume of nitrogen at ordinary temperatures does not produce any noticeable ammonia, regardless of how long the gases remain together. The rate of ammonia formation is too slow because the temperature is too low to furnish the necessary activation energy (see Air unit, page 437).

The major breakthrough that led to workable, profitable ammonia production by the Haber–Bosch process was the discovery of a suitable catalyst. The catalyst made it possible to produce ammonia at lower temperatures (450–500 °C), where it is more stable. The first catalyst used was metallic iron, but now it is a mixture of iron oxides and aluminum oxide.

Commercial ammonia production involves more than allowing nitrogen gas and hydrogen gas to react in the presence of a catalyst. First, of course, the reactants must be obtained. Nitrogen gas is taken from the air, and hydrogen is obtained from natural gas (mainly methane, CH_4). Consequently, an ammonia plant in Riverwood would require construction of a natural gas pipeline.

Sulfur-containing compounds are removed from the natural gas by passing it through absorptive beds. Next, methane reacts with steam to produce hydrogen gas:

$$CH_4 + H_2O \rightarrow 3\,H_2 + CO$$

In modern ammonia plants, this endothermic reaction takes place at 200–600 °C, at pressures of 200–900 atm. The ratio of methane (CH_4) to steam must be controlled carefully to prevent the formation of a variety of carbon compounds.

Carbon monoxide, a product of the hydrogen-generating reaction shown above, is converted to carbon dioxide, with the formation of additional hydrogen gas:

$$CO + H_2O \rightarrow H_2 + CO_2$$

The hydrogen gas is then purified by removal of carbon dioxide and any unreacted methane from the CO-producing step.

▶ *The carbon dioxide can be removed in a variety of ways, including allowing it to react with calcium oxide (lime) to form solid calcium carbonate, or by dissolving carbon dioxide gas at high pressure in water.*

Steam turbines compress the hydrogen and nitrogen to high pressures (150 to 300 atm). Ammonia forms as the gases at 500 °C flow over a catalyst of reduced iron oxide (Fe_3O_4) and aluminum oxide:

$$N_2 + 3\,H_2 \rightleftharpoons 2\,NH_3$$

Unreacted nitrogen and hydrogen gas are recycled, mixed with new supplies of reactants, and passed through the reaction chamber again.

Before the Haber–Bosch synthesis was developed in 1914, nitrogen-containing fertilizers came either from animal waste or from nitrate compounds. Large quantities of such compounds came from nitrate beds in South American deserts, in Chile. At the turn of the twentieth century, there had been speculation that this nitrate source would be used up by about 1930, which would have led to an agricultural crisis and the spectre of world famine. However, the synthetic manufacturing of fertilizers derived from ammonia by the Haber–Bosch process removed agriculture's sole dependency on natural nitrate sources.

The impact that commercial ammonia for fertilizer has had on agriculture and world food supplies cannnot be overemphasized. World ammonia production has increased dramatically since 1950, as farmers worldwide have increased their use of fertilizer to meet the food needs of growing populations. The U.S. chemical industry annually produces over 18 million tons of ammonia.

CHEMQUANDARY

WHAT PRICE SURVIVAL?

A significant portion of the world's population depends on food grown and preserved with the aid of products of the chemical industry—such as fuel, fertilizers, fungicides, and pesticides. Without these materials, world access to food would be insufficient to support all of the world's current human inhabitants.

In what ways might the goal of a pollution-free and chemically safer environment conflict with the goal of feeding and housing a rapidly growing world population?

B.6 NITROGEN'S OTHER FACE

The Haber–Bosch process made inexpensive ammonia commercially available. Ammonia itself can react with nitric acid to produce ammonium nitrate, an important substitute for natural nitrates.

$$NH_3 \quad + \quad HNO_3 \quad \rightarrow \quad NH_4NO_3$$
$$\text{ammonia} \quad + \quad \text{nitric acid} \quad \rightarrow \quad \text{ammonium nitrate}$$

The widespread availability of ammonia and nitrates was important not only in agriculture, but also in warfare. Because almost all chemical explosives involve nitrogen-containing compounds, ammonia can be converted to explosives. The Haber–Bosch process provided an independent source of ammonia from which fertilizers and military munitions could be made. Making military explosives in this way allowed Germany to continue fighting in World War I even after its shipping connections to Chilean nitrate deposits were cut off by the British Navy.

Explosives also have important nonhostile uses. Air bags in automobiles are one such modern application. The air bag inflates like a big "pillow" during a collision to reduce injuries to the driver and passengers. The uninflated air bag assembly contains solid sodium azide, NaN_3. In a collision, sensors initiate a sequence of events that rapidly decompose the sodium azide to form nitrogen gas. The nitrogen inflates the driver's air bag fully (to about 50 L) within 50 milliseconds (0.050 seconds) after the collision begins (60 milliseconds for a 150-L passenger air bag).

The chemical forces released by explosives also blast road-cuts for interstate highways. To cut through the stone faces of hills and mountains, road crews drill holes, drop in explosive canisters, and detonate them. The

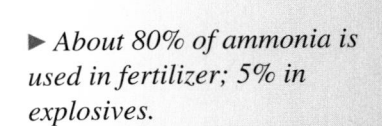

▶ *About 80% of ammonia is used in fertilizer; 5% in explosives.*

Sodium azide is decomposed to form nitrogen gas when an air bag deploys.

The Chemistry of Some Nitrogen-based Products

One use for explosives—the demolition of the Hotel Madison.

explosion is generally caused by rapid formation of gaseous products from liquid or solid reactants. A gas takes up more than a thousand times the volume of the same number of molecules of solid or liquid explosive, such as sodium azide, dynamite, or nitroglycerin.

Working with loud, dangerous explosives caused considerable trouble in the 1880s for a family named Nobel. It also made and lost fortunes for them several times. The father and his four sons were all interested in explosives, but Alfred, one of the sons, was the most persistent experimenter.

In 1846, the powerful explosive nitroglycerin (see page 550) was invented. However, it was too sensitive to be useful—one never knew when it was going to explode. The Nobels built a laboratory in Stockholm where they explored ways to control this unstable substance. Carelessness, as well as ignorance of nitroglycerin's properties, led to many destructive explosions. Alfred's brother, Emil, was killed in one of them. The city of Stockholm finally insisted that Alfred take his experimenting elsewhere. Grimly determined to continue and to make nitroglycerin less dangerous, Alfred rented a barge, carrying out his experiments in the middle of a lake. He finally discovered that adsorbing the oily nitroglycerin on finely divided sand (diatomaceous earth) made it stable enough to be transported and stored, but it would still explode when activated by a blasting cap. This new, more stable form of nitroglycerin was called dynamite.

A new era had begun. At first, dynamite served peaceful uses in mining and in road and tunnel construction. By the late 1800s, however, dynamite was also used in warfare.

The military use of his invention caused Alfred Nobel considerable anguish and motivated him to use his fortune to benefit humanity. In his will, he left money for annual prizes in physics, chemistry, physiology and medicine, literature, and peace. (The Swedish parliament later added economics to the award categories.) The Nobel prizes, first awarded in 1901, are still regarded as the highest honor scientists can receive. Recent Nobel laureates in chemistry are listed in Table 2, along with their contributions to chemical science.

A Nobel medallion.

Table 2	Nobel Laureates in Chemistry 1990–96	
Year	Awardees	Contributions
1996	Richard E. Smalley, United States, Rice University Robert F. Curl, United States, Rice University Harold W. Kroto, United Kingdom, University of Sussex	Discovered fullerenes, a new class of carbon allotropes ("buckyballs").
1995	Paul Crutzen, Germany, Max Planck Institute Mario Molina, United States, Massachusetts Institute of Technology F. Sherwood Rowland, United States, University of California–Irvine	Discovered and evaluated roles of nitrogen oxides and chlorofluorocarbons in stratospheric ozone depletion.
1994	George Olah, United States, University of Southern California	Developed classes of "super acids" to use in characterizing unstable chemical species.
1993	Michael Smith, Canada, University of British Columbia Kary B. Mullis, United States, University of California–San Diego	Developed site-directed ways of reprogramming cellular DNA; DNA-based chemistry.
1992	Rudolph Marcus, United States, California Institute of Technology	Contributed a major theory of electron-transfer reactions.
1991	Richard R. Ernst, Switzerland, Federal Institute of Technology, Zurich	Improved technology for nuclear magnetic resonance, used in medical diagnosis and in revealing structure of complex molecules.
1990	Elias Corey, United States, Harvard University	Established logical approaches to synthesizing complex organic molecules, including drugs.

The Chemistry of Some Nitrogen-based Products

The EKS Company produces several chemicals in its Explosives Division. Names and formulas of some of these explosives are shown in Figure 5; their formulas differ greatly, but they all contain nitrogen atoms. Most have nitrogen in a positive oxidation state and carbon in a negative oxidation state within the same molecule. This creates conditions for a very rapid transfer of electrons from carbon to nitrogen, accompanied by release of vast quantities of energy. The driving force of this type reaction is the formation of N_2, a very stable molecule, as you know.

Reactions of explosives are not reversible and do not reach equilibrium; they go to completion. Many different products are possible. Here are two examples of explosive reactions:

▶ *Explosions are rapid, exothermic oxidation–reduction reactions that release large amounts of gas.*

$$4\ C_3H_5(NO_3)_3(l) \rightarrow 12\ CO_2(g) + 6\ N_2(g) + 10\ H_2O(g) + O_2(g)$$
nitroglycerin

$$4\ C_7H_5N_3O_6(s) + 21\ O_2(g) \rightarrow 28\ CO_2(g) + 10\ H_2O(g) + 6\ N_2(g)$$
TNT, trinitrotoluene

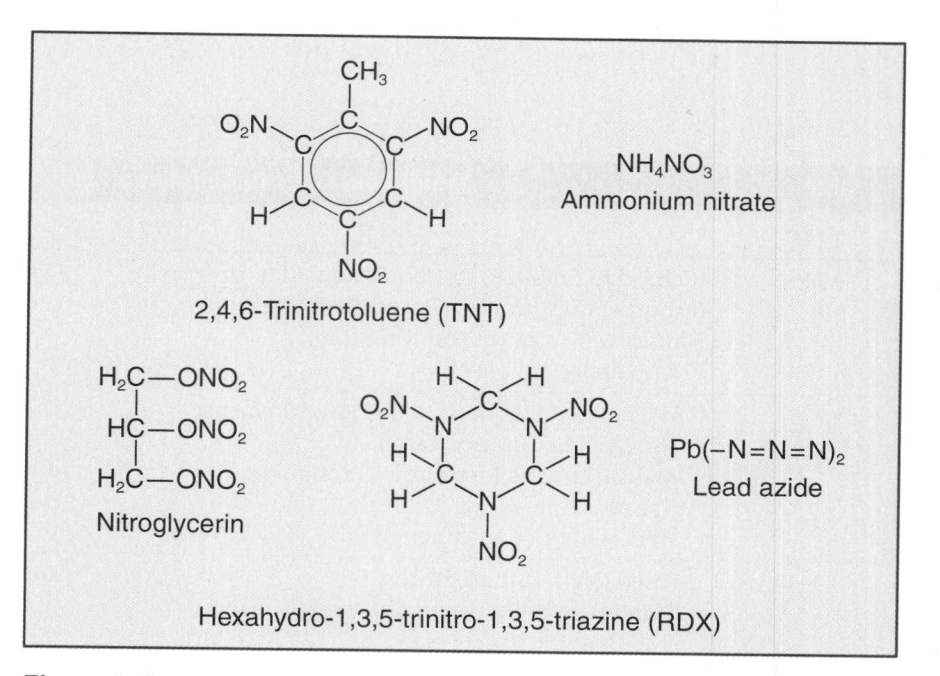

Figure 5 Some explosives produced by the EKS Company.

Refer to the equations above in answering the following *Your Turn.*

YOUR TURN

Chemistry of Explosives

1. How many total moles of gas are formed in the explosion of one mole of TNT?

2. Assume that a gas sample at a given temperature occupies 1,000 times as much space as the same number of moles of solid. Also assume that one mole of TNT occupies "one unit" of volume.

   a. Using the value you found in answering Question 1, how many "units" of gas volume are formed when one mole of TNT explodes? (Assume temperature remains constant.)

   b. By what factor would the total volume (at constant temperature) increase in a TNT explosion?

3. In fact, when TNT explodes, the rise in temperature causes the volume to increase eight times more than the factor calculated in Question 2b.

   a. By what combined factor, then, does volume increase in an actual TNT explosion?

   b. How does this help explain the destructive power of such an explosion?

4. Which of these equations might represent possible explosive reactions? Why?

   a. $C_5H_{12}(l) + 8\ O_2(g) \rightarrow 5\ CO_2(g) + 6\ H_2O(g) +$ energy

   b. $CaCO_3(s) +$ energy $\rightarrow CaO(s) + CO_2(g)$

   c. $C_3H_6N_6O_6(s) \rightarrow 3\ CO(g) + 3\ H_2O(g) + 3\ N_2(g) +$ energy

   d. $2\ NaN_3(s) \rightarrow 2\ Na(l) + 3\ N_2(g) +$ energy

B.7 YOU DECIDE:
FOOD OR ARMS

Substances produced from ammonia are useful in both war and peace. Because ammonia is the starting material for explosives and for fertilizers, it can be used either for destructive or constructive purposes.

1. Assume it is 1917, and World War I has started. A group seeks to ban the production of ammonia entirely, because it is used to produce destructive military explosives. The group asks you to join. What is your decision? Explain your reasoning.

2. If the Haber–Bosch process for ammonia production had been banned after World War I, how might the world today be different?

3. Now apply the same reasoning to the issue of nuclear energy today:

   a. A group seeks to ban the production of nuclear energy entirely, because it is used to produce destructive military explosives. The group asks you to join. What is your decision? Explain your reasoning.

   b. If the technology for producing nuclear energy had been banned after World War II, how might the world today be different?

You have learned how the EKS Nitrogen Products Company produces ammonia, a product that is sometimes used to serve destructive purposes. How will Riverwood citizens feel about that as they consider whether to invite a chemical company to build a plant in the town? Will it cause the citizens to favor the WYE Metals Corporation over the EKS Nitrogen Products Company? Should either company be invited?

To help you answer these questions, it's time to learn more about the WYE Metals Corporation. It specializes in electrochemical processes—chemical changes that produce or are caused by electrical energy.

? PART B: SUMMARY QUESTIONS

1. How does the concept of limiting reactants relate to the use of fertilizers?

2. Molecular nitrogen constitutes 78% of all the molecules in the atmosphere, yet nitrogen can be a limiting reactant for plants. Explain.

3. A newspaper story states that ". . . oxygen is needed for all oxidation reactions . . ." On what chemical basis would you agree or disagree with this statement?

4. a. What is meant by oxidation state?

   b. How is oxidation state related to oxidation and to reduction?

   c. Illustrate how this concept applies to this equation:

   $$4\,NH_3 + 7\,O_2 \rightarrow 4\,NO_2 + 6\,H_2O$$

5. The following questions are based on this equation:

   $$N_2(g) + 3\,H_2(g) \rightleftharpoons 2\,NH_3(g) + energy$$

   a. What are the sources of the raw materials for this reaction?

   b. Is the forward reaction endothermic or exothermic?

   c. In light of your answer to Question 5b, would you expect the forward reaction to be favored at high or low temperatures?

   d. What is the disadvantage of running the reaction at these temperatures?

   e. What role does a catalyst play in this reaction?

EXTENDING YOUR KNOWLEDGE
(OPTIONAL)

- Visit a garden shop or hardware store and read the labels on various fertilizer preparations. What chemical substances are commonly found in these fertilizers? How are the quantities of these substances in various fertilizers reported? How do the compositions of various fertilizers vary for different uses? Can you find any fertilizers made up of only one chemical compound? If so, what is the name and formula of the compound? For what uses is that fertilizer recommended? Why?

- Scientists have long debated whether they are responsible for all the consequences of their discoveries. Do you think scientists have more responsibility than other citizens in these decisions? Should scientists try to control society's uses of their discoveries, helping to ensure that the discoveries are not used for destructive or evil purposes? Or, is this a matter for all members of society to decide? If so, how can citizens become well enough informed to make wise decisions?

THE CHEMISTRY OF ELECTRO-CHEMISTRY

The WYE Metals Corporation wants to use its long experience with electrochemical processses to produce aluminum metal at its Riverwood plant using oxidation–reduction methods. That might sound like an easy task, as aluminum is the most abundant metallic element in the Earth's crust (Resources unit, page 133). However, aluminum is present in the Earth's crust not as aluminum metal, but as the ore bauxite, in which Al^{3+} ions are tied up chemically with silicon and oxygen atoms. Thus, the aluminum must be extracted from bauxite. This is done by electrolysis—using an electric current to reduce Al^{3+} ions to aluminum metal.

The electrolysis process requires so much electricity that electrical costs are an important factor in plant location. The hydroelectric plant at the Snake River Dam produces considerably more electrical power than is needed to serve Riverwood and surrounding communities. To encourage WYE to consider locating in Riverwood, power company officials have offered WYE large quantities of electrical power at very competitive rates.

The following laboratory activities and discussions provide background on electrochemistry. This information will help you understand how the proposed new plant will operate.

C.1 LABORATORY ACTIVITY: VOLTAIC CELLS

GETTING READY

In the Resources unit (page 142) and earlier in this unit, you learned that some metals release electrons (become oxidized) more readily than others—that is, they are more active. The relative tendencies for metals in contact with water to release electrons can be summarized in an activity series of the metals (see Table 3, page 555). A metal higher in the activity series will give up electrons more readily than a metal that is lower. For example, aluminum is oxidized (loses electrons) easier than is iron.

We can make use of the differing tendencies of metals to lose electrons to obtain electrical energy from a chemical reaction. We do this by constructing a *voltaic cell* in which electrons flow spontaneously through a wire connecting the two metals. The flow of electrons is called an *electric current*. In this laboratory activity you will study several voltaic cells. Each voltaic cell consists of two half-cells. A *half-cell* consists of a metal dipping into a solution of that metal's ions, for example, a piece of copper metal immersed in a solution of Cu^{2+} ions.

Table 3	Activity Series of Common Metals		
Metal	**Products of Metal Reactivity**		
$Li(s)$	$\rightarrow$ $Li^+(aq)$	+	e^-
$Na(s)$	$\rightarrow$ $Na^+(aq)$	+	e^-
$Mg(s)$	$\rightarrow$ $Mg^{2+}(aq)$	+	$2\,e^-$
$Al(s)$	$\rightarrow$ $Al^{3+}(aq)$	+	$3\,e^-$
$Mn(s)$	$\rightarrow$ $Mn^{2+}(aq)$	+	$2\,e^-$
$Zn(s)$	$\rightarrow$ $Zn^{2+}(aq)$	+	$2\,e^-$
$Cr(s)$	$\rightarrow$ $Cr^{3+}(aq)$	+	$3\,e^-$
$Fe(s)$	$\rightarrow$ $Fe^{2+}(aq)$	+	$2\,e^-$
$Ni(s)$	$\rightarrow$ $Ni^{2+}(aq)$	+	$2\,e^-$
$Sn(s)$	$\rightarrow$ $Sn^{2+}(aq)$	+	$2\,e^-$
$Pb(s)$	$\rightarrow$ $Pb^{2+}(aq)$	+	$2\,e^-$
$Cu(s)$	$\rightarrow$ $Cu^{2+}(aq)$	+	$2\,e^-$
$Ag(s)$	$\rightarrow$ $Ag^+(aq)$	+	e^-
$Au(s)$	$\rightarrow$ $Au^{3+}(aq)$	+	$3\,e^-$

When two metals of differing electron-releasing tendencies are connected in a voltaic cell, an ***electrical potential*** (that is, a difference in electrical energy) is created between the metals. Electrical potential, which is measured in volts (V), is somewhat like water pressure in a pipe. It represents the "push" that drives electrons through the wire connecting the two metals. The greater the difference in activity of the metals, the greater the "electron pressure," or electrical potential, of the cell.

To provide both electrical conduction inside the cell and a pathway for electrons leaving the wire or external circuit, a voltaic cell is prepared by immersing each metal in a solution of its ions. In this laboratory activity, the half-cells make electrical contact through the saturated filter paper.

Prepare an appropriate data table for the following procedures.

PROCEDURE

1. Put on your protective goggles.
2. Add 1 mL of 0.1 M $Cu(NO_3)_2$ to each of three wells on the wellplate.
3. Add 1 mL of 0.1 M $Zn(NO_3)_2$ to a well adjacent to one containing $Cu(NO_3)_2$.
4. Add 1 mL of 0.1 M $Mg(NO_3)_2$ to another well adjacent to one containing $Cu(NO_3)_2$.
5. Add 1 mL of 0.1 M $Fe(NO_3)_2$ to a third well adjacent to one containing $Cu(NO_3)_2$.

You will use copper metal and zinc metal strips of different widths—0.25 cm (narrow), 0.50 cm (medium), and 1.0 cm (wide). Make sure that you use the proper size metal strip.

6. Place one medium copper metal strip into each $Cu(NO_3)_2$ well.

7. Place one medium zinc metal strip into the $Zn(NO_3)_2$ well.

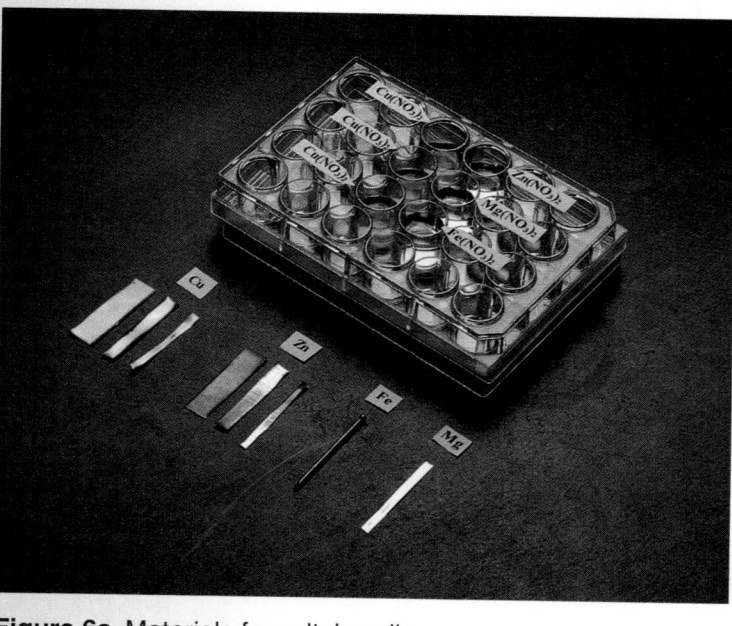

Figure 6a Materials for voltaic cells.

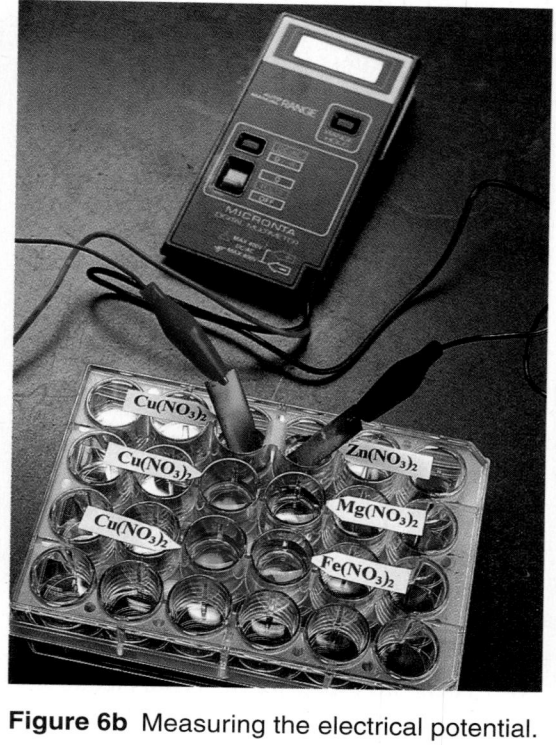

Figure 6b Measuring the electrical potential.

8. Place a magnesium metal strip into the $Mg(NO_3)_2$ well, and an iron strip (nail) into the $Fe(NO_3)_2$ well.

9. Use a piece of filter paper saturated with KCl solution to connect the solutions in the $Cu(NO_3)_2$ and $Zn(NO_3)_2$ wells. In doing so, do not allow the metals to touch.

10. Make similar connections between the $Cu(NO_3)_2$ and $Mg(NO_3)_2$ wells, and the $Cu(NO_3)_2$ and $Fe(NO_3)_2$ wells. In doing so, do not allow the metals to touch.

11. Attach an alligator clip from the voltmeter to the copper metal strip in the $Cu(NO_3)_2$ well. Lightly touch the second alligator clip attached to the voltmeter to the zinc metal strip in the $Zn(NO_3)_2$ well. If the needle deflects in the direction of a positive potential, attach the clip to the zinc metal. If the needle deflects in the negative potential direction, reverse the clip connections to the metal strips (see Figure 6b).

12. Record the reading from the voltmeter in the data table.

13. Repeat Steps 11 and 12 for the copper and magnesium strips, and then for the copper and iron strips.

14. Repeat the measurements using the narrow pieces of copper metal and zinc metal in their respective solutions.

15. Repeat the measurements using the wide pieces of copper metal and zinc metal in their respective solutions.

16. Wash your hands thoroughly before leaving the laboratory.

QUESTIONS

1. a. Rank the three voltaic cells in order of decreasing electrical potential.

   b. Explain the observed order in terms of the activity series.

2. Use Table 3 (page 555) to predict whether the electrical potential of cells composed of these metal pairings would be higher or lower than that of the Zn–Cu cell:

   a. Zn and Cr   b. Zn and Ag   c. Sn and Cu

3. How did decreasing the size of the zinc and copper electrodes affect the measured electrical potential? Did increasing the size have any effect?

4. Would a Ag–Au cell be a practical device? Why or why not?

C.2 ELECTROCHEMISTRY

In the voltaic cells you made in the laboratory activity, each metal and the solution of its ions was a half-cell. In the zinc–copper cell, oxidation (electron loss) occurred in the half-cell with zinc metal immersed in zinc nitrate solution. Reduction (electron gain) took place in the half-cell composed of copper metal in copper(II) chloride solution. The activity series helps us predict that zinc is more likely to be oxidized (lose electrons) than copper. The **half-reactions** (individual electron-transfer steps) for this cell are:

$$\text{Oxidation: } Zn(s) \rightarrow Zn^{2+}(aq) + 2\,e^-$$
$$\text{Reduction: } Cu^{2+}(aq) + 2\,e^- \rightarrow Cu(s)$$

The electrode at which oxidation takes place is the **anode.** Reduction occurs at the **cathode.**

The overall reaction in the zinc–copper voltaic cell is the sum of the two half-reactions, added so that electrons cancel; the charges are balanced:

$$Zn(s) \rightarrow Zn^{2+}(aq) + 2e^-$$
$$Cu^{2+}(aq) + 2e^- \rightarrow Cu(s)$$
$$\overline{Zn(s) + Cu^{2+}(aq) \rightarrow Zn^{2+}(aq) + Cu(s)}$$

Because a barrier separates the reactants (Zn and Cu^{2+}), electrons released by zinc in the zinc–copper voltaic cell must go through the wire to reach copper ions.

The greater the difference in reactivity of the two metals used in a voltaic cell, the greater the tendency to transfer electrons, and the greater the cell's electrical potential. For example, a cell containing zinc, gold, and their ions generates a larger electrical potential than does a zinc–copper cell.

Voltaic cells represent a simple way to convert chemical energy to electrical energy in small portable containers. Chemists have packaged various combinations of metals and ions to make useful commercial cells. In ordinary dry cells—often called batteries—zinc is the anode, and a graphite rod and manganese dioxide (MnO_2) is the cathode, along with an ionic solution of ammonium and zinc chlorides. In alkaline batteries,

▶ *These terms can be kept straight by remembering: anode and its process (oxidation) both begin with a vowel; cathode and its process (reduction) both start with a consonant.*

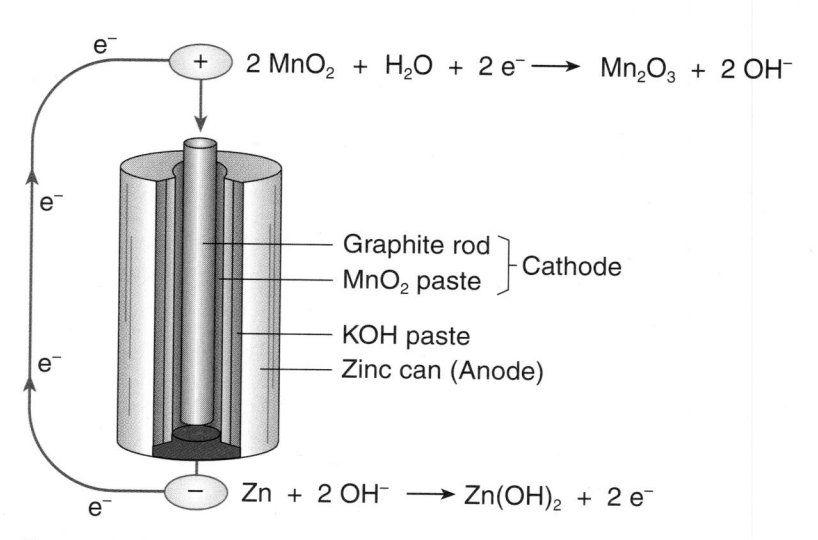

$$2 \, MnO_2 + H_2O + 2 \, e^- \longrightarrow Mn_2O_3 + 2 \, OH^-$$

Graphite rod ⎤
MnO₂ paste ⎦ Cathode

KOH paste
Zinc can (Anode)

$$Zn + 2 \, OH^- \longrightarrow Zn(OH)_2 + 2 \, e^-$$

Figure 7 An alkaline battery.

Despite the differences in size, each of these dry cell batteries generates 1.5 V, because they are all based on the same chemistry.

commonly used in portable tape and CD players, zinc is the anode and a graphite rod and MnO_2 paste is the cathode. A potassium hydroxide (KOH) paste is the alkaline (basic) material between the electrodes (Figure 7). Alkaline batteries produce 1.54 V by the two half-reactions:

Oxidation: $Zn(s) + 2 \, OH^-(aq) \rightarrow Zn(OH)_2(s) + 2 \, e^-$
Reduction: $2 \, MnO_2(s) + H_2O(l) + 2 \, e^- \rightarrow Mn_2O_3(s) + 2 \, OH^-(aq)$

Adding the two half-reactions yields the overall cell reaction:

$$Zn(s) + 2 \, MnO_2(s) + H_2O(l) \rightarrow Zn(OH)_2(s) + Mn_2O_3(s)$$

Nickel–cadmium (NiCad) batteries are rechargable—when this battery discharges, the reactants Cd and NiO_2 are converted to $Cd(OH)_2(s)$ and $Ni(OH)_2(s)$. These solid products cling to the electrodes, allowing them to be converted back to reactants when the battery is placed in a recharging circuit.

An automobile battery, known as a lead storage battery, is composed of a series of electrochemical cells. As illustrated in Figure 8 (page 559), each cell consists of lead plates that serve as one electrode and lead dioxide plates (PbO_2) that serve as the other electrode. All plates are immersed in sulfuric acid. When you turn the car's ignition key, an electrical circuit is completed. Chemical energy in the lead storage battery is converted to electrical energy that runs the starter motor, and the lead electrode is oxidized (it loses electrons). These electrons travel through the circuitry to the lead dioxide electrode, reducing it (the electrode gains electrons). As the electrons travel between the electrodes, they energize the car's electrical systems. Such batteries are also used for motorcycles, self-starting lawn mowers, and golf carts.

The oxidation and reduction half-reactions in a discharging car battery are:

Oxidation: $Pb(s) + SO_4^{2-}(aq) \rightarrow PbSO_4(s) + 2 \, e^-$
Reduction: $PbO_2(s) + SO_4^{2-}(aq) + 4 \, H^+(aq) + 2 \, e^- \rightarrow$
$\qquad\qquad PbSO_4(s) + 2 \, H_2O(l)$

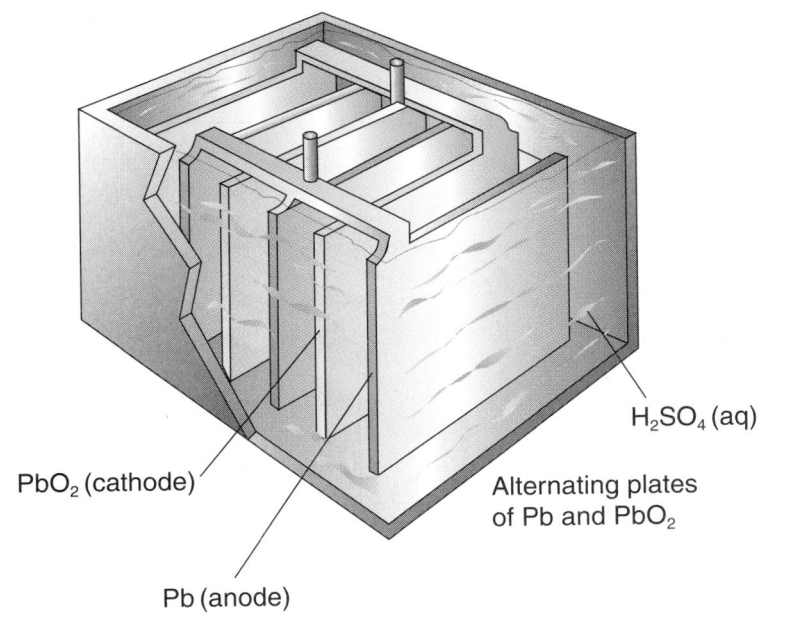

PbO$_2$ (cathode)

Pb (anode)

H$_2$SO$_4$ (aq)

Alternating plates
of Pb and PbO$_2$

Figure 8 One cell of a lead storage battery.

The overall reaction is the sum of the two half-reactions:

$$Pb(s) + SO_4^{2-}(aq) \rightarrow PbSO_4(s) + 2e^-$$
$$PbO_2(s) + SO_4^{2-}(aq) + 4\,H^+(aq) + 2e^- \rightarrow PbSO_4(s) + 2\,H_2O(l)$$

$$\overline{PbO_2(s) + Pb(s) + 4\,H^+(aq) + 2\,SO_4^{2-}(aq) \rightarrow 2\,PbSO_4(s) + 2\,H_2O(l)}$$

If an automobile battery is used too long without recharging, it runs down because the lead sulfate formed eventually coats both electrodes and reduces their ability to produce current. To maintain the battery's charge, a generator or alternator converts some mechanical energy from the car's running engine to electrical energy, which causes electrons to move in the opposite direction through the battery, reversing the direction of the battery's chemical reaction and recharging the battery (Figure 9, page 560).

Lead storage batteries can be quite dangerous when they are being rapidly charged from an outside source of electric energy. Hydrogen gas, formed by the reduction of H$^+$, is released at the lead electrode and can be ignited by a spark or flame, causing an explosion. This is one of the reasons why you need to be careful when "jumping" a battery.

It's estimated that about 70% of used car batteries are currently recycled in this country, leading to the recovery of substantial quantities of lead metal, sulfuric acid, and battery casing polymer.

Although the lead storage battery provides adequate electric current for an automobile, it is too heavy to be used in the electric cars now being produced. Electrochemical research has developed several experimental batteries for such cars, one of which uses the reaction of aluminum with oxygen from the air. The three major U.S. auto producers—Chrysler, Ford, and General Motors—have a joint effort to produce a commercial electric car. In January 1996, General Motors announced plans to start large-scale sales of EV-1 electric passenger cars in California and

▶ *Such recycling reduces potentially hazardous waste disposal problems.*

DISCHARGING

PbO$_2$ (cathode):
PbO$_2$(s) + SO$_4^{2-}$(aq) + 4 H$^+$(aq) + 2 e$^-$ ⟶
    PbSO$_4$(s) + 2 H$_2$O(l)

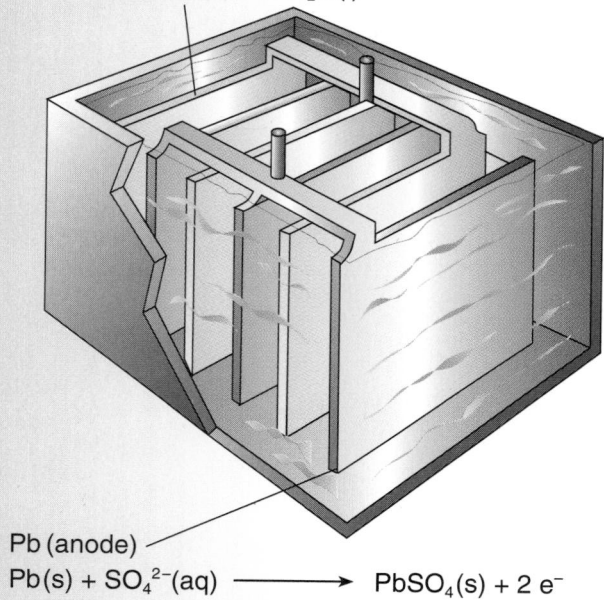

Pb (anode)
Pb(s) + SO$_4^{2-}$(aq) ⟶ PbSO$_4$(s) + 2 e$^-$

CHARGING

PbO$_2$ (anode):
PbSO$_4$(s) + 2 H$_2$O(l) ⟶
    PbO$_2$(s) + SO$_4^{2-}$(aq) + 4 H$^+$(aq) + 2 e$^-$

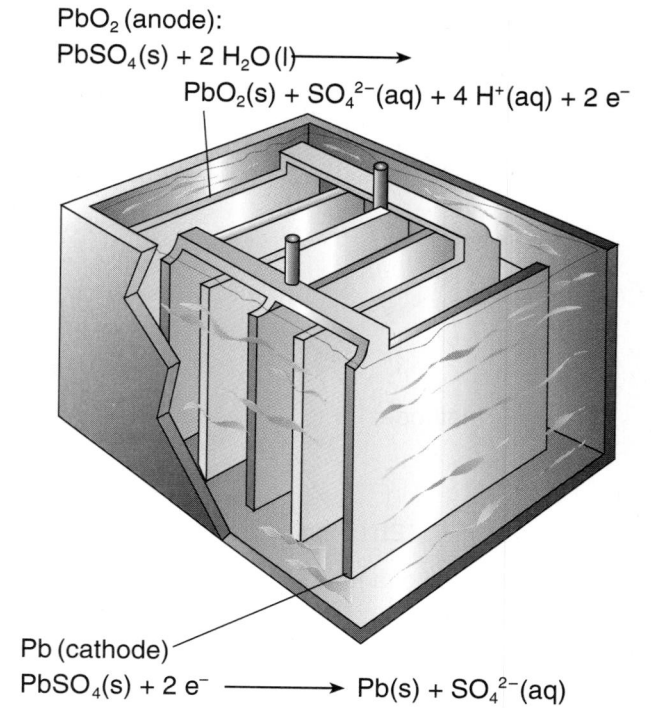

Pb (cathode)
PbSO$_4$(s) + 2 e$^-$ ⟶ Pb(s) + SO$_4^{2-}$(aq)

Figure 9 Discharging and charging a car battery.

The EV-1 battery-powered electric car.

Arizona. A network of recharging stations is planned, possibly in shopping mall parking lots.

This multimillion-dollar venture to develop an electric car is motivated, in part, by California legislation to improve air quality. The legislation requires all cars in the Los Angeles area to be powered by electricity or other clean fuels by 2007. It requires 10% of all cars sold in California by 2003 to have "zero emissions." This action is part of the strategy to move federal regulations more under states' control.

▶ *"Jumping" a run-down battery can be extremely dangerous unless the proper electrical connections are made.*

Jump-starting a car battery.

In some electric cars, the energy for the battery comes from the sun. These so-called "solar" cars are powered by energy coming from solar (photovoltaic) cells that convert sunlight into electrical energy for storage in a battery.

CHEMQUANDARY

ELECTRIC CARS—TRULY "ZERO EMISSIONS"?

Electric cars are thought to reduce air pollution because they are powered by electricity, not by the burning of fossil fuels. However, how the electricity is generated to recharge the car's batteries also must be considered. As such, are electric cars truly emission-free? In what ways do they contribute to air pollution?

YOUR TURN

Getting a Charge from Electrochemistry

Consider the following sample questions about a voltaic cell in which lead (Pb), silver (Ag), and solutions of lead nitrate, $Pb(NO_3)_2$, and silver nitrate, $AgNO_3$, are appropriately arranged:

a. Predict the direction of electron flow in the wire connecting the two metals.

b. Write equations for the two half-reactions.

c. Which metal is the anode and which the cathode?

Example: Here's how answers to these questions can be expressed for a lead–silver cell:

a. Table 3 (page 555) shows that lead is the more active metal. Therefore, electrons should flow from lead to silver.

b. One half-reaction involves forming Pb^{2+} from Pb, as shown in Table 3. The other half-reaction produces Ag from Ag^+, which can be written by reversing the Table 3 equation.

$$Pb \rightarrow Pb^{2+} + 2\,e^-$$
$$Ag^+ + e^- \rightarrow Ag$$

c. In the cell reaction, each Pb atom loses two electrons. Lead metal is therefore oxidized. By definition, then, lead metal is the anode. Each Ag^+ ion gains one electron. This is a reduction reaction. Since reduction takes place at the cathode, silver metal must be the cathode.

Now try these questions on your own:

1. Predict the direction of electron flow in a voltaic cell made from each pair of metals in solutions of their ions.

   a. Al and Sn   b. Pb and Mg   c. Cu and Fe

2. We design a voltaic cell using the metals tin and cadmium. The overall equation for the reaction is:

$$Sn^{2+}(aq) + Cd(s) \rightarrow Cd^{2+}(aq) + Sn(s)$$

a. Write the half-reactions.

b. Which metal, Sn or Cd, loses electrons more easily?

3. Sketch a voltaic cell composed of a Ni–Ni(NO$_3$)$_2$ half-cell and a Cu–Cu(NO$_3$)$_2$ half-cell.

4. In each of these cells, identify the anode and the cathode. (Assume that appropriate ionic solutions are used with each.)

a. Cu–Zn cell   c. Mg–Mn cell

b. Al–Zn cell   d. Au–Ni cell

C.3 LABORATORY ACTIVITY: ELECTROPLATING

GETTING READY

In Laboratory Activity C.1: Voltaic Cells, all the reactions you observed were spontaneous. Materials were arranged so electrons would spontaneously move through wires. Voltaic cells involving such reactions generate electrical energy to run motors and perform other useful work.

By contrast, electrical energy can be used to "force" nonspontaneous (or difficult) oxidation–reduction reactions to occur. Charging a lead storage battery in an automobile is one example. Electroplating is another.

In electroplating, electrical energy from a battery or other source provides electrons to convert metal ions to atoms. Electroplating deposits a thin layer of metal on another surface, protecting the surface or improving its appearance. Coating inexpensive jewelry with a very thin layer of gold or silver can make it more attractive.

An electroplating cell used to accomplish these chemical changes consists of two electrodes (anode and cathode), an appropriate solution of ions, and a source of electricity. In this laboratory activity you will electroplate an object of your choice. The object will become one of the electrodes; a piece of copper will serve as the other electrode.

PROCEDURE

Part 1: Preliminary Electroplating Test

1. Add 0.50 M CuSO$_4$ to a corner well of a 24-well wellplate until the well is 3/4 full.

2. Obtain a battery clip and a graphite electrode/paper clip assembly. Wedge the end of the positive lead of the battery clip (red wire) into one of the paper clips as shown in Figure 10. Wedge the end of the

► Converting metal ions to neutral metal atoms is reduction.

► Electroplating is one particular kind of electrolysis, a technique used in the Resources unit, page 143.

negative lead (black wire) into the other paper clip (See Figure 10.) Attach the battery clip to a 9-volt battery, being careful not to allow the graphite electrodes to touch each other.

3. Quickly (1–2 s) place the electrodes into the copper (II) sulfate solution, then withdraw them. Examine them for evidence of a chemical reaction on their surface.

4. Repeat Step 3 four or five times, making observations of the electrodes while they are in solution. Record your observations, noting any changes at specific electrodes.

Part 2: *Electroplating a Paper Clip*

1. Obtain a large paper clip from your teacher and clean the paper clip with a piece of steel wool. Based on your observations from Part 1, decide which battery clip lead, red (+) or black (−), should be wedged into an end of the paper clip in order to copper plate it. Wedge the correct lead into the paper clip.

2. Obtain an S-shaped copper wire/small paper clip assembly. Squeeze the S-curve together enough so that it will clip securely to the side of the well containing the 0.50 M copper(II) sulfate solution. Place the exposed end of the copper wire into the 0.50 M $CuSO_4$ solution as shown in Figure 11. Wedge the other battery clip lead into the small paper clip attached to the copper wire (see Figure 11).

3. For 1–2 s dip the end of the large paper clip not attached to the lead into the 0.50 M $CuSO_4$ solution. Be sure to keep the large paper clip away from the copper wire in solution. Observe what happens to the portion of the large paper clip that is in solution for 1–2 s. Remove it and examine its surface. Repeat this step 4–5 more times, recording your observations. Did the same reactions happen as in Part 1?

4. Switch the connections so that the connection that had been on the large paper clip is now on the small paper clip, and vice versa. Repeat Step 3, removing the large paper clip from the solution to examine the clip. Record your observations.

5. Switch the leads once more and repeat Step 3 again.

6. Show your electroplated paper clip to your teacher.

7. Transfer the copper(II) sulfate solution into a recovery beaker so that the solution can be reused. Rinse the wellplate with distilled or deionized water.

8. Wash your hands thoroughly before leaving the laboratory.

QUESTIONS

1. a. Which electrode was the anode?

   b. Which was the cathode?

2. In which direction did electrons flow during the reaction?

3. Why were you instructed not to let the large paper clip touch the S-shaped copper wire?

4. Why was a copper wire used in Part 2 rather than a graphite electrode?

Figure 10 A preliminary electroplating test.

Figure 11 Electroplating a paper clip.

The Chemistry of Electrochemistry

563

A Career That's Good from Start to Finish

Gene had just pulled out of the parking lot when his cellular phone rang. It was a "code call"—a serious problem at Ardmore Manufacturing. "Well," he thought to himself, "so much for lunch with Pat."

Gene Kropp is a Technical Field Engineer and Salesman for Technic Inc. in Rhode Island. Gene sells equipment and supplies to electroplating manufacturers across a wide geographic area. Usually his days are fairly predictable: appointments, meetings, and trips to suppliers. Gene visits four or five customers, dropping off chemicals and equipment and taking new orders. Gene also visits prospective customers to develop new business.

But sometimes he gets a code call, an emergency call from a customer who is having electroplating system trouble. Gene listens to the customer describe the problem. Then, using the process of elimination and a few "tricks of the trade," he helps get things up and running again.

Electroplating is the process of attaching a metal (such as chromium, silver, or gold) to the surface of an article to protect it and make it more attractive. Gene's knowledge of chemistry in general (and silver and other metals in particular) helps him meet clients' needs and solve their problems.

Gene attended a two-year college and earned a degree in general studies with a concentration in chemistry. Since then, Gene has worked in various positions in the electroplating field, adding to his knowledge with additional coursework and on-the-job training.

Technic Inc.

Call Log

Date: March 3

To Do:	Call Barbara @ Robinson Electroplating- set up plant visit
	Call Bill Smith; ask about new valve
8:30 am	Robinson Electroplating-meet new plant manager, inspect equipment
	Visit Bill in purchasing; drop off bid on replacing backup pump
10:00	Drop off chemicals at headlamp plant; visit Bill in lab
10:35	**Code Call** from Ardmore/Tom D. says pump has quit
10:50	Ordered new pump from L-Way, will pick up on way to Ardmore
	Called service, will have Joe and engineer meet me at Ardmore
11:00	Left for Ardmore
~~12:30~~	~~Lunch with Pat~~ cancelled

Photograph courtesy of Technic Inc.

5. Write the half-reaction for
   a. the oxidation.
   b. the reduction.

6. Diagram and label the parts of an electroplating cell for
   a. plating a ring with gold.
   b. plating silver onto spoons.
   c. purifying copper metal.

C.4 INDUSTRIAL ELECTROCHEMISTRY

When a solution of sodium chloride in water (brine) is electrolyzed, chlorine gas, Cl_2, is produced at the anode and hydrogen gas, H_2, at the cathode (Figure 12). Positive sodium ions remain in solution; electrolysis leaves negative hydroxide ions in solution, replacing the chloride ions:

$$2\,Na^+(aq) + 2\,Cl^-(aq) + 2\,H_2O(l) \rightarrow$$
$$2\,Na^+(aq) + 2\,OH^-(aq) + H_2(g) + Cl_2(g)$$

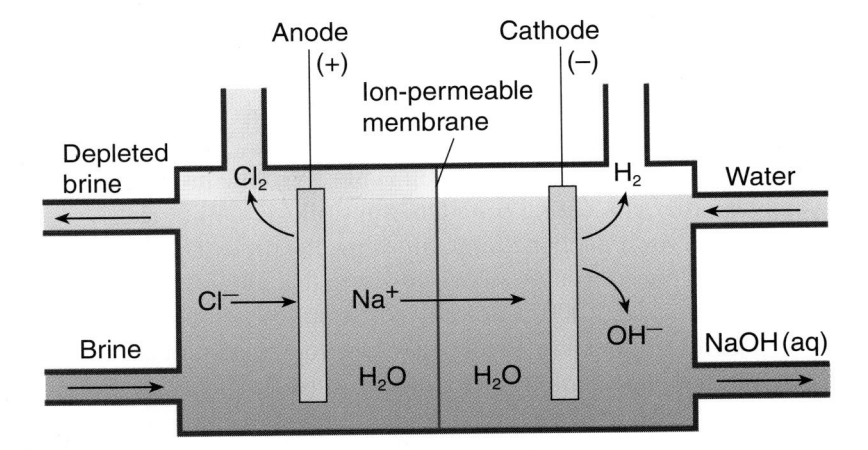

From: Kotz, J. et al.: *The Chemical World,* Saunders College Publishing, Philadelphia, PA, 1994.

Figure 12 Electrolysis of aqueous NaCl.

This electrolysis reaction, called the chlor–alkali process, generates three important, high-volume industrial chemicals: hydrogen gas, chlorine gas, and sodium hydroxide. More than 12 million tons of chlorine and about the same quantity of sodium hydroxide are produced by this method annually by the U.S. chemical industry.

The hydrogen gas and chlorine gas can be used without additional purification, because no other gases are released at the anode and cathode. The sodium hydroxide, on the other hand, must be purified, since it remains in solution with unreacted sodium chloride.

In earlier years, the large industrial electrolysis cells for the chlor-alkali process used metallic mercury as the cathode. Before the environmental hazards of mercury were discovered, it was dumped into nearby rivers and lakes during maintenance and cleaning of the mercury cells. However, mercury is no longer used in chlor–alkali cells. Responding to

▶ *Electrolysis uses electrical energy to cause chemical changes.*

▶ *Chlorine and sodium hydroxide are among the top industrial chemicals (by mass) produced in the United States.*

The Washington Monument; the aluminum cap is about four inches high.

▶ *There is a curious coincidence associated with Hall's work. Within two months of the time Hall came up with his process, Paul-Louis Héroult, a young French scientist working in Paris, independently developed the same process. In recognition of both young men's accomplishment, the production of aluminum in this way is called the Hall-Héroult process. Remarkably, these two men, linked through their common discovery, also shared the same birth year (1863) and died the same year (1914).*

the need to replace mercury in the electrolytic cells, research chemists and chemical engineers modified the chlor–alkali process considerably. They developed a new permeable polymer membrane to separate the anode and cathode compartments. Also, metallic iron replaced mercury as the cathode, and new metals substitute for graphite as the anode.

The industrial production of aluminum, such as done by the WYE Metals Corporation, is another major process that uses electrical energy to produce chemical change. Aluminum was first isolated in metallic form in the 1820s by an expensive and potentially dangerous task, using metallic sodium or potassium to reduce Al^{3+} in aluminum compounds. It was very difficult to reduce the Al^{3+} in its ores to the metallic state. Therefore, for more than 60 years, aluminum metal was very expensive, despite the fact that aluminum is the most plentiful metal ion in the Earth's crust. The prized aluminum metal was even used in jewelry, including the French crown jewels and the Danish crown. In 1884, a 2.8-kg (6-lb) aluminum cap was installed on top the Washington Monument in Washington, D.C., as ornamentation and the tip of a lighting rod system. At that time, the aluminum cap cost considerably more than the same mass of silver.

The challenge for industrial users was how to produce aluminum at a much lower cost. Aluminum ion, Al^{3+}, is very stable. No common substances give up electrons easily enough to reduce aluminum ions to aluminum metal. For example, carbon, an excellent reducing agent for metal compounds such as iron oxide or copper sulfide, simply doesn't work with aluminum compounds.

A young man named Charles Martin Hall solved the problem. In 1886, one year after graduating from Oberlin College (Ohio), Hall devised a method for reducing aluminum using electricity (Figure 13, page 567). Hall's breakthrough was discovering that aluminum oxide (Al_2O_3), which melts at 2,000 °C, dissolved in molten cryolite (Na_3AlF_6) at 950 °C. Thus, he found a relatively easy way to get aluminum ions from aluminum ore into "solution" for electrolysis. Hall's discovery became the basis of a rapidly growing and important aluminum industry; he founded the Aluminum Corporation of America (ALCOA) and was a multimillionaire at the time of his death in 1914.

In the Hall–Héroult process, aluminum oxide (bauxite) is dissolved in molten cryolite (Na_3AlF_6) at 950 °C in a large steel tank lined with carbon. The carbon tank lining is made the negative electrode by a source of direct current. This carbon cathode transfers electrons to aluminum ions, reducing them to molten metal. Molten aluminum sinks to the bottom, where it is drawn off periodically. The aluminum is then formed into various shapes and used to manufacture a wide variety of products, from stepladders and baking foil to airplane parts.

The positive electrode (anode), which is also made of carbon, is oxidized during the reaction. As the tips of the carbon-rod anodes slowly burn away, the carbon rods are lowered deeper into the molten cryolite bath. The half-reactions for making aluminum are:

$$4\ Al^{3+}(melt) + 12\ e^- \rightarrow 4\ Al(l)$$
$$3\ C(s) + 6\ O_2(melt) \rightarrow 3\ CO_2(g) + 12\ e^-$$

Ions from cryolite carry the electric current in the molten mixture.

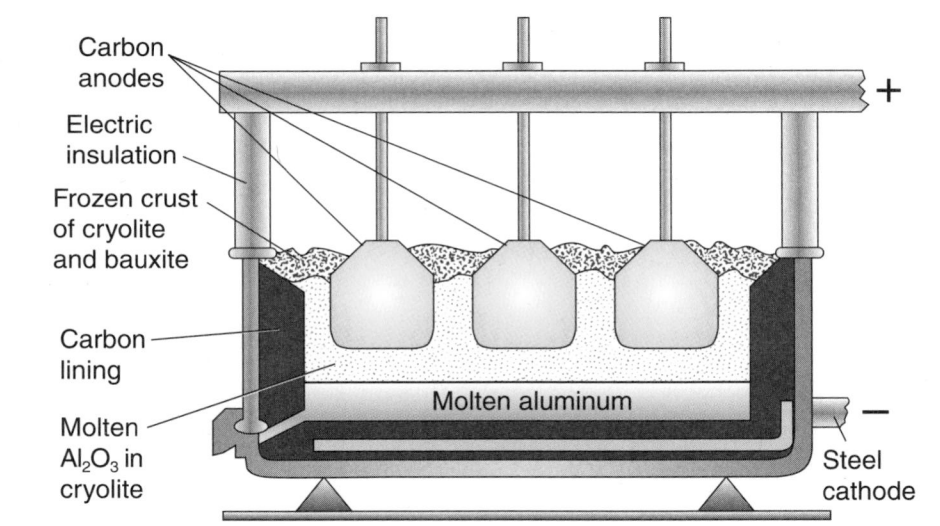

Carbon
anodes

Electric
insulation

Frozen crust
of cryolite
and bauxite

Carbon
lining

Molten aluminum

Molten
Al_2O_3 in
cryolite

Steel
cathode

+

−

Figure 13 The Hall–Héroult process for aluminum manufacture.

Molten aluminum flows
from furnaces into molds
in a factory.

The Hall–Héroult process had a sudden and long-lasting impact on aluminum production. Prior to the process, only about 1,000 kg (about 2,000 pounds) of aluminum were produced annually worldwide (Figure 14, page 568). In 1884, a pound of aluminum metal sold for about $12, a significant sum in those times. Within three years of their discovery, Hall and Héroult each had commercial operations producing aluminum, and a million tons per year were produced worldwide by the turn of the century. The price of aluminum plummeted so that five years after large-scale commercial production started, aluminum sold for 70 cents a pound. What once was a metal affordable only by the very rich became commonplace.

The U.S. chemical industry produces more than 5 billion kilograms (over 10 billion pounds) of aluminum each year, requiring about 7.15×10^9 worth of electrical energy ($1.43 per kilogram of aluminum). Because aluminum plants use such large quantities of electricity, they are often located near sources of hydroelectric power. Many are in the Pacific Northwest. Electricity from hydropower generally costs less than electricity from thermal power plants (those that produce steam to spin the generator

Molten aluminum.

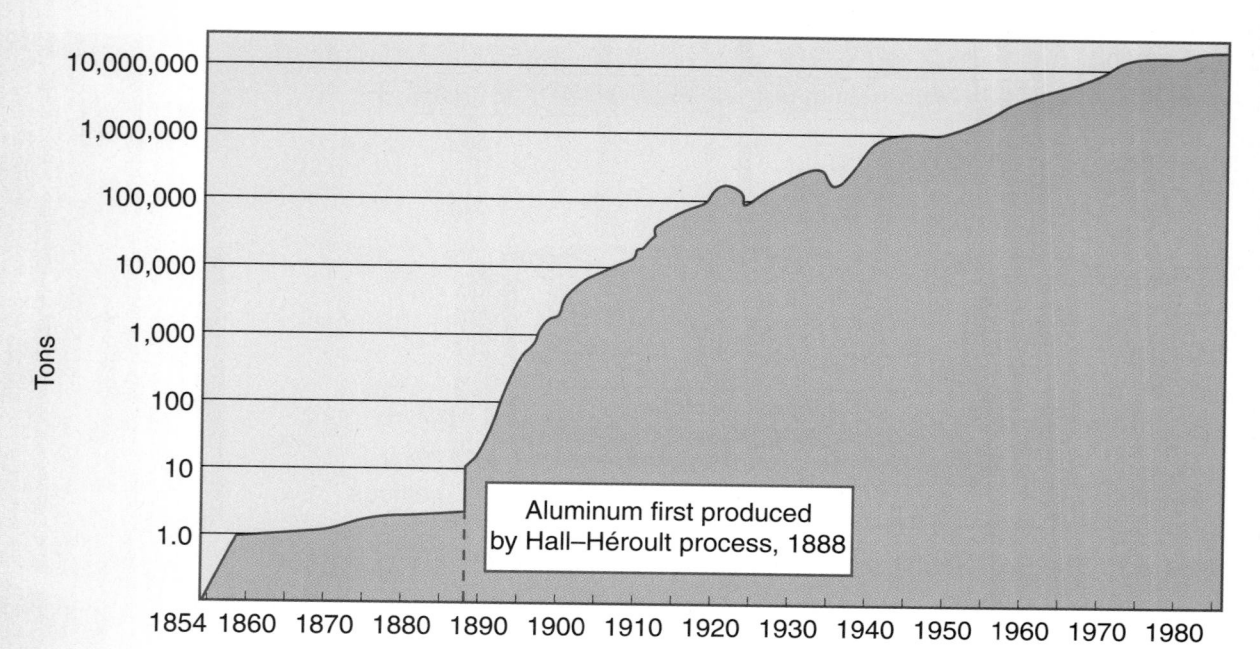

From: *Encyclopedia of Physical Science and Technology,* Harcourt Brace & Company, Orlando, FL.

Figure 14 Worldwide aluminum production.

turbines). The WYE Metals Corporation is interested in Riverwood because of cheap hydroelectric power and proximity to aluminum ore sources.

We now have considered the chemical processes by which EKS and WYE make their products. To further help you better understand the choices facing Riverwood, we will next turn to consider some general aspects of the chemical industry.

Removing aluminum bars from their molds.

Transporting bauxite ore by barge.

? PART C: SUMMARY QUESTIONS

1. Does the electrical potential produced in a voltaic cell depend on

   a. the size of the cell (the mass of metal forming each electrode)?

   b. the specific metals used?

2. For highest electrical potential, should a voltaic cell's two metals be close together or far apart on the activity series? Explain.

3. The following questions refer to this equation:

$$PbO_2(s) + Pb(s) + 4\,H^+(aq) + 2\,SO_4^{2-}(aq) \rightarrow$$
$$2\,PbSO_4(s) + 2\,H_2O(l)$$

   a. Does this equation represent the charging or discharging of a lead storage battery?

   b. Electrical energy or work can be extracted from this reaction. What does that imply about the reverse reaction?

   c. Would the reverse reaction be endothermic or exothermic?

   d. Would the reverse reaction represent battery charging or discharging?

   e. Why can the condition of a lead storage battery be tested with a hydrometer, a device that measures liquid density? (*Hint:* Sulfuric acid solutions have a greater density than liquid water.)

   f. In a fully charged condition, will the density of the battery liquid be greater or less than in a discharged state? Why?

   g. Identify the species being oxidized and the species being reduced.

   h. Under what conditions may a lead storage battery produce hydrogen gas?

   i. Why does this pose a hazard?

4. Magnesium and aluminum are both active metals. Would a voltaic cell using these two metals produce a significant electrical potential? Why?

5. Use Table 3 to predict the direction of electron flow in a voltaic cell composed of each pair of metals and their ions:

   a. Al and Cr

   b. Mn and Cu

   c. Fe and Ni

6. Compare chemical reactions in voltaic cells with those in electroplating cells. Consider

   a. energy released or absorbed.

   b. spontaneity vs. nonspontaneity.

   c. direction of electron flow.

7. Iron and copper were available to humans as free (chemically uncombined) metals long before aluminum was—even though aluminum is much more abundant in Earth's crust.

   a. Why were free iron and copper available first?

   b. Why is aluminum oxide reduction more difficult than iron oxide reduction?

   c. How is the aluminum oxide reduction carried out?

8. a. Write the overall equation for making aluminum by the Hall–Héroult process.

   b. How many moles of electrons are needed to reduce enough Al^{3+} ions to produce a 378-g roll of "kitchen" aluminum foil?

EXTENDING YOUR KNOWLEDGE (OPTIONAL)

Prepare a brief report on an electrolysis reaction or electroplating process that is used to produce some item you encounter daily.

D THE CHEMICAL INDUSTRY AND RIVERWOOD— A CLOSER LOOK

Riverwood citizens discussing whether their town is to be the home of either the EKS Nitrogen Products Company or the WYE Metals Corporation manufacturing facility will want answers to some questions. How are chemical products manufactured, sold, and distributed? What employment opportunities are available in a chemical company? In what ways do chemical companies comply with environmental regulations?

Loading a tank with sulfuric acid.

The chemical industry, loosely defined, is the nation's largest industry. John Winthrop, a Mayflower passenger, started the U.S. chemical industry in 1635 in Massachusetts with the production of two very practical materials: alum—potassium aluminum sulfate, $KAl(SO_4)_2 \cdot 12 H_2O$, and saltpeter—potassium nitrate, KNO_3. Both substances were needed by colonial settlers to preserve meat and produce. From such humble beginnings, the chemical industry has become a colossus that includes many of today's leading industries. Petroleum, pharmaceuticals, tires, clothing, paint, and even processed foods are just a few products for which the chemical industry is wholly or partly responsible. Even if the large petroleum and food industries are excluded, the remaining chemical industries generate roughly $1 of every $10 earned in manufacturing.

Sulfuric acid has traditionally been the substance produced in the largest quantity annually by the U.S. chemical industry. This is principally because sulfuric acid acts as an intermediate—a substance used to synthesize consumer products or other chemicals (Figure 15). More than half of the sulfuric acid is used to produce phosphate fertilizers, particularly ammonium monohydrogen phosphate, $(NH_4)_2HPO_4$.

The "top 25" chemicals and their recent production are given in Table 4 (page 572). Nitrogen gas, second in total U.S. quantity production, is a major raw material for the EKS Company. Nitrogen and oxygen gas—third-ranked in production—are obtained by low-temperature distillation from liquefied air. Ammonia, an important EKS product, is among the nation's top 10 chemicals in terms of quantity produced. Nitric acid, another EKS product, is among the top 15. Petroleum and natural gas are used to produce tremendous quantities of consumer products, from solvents to pharmaceuticals. Natural gas and petroleum are sources for many basic chemical industry materials, including ethylene, fourth in total production. Ethylene is the starting material for almost 40% of all organic chemicals produced, such as polyethylene. Industrial raw materials also are extracted from the Earth's crust (minerals, precious metals, sulfur), the oceans (magnesium, bromine), and the atmosphere (nitrogen, oxygen).

▶ *More than 1 million people are employed in the United States in producing industrial chemicals and allied products.*

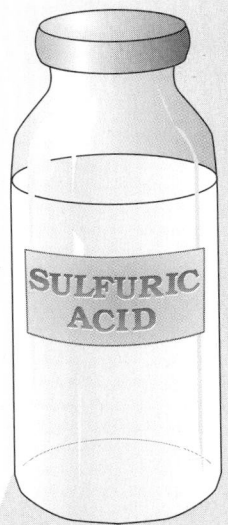

Plastics
Pharmaceuticals
Food processing
Agriculture
Textiles
Rubber
Paints, dyes, pigments
Petroleum
Pulp and paper manufacturing
Glass and ceramics
Ferrous metallurgy
Nonferrous metallurgy
Water and sewage treatment
Cleaning and refrigeration
Explosives

Figure 15 Uses of sulfuric acid.

YOUR TURN

The Top of the Chart

1. Which chemicals listed in Table 4 are inorganic substances? Explain your answer.

2. Which chemicals listed in Table 4 are organic substances? Explain your answer.

3. What is the manufacturing relationship among ammonia, nitric acid, and ammonium nitrate?

4. How are chlorine and sodium hydroxide manufacturing associated?

Table 4	Top 25 Chemicals Produced in the United States in 1994		
Rank	Name	Formula	Billions of Pounds
1	Sulfuric acid	H_2SO_4	89.2
2	Nitrogen	N_2	67.5
3	Oxygen	O_2	49.7
4	*Ethene (ethylene)	C_2H_4	48.5
5	Calcium oxide (lime)	CaO	38.4
6	Ammonia	NH_3	37.9
7	*Propene (propylene)	C_3H_6	28.8
8	Sodium hydroxide	$NaOH$	25.8
9	Phosphoric acid	H_3PO_4	25.3
10	Chlorine	Cl_2	24.2
11	Sodium carbonate	Na_2CO_3	20.6
12	*1,2-Dichloroethane (ethylene dichloride)	$C_2H_4Cl_2$	18.7
13	Nitric acid	HNO_3	17.6
14	Ammonium nitrate	NH_4NO_3	17.6
15	*Urea	$(NH_2)_2CO$	16.1
16	*Vinyl chloride	C_2H_3Cl	14.8
17	*Benzene	C_6H_6	14.7
18	*Methyl tert-butyl ether (MTBE)	$C_5H_{12}O$	12.3
19	*Ethylbenzene	C_8H_{10}	11.9
20	*Styrene	C_8H_8	11.3
21	Carbon dioxide	CO_2	11.0
22	*Methanol	CH_3OH	10.8
23	*Xylene	C_8H_{10}	9.1
24	*Terphthalic acid	$C_8H_6O_4$	8.6
25	*Formaldehyde	H_2CO	7.9

Total Organics	279.2
Total Inorganics	450.2
Grand Total	729.4

Reprinted in part from *Chemical & Engineering News,* April 10, 1995, *73* (15), p.17. Copyright 1995 American Chemical Society.

*Organic

D.1 FROM RAW MATERIALS TO PRODUCTS

Many chemical reactions you have carried out in *ChemCom* laboratory activities are basically the same reactions used in industry to synthesize chemical products. However, chemical reactions for industry must be scaled up to produce large quantities of high-quality products at low cost.

Four factors become crucial in scaling up reactions: engineering, profitability, waste, and safe operation. Sometimes an industrial reaction is run as a batch process, a single "run" of converting reactants to products, such as you did in the laboratory activity to produce esters (Petroleum unit, page 220). When more products are needed, additional batches are run. Early production of nylon was a batch process. More commonly and less expensively, industrial reactions are run as continuous processes where reactants flow steadily into reaction chambers and product flows out continuously. Rate of flow, time, temperature, and catalyst composition must all be carefully controlled.

Chemical engineers face many challenges in designing manufacturing systems for industry. In your classroom laboratory, the small quantity of heat generated by a reaction in a test tube may seem unimportant. But, in industrial processes where thousands of liters may react in huge vats, that heat must be anticipated and carefully managed. Otherwise, the reaction temperature can rise, creating potentially dangerous and costly situations.

The need to generate profits for their company heightens the challenge for chemical engineers. Because few chemical reactions produce 100% of the product sought, profitability can often be enhanced by modifying reaction conditions to increase the yield of desired material. For some products, a difference of one penny per liter in production costs can literally mean the difference between profit or loss.

Industry also faces the problem of dealing with unwanted materials resulting from chemical processes. These substances can quickly accumulate when reactions occur on an industrial scale. A major problem for the Environmental Protection Agency (EPA) is managing the cleanup of hundreds of chemical waste dumps. These dumps are legacies of an earlier time when there were fewer people. In proportion, the United States seemed larger than it does now. In those days, unwanted materials released into the air, rivers, and the ground seemed to disappear. It was simply a case of "out of sight, out of mind."

When EPA put an end to such waste releases, many chemical industries discovered that some previously unwanted materials or products could, with a little additional processing, become valuable commodities. Often such former "waste" compounds can be used as intermediates to make other substances. Thus, instead of polluting the environment, these wastes-turned-resources provide a new source of income. Additionally, recently developed catalysts and new or modified processes have allowed manufacturers to increase the efficiency of producing certain materials while decreasing the amounts of starting materials (reactants).

Chemical manufacturers also have come to learn that pollution prevention pays off. For example, the 3M Corporation, a major producer and user of chemical substances, has maintained a pollution prevention program for more than 20 years. During that time, 3M has saved over a half a billion dollars by eliminating more than 135,000 tons of air pollutants, nearly 17,000 tons of water pollutants, and 430,000 tons of solid and semi-solid wastes. Most of these pollution-prevention suggestions and their implementation came from 3M employees. By the year 2000, 3M intends to reduce all plant discharges into air, water, or soil to 10% of 1987 levels. Its post-2000 goal is to approach zero pollution.

D.2 INDUSTRY AS A SOCIAL PARTNER

In recent years, U.S. industries and citizens have recognized their joint responsibility to ensure that chemical products are manufactured with a net benefit and minimum hazard to society. The EPA has initiated a "Design for Environment" program that is having an impact on the ways chemical process, printing, and dry-cleaning industries do business. For example, many of the nation's 34,000 commercial dry-cleaning shops have significantly reduced the use of perchloroethylene, a toxic dry-cleaning solvent. Dry cleaners have shifted to a much safer and cost-effective process using controlled application of water as the cleaning solvent.

Over the past decade, the chemical industry has begun to develop new synthesis methods, based on using safe starting materials that replace toxic or environmentally unsafe substances. This new "green chemistry" focuses on preventing environmental pollution directly at the point of manufacturing. In this approach, and in others such as Responsible Care (page 527), the chemical industry is trying to work as a social partner to sustain development and international trade without damaging the environment.

New "benign by design" chemical processes use more environmentally benign reactants and create waste products that do less damage to air and water. For example, a process has been developed using D-glucose, found in ordinary table sugar, to replace benzene, a known carcinogen. The D-glucose serves as the chemical feedstock to make reactants that eventually produce nylon and various medicines. Also, nontoxic food dyes have been demonstrated in some processes to be effective substitutes for catalysts composed of toxic metals such as lead, chromium, or cadmium.

Phosgene, a toxic gas, can be replaced by carbon dioxide in the manufacture of isocyanates. Isocyanates are substances used to make polyurethanes, materials used widely in the manufacture of seat cushions, insulation, and contact lenses. In addition, "green chemistry" seeks ways to synthesize industrial chemicals in water solution rather than in toxic solvents, and to use materials that can be recycled and reused to reduce waste disposal problems significantly. To succeed commercially, these newly developed processes must also be cost effective.

The chemical industry also has the responsibility to make products in ways that are as hazard-free as possible, in workplaces that are as safe as possible. Industry is obliged to deal honestly with the public to ensure that risks and benefits of chemical operations are clearly known. The National Safety Council ranked the U.S. chemical industry as the safest of all manufacturing industries four times during the 1980s. In 1992, the chemical industry's occupational injury and illnesses rate was less than 50% that of U.S. manufacturing industries as a whole. Chemical companies also need to assure consumers that the products are safe when used as intended by the manufacturer.

Chemical industries must comply with certain laws and regulations, as well as with voluntary standards met by many manufacturers. Often, compliance with voluntary or mandatory requirements is monitored by independent, outside organizations. An international initiative, ISO 9000, calling for voluntary quality management systems, has been developed by the International Organization for Standardization (ISO) in Geneva, Switzerland, in conjunction with representatives from the United States

► *A chemical feedstock is a starting material from which many other materials can be made. For example, crude oil is a chemical feedstock for fuels, plastics, and many other substances.*

and 73 other countries. The five ISO 9000 standards supply guidance to corporations in setting up quality management systems.

In addition, there is a worldwide movement to improve environmental quality. The governments of various nations around the world have passed laws to regulate environmental pollution, although the laws frequently are not consistent from country to country. A program known as ISO 14000 has been developing an international series of standards to help manufacturers and organizations consider the impact of their operations, products, and services on the environment. Work on ISO 14000 standards is being carried out by representatives from more than 110 countries, including the United States. The goal is to develop one set of internationally accepted environmental management system standards rather than many, sometimes conflicting, national standards.

No activity of the chemical industry or the government can completely eliminate the risks involved in manufacturing chemical substances, any more than the risks of traveling in a car can completely be eliminated. Knowing the risks, continuing to explore them, and dealing with them prudently are essential.

These concerns are not the sole responsibility of the chemical industry. As users of the industry's products, we share this responsibility. Having studied some basic concepts about the manufacturing of chemical products, you can use such knowledge to make decisions involving risks and benefits to you and to the environment. You will be asked to provide information to help the Riverwood Town Council make a decision at its next meeting—will a chemical plant come to Riverwood?

D.3 JOBS IN CHEMICAL COMPANIES

The corporate managers of the EKS Nitrogen Products Company and the WYE Metals Corporation are eager to set up operation in Riverwood. In remarks to the town council, Jill Mulligan of EKS and Nancy Belski of WYE spoke about how their respective companies are organized (Figure 16 and Figure 17, page 576). Parallel jobs and services are found in both firms. Analytical Department chemists at EKS or WYE test the purity of raw materials purchased from other companies and the substances produced from them. This department also performs services related to quality control to ensure the uniformity and properties of each product. Trained personnel in the Environmental Departments monitor the plant's wastes and their potential effects on the environment. Chemists and chemical engineers in the Research Departments work to develop new products and improve old ones, to reduce operating costs and energy use, and to minimize by-products. Public Relations or Public Information Departments work with the news media, the government, and the general public. This is the department Riverwood citizens would contact if they have a concern about the company's operations. Corporate Management oversees the work of all divisions and handles personnel, company policy, and finances.

Anticipating a favorable response from the town council and the community, each company has placed announcements (page 577) in the *Riverwood News* indicating the kinds of employment opportunities to be made available.

▶ *U.S. chemical companies invest about 5% of their sales revenue in research and development (R&D) activities.*

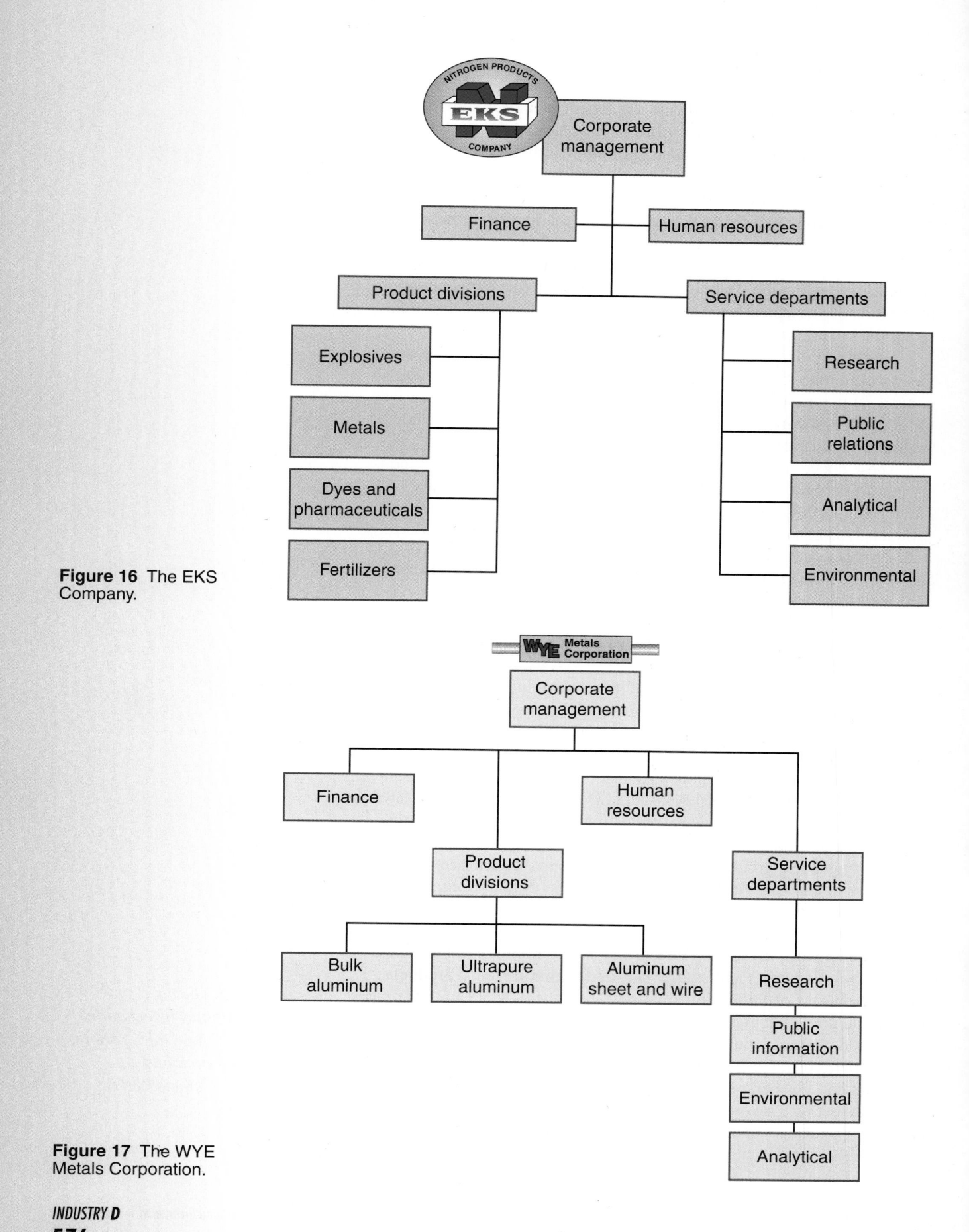

Figure 16 The EKS Company.

Figure 17 The WYE Metals Corporation.

Advertisements in the *Riverwood News:*

EKS NITROGEN PRODUCTS COMPANY
AN EXCEPTIONAL PLACE TO WORK

The **EKS Nitrogen Products Company** hopes to establish a chemical manufacturing plant in Riverwood in the near future. If selected, the company will have vacancies for these technical and management positions:

Chemical technicians	**Chemists**
Chemical engineers	**Department managers**
Environmental specialists	**Personnel officers**
Public relations specialists	**Financial analysts**
Purchasing agents	

Competitive wages, excellent benefits, health care insurance, three weeks paid vacation annually.

Application materials can be obtained from the Riverwood Chamber of Commerce.

An announcement will be made soon regarding opportunities for control operators and production, clerical, and maintenance staff positions.

The EKS Nitrogen Products Company is an Equal Opportunity employer.

WYE Metals Corporation

Get to Know Why WYE is a Great Place to Work

The WYE Metals Corporation announces plans to hire qualified individuals for technical, staff, and administrative positions for its proposed facility. The following positions will be available pending the approval of WYE to build a manufacturing facility in Riverwood:

Chemical technicians	*Metallurgists*
Chemists	*Chemical engineers*
Production supervisors	*Environmental engineers*
Public information officers	*Personnel officers*
Purchasing agents	*Fiscal supervisors*

The WYE Metals Corporation provides a friendly working atmosphere, excellent salaries, broad health care benefits, an outstanding retirement plan, and three weeks of paid vacation annually.

Applications can be obtained from the Riverwood Chamber of Commerce.

Control operators and production, clerical, and maintenance staffs will also be needed. An announcement for those positions will be made soon.

The WYE Metals Corporation is an Equal Opportunity employer.

Imagine that you work in the Personnel Department of either the EKS Nitrogen Products Company or WYE Metals Corporation. You have been assigned to write a position description to be used in hiring a chemical technician, an important laboratory position with your company.

Answer the following questions to help you with your assignment. Reference sources other than your textbook may be consulted.

1. What is the the title of the position?

2. What educational background is needed for the position?

3. What minimum requirements are there for the position?

4. What will be the specific duties and responsibilities of the chemical technician?

5. You and your classmates will compare the descriptions you have written for the chemical technician job. How are the chemical technician position descriptions for each company

   a. similar?

   b. different? If different, account for the differences.

? PART D: SUMMARY QUESTIONS

1. Describe the nature of the partnership that should exist between the chemical industry and society.

2. Provide evidence to support or refute the claim that the chemical industry plays a major role in our economy.

3. a. Identify the top four chemicals in terms of quantity produced annually in the United States.

   b. List two uses of each.

4. Briefly describe several major factors that must be considered when a reaction is scaled up from a laboratory level to an industrial level.

5. What role is played by each of these within a chemical company?

   a. Analytical department

   b. Environmental department

   c. Public relations (information) department

   d. Corporate management

   e. Research department

6. Unwanted materials from chemical processing plants can often be regarded as "resources out of place." Explain why.

7. Why does "green chemistry" represent a change from "doing business as usual."

8. What does ISO 14000 seek to do that is not being done now?

EXTENDING YOUR KNOWLEDGE (OPTIONAL)

• Learn more about the printing and dry-cleaning businesses in your community to find out in what ways they have changed their operations to be more environmentally "friendly."

• The Toxics Release Inventory (TRI) is available from the EPA as a book or on computer disks. If your school has a copy, you can use the TRI to check on how well your state or metropolitan area has done in reducing emissions from manufacturing facilities.

E PUTTING IT ALL TOGETHER: A CHEMICAL PLANT FOR RIVERWOOD—CONFRONTING THE FINAL CHOICES

Final Meeting to Decide Future Chemical Plant

By Gary Franzen
Staff writer of *Riverwood News*

After months of study and discussion, the Riverwood Town Council is prepared to act on separate proposals from the EKS Nitrogen Products Company and the WYE Metals Corporation to locate a chemical plant in Riverwood. At tonight's town council special meeting, the council will decide which, if either, plant should be invited to Riverwood. The meeting starts at 7:30 P.M. in Town Hall. Mayor Cisko said, "I'm very pleased with the turnout we've had for the town meetings held on this issue. I want to emphasize that tonight's council meeting is open to the public. I encourage all community members to attend and express their views about a chemical plant in Riverwood."

At the request of the council, both companies have prepared comprehensive summaries to inform cit-izens of their plans for a Riverwood plant. These summaries have been widely distributed throughout the community and were circulated at the previous town meetings on the subject. Additional copies can be obtained from the Mayor's office.

Editorial:
Plan to Attend Tonight's Council Meeting

Tonight's meeting of the town council deserves special attention because it will impact this community and affect all Riverwood citizens for many years to come. After extended discussions with all parties concerned, the council faces two decisions, equally important. The first is whether to locate a chemical plant in Riverwood. If that decision is yes, the next question is which of the two competing bids should be accepted. The WYE Metals Corporation and the EKS Nitrogen Products Company each have made strong cases for selecting their firm. Their discussions with us have been friendly and candid. As members of the Responsible Care initiative, they have answered questions openly about their current operations and those proposed for our community, especially in terms of the criteria for new businesses established by the council. However, questions may still remain among citizens who may have to live with a chemical plant in their vicinity.

We urge you to attend and participate in tonight's meeting. It is likely that no other town council decisions in the near term will be as important to the future of our community as those made tonight.

E.1 YOU DECIDE: ASSET OR LIABILITY

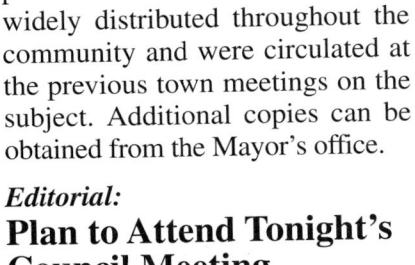

If you were a citizen of Riverwood, what would be your choice—should a chemical plant be invited to come and, if so, which one? On what basis should that choice be made?

To help clarify these questions, you will participate in a class debate regarding some benefits and risks involved in building such a chemical plant in

Riverwood. You might wish to consider the following factors, positive as well as negative, in your discussion.

POSITIVE FACTORS

The local economy will be improved. Either plant will employ about 200 local people, working in three shifts. This will add about $8 million to Riverwood's economy each year. In addition, each plant employee will indirectly provide jobs for another four people in local businesses. This is quite desirable, because 10% of Riverwood's labor force of 21,000 is currently unemployed.

Farming costs will be lowered. Fertilizer for farms near Riverwood is now trucked from a fertilizer plant 200 miles away. These transport costs increase farmers' expenses by $15 for each ton of ammonia-based fertilizer. Each year 700 tons of fertilizer are spread on farms in the Riverwood area. Local farmers thus stand to save about $10,000 annually in transportation costs.

New local industries could result. With direct access to aluminum metal locally, several Riverwood entrepreneurs are considering starting a factory to make aluminum window frames and studding for home construction. Without significant transportation costs for raw materials, the proposed company could establish itself in the market by selling its products at lower prices. The projected aluminum fabrication factory would likely employ 30-45 employees, mainly shop workers and drivers, that would add $700,000 annually to the local economy.

Ammonia is a commercial refrigerant used to produce ice in large quantities, and ready access to ammonia supplies could lead to the development of a commercial ice company in Riverwood. Such a company might have a workforce of 25–35 individuals, principally drivers and some plant workers, which might add $500,000 annually to Riverwood's economy as well as providing ice for the region.

Riverwood air quality will be enhanced. A natural gas transmission company will build a line to deliver natural gas (mainly methane, CH_4) to a Riverwood ammonia plant. Town residents, who have traditionally burned fuel oil in their home furnaces, could convert to natural gas. If all 11,000 Riverwood homes and businesses burned natural gas rather than oil, emissions of sulfur dioxide and particulate matter would decrease.

The tax base will be improved. The aluminum or ammonia plant will be taxed at the current commercial rate in Riverwood. This will provide a large increase in revenues for the community.

NEGATIVE ASPECTS

Injuries and accidents pose a threat. Ammonia is manufactured at high pressure and high temperature. An accident at the plant could injure or kill nearby workers. For each 100 U.S. workers in ammonia-based fertilizer plants, several reported cases of work-related injury or illness have been reported annually. A few worker-related deaths have also been reported in such plants. Aluminum production illness and injury rates annually are about 20 per 100 full-time employees. Infrequent deaths at such plants have also been reported. However, the injury and accident rates in aluminum and in ammonia-based fertilizer production are both lower than the rates in motor vehicle manufacturing, meat packing, and steel foundries.

Production involves chemical hazards. Ammonia, a gas at ordinary temperature and pressure, is extremely toxic at high concentrations. It is often shipped in tanker trucks. An accident on the roadway or at the plant releasing large amounts of ammonia could create a health hazard. A 1994 explosion at an Iowa nitrogen fertilizer plant near Sioux City killed four employees and seriously injured two more. The blast, suspected to be caused by explosion of ammonium nitrate at a nearby production facility, tore open a 15,000-ton anhydrous ammonia storage tank, releasing ammonia vapor. Many of the other 24 employees on that shift suffered from ammonia inhalation. Although such accidents are rare, injuries and illnesses associated with storing and shipping ammonia are reported each year.

Aluminum metal itself is nontoxic, but its production has its own dangers. High temperatures are used and molten aluminum is drawn from large electrolytic cells. As the carbon anodes burn off in the cryolite bath, toxic carbon monoxide gas is produced and must be handled properly. Injuries are also associated with mining and transporting aluminum ore.

Water quality could be threatened. If leaks occurred during production so that ammonia entered the Snake River, the resulting water solution would become basic—that is, it would have a pH higher than 7. If the plant were to malfunction, excessive release of ammonia-laden wastewater could kill aquatic life in the Snake River.

The aluminum or ammonia markets might decline. Aluminum is important commercially because it is inexpensive, corrosion-resistant, and has low density (low mass for a given volume). The market for aluminum is driven by principal uses where such properties are important—building trades, transportation, and packaging containers. If aluminum substitutes were developed in any of these major areas, demand for aluminum would weaken. These substitutes might come from new alloys or by development of new materials such as different plastics or ceramics. If demand for aluminum fell, the Riverwood plant would benefit the community only relatively briefly.

The fertilizer industry is among the largest consumers of ammonia. Current fertilizer-intensive agricultural methods have created controversy. In some cases crop yields have declined despite application of increased amounts of synthetic fertilizer. Some farmers have elected to use less synthetic fertilizer; ammonia demand may decline in coming years. This could hurt Riverwood's future economy, even though the plant would provide short-term economic assistance.

E.2 THE TOWN COUNCIL MEETING

In this final *Chemistry in the Community* activity, we come full circle, returning to the type of activity that closed the first unit (Water), a debate before the Riverwood Town Council. You will be assigned to one of six "special interest" groups who will testify as the council tries to decide whether to approve an industrial chemical plant to be located in the town.

Each group will have planning time to prepare and organize the group's presentation to the council. Select a group spokesperson. You will be given some suggestions about information to consider including in your group's presentation. These suggestions are merely a starting point; you should identify other points that you intend to stress.

Also consider the pros and cons noted above. Prepare written notes. You may find it useful to get information from other groups. What other sources of information would be most useful?

Following the debate, the town council will make appropriate motions, and discuss and vote on them. Several options are available. One option would be to vote down having either chemical manufacturing plant in Riverwood. Another would be to approve such a plant and identify which firm to invite first. Yet another option might be to call for a local referendum on the matter and give the townspeople the choice of whether to invite a chemical company to Riverwood. If such a measure were approved, an additional ballot item would be needed to determine which company should be invited first.

MEETING RULES AND PENALTIES FOR RULE VIOLATIONS

1. The order of presentations is decided by council members and announced at the start of the meeting.
2. As in the Water unit, each group will have two minutes for its presentation. Time cards will notify the speaker of time remaining.
3. If a member of another group interrupts a presentation, the offending group will be penalized 30 seconds for each interruption, to a maximum of one minute. If the group has already made its presentation, it will forfeit its rebuttal time.

INFORMATION FOR TOWN COUNCIL MEMBERS

The council is responsible for conducting the meeting in an orderly manner. Therefore, be prepared to:

1. Decide and announce the order of presentations at the meeting. Groups presenting factual information should be heard before those voicing opinions.
2. Explain the meeting rules and the penalties for violating those rules.
3. Recognize each special interest group at its assigned presentation time.
4. Enforce the two-minute presentation time limit. One suggestion: Prepare time cards with "one minute," "30 seconds," and "time is up." Place these cards in the speaker's line of sight to serve as useful warnings.

INFORMATION FOR EKS NITROGEN PRODUCTS COMPANY OFFICIALS

The following may help your presentation:

The EKS Company is a multinational organization with a long record of safely manufacturing and handling a wide variety of industrial chemicals. EKS helped to develop chemical industry standards for limiting environmental pollution and agrees to meet the council's criteria for any new industries. It is a member of Responsible Care.

Ammonia production is potentially dangerous because of the high temperatures and pressures used. However, long-standing proper engineering and safety measures make ammonia production safer than many nonchemical manufacturing processes.

INFORMATION FOR WYE METALS CORPORATION OFFICIALS

The following may help your presentation:

The WYE Corporation is a world leader in aluminum production. Its plants are modern, efficient, and safe. It has been cited as an industry model for its environmentally sound practices and community outreach. WYE has agreed to abide by the council's criteria for a new industry. It is also a member of Responsible Care.

Aluminum production relies on adequate raw materials and abundant, inexpensive electrical energy. The Riverwood plant would be similar in design and operation to several other WYE aluminum production facilities, known for their safe, efficient operation.

INFORMATION FOR RIVERWOOD INDUSTRIAL DEVELOPMENT AUTHORITY (RIDA) MEMBERS

The following may help your presentation:

Because of high unemployment, Riverwood needs to attract additional employers. EKS or WYE would each provide a significant increase in new jobs, which will help revive the local economy. A new chemical manufacturing plant might encourage other industries to locate in Riverwood.

RIDA will continue its efforts to bring additional industries to Riverwood. Riverwood has abundant electrical power, excellent railroad and highway access, and favorable tax rates.

INFORMATION FOR RIVERWOOD ENVIRONMENTAL LEAGUE MEMBERS

The following may help your presentation:

League members are concerned about the potential problems of air and water pollution that might be caused by either EKS or WYE. They are concerned about how closely either plant would comply with federal, state, and local policies governing the manufacturing of its products.

Although each company has a strong safety record, accidents occur in chemical manufacturing, some with significant environmental impact. Other types of industries might offer fewer potential problems while creating as many jobs as EKS or WYE.

INFORMATION FOR MEMBERS OF CITIZENS OPPOSED TO POTENTIAL EXPENSES (COPE)

The following may help your presentation:

COPE opposes locating either EKS or WYE in Riverwood. Such an industry could become larger than any other company in the history of

Riverwood, and would dominate the economy and workforce of the town. COPE believes that no chemical manufacturing plant should be located within a municipality.

COPE members also feel that if EKS or WYE were to locate in Riverwood, neither company would pay sufficient taxes to offset demands it would make for local services.

INFORMATION FOR MEMBERS OF RIVERWOOD TAXPAYERS ASSOCIATION

The following may help your presentation:

What level of new taxes will be needed to upgrade the local water treatment plant to meet growing industrial needs?

Who will pay for construction and installation of new road signs and, perhaps, even new roads?

Will WYE or EKS pay a fair proportion of taxes compared to other Riverwood businesses?

E.3 LOOKING BACK AND LOOKING AHEAD

During your study of chemistry in this course, you have learned that all changes in matter and energy on our planet involve chemistry. Beyond this, you have explored contributions of chemistry and chemical technology to the growth and well-being of human communities, as well as to problems these activities have created.

To maximize benefits and minimize chemical technology's costs (and risks), individuals must combine an understanding of chemistry, other sciences, and technology with their own values in making informed personal and societal decisions. Chemistry will help improve modern life as long as citizens continue asking reasoned questions, weighing opinions and options in light of the best available information, and making intelligent decisions.

We hope we have encouraged you to continue a lifelong exploration of science-related issues. Whether you decide to study more chemistry or not, many opportunities are available to further your "chemical common sense."

Look ahead to your future to decide what kind of world you want—and how you, science, technology, and other components of society can contribute toward attaining it. As we have suggested, the challenge is great. But, so are the potential rewards.

APPENDIX: COMPOSITION OF FOODS

	Measure	Mass	Energy	Protein	Carbohydrate	Fat	VITAMINS Vitamin A	Vitamin B_1	Vitamin B_2	Vitamin B_6	Vitamin B_{12}	Niacin	Vitamin C	MINERALS Calcium	Iron	Potassium	Sodium
		g	Cal	g	g	g	IU	mg	mg	mg	μg	mg	mg	mg	mg	mg	mg
BEVERAGES																	
Club soda	12 oz	355	0	0	0	0	0	0	0	0	—	0	0	18	—	1	78
Coffee	6 oz	180	3	—	.54	.01	0	0	.02	—	—	1.26	0	13	.02	117	2
Cola drink	12 oz	369	159	0	40	0	0	0	0	0	—	0	0	11	.18	7	20
Fruit-flavored drink	12 oz	372	179	0	45.8	0	0	0	0	0	—	0	0	15	.26	37	52
Diet drink	12 oz	355	0	0	.43	0	0	0	0	0	—	0	0	14	.46	6	76
Tea	8 oz	240	0	—	.06	.01	—	0	—	—	—	—	—	5	.1	58	19
BAKERY AND GRAINS																	
Breads																	
Bagel	1	100	296	11	56	2.57	—	.38	.29	.044	—	3.53	0	42	2.65	74	360
Biscuit, enr*	2"	28	103	2.1	12.8	4.8	t†	.06	.06	—	—	.5	t	34	.4	33	175
Cornbread, whole ground	2"	45	93	3.3	13.1	3.2	68	.06	.09	—	—	.3	t	54	.5	71	283
English muffin, enr	1	27	130	4	1	t	.23	.136	—	—	2	t	20	1.08	—	—
French or Vienna, enr	1 slice	20	58	1.8	11.1	.64	t	.06	.04	.01	0	.5	t	9	.4	18	116
Mixed grain	1 slice	100	257	9.96	46.6	3.72	t	.39	.38	.103	0	4.16	t	104	3.26	218	412
Pita, whole wheat	1	42	140	6	24	2	t	.2	.068	—	—	2.4	t	40	1.8	—	—
Pumpernickel	1 slice	32	79	2.9	17	.4	—	.07	.04	.05	0	.4	t	27	.8	145	182
Roll, dinner, enr	1	38	113	3.1	20.1	2.2	t	.11	.07	—	—	.8	t	28	.7	36	192
Roll, hamburger/hot dog, enr	1	40	119	3.3	21.2	2.2	t	.11	.07	—	—	.9	t	30	.8	38	202
Roll, whole wheat	1	35	90	3.5	18.3	1	t	.12	.05	—	—	1.1	t	37	.8	102	197
Rye	1 slice	23	56	2.1	12	.3	—	.04	.02	.02	0	.3	—	17	.4	33	128
Wheat	1 slice	100	255	9.55	47	4.1	t	.46	.32	.109	0	4.52	t	126	3.49	138	539
White, enr	1 slice	23	62	2	11.6	.79	t	.06	.05	.009	t	.6	t	20	.6	24	117
Whole wheat	1 slice	23	56	2.4	11	.7	t	.06	.02	.04	0	.6	t	23	.5	63	121
Cracker, graham	1	14.2	55	1.1	10.4	1.3	0	.01	.03	—	—	.2	0	6	.2	55	95
Cracker, soda	1	2.8	12.5	.26	2	.37	0	t	.001	—	0	.03	0	.6	.04	3.4	31
Muffin, bran, enr	1	40	104	3.1	17.2	3.9	90	.06	.1	—	—	1.6	t	57	1.5	172	179
Muffin, whole wheat	1	40	103	4	20.9	1.1	t	.14	.05	—	—	1.2	t	42	1	117	226
Pancake, plain, enr	4"	27	62	1.9	9.2	1.9	30	.05	.06	—	—	.4	—	27	.4	33	115
Pancake, whole wheat	4"	45	74	3.4	8.8	3.2	80	.09	.07	—	—	.4	t	50	.54	—	—
Pizza, cheese, 14"	⅛	65	153	7.8	18.4	5.4	410	.04	.13	—	—	.7	5	144	.7	85	456
Pretzel, twisted	1	16	62	1.6	12.1	.7	0	.003	.008	.003	t	.1	0	4	.2	21	269
Taco shell	1	100	453	8.79	65.7	19.5	—	.29	.15	—	0	1.72	—	142	2.6	—	—
Tortilla, corn	6"	30	63	1.5	13.5	.6	6	.04	.015	.022	0	.3	0	60	.9	—	—
Waffle, enr	5½"	75	209	7	28	7.4	248	.13	.19	—	—	1	t	85	1.3	109	356
Cereals																	
All-Bran	⅓ c	28	71	4	21	1	376	.37	.43	.51		5	15	23	4.52	350	320
Alpha Bits	1 c	28	111	2	25	1	376	.37	.43	.51		5	0	8	2.7	110	180

*enr = enriched
†t = trace
Source: United States Department of Agriculture

BAKERY AND GRAINS (cont.)

	Measure	Mass	Energy	Protein	Carbohydrate	Fat	VITAMINS Vitamin A	Vitamin B$_1$	Vitamin B$_2$	Vitamin B$_6$	Vitamin B$_{12}$	Niacin	Vitamin C	MINERALS Calcium	Iron	Potassium	Sodium
		g	Cal	g	g	g	IU	mg	mg	mg	µg	mg	mg	mg	mg	mg	mg
Apple Jacks	1 c	28	109	2	26	<1	376	.37	.43	.51		5	15	3	4.52	23	124
Bran Flakes, 40%, Post	1 c	47	152	5.3	37.3	.8	2072	.6	.7	.8	2.5	8.3	—	21	7.5	251	431
Corn Flakes, Kellogg's	1¼ c	28	110	2.3	24.4	.1	1250	.4	.4	.5	—	5	15	1	1.8	26	351
Corn, puffed, Kix	1½ c	28.4	110	2.5	23.4	.7	1250	.4	.4	.5	1.5	5	15	35	8.1	44	339
C.W. Post, plain	1 c	97	431	9	69	15	1283	1.26	1.46	1.75		17.1	0	47	15.4	197	166
C.W. Post, with raisins	1 c	103	445	9	74	15	1362	1.34	1.55	1.85		18.1	0	50	16.4	260	160
Cap'n Crunch	1 c	37	156	2	30	3	5	.66	.71	1		8.62	0	6	9.81	48	278
Cap'n Crunchberries	1 c	38	160	2	31	3	5	.65	.74	1.02		8.92	0	12	9.92	54	268
Cocoa Krispies	1 c	36	139	2	32	1	476	.47	.54	.65		6.34	19	6	2.27	53	275
Cocoa Pebbles	1 c	31	128	1	27	2	415	.41	.47	.56		5.52	0	5	1.97	52	149
Corn Chex	1 c	28	111	2	25	1	14	.37	.07	.51		5	15	3	1.79	23	271
Corn Flakes, Post Toasties	1¼ c	28	108	2	24	<1	370	.36	.42	.5		4.93	0	1	.74	32	293
Crispy Wheat 'N Raisins	1 c	43	150	3	35	1	569	.56	.64	.77		7.57	0	71	6.84	173	204
Froot Loops	1 c	28	111	2	25	1	376	.37	.43	.51		5	15	3	4.52	26	144
Frosted Mini-Wheats	4 ea	31	111	3	26	<1	410	.4	.46	.56		5.46	16	10	1.95	105	9
Frosted Rice Krispies	1 c	28	108	1	26	<1	376	.37	.43	.51		5	15	1	1.79	21	239
Fruity Pebbles	1 c	32	131	1	28	2	429	.42	.49	.58		5.71	0	4	2.04	25	180
Golden Grahams	1 c	39	150	2	33	1	517	.51	.59	.7		6.87	21	24	6.21	86	386
Granola, homemade	½ c	61	297	7	34	17	2	.37	.15	.21		1.07	1	38	2.42	306	6
Granola, low-fat	⅓ c	31	120	3	25	2	150	.37	.42	.5		4.99	—	—	1.8	95	60
Granola, Nature Valley	1 c	113	503	11.5	75.5	19.6	—	.39	.19	—	—	.83	—	71	3.78	389	232
Honey BucWheat Crisp	¾ c	28	110	3	23	1	682	.68	.77	1.4		9	27	40	8.12	106	270
HoneyComb	1 c	22	86	1	20	<1	291	.29	.33	.4		3.87	0	4	2.09	70	123
Honey & Nut Corn Flakes	1 c	38	151	2	31	2	501	.49	.57	.68		6.67	20	5	2.39	48	300
Honey Nut Cheerios	1 c	33	126	4	27	1	441	.43	.5	.6		5.87	18	23	5.3	116	302
Honey Smacks	1 c	38	140	3	33	1	501	.49	.57	.68		6.67	20	4	2.39	56	100
King Vitaman	1 c	19	77	1	16	1	644	.83	.95	1.06		11.6	30	2	11.4	23	145
Life	1 c	43	158	8	31	1	9	.93	.97	.08		11.3	0	150	11.3	192	224
Lucky Charms	1 c	32	125	3	26	1	424	.42	.48	.58		5.63	17	36	5.09	66	227
Nutri-Grain, corn	1 c	42	160	3	35	1	556	.55	.63	.76		7.39	22	1	.89	98	276
Nutri-Grain, wheat	1 c	44	158	4	37	<1	583	.57	.66	.79		7.74	23	12	1.24	119	299
100% Bran	1 c	66	177	8	48	3	0	1.58	1.78	2.11		20.9	63	46	8.12	822	457
100% Natural cereal, plain	½ c	57	267	7	36	12	3	.17	.31	.1		1.3	0	99	1.68	281	24
100% Natural with apples & cinnamon	1 c	104	478	11	70	20	6	.33	.57	.11		1.88	1	157	2.9	515	52
Oats, puffed, Cheerios	1¼ c	28.4	111	4.3	19.6	1.8	1250	.4	.4	.5	1.5	5	15	48	4.5	101	307

BAKERY AND GRAINS (cont.)

	Measure	Mass	Energy	Protein	Carbohydrate	Fat	VITAMINS Vitamin A	Vitamin B_1	Vitamin B_2	Vitamin B_6	Vitamin B_{12}	Niacin	Vitamin C	MINERALS Calcium	Iron	Potassium	Sodium
		g	Cal	g	g	g	IU	mg	mg	mg	µg	mg	mg	mg	mg	mg	mg
Oatmeal, cooked	1 c	234	145	6	25.2	2.4	38	.26	.05	.047	—	.3	—	20	1.59	132	1
Product 19	1 c	33	125	3	27	<1	1746	1.75	1.98	2.34		23.3	70	4	21	51	378
Quisp	1 c	30	124	1	25	2	5	.54	.76	.91		5.79	0	9	6.33	45	240
Raisin Bran, Kellogg's	1 c	49	152	5	37	1	498	.49	.59	.69		6.66	0	17	22.2	254	271
Raisin Bran, Post	1 c	56	171	5	42	1	741	.73	.84	1.01		9.86	0	26	8.9	344	365
Rice Chex	¾ c	19	75	1	17	1	1	.25	.01	.34		3.34	10	3	1.2	22	158
Rice Krispies, Kellogg's	1 c	29	114	2	25	<1	384	.38	.43	.52		5.1	15	4	1.83	30	348
Rice, puffed	1 c	14	56	1	13	<1	0	.01	.01	.01		.42	0	1	.15	16	<1
Shredded Wheat	1 c	43	152	5	34	1	0	.11	.12	.11		2.24	0	16	1.8	153	4
Special K	1 c	21	83	4	16	<1	282	.28	.32	.38		3.75	11	6	3.39	37	199
Sugar Corn Pops	1 c	28	108	1	26	<1	376	.37	.43	.51		5	15	1	1.79	17	103
Sugar Frosted Flakes	1 c	35	133	2	32	<1	463	.45	.52	.63		6.16	19	1	2.21	22	283
Super Golden Crisp	1 c	33	123	2	30	<1	437	.43	.49	.59		5.81	0	7	2.08	123	29
Total, wheat, with added calcium	1 c	33	116	3	26	1	1746	1.75	1.98	2.34		23.3	70	281	21	123	326
Trix	1 c	28	109	2	25	<1	376	.37	.43	.51		5	15	6	4.52	27	181
Wheat Chex	1 c	46	168	5	38	1	0	.6	.17	.83		8.1	24	18	7.31	173	308
Wheat, Cream of, cooked	1 c	251	134	3.8	27.7	.5	—	.2	.1	—	—	1.5	—	51	10.3	43	2
Wheat, flakes, Total	1 c	33	116	3.3	26	.7	5820	1.7	2	2.3	7	23.3	70	56	21	123	409
Wheaties	1 c	29	101	3	23	<1	384	.38	.43	.52		5.1	102	44	4.61	108	276
Wheat, puffed	1 c	12	44	1.8	9.5	.1	—	.02	.03	.02	—	1.3	—	3	.57	42	0
Wheat, shredded	1 large	23.6	83	2.6	18.8	.3	—	.07	.06	.06	—	1.08	—	10	.74	77	0

Desserts and Sweets

	Measure	Mass	Energy	Protein	Carbohydrate	Fat	VITAMINS Vitamin A	Vitamin B_1	Vitamin B_2	Vitamin B_6	Vitamin B_{12}	Niacin	Vitamin C	MINERALS Calcium	Iron	Potassium	Sodium
Apple or brown betty	1 c	215	325	3.4	64	7.5	220	.13	.09	—	—	.9	2	39	1.3	215	329
Cake, angel food	1/10	45	121	3.2	27	.1	0	.004	.06	—	—	.1	0	4	.1	40	127
Cake, cheese	1/10	100	302	5.42	28.5	19.2	254	.03	.13	.064	.495	.46	5	56	48	98	222
Cake, devil's food	2 × 3 × 2"	45	165	2.2	23.4	7.7	68	.009	.045	—	—	.1	t	33	.4	63	132
Cake, gingerbread	3 × 3 × 2"	117	371	4.4	60.8	12.5	110	.14	.13	—	—	1.1	0	80	2.7	531	277
Cake, pound	3 × 3 × ½"	30	142	1.7	14.1	8.8	84	.009	.03	—	—	.1	0	6	.2	18	33
Cake, sponge	1/10	50	149	3.8	27	2.85	225	.03	.07	—	—	.1	t	15	.6	44	84
Cake, white	1/10	50	188	2.3	27	8	15	.005	.04	.025	—	.1	t	32	.1	38	162
Candy, caramel	1	5	20	.2	3.88	.536	t	.002	.009	—	—	.018	t	7.5	.071	9.64	11.4
Candy, chocolate bar	1 oz	28	147	2.2	16	9.2	80	.02	.1	—	—	.1	t	65	.3	109	27
Candy, chocolate fudge	1" cube	21	84	.6	15.8	.3	t	t	.02	—	—	t	t	16	.2	31	40
Candy, peanut brittle	1 oz	25	119	1.6	23	2.6	0	.05	.01	—	—	1	0	10	.7	43	9
Chocolate, baking	1 oz	28	143	3	8.2	15	20	.01	.07	—	—	.4	0	22	1.9	235	1
Chocolate, semisweet	1 oz	28	144	1.2	16.2	10.1	10	t	.02	—	—	.1	t	9	.7	92	1
Chocolate syrup	1 T	18.7	46	.45	11.7	.49	t	.005	.015	—	—	.1	0	3	.3	53	10
Cookie, brownies	2 × 2 × ¾"	30	146	2	15.3	9.4	60	.05	.03	—	—	.2	t	12	.6	57	75
Cookie, chocolate chip, 2⅓"	1	10	51	.55	6	3	10	.01	.01	—	—	.1	t	3.5	.2	11.8	34.8

BAKERY AND GRAINS (cont.)

	Measure	Mass	Energy	Protein	Carbohydrate	Fat	VITAMINS Vitamin A	Vitamin B₁	Vitamin B₂	Vitamin B₆	Vitamin B₁₂	Niacin	Vitamin C	MINERALS Calcium	Iron	Potassium	Sodium
		g	Cal	g	g	g	IU	mg	mg	mg	μg	mg	mg	mg	mg	mg	mg
Cookie, fig bar	1	14	50	.55	10.5	.775	15	.005	.01	—	—	.05	t	11	.15	27.8	35.3
Cookie, gingersnap	1	7	29.4	.39	5.59	.62	5	.003	.004	—	—	.03	t	5.1	.16	32.3	40
Cookie, macaroon	1	14	67	.7	9.3	3.2	0	.006	.02	—	—	.1	0	4	.1	65	5
Cookie, oatmeal and raisin, 3"	1	14	63	.9	10.3	2.2	7	.015	.011	—	—	.1	t	3	.4	52	23
Custard, baked	1 c	265	305	14.3	29.4	14.6	930	.11	.5	—	—	.3	1	297	1.1	387	209
Danish pastry	1	42	179	3.1	19.4	10	130	.03	.06	—	—	.3	t	21	.4	48	156
Doughnut, cake	1	32	125	1.5	16.4	6	26	.05	.05	—	—	.4	t	13	.4	29	160
Doughnut, raised	1	30	124	1.9	11.3	8	18	.05	.05	—	—	.4	0	11	.5	24	70
Eclair, custard	1	100	239	6.2	23.2	13.6	340	.04	.16	—	—	.1	t	80	.7	122	82
Honey	1 T	21	64	.1	17.8	0	0	.002	.014	.004	0	.1	t	1	.1	11	1
Jams	1 T	20	54	.1	14	t	t	t	.01	.005	0	t	0	4	.2	18	2
Jellies	1 T	18	49	t	12.7	t	t	t	.01	—	—	t	1	4	.3	14	3
Molasses, blackstrap	1 T	20	43	0	11	—	—	.02	.04	.054	—	.4	—	137	3.2	585	19
Molasses, light	1 T	20	50	0	13	—	—	.01	.01	—	—	t	—	33	.9	183	3
Pie, apple, 9"	⅙	160	410	3.4	61	17.8	48	.03	.03	—	0	.6	2	1	.5	128	482
Pie, chocolate cream	⅙	100	264	4.58	29.4	15.1	264	.1	.17	.047	.366	.72	t	84	1.08	142	273
Pie, lemon meringue	⅙	140	357	5.2	52.8	14.3	238	.05	.11	—	—	.3	4	20	.7	70	395
Pie, pecan	⅙	160	668	8.2	82	36.6	256	.25	.11	—	—	.5	t	75	4.5	197	354
Pie, pumpkin	⅙	150	317	6	36.7	16.8	3700	.04	.15	—	—	.8	t	76	.8	240	321
Pie crust, enr	9"	135	675	8.2	59.1	45.2	0	.27	.19	—	—	2.4	0	19	2.3	67	825
Pudding, bread, w/raisins	1 c	265	496	14.8	75.3	16.2	800	.16	.5	—	—	1.3	3	289	2.9	570	533
Pudding, chocolate	1 c	260	385	8.1	66.8	12.2	390	.05	.36	—	—	.3	1	250	1.3	445	146
Pudding, rice, w/raisins	1 c	265	387	9.5	70.8	8.2	290	.08	.37	—	—	.5	t	260	1.1	469	188
Pudding, tapioca cream	1 c	165	221	8.3	28.2	8.4	480	.07	.3	—	—	.2	2	173	.7	223	257
Sugar, cane	1 T	12	46	0	11.9	0	0	0	0	—	—	0	0	0	t	t	t
Sugar, brown, packed	1 c	220	821	0	212	0	0	.02	.07	—	—	.4	0	187	7.5	757	66
Sugar, confectioner's	1 c	120	462	0	119	0	0	0	0	—	—	0	0	0	.1	4	1
Sugar, raw	1 T	14	14	.06	12.7	.07	t	.003	.016	—	—	.04	.3	7	.6	—	—
Syrup, corn	1 T	20	57	0	14.8	0	0	0	.002	—	—	t	.1	9	.8	—	—
Syrup, maple	1 T	20	50	0	12.8	0	0	—	—	—	—	—	0	33	.2	26	3
Grains																	
Bran, wheat	1 c	57	121	9	35.4	2.6	0	.41	.2	.468	0	12	0	67.8	8.49	639	5.13
Cornmeal, whl grd	1 c	118	427	10.6	88	4	566	.35	.09	.29	0	2.2	—	20	2.1	293	1
Cornstarch	1 T	8	29	t	7	.05	0	0	.006	t	—	.002	0	0	.04	.32	.32
Macaroni, enr, ckd	1 c	140	151	4.8	32.2	1	0	.2	.11	.029	0	1.5	0	11	1.2	85	1
Noodles, egg, enr, cooked	1 c	160	200	6.6	37.3	2.4	112	.22	.13	.04	t	1.9	0	16	1.4	70	3
Pasta, whole wheat	4 oz	113	400	20	78	1	200	.6	.85	—	—	8	10.8	20	5.4	—	—
Popcorn	1 c	14	54	1.8	10.7	.7	—	.055	.02	.03	0	.3	0	2	.4	33.6	t
Rice, brown	1 c	196	704	14.8	152	3.6	0	.68	.08	1	0	9.2	0	64	3.2	420	16
Rice, white enr	1 c	195	708	13.1	157	1.5	0	.86	.06	.3	0	6.8	0	47	5.7	179	10
Rice, wild	1 c	160	565	22.6	121	1.1	0	.72	1.01	—	0	9.9	0	30	6.7	352	11

BAKERY AND GRAINS (cont.)	Measure	Mass	Energy	Protein	Carbohydrate	Fat	VITAMINS Vitamin A	Vitamin B$_1$	Vitamin B$_2$	Vitamin B$_6$	Vitamin B$_{12}$	Niacin	Vitamin C	MINERALS Calcium	Iron	Potassium	Sodium
		g	Cal	g	g	g	IU	mg	mg	mg	µg	mg	mg	mg	mg	mg	mg
Spaghetti, enr, ckd	1 c	140	155	4.8	32.2	.6	0	.2	.11	—	0	1.5	0	11	1.3	85	1
Tapioca	1 c	152	535	.9	131	.3	0	0	.15	—	—	0	0	15	.6	27	5
DAIRY AND EGGS **Cheese**																	
Bleu	1 oz	28	100	6.07	.66	8.15	204	.008	.108	.047	.345	.288	0	150	.09	73	396
Brick	1 oz	28	105	6.59	.79	8.41	307	.004	.1	.018	.356	.033	0	191	.12	38	159
Brie	1 oz	28	95	5.88	.13	7.85	189	.02	.147	.067	.468	.108	0	52	.14	43	178
Cheddar	1 oz	28	114	7.06	.36	9.4	300	.008	.106	.021	.234	.023	0	204	.19	28	176
Cheese food, American	1 oz	28	94	5.57	2.36	6.93	200	.009	.126	.04	.363	.021	0	141	.24	103	274
Cheese food, Swiss	1 oz	28	92	6.2	1.28	6.84	243	.004	.113	—	.652	.029	0	205	.17	81	440
Cheese spread, American	1 oz	28	82	4.65	2.48	6.02	223	.014	.122	.033	.113	.037	0	159	.09	69	381
Colby	1 oz	28	112	6.74	.73	9.1	293	.004	.106	.022	.234	.026	0	194	.22	36	171
Cottage, creamed	1 c	210	217	26.2	5.63	9.45	342	.044	.342	.141	1.3	.265	t	126	.29	177	850
Cottage, dry	1 c	145	123	25	2.68	.61	44	.036	.206	.119	1.19	.225	0	46	.33	47	19‡
Cottage, lowfat, 2%	1 c	226	203	31	8.2	4.36	158	.054	.418	.172	1.6	.325	t	155	.36	217	918
Cream	1 oz	28	99	2.14	.75	9.89	405	.005	.056	.013	.12	.029	0	23	.34	34	84
Feta	1 oz	28	75	4	1.16	6	—	—	—	—	—	—	0	140	.18	18	316
Gouda	1 oz	28	101	7.07	.63	7.78	183	.009	.095	.023	—	.018	0	198	.07	34	232
Gruyère	1 oz	28	117	8.45	.1	9.17	346	.017	.079	.023	.454	.03	0	287	—	23	95
Monterey jack	1 oz	28	106	6.94	.19	8.58	269	8.58	.111	—	—	—	0	212	.2	23	152
Mozzarella	1 oz	28	80	5.5	.63	6.12	225	.004	.069	.016	.185	.024	0	147	.05	19	106
Mozzarella, part skim	1 oz	28	72	6.88	.78	4.51	166	.005	.086	.02	.232	.03	0	183	.06	24	132
Muenster	1 oz	28	104	6.64	.32	8.52	318	.004	.091	.016	.418	.029	0	203	.12	38	178
Parmesan	1 oz	28	111	10	.91	7.32	171	.011	.094	.026	—	.077	0	336	.23	26	454
Parmesan, grated	1 T	5	23	2	.19	1.5	35	.002	.019	.005	—	.016	0	69	.05	5	93
Processed, American	1 oz	28	106	6.28	.45	8.86	343	.008	.1	.02	.197	.02	0	174	.11	46	406
Processed, Swiss	1 oz	28	95	7	.6	7.09	229	.004	.078	.01	.348	.011	0	219	.17	61	388
Provolone	1 oz	28	100	7.25	.61	7.55	231	.005	.091	.021	.415	.044	0	214	.15	39	248
Ricotta	1 c	246	428	27.7	7.48	31.9	1205	.032	.48	.106	.831	.256	0	509	.94	257	207
Ricotta, part skim	1 c	246	340	28	12.6	19.4	1063	.052	.455	.049	.716	.192	0	669	1.08	308	307
Romano	1 oz	28	110	9	1	7.64	162	—	.105	—	—	.022	0	302	—	—	340
Roquefort	1 oz	28	105	6.11	.57	8.69	297	.011	.166	.035	.182	.208	0	188	.16	26	513
Swiss	1 oz	28	107	8.06	.96	7.78	240	.006	.103	.024	.475	.026	0	272	.05	31	74
Cream Half and half	1 c	242	315	7.16	10.4	27.8	1050	.085	.361	.094	.796	.189	2.08	254	.17	314	98
Whipping, light	1 c	239	699	5.19	7.07	73.8	2694	.057	.299	.067	.466	.1	1.46	166	.07	231	82
Whipping, heavy	1 c	238	821	4.88	6.64	88	3499	.052	.262	.062	.428	.093	1.38	154	.07	179	89
Whipped, pressurized	1 c	60	154	1.92	7.49	13.3	548	.022	.039	.025	.175	.042	0	61	.03	88	78
Sour cream	1 c	230	493	7.27	9.82	48.2	1817	.081	.343	.037	.69	.154	1.98	268	.14	331	123
Frozen Desserts Ice cream	1 c	133	269	4.8	31.7	14.3	543	.052	.329	.061	.625	.134	.7	176	.12	257	116
Ice cream, rich	1 c	148	349	4.13	31.9	23.6	897	.044	.283	.053	.537	.115	.61	151	.1	221	108
Ice milk	1 c	131	184	5.16	28.9	5.63	214	.076	.347	.085	.875	.118	.76	176	.18	265	105
Sherbet	1 c	193	270	2.16	58.7	3.82	185	.033	.089	.025	.158	.131	3.86	103	.31	198	88

‡unsalted

DAIRY AND EGGS (cont.)	Measure	Mass	Energy	Protein	Carbohydrate	Fat	VITAMINS Vitamin A	Vitamin B_1	Vitamin B_2	Vitamin B_6	Vitamin B_{12}	Niacin	Vitamin C	MINERALS Calcium	Iron	Potassium	Sodium
		g	Cal	g	g	g	IU	mg	mg	mg	µg	mg	mg	mg	mg	mg	mg
Milk																	
Whole	1 c	244	150	8.03	11.37	8.15	307	.093	.395	.102	.871	.205	2.29	291	.12	370	120
Lowfat, 2%	1 c	244	121	8.12	11.7	4.68	500	.095	.403	.105	.888	.21	2.32	297	.12	377	122
Skim	1 c	245	86	8.35	11.8	.44	500	.088	.343	.098	.926	.216	2.4	302	.1	406	126
Buttermilk	1 c	245	99	8.11	11.7	2.16	81	.083	.377	.537	.083	.142	2.4	285	.12	371	257
Whole, dry	1 c	128	635	33.6	49	34.2	1180	.362	1.54	.387	4.16	.827	11	1168	.6	1702	475
Nonfat, dry	1 c	120	435	43.4	62.3	.92	43	.498	1.86	.433	4.84	1.14	8.11	1508	.38	2153	642
Nonfat, dry, instant	1 c	68	244	23.8	35.5	.49	18.3	.281	1.18	.235	2.71	.606	3.79	837	.21	1160	373
Chocolate	1 c	250	208	7.92	25.8	8.48	302	.092	.405	.1	.835	.313	2.28	280	.6	417	149
Malted, powder	1 T	21	86	2.74	15.2	1.78	68	.111	.142	.078	.164	1.07	0	56	.16	159	96
Yogurt																	
Plain	1 c	227	139	7.88	10.5	7.38	279	.066	.322	.073	.844	.17	1.2	274	.11	351	105
Plain, skim	1 c	227	127	13	17.4	.41	16	.109	.531	.120	1.39	.281	1.98	452	.2	579	174
Fruit, lowfat	1 c	227	225	9.04	42.3	2.61	111	.077	.368	.084	.967	.195	1.36	314	.14	402	121
Eggs																	
Whole	1 lge	50	79	6.07	.6	5.58	260	.044	.15	.06	.773	.031	0	28	1.04	65	69
White	1 lge	33	16	3.35	.41	t	0	.002	.094	.001	.021	.029	0	4	.01	45	50
Yolk	1 lge	17	63	2.79	.04	5.6	313	.043	.074	.053	.647	.012	0	26	.95	15	8
Whole, dried	1 T	5	30	2.29	.24	2.09	98	.015	.059	.02	.5	.012	0	11	.39	24	26
FAST FOODS AND SANDWICHES **Sandwiches** Egg salad:																	
On white bread, soft	1 ea	111	361	9	29	23	73	.29	.36	.2		2.26	0	81	2.21	115	521
On part whole wheat	1 ea	111	358	9	28	24	73	.26	.32	.22		2.35	0	79	2.36	161	516
On whole wheat	1 ea	125	385	11	33	24	73	.27	.31	.3		2.71	0	71	2.81	225	588
Ham: On rye bread	1 ea	116	241	16	20	10	6	.78	.28	.38		4.66	17	38	1.72	303	1289
On white bread, soft	1 ea	122	260	17	23	11	6	.83	.3	.38		4.96	17	57	1.97	293	1277
On part whole wheat	1 ea	122	257	17	22	11	6	.8	.27	.39		5.03	17	56	2.09	329	1274
On whole wheat	1 ea	136	284	19	27	12	6	.83	.27	.46		5.45	18	50	2.5	389	1364
Ham & cheese: On white bread, soft	1 ea	151	388	21	29	21	87	.78	.4	.36		4.84	14	237	2.34	311	1582
On part whole wheat	1 ea	151	384	22	27	21	87	.75	.37	.38		4.92	14	236	2.49	356	1578
On whole wheat	1 ea	165	411	24	33	22	88	.76	.36	.45		5.29	14	228	2.93	419	1655
Ham & swiss on rye	1 ea	145	368	23	25	19	77	.73	.38	.37		4.52	14	309	1.99	313	1289
Ham salad: On white bread, soft	1 ea	125	344	10	34	18	8	.52	.26	.18		3.49	4	67	2.11	159	898
On part whole wheat	1 ea	125	340	10	33	19	8	.49	.23	.2		3.57	4	66	2.27	206	893
On whole wheat	1 ea	139	367	12	39	19	8	.51	.21	.27		3.94	4	57	2.71	269	967
Patty melt: Ground beef & cheese on rye	1 ea	177	583	36	25	37	137	.28	.47	.33		6.56	<1	224	3.83	377	892

FAST FOODS AND SANDWICHES (cont.)	Measure	Mass	Energy	Protein	Carbohydrate	Fat	VITAMINS Vitamin A	Vitamin B₁	Vitamin B₂	Vitamin B₆	Vitamin B₁₂	Niacin	Vitamin C	MINERALS Calcium	Iron	Potassium	Sodium
		g	Cal	g	g	g	IU	mg	mg	mg	μg	mg	mg	mg	mg	mg	mg
Peanut butter & jelly sandwich: On white bread, soft	1 ea	100	345	10	47	14	<1	.3	.22	.15		5.55	<1	74	2.36	263	309
On part whole wheat	1 ea	100	341	11	46	14	<1	.27	.19	.17		5.63	<1	73	2.52	309	305
On whole wheat	1 ea	114	368	13	51	15	<1	.28	.18	.24		6.03	<1	64	2.97	374	375
Reuben, grilled: Corned beef, swiss cheese, sauerkraut on rye	1 ea	233	496	29	29	29	128	.25	.37	.3		3.43	12	347	3.96	337	1655
Roast beef: On a bun	1 ea	150	374	23	36	15	22	.4	.33	.28		6.33	2	58	4.56	341	855
On white bread, soft	1 ea	122	314	22	26	13	9	.26	.28	.32		5.29	10	57	3.21	341	1257
On part whole wheat	1 ea	122	311	23	25	13	9	.24	.25	.33		5.36	10	56	3.33	377	1254
On whole wheat	1 ea	136	339	25	30	14	9	.25	.25	.4		5.78	10	50	3.77	438	1343
Tuna salad: On white bread, soft	1 ea	116	310	13	33	14	21	.28	.23	.13		5.63	1	71	2.26	160	558
On part whole wheat	1 ea	116	306	13	32	14	21	.25	.19	.15		5.71	1	69	2.42	206	553
On whole wheat	1 ea	130	333	15	37	15	21	.26	.18	.22		6.1	1	60	2.86	270	625
Turkey: On white bread, soft	1 ea	122	270	19	22	11	9	.24	.22	.33		7.32	0	55	1.67	242	1252
On part whole wheat	1 ea	122	267	19	21	11	9	.21	.19	.35		7.39	0	53	1.8	279	1249
On whole wheat	1 ea	136	294	21	26	12	9	.22	.19	.41		7.88	0	47	2.19	336	1339
Turkey ham: On rye bread	1 ea	116	239	16	19	10	6	.2	.29	.23		3.81	0	40	3.03	287	1004
On white bread, soft	1 ea	122	258	16	23	11	6	.24	.31	.23		4.11	0	59	3.29	276	991
On part whole wheat	1 ea	122	255	17	22	11	6	.22	.29	.25		4.17	0	58	3.42	313	987
On whole wheat	1 ea	136	282	19	27	12	6	.23	.28	.31		4.57	0	52	3.86	371	1069
FAST FOOD RESTAURANTS Arby's Bac'n cheddar deluxe	1 ea	226	526	27	33	36	100	.38	.51	—		8	9	150	4.5	422	1672
Roast beef sandwiches: Regular	1 ea	147	353	22	32	15	1	.23	.43	.2		7	1	80	3.6	368	588
Junior	1 ea	86	218	12	22	8	—	.15	.25	.1		4	—	40	1.8	197	345
Super	1 ea	234	501	25	50	22	150	.38	.6	.3		9	4	100	4.5	503	798
Beef 'n cheddar	1 ea	197	455	26	28	27	80	.38	.51	.22		8	0	60	3.6	335	955
Chicken breast sandwich	1 ea	184	493	23	48	25	—	.23	.51	.38		8	5	80	3.6	330	1019
Ham 'n cheese sandwich	1 ea	156	292	23	19	14	50	.15	.26	.31		6	24	200	2.7	312	1350
Turkey sandwich, deluxe	1 ea	197	375	24	32	17	60	.23	.43	.52		12	5	80	2.7	346	1047

Source: Arby's Inc. for the basic nutrients. Values for some nutrients from known values of major ingredients.

FAST FOOD RESTAURANTS (cont.)	Measure	Mass	Energy	Protein	Carbohydrate	Fat	VITAMINS Vitamin A	Vitamin B₁	Vitamin B₂	Vitamin B₆	Vitamin B₁₂	Niacin	Vitamin C	MINERALS Calcium	Iron	Potassium	Sodium
		g	Cal	g	g	g	IU	mg	mg	mg	μg	mg	mg	mg	mg	mg	mg
Milk shakes: Chocolate	1 ea	340	451	10	76	12	40	.06	.85	.14		.8	5	250	2.7	410	341
Jamocha	1 ea	326	368	9	59	10	60	.06	.77	.14		5	2	250	2.7	525	262
Vanilla	1 ea	312	330	10	46	11	100	.23	.85	.14		4	2	300	2.7	686	281
Burger King Croissant sandwiches: Egg, bacon, & cheese	1 ea	119	364	15	19	24	151	.32	.3	.11		2.02	2	137	2.02	—	725
Egg, sausage, & cheese	1 ea	163	547	21	23	41	154	.37	.33	.12		4.1	<1	149	2.97	—	1009
Egg, ham, & cheese	1 ea	145	348	19	19	21	151	.49	.32	.22		3.02	10	137	2.22	—	969
Whopper sandwiches: Whopper	1 ea	265	603	26	44	35	59	.32	.4	.34		6.87	14	78	4.81	—	849
Whopper with cheese	1 ea	289	694	31	46	43	84	.33	.47	.32		6.88	14	206	4.82	—	1156
Double beef	1 ea	351	844	46	45	53	60	.34	.56	—		10	14	91	7.3	—	933
Double beef & cheese	1 ea	374	933	51	47	61	85	.35	.63	—		9.97	14	221	7.28	—	1241
Hamburger deluxe	1 ea	136	339	15	28	19	30	.23	.25	.14		3.94	6	39	2.76	—	489
Cheeseburger deluxe	1 ea	158	408	19	30	24	89	.24	.3	—		4.19	6	110	2.93	—	682
Hamburger	1 ea	109	275	15	28	11	15	.23	.25	—		4.04	3	37	2.73	—	510
Cheeseburger	1 ea	120	315	17	28	15	69	.23	.29	—		3.97	3	101	3.77	—	656
Double cheeseburger with bacon	1 ea	159	512	32	26	31	84	.31	.42	—		5.96	1	167	3.78	—	743
Chicken sandwich	1 ea	230	688	26	56	40	13	.45	.31	—		10	1	79	3.31	—	1423
Ham & cheese sandwich	1 ea	230	471	24	44	23	85	.87	.42	—		6	7	195	3.2	—	1534
French fries (salted)	1 ea	74	227	3	24	13	0	.07	.2	—		5	3	6	.33	—	161
Onion rings	1 ea	79	277	4	31	16	0	.05	.03	—		.66	<1	114	.73	—	514
Milk shakes: Chocolate	1 ea	273	313	9	47	10	58	.12	.53	—		.12	0	250	1.54	—	190
Milk shakes: Vanilla	1 ea	273	321	9	49	10	77	.11	.55	—		.12	0	284	—	—	205
Fried apple pie	1 ea	125	311	3	44	14	4	.27	.16	—		.6	5	15	1.2	—	412

Source: Burger King Corporation

Jack in the Box Breakfast items: Breakfast Jack sandwich	1 ea	126	307	18	30	13	90	47	.41	—		3	—	170	3.1	—	871
Sausage crescent	1 ea	156	584	22	28	43	110	.6	.51	—		4.6	—	170	2.9	—	1012
Supreme crescent	1 ea	146	547	20	27	40	110	.65	.54	—		4.2	—	150	2.7	—	1053
Pancake platter	1 ea	231	612	15	87	22	80	.03	.85	—		7	6	100	1.8	—	888
Scrambled egg platter	1 ea	249	653	21	58	37	164	—	.77	—		5.85	11	175	5.73	—	1239
Sandwiches: Hamburger	1 ea	98	267	13	28	11	—	.15	.26	—		2	—	150	1.8	—	556

FAST FOOD RESTAURANTS (cont.)	Measure	Mass	Energy	Protein	Carbohydrate	Fat	VITAMINS Vitamin A	Vitamin B$_1$	Vitamin B$_2$	Vitamin B$_6$	Vitamin B$_{12}$	Niacin	Vitamin C	MINERALS Calcium	Iron	Potassium	Sodium
		g	Cal	g	g	g	IU	mg	mg	mg	µg	mg	mg	mg	mg	mg	mg
Sandwiches: Cheeseburger	1 ea	113	318	16	33	14	40	.23	.23	—		3.03	—	252	2.72	—	753
Jumbo Jack burger	1 ea	205	539	24	39	31	—	.33	.27	—		1.66	—	129	2.86	—	677
Jumbo Jack burger with cheese	1 ea	246	688	32	47	41	—	.37	.45	—		1.63	—	274	3.86	—	1108
Chicken supreme	1 ea	228	597	25	44	36	74	.36	.3	—		10.2	6	223	2.7	—	1368
Double cheeseburger	1 ea	149	467	21	33	27	200	.15	.34	—		6	—	400	2.7	—	842
Tacos: Regular	1 ea	81	191	8	16	11	57	.07	.17	.13		1	<1	100	1.1	257	460
Tacos: Super	1 ea	135	288	12	21	17	85	.12	.08	.18		1.4	2	150	1.6	347	765
Taco salad	1 ea	402	503	34	28	31	270	.29	.53	—		5.8	9	410	3.8	—	1600
French fries	1 ea	109	351	4	45	17	—	.18	.03	—		3.8	26	—	1.3	—	194
Hash browns	1 ea	62	170	1	15	12	0	.05	—	—		1.09	8	—	.39	—	339
Onion rings	1 ea	108	398	5	40	24	—	.3	.18	—		2.73	3	31	2.31	—	473
Milk shakes: Chocolate	1 ea	322	330	11	55	7	—	.15	.6	—		.4	—	350	.72	—	270
Strawberry	1 ea	328	320	10	55	7	—	.15	.43	—		.4	—	350	.36	—	240
Vanilla	1 ea	317	320	10	57	6	—	.15	.34	—		.4	—	350	—	—	230

Source: Jack in the Box Restaurant, Inc.

McDonald's Sandwiches: Big Mac	1 ea	215	500	25	42	26	60	.45	.43	.22		7	1	250	3.6	—	890
McChicken	1 ea	187	415	19	39	19	20	.9	.17	—		9	2	150	2.7	—	83
McLean Deluxe with Cheese	1 ea	219	370	24	35	14	150	.38	.34	—		7	6	200	3.6	—	890
Quarter-Pounder	1 ea	166	410	23	34	20	40	.38	.26	.32		7	4	150	3.6	—	645
Quarter-Pounder with Cheese	1 ea	194	510	28	34	28	150	.38	.34	.32		7	4	300	3.6	—	1110
Filet-O-Fish	1 ea	142	373	14	38	18	20	.3	.14	.1	—	9.06	<1	151	1.81	—	735
Hamburger	1 ea	102	255	12	30	9	40	.3	.17	—		4	2	100	2.7	—	490
Cheeseburger	1 ea	116	305	15	30	13	80	.3	.26	—		4	2	200	2.7	—	725
French fries, small serving	1 ea	68	220	3	26	12	0	.15	0	.18		2	9	10	.36	—	110
Chicken McNuggets	6 ea	112	270	20	17	15	0	.12	.14	.36		8	0	13	1.08	—	580
Sauces (packet): Hot mustard	1 ea	30	70	0	8	4	2	.01	.01	—		.15	<1	20	.22	—	250
Barbecue	1 ea	32	50	0	12	<1	40	.01	.01	—		.17	2	13	.36	—	340
Sweet & sour	1 ea	32	60	0	14	<1	60	0	.01	—		.08	1	11	.17	—	190
Low-fat (frozen yogurt) milk shakes: Chocolate	1 ea	293	324	12	66	2	60	.12	.51	.1		.4	0	352	.84	—	242
Strawberry	1 ea	293	320	11	67	1	60	.12	.51	.11		.4	0	352	.09	—	171
Vanilla	1 ea	293	291	11	60	1	60	.12	.51	—		.31	0	352	.1	—	171
Low-fat (frozen yogurt) sundaes: Hot caramel	1 ea	168	270	7	59	3	60	.09	.34	—		.27	0	200	.08	—	180

FAST FOOD RESTAURANTS (cont.)	Measure	Mass g	Energy Cal	Protein g	Carbohydrate g	Fat g	VITAMINS Vitamin A IU	Vitamin B$_1$ mg	Vitamin B$_2$ mg	Vitamin B$_6$ mg	Vitamin B$_{12}$ µg	Niacin mg	Vitamin C mg	MINERALS Calcium mg	Iron mg	Potassium mg	Sodium mg
Low-fat (frozen yogurt) sundaes: Hot fudge	1 ea	168	240	7	50	3	40	.09	.34	—		.29	0	250	.36	—	170
Strawberry	1 ea	168	210	6	49	1	40	.06	.34	—		.25	1	200	.16	—	95
Vanilla	1 ea	80	100	4	21	1	19	.03	.16	—		.38	0	95	.23	—	76
Pie, apple	1 ea	83	220	2	31	10	10	.12	.09	.03		1.02	1	6	.94	66	175
Muffins (fat-free): Blueberry	1 ea	75	170	3	40	0	—	.12	.14	—		.8	1	80	.72	—	220
Apple bran	1 ea	85	204	6	45	0	1	.17	.19	—		2.27	1	45	1.22	—	227
Cookies: McDonaldland	1 ea	56	262	4	42	8	0	.13	.15	.03		1.81	0	18	1.63	—	271
Cookies: Chocolaty chip	1 ea	56	293	4	37	13	0	.13	.15	—		1.78	0	18	1.6	—	249
Breakfast items: English muffin with spread	1 ea	59	177	5	28	5	31	.24	.29	.03		2.45	1	96	1.49	65	362
Egg McMuffin	1 ea	138	286	18	29	11	102	.48	.34	.08		3.79	0	256	2.76	—	726
Hotcakes with margarine & syrup	1 ea	176	445	8	75	12	40	.3	.34	.11		3.03	0	101	1.81	—	693
Scrambled eggs	1 ea	100	140	12	1	10	100	.07	.26	—		.05	0	60	1.8	—	290
Pork sausage	1 ea	48	179	8	0	17	0	.26	.11	—		2.23	0	8	.8	—	346
Hash brown potatoes	1 ea	53	130	1	15	7	0	.06	.02	—		.8	1	6	.27	—	330
Sausage McMuffin	1 ea	117	299	13	23	17	35	.46	.22	.13		4.33	0	173	2.34	—	667
Sausage McMuffin with egg	1 ea	167	452	22	28	26	105	.56	.45	.21		5.25	0	263	3.78	—	966
Biscuit with biscuit spread	1 ea	75	260	5	32	13	0	.23	.1	.03		1.65	0	80	1.44	—	730
Biscuit with sausage	1 ea	123	438	12	33	29	0	.47	.18	.21		4.17	0	83	1.88	—	1084
Biscuit with sausage & egg	1 ea	180	519	19	34	34	62	.46	.36	.21		4.1	0	103	3.7	—	1244
Biscuit with bacon, egg, cheese	1 ea	156	449	15	34	26	102	.39	.35	.13		2.04	0	204	2.75	—	1238
Salads: Chef salad	1 ea	283	182	18	9	10	1067	.32	.28	—		4.27	22	160	1.54	—	427
Garden salad	1 ea	213	56	5	7	2	1014	.1	.11	—		.45	24	45	1.62	—	79
Chunky chicken salad	1 ea	250	147	24	7	4	1666	.22	.17	—		8.82	26	39	1.06	—	225

Source: McDonald's Corporation

Pizza Hut Pan Pizza: Cheese	2 pce	205	492	30	57	18	90	.56	.6	.17		5.2	7	630	5.4	320	940
Pepperoni	2 pce	211	540	29	62	22	100	.63	.49	.17		5.4	8	520	6.3	405	1127
Supreme	2 pce	255	589	32	53	30	120	.81	.8	.31		6	10	500	5	580	1363
Super Supreme	2 pce	257	563	33	53	26	120	.75	.66	—		6.4	11	540	6.7	532	1447
Thin 'N Crispy: Cheese Pizza	2 pce	148	398	28	37	17	70	.39	.39	.16		4.8	5	660	3.2	261	867
Pepperoni Pizza	2 pce	146	413	26	36	20	70	.42	.43	—		5.2	6	450	3.2	287	986
Supreme Pizza	2 pce	200	459	28	41	22	100	.6	.49	—		5.4	10	430	5.9	544	1328

FAST FOOD RESTAURANTS (cont.)	Measure	Mass	Energy	Protein	Carbohydrate	Fat	VITAMINS Vitamin A	Vitamin B_1	Vitamin B_2	Vitamin B_6	Vitamin B_{12}	Niacin	Vitamin C	MINERALS Calcium	Iron	Potassium	Sodium
		g	Cal	g	g	g	IU	mg	mg	mg	µg	mg	mg	mg	mg	mg	mg
Thin 'N Crispy: Super Supreme Pizza	2 pce	203	463	29	44	21	100	.59	.44	—		5.4	8	460	4.9	463	1336
Hand Tossed: Cheese Pizza	2 pce	220	518	34	55	20	100	.48	.49	—		5.4	10	750	5.4	396	1276
Pepperoni Pizza	2 pce	197	500	28	50	23	100	.54	.53	—		5.6	7	440	5	415	1267
Supreme Pizza	2 pce	239	540	32	50	26	110	.69	.53	—		7.2	12	480	8.1	578	1470
Super Supreme Pizza	2 pce	243	556	33	54	25	110	.71	.58	—		7.4	12	440	6.8	516	1648
Personal Pan Pizza: Pepperoni	2 pce	256	675	37	76	28	120	.56	.66	.2		8.2	10	730	5.8	408	1335
Supreme	2 pce	264	647	33	76	28	120	.59	.66	.32		8	11	520	6.7	487	1313
Source: Pizza Hut																	
Taco Bell Burritos: Bean with red sauce	1 ea	191	414	14	58	13	46	.03	1.87	.29		1.84	49	136	3.22	459	1064
Beef with red sauce	1 ea	191	457	23	44	19	67	.37	1.98	.3		3.19	2	106	3.46	352	1215
Beef & bean with red sauce	1 ea	191	393	17	44	15	77	.47	.4	.57		2.98	2	107	2.07	426	1095
Supreme with red sauce	1 ea	241	475	19	52	21	118	.39	2	.33		2.73	24	145	3.4	473	1116
Enchirito with red sauce	1 ea	213	382	20	31	20	100	.26	.42	1		2.3	28	269	2.84	423	1243
Tacos: Taco	1 ea	78	183	10	11	11	24	.05	.14	.12		1.2	1	84	1.07	159	276
Taco Bellgrande	1 ea	163	355	18	18	23	40	.11	.29	.21		2.02	5	182	1.9	334	472
Soft Taco	1 ea	92	225	12	18	12	30	.39	.22	.1		2.74	1	116	2.27	196	554
Tostada with red sauce	1 ea	156	243	9	27	11	95	.06	.17	.26		.63	45	179	1.53	401	596
Mexican Pizza	1 ea	223	575	21	40	37	215	.32	.33	1.11		2.96	31	257	3.74	408	1031
Taco salad with salsa	1 ea	595	939	36	60	62	407	.52	.77	.57		4.88	78	405	7.22	1066	1307
Nachos: Regular	1 ea	107	349	8	38	19	88	.17	.16	.19		.69	2	193	.91	161	403
Nachos: Bellgrande	1 ea	287	649	22	61	35	40	.1	.34	—		2.17	58	297	3.48	674	997
Pintos & cheese with red sauce	1 ea	128	190	9	19	9	87	.05	.15	.21		.4	52	156	1.42	384	642
Taco sauce, packet	1 ea	4	1	<1	<1	<1	6	0	<.01	<.01		.02	<1	1	.02	4	42
Salsa	1 ea	10	18	1	4	<1	7	.02	.14	—		0	2	36	.6	376	376
Source: Taco Bell Corporation																	
Wendy's Hamburgers: Single on white bun, no toppings	1 ea	119	350	21	29	16	0	.38	.34	—		6	—	100	4.5	265	420
Double on white bun, no toppings	1 ea	197	560	41	32	34	0	.22	.43	.47		9	<1	48	6.3	431	575
Big Classic	1 ea	241	470	26	36	25	60	.3	.25	—		5	12	40	4.5	470	900

FAST FOOD RESTAURANTS (cont.)	Measure	Mass	Energy	Protein	Carbohydrate	Fat	VITAMINS Vitamin A	Vitamin B$_1$	Vitamin B$_2$	Vitamin B$_6$	Vitamin B$_{12}$	Niacin	Vitamin C	MINERALS Calcium	Iron	Potassium	Sodium
		g	Cal	g	g	g	IU	mg	mg	mg	μg	mg	mg	mg	mg	mg	mg
Cheeseburgers: Bacon cheeseburger	1 ea	147	460	29	23	28	82	.26	.28	.23		5.7	1	136	3.6	332	860
Double with lettuce & tomato	1 ea	215	548	30	32	33	111	.34	.35	.25		5.29	5	177	4	430	864
Double with all toppings	1 ea	291	735	48	27	47	112	.36	.53	.46		10	5	180	5.4	620	883
Baked potatoes: Plain	1 ea	250	250	6	52	<1	0	.27	.1	.7		3.82	36	40	2.7	1360	60
With bacon & cheese	1 ea	350	570	19	57	30	150	.22	.17	.87		4.64	36	200	3.7	1380	180
With broccoli & cheese	1 ea	365	500	13	54	25	350	.3	.25	.86		4	90	250	3.6	1550	2
With cheese	1 ea	350	590	16	55	34	200	.22	.25	.8		3.3	36	350	3.6	1380	2
With chili & cheese	1 ea	400	510	22	63	20	172	.3	.26	.9		4.1	36	250	6.13	1590	810
With sour cream & chives	1 ea	310	460	6	53	24	100	.22	.14	.79		3	36	40	2.7	1420	230
Chili	1 ea	256	230	21	16	9	200	.12	.17	—		3	9	60	4.5	565	960
French fries	1 ea	106	306	4	38	15	0	.15	.04	.26		2.96	12	13	1.02	689	105
Frosty dairy dessert	1 c	216	354	7	53	13	143	.11	.45	.12		.31	<1	257	.86	518	194
Chocolate chip cookies	1 ea	64	320	3	40	17	0	.06	.07	.03		.4	0	10	1.09	100	235

Source: Wendy's International

FATS AND OILS Fats																	
Butter	1 T	14.1	101	.12	.008	11.5	433	t	.004	t	—	.006	0	3.37	.022	3.62	117
Margarine	1 T	14.1	101	0	0	11.4	465	0	.006	0	.012	.003	.024	4.23	—	5.97	133
Vegetable shortening	1 T	12.8	115	0	0	12.8	—	—	—	—	—	—	—	—	—	—	—
Oils Corn	1 T	13.6	120	0	0	13.6	—	—	—	—	—	—	—	—	—	—	—
Olive	1 T	13.5	119	0	0	13.5	—	—	—	—	—	—	—	.02	.05	—	0
Peanut	1 T	13.5	119	0	0	13.5	—	—	—	—	—	—	—	.01	0	0	.01
Safflower	1 T	13.6	120	0	0	13.6	—	—	—	—	—	—	—	—	—	—	—
Sesame	1 T	13.6	120	0	0	13.6	—	—	—	—	—	—	—	—	—	—	—
Soybean	1 T	13.6	120	0	0	13.6	—	—	—	—	—	—	—	.01	0	—	0
Sunflower	1 T	13.6	120	0	0	13.6	—	—	—	—	—	—	—	.03	0	—	.01

FRUITS AND FRUIT JUICES																	
Apple	1	150		.27	21	.49	74	.023	.019	.066	0	.106	7.8	10	.25	159	1
Apple, dried	10 rings	64	155	.59	42	.2	0	0	.102	.08	0	.593	2.5	9	.9	288	56
Apple juice	1 c	248	116	.15	29	.28	2	.052	.042	.074	0	.248	2.3	16	.92	296	7
Applesauce, unsw	1 c	244	106	.4	27.5	.12	70	.032	.061	.063	0	.459	2.9	7	.29	183	5
Apricot	3	114	51	1.48	11.7	.41	2769	.032	.042	.057	0	.636	10.6	15	.58	313	1
Apricot, dried	10 halves	35	83	1.28	21.6	.16	2534	.003	.053	.055	0	1.05	.8	16	1.65	482	3
Apricot nectar	1 c	251	141	.92	36	.22	3304	.023	.035	—	0	.653	1.4	17	.96	286	9
Avocado	1	272	324	3.99	14.8	30.8	1230	.217	.245	.563	0	3.86	15.9	22	2.05	1204	21
Banana	1	175	105	1.18	26.7	.55	92	.051	.114	.659	0	.616	10.3	7	.35	451	1
Blackberries	1 c	144	74	1.04	18.3	.56	237	.043	.058	.084	0	.576	30.2	46	.83	282	0
Blueberries	1 c	145	82	.97	20.5	.55	145	.07	.073	.052	0	.521	18.9	9	.24	129	9
Cherries	1 c	145	104	1.74	24	1.39	310	.073	.087	.052	0	.58	10.2	21	.56	325	1
Cranberries	1 c	95	46	.37	12	.19	44	.029	.019	.062	0	.095	12.8	7	.19	67	1

FRUITS AND FRUIT JUICES (cont.)	Measure	Mass	Energy	Protein	Carbohydrate	Fat	VITAMINS Vitamin A	Vitamin B_1	Vitamin B_2	Vitamin B_6	Vitamin B_{12}	Niacin	Vitamin C	MINERALS Calcium	Iron	Potassium	Sodium
		g	Cal	g	g	g	IU	mg	mg	mg	µg	mg	mg	mg	mg	mg	mg
Dates	10	83	228	1.63	61	.37	42	.075	.083	.159	0	1.82	0	27	.96	541	2
Figs	1	65	47	.48	12.2	.19	91	.038	.032	.072	0	.256	1.3	22	.23	148	1
Figs, dried	10	189	477	5.7	122	2.18	248	.133	.165	.419	0	1.3	1.6	269	4.18	1332	20
Grapefruit	1/2	241	38	.75	9.7	.12	149	.043	.024	.05	0	.3	41.3	14	.1	167	0
Grapefruit juice	1 c	247	96	1.24	22.7	.25	—	.099	.049	—	0	.494	93.9	22	.49	400	2
Grapes, slip skin	1 c	153	58	.58	15.7	.32	92	.085	.052	.1	0	.276	3.7	13	.27	176	2
Grapes, adherent skin	1 c	160	114	1.06	28.4	.92	117	.147	.091	.176	0	.48	17.3	17	.41	296	3
Grape juice	1 c	253	155	1.41	37.8	.19	20	.066	.094	.164	0	.663	.2	22	.6	334	7
Guava	1	112	45	.74	10.7	.54	713	.045	.045	.129	0	1.08	165	18	.28	256	2
Kiwi fruit	1	88	46	.75	11.3	.34	133	.015	.038	—	0	.38	74.5	20	.31	252	4
Lemon juice	1 T	15.2	3	.06	.99	.04	2	.006	.001	.007	0	.03	3.8	2	.02	15	3
Lime juice	1 T	15.4	4	.07	1.39	.02	2	.003	.002	.007	0	.015	4.5	1	0	17	0
Lychee	1	16	6	.08	1.59	.04	—	.001	.006	—	0	.058	6.9	0	.03	16	0
Mango	1	300	135	1.06	35	.57	8060	.12	.118	.277	0	1.21	57.3	21	.26	322	4
Melon, cantaloupe	1/2	477	94	2.34	22.3	.74	8608	.096	.056	.307	0	1.53	112	28	.57	825	23
Melon, casaba	1/10	245	43	1.48	10	.16	49	.098	.033	—	0	.656	26.2	8	.66	344	20
Melon, honeydew	1/10	226	46	.59	11.8	.13	52	.099	.023	.076	0	.774	32	8	.09	350	13
Mulberries	1 c	140	61	2.02	13.7	.55	35	.041	.141	—	0	.868	51	55	2.59	271	14
Nectarine	1	150	67	1.28	16	.62	1001	.023	.056	.034	0	1.34	7.3	6	.21	288	0
Orange	1	180	62	1.23	15.4	.16	269	.114	.052	.079	0	.369	69.7	52	.13	237	0
Orange juice	1 c	248	111	1.74	25.8	.5	496	.223	.074	.099	0	.992	124	27	.5	496	2
Papaya	1	454	117	1.86	30	.43	6122	.082	.097	.058	0	1.02	187	72	.3	780	8
Passion fruit	1	35	18	.4	4.21	.13	126	—	.023	—	0	.27	5.4	2	.29	63	5
Peach	1	115	37	.61	9.65	.08	465	.015	.036	.016	0	.861	5.7	5	.1	171	0
Peach, dried	10 halves	130	311	4.69	79.7	.99	2812	.003	.276	.087	0	5.68	6.3	37	5.28	1295	9
Peach nectar	1 c	249	134	.67	34.6	.05	643	.007	.035	—	0	.717	13.1	13	.47	101	17
Pear	1	180	98	.65	25	.66	33	.033	.066	.03	0	.166	6.6	19	.41	208	1
Pear, dried	10 halves	175	459	3.28	122	1.1	6	.014	.254	—	0	2.4	12.3	59	3.68	932	10
Pear nectar	1 c	250	149	.27	39.4	.03	1	.005	.033	—	0	.32	2.7	11	.65	33	9
Pineapple	1 c	155	77	.6	19.2	.66	35	.143	.056	.135	0	.651	23.9	11	.57	175	2
Pineapple juice	1 c	250	139	.8	34.4	.2	12	.138	.055	.24	0	.643	26.7	42	.65	334	2
Plum	1	70	36	.52	8.59	.41	213	.028	.063	.053	0	.330	6.3	2	.07	113	0
Prune	10	97	201	2.19	52.7	.43	1669	.068	.136	.222	0	1.64	2.8	43	2.08	626	3
Prune juice	1 c	256	181	1.55	44.6	.08	9	.041	.179	—	0	2.01	10.6	30	3.03	706	11
Raisins, packed	1 c	165	488	4.16	12.9	.9	0	.185	.3	.31	0	1.83	9	46	4.27	1362	47
Raspberries	1 c	123	61	1.11	14.2	.68	160	.037	.111	.07	0	1.1	30.8	27	.7	187	0
Rhubarb	1 c	122	26	1.09	5.53	.24	122	.024	.037	.029	0	.366	9.8	105	.27	351	5
Strawberries	1 c	149	45	.91	10.4	.55	41	.03	.098	.088	0	.343	84.5	21	.57	247	2
Tangerine	1	116	37	.53	9.4	.16	773	.088	.018	.056	0	.134	26	12	.09	132	1
Tangerine juice	1 c	247	106	1.24	25	.49	1037	.148	.049	—	0	.247	76.6	44	.49	440	2
Watermelon	1 c	160	50	.99	11.5	.68	585	.128	.032	.23	0	.32	15.4	13	.28	186	3
MEATS **Beef** Chuck roast	1 lb	454	1164	83	0	90	130	.485	.794	1.7	13.7	14.6	0	32	9.44	1374	266
Corned, brisket	1 lb	454	896	66.58	.63	67.6	—	.195	.712	1.32	8.07	16.6	0	30	7.66	1348	5519
Dried	1 oz	28	47	8.25	.44	1.11	—	.02	.09	—	.52	1.06	—	2	1.28	126	984
Flank steak	1 lb	454	888	87.4	0	57	50	.499	.680	1.87	13.4	20.6	0	22	8.9	1585	321
Ground beef, lean	4 oz	113	298	20	0	23.4	22.5	.057	.237	.28	2.64	5.1	0	9	1.99	295	78
Ground beef, regular	4 oz	113	351	18.8	0	30	40	.043	.171	.27	2.99	5.06	0	10	1.96	258	77

	Measure	Mass	Energy	Protein	Carbohydrate	Fat	VITAMINS Vitamin A	Vitamin B_1	Vitamin B_2	Vitamin B_6	Vitamin B_{12}	Niacin	Vitamin C	MINERALS Calcium	Iron	Potassium	Sodium
		g	Cal	g	g	g	IU	mg	mg	mg	μg	mg	mg	mg	mg	mg	mg
Liver	4 oz	113	161	22.6	6.58	4.34	39941	.292	3.14	1.06	78.2	14.4	25.3	6	7.71	365	82
Pastrami	1 oz	28	99	4.9	.86	8.27	—	.027	.048	.05	.5	1.44	.9	2	.54	65	348
Porterhouse steak	1 lb	454	1289	78.8	0	105.6	300	.44	.739	1.62	11.85	15.3	0	29	7.83	1305	222
Rib roast	1 lb	454	1503	72.8	0	132	310	.349	.576	1.39	12.45	12.4	0	39	7.63	1180	241
Round steak	1 lb	454	1093	88	0	79.5	110	.435	.748	2.02	12.21	15.97	0	23	8.5	1434	232
Short ribs	1 lb	454	1761	65.3	0	164	—	.322	.535	1.34	11.6	11.6	0	41	7.03	1053	224
Sirloin steak	1 lb	454	1179	82.7	0	91.5	220	.503	.88	1.71	12.58	13.9	0	34	10.2	1331	234
Smoked, chopped	1 oz	28	38	5.7	.53	1.25	—	.024	.05	.1	.49	1.3	0	—	.81	107	357
T-bone steak	1 lb	454	1394	76	0	118.5	300	.422	.712	1.58	11.6	14.8	0	30	7.58	1248	217
Tenderloin	1 lb	454	1095	84.1	0	81.6	—	.54	.97	1.74	12	13.9	0	30	10.9	1422	223
Lamb																	
Leg	1 lb	454	845	67.7	0	61.7	—	.59	.82	1.05	8.2	19	—	.39	5.1	1083	237
Chops	1 lb	454	1146	63.7	0	97	—	.57	.79	1.05	8.2	18.5	—	35	4.7	1019	223
Liver	1 lb	454	617	95.3	13.2	19.6	229070	1.81	14.9	1.36	472	76.5	152	45	49.4	916	236
Shoulder	1 lb	454	1082	59	0	92	—	.53	.73	1.05	8.2	17.1	—	35	3.9	942	206
Pork																	
Bacon	1 lb	454	2523	39	.42	261	0	1.67	.472	.64	4.2	12.6	0	34	2.7	631	3107
Bacon, Canadian style	1 lb	454	714	93.6	7.61	31.6	0	3.4	.78	1.77	3.02	28.2	0	36	3.07	1560	6391
Ham, boneless	1 lb	454	827	79.6	14.1	47.9	0	3.9	1.14	1.52	3.75	23.8	—	.32	4.5	1508	5974
Leg	1 lb	454	1182	77.4	0	94.4	.30	3.24	.889	1.81	2.79	20.5	3.2	25	.387	1405	214
Loin, chop	1 chop	151	345	20	0	28.7	9	.948	.294	.45	.86	5.33	.8	7	.85	346	63
Shoulder	1 lb	454	1249	73	0	103	30	3.08	1.19	1.26	3.25	16.8	3	24	4.59	1325	286
Spareribs	1 lb	454	804	48	0	66.3	30	1.74	.768	1.18	2.45	13.6	—	19	2.78	728	212
Veal																	
Breast	1 lb	454	828	65.6	0	61	—	.48	.87	1.22	5.7	22	—	39	9.7	1050	230
Chuck	1 lb	454	628	70.4	0	36	—	.52	.94	1.22	5.7	23.6	—	40	10.5	1126	246
Cutlet	1 lb	454	681	72.3	0	41	—	.53	.96	1.22	5.7	24.2	—	41	10.9	1157	253
Liver	1 lb	454	635	87.1	18.6	21.3	102060	.9	12.3	3.04	272	51.8	161	36	39.9	1275	331
Rib roast	1 lb	454	723	65.7	0	49	—	.48	.87	1.22	5.7	22	0	38	9.8	1051	230
Rump roast	1 lb	454	573	68	0	31	—	.5	.9	1.22	5.7	22.8	—	38	10	1090	238
Luncheon and Sausage																	
Bologna, beef	1 oz	28	89	3.31	.55	8.04	—	.016	.036	.05	.4	.746	t	3	.4	44	284
Bologna, beef and pork	1 oz	28	89	3.31	.79	8.01	—	.049	.039	.05	.38	.731	t	3	.43	51	289
Bologna, pork	1 oz	28	70	4.34	.21	5.63	—	.148	.045	.08	.26	1.1	t	3	.22	80	336
Bratwurst, cooked	1 link	85	256	12	1.76	22	—	.429	.156	.18	.81	2.72	1	38	1.09	180	473
Brotwurst	1 oz	28	92	4.04	.84	7.88	—	.071	.064	.04	.58	.936	t	14	.29	80	315
Frankfurter, beef	1	45	145	5.08	1.08	13.2	—	.023	.046	.05	.74	1.13	t	6	.6	71	461
Frankfurter, beef and pork	1	45	144	5.08	1.15	13.1	—	.09	.054	.06	.58	1.18	t	5	.52	75	504
Italian sausage, cooked	1 link	67	216	13.4	1	17.2	—	.417	.156	.22	.87	2.79	1	16	1	204	618
Kielbasa	1 oz	28	88	3.76	.61	7.7	—	.065	.061	.05	.46	.816	t	12	.41	77	305
Pepperoni	1 slice	5.5	27	1.15	.16	2.42	—	.018	.014	.01	.14	.273	.018	1	.08	19	112
Polish sausage	1 oz	28	92	4	.46	8.14	—	.142	.042	.05	.28	.976	0	3	.41	67	248
Pork and beef sausage	1 link	13	52	1.79	.35	4.71	—	.096	.019	.01	.06	.438	—	—	.15	—	105
Pork sausage	1 link	28	118	3.31	.29	11.4	—	.155	.046	.07	.32	.804	0	5	.26	58	228
Salami, hard	1 slice	10	42	2.29	.26	3.44	—	.06	.029	.05	.19	.487	t	1	.15	38	186
Summer sausage	1 slice	23	80	3.69	.53	6.88	—	.039	.069	.07	1.06	.94	t	2	.47	53	334

	Measure	Mass	Energy	Protein	Carbohydrate	Fat	VITAMINS Vitamin A	Vitamin B_1	Vitamin B_2	Vitamin B_6	Vitamin B_{12}	Niacin	Vitamin C	MINERALS Calcium	Iron	Potassium	Sodium
		g	Cal	g	g	g	IU	mg	mg	mg	µg	mg	mg	mg	mg	mg	mg
MEATS (cont.)																	
Vienna sausage	1	16	45	1.65	.33	4.03	—	.014	.017	.02	.16	.258	0	2	.14	16	152
NUTS AND SEEDS																	
Almonds	1 c	142	849	26.4	27.7	77	0	.34	1.31	.142	0	5	t	332	6.7	1098	6
Brazil nuts	1 c	140	916	20	15.3	93.7	t	1.34	.17	.238	0	2.2	14	260	4.8	1001	1
Cashews	1 c	140	785	24.1	41	64	140	.6	.35	.325	0	2.5	0	53	5.3	650	21
Coconut, shredded	1 c	80	277	2.8	7.5	28.2	0	.04	.02	.035	0	.4	2	10	1.4	205	18
Hazelnuts	1 c	135	856	17	22.5	84.2	144	.62	.738	.735	0	1.2	t	282	4.6	950	3
Macadamia nuts	1 c	134	940	11	18.4	98.8	0	.469	.147	—	0	2.87	—	94	3.23	493	6
Peanuts	1 c	144	838	37.7	29.7	70.1	t	.46	.19	.576	0	24.6	0	104	3.2	1009	7
Peanut butter	1 T	15	86	3.9	3.2	8.1	0	.018	.02	.05	0	2.4	0	11	.3	123	18
Pecans	1 c	108	742	9.9	15.8	76.9	140	.93	.14	.183	0	1	2	79	2.6	651	t
Pine nuts	1 oz	28	180	3.7	5.8	14.3	10	.36	.07	—	0	1.3	t	3	1.5	170	1
Pistachios, shelled	1 c	128	739	26	31.7	61.9	299	1.05	.223	—	0	1.38	—	173	8.67	1399	7
Pumpkin seeds & squash	1 c	140	774	40.6	21	65.4	100	.34	.442	—	0	3.4	0	71	15.7	—	24
Sesame seeds	1 c	150	873	27.3	26.4	80	99	.27	.2	.126	0	8.1	0	165	3.6	610	59
Sunflower seeds	1 c	145	812	34.8	28.9	68.6	70	2.84	.33	1.8	0	7.8	—	174	10.3	1334	4
Tahini	1 T	15	89	2.55	3.18	8.06	—	.183	.071	—	0	.818	0	64	1.34	62	17
Walnuts	1 c	100	651	14.8	15.8	64	30	.33	.13	.73	0	.9	2	99	3.1	450	2
POULTRY **Chicken**																	
Light meat	1 lb	454	216	23.5	0	12.8	115	.068	.1	.56	.39	10.3	1.1	13	.92	237	76
Dark meat	1 lb	454	379	26.7	0	29.3	273	.098	.234	.39	.47	8.33	3.4	18	1.57	285	117
Light meat, no skin	1 lb	454	100	20.4	0	1.45	25	.06	.081	.48	.34	9.33	1.1	10	.64	210	60
Dark meat, no skin	1 lb	454	136	21.9	0	4.7	78	.084	.201	.36	.39	6.8	3.4	13	1.12	241	93
Breast	1/2	181	250	30.2	0	13.4	121	.091	.123	.77	.5	14.3	1.5	16	1.07	319	91
Drumstick	1	110	117	14	0	6.34	69	.054	.13	.22	.25	3.97	2	8	.75	151	61
Thigh	1	120	199	16.2	0	14.3	136	.058	.144	.24	.28	5.1	2.1	9	.93	181	71
Wing	1	90	109	8.98	0	7.82	72	.024	.043	.17	.15	2.9	.3	6	.47	76	36
Canned, boned	5 oz	142	234	30.9	0	11.3	—	.021	.183	.5	.42	8.98	2.8	20	2.25	196	714
Duck Domesticated	1 lb	454	1159	33	0	113	483	.565	.603	.55	.73	11.3	8	30	6.89	600	181
Goose Domesticated	1 lb	454	1187	50.7	0	107	176	.272	.784	1.24	—	11.5	—	38	8	985	234
Turkey Light meat	1 lb	454	286	39	0	13.2	12	.101	.207	.86	.75	9.24	0	23	2.18	489	106
Dark meat	1 lb	454	243	28.7	0	13.3	8	.111	.307	.49	.58	4.34	0	26	2.57	396	108
Canned, boned	5 oz	142	231	33.6	0	9.74	0	.02	.243	—	—	9.4	2.8	17	2.64	—	663
SALAD DRESSINGS AND SAUCES **Salad Dressings** Bleu cheese	1 T	15.3	77	.7	1.1	8	32	0	0	—	—	0	.3	12.4	0	—	—
French	1 T	15.6	67	.1	2.7	6.4	—	—	—	—	—	—	—	1.7	.1	12.3	213
Italian	1 T	14.7	68.7	.1	1.5	7.1	—	—	—	—	—	—	—	1	0	2	116
Mayonnaise	1 T	13.8	99	.2	.4	11	39	0	—	—	—	—	0	2	.1	5	78.4
Russian	1 T	15.3	76	.2	1.6	7.8	106	.01	.01	—	—	1	1	3	.1	24	133
Thousand Island	1 T	15.6	58.9	.1	2.4	5.6	50	—	—	—	—	—	—	2	.1	18	109

SALAD DRESSINGS AND SAUCES (cont.)	Measure	Mass	Energy	Protein	Carbohydrate	Fat	VITAMINS Vitamin A	Vitamin B₁	Vitamin B₂	Vitamin B₆	Vitamin B₁₂	Niacin	Vitamin C	MINERALS Calcium	Iron	Potassium	Sodium
		g	Cal	g	g	g	IU	mg	mg	mg	μg	mg	mg	mg	mg	mg	mg
Sauces																	
Barbecue	1 c	250	188	4.5	32	4.5	2170	.075	.05	.188	0	2.25	17.5	48	2.25	435	2038
Catsup	1 T	15	16	.3	3.8	.1	210	.01	.01	.016	0	2	2	3	.1	54	156
Horseradish, prepared	1 T	15	6	.2	1.4	t	—	—	—	.022	0	—	—	9	.1	44	14
Mustard	1 T	15	15	.9	.9	.9	—	—	—	—	—	—	—	18	.3	21	195
Soy	1 T	18	11	1.56	1.5	0	0	.009	.023	.031	0	605	0	3	.49	64	1029
Tartar	1 T	14	31	.1	.9	3.1	30	t	t			t	t	3	.1	11	99
Vinegar	1 T	15	2	t	.9	0	—	—	—	t	0	—	—	1	.1	15	t
SEAFOOD																	
Abalone	3 oz	85	89	14.5	5	.64	—	.16	.12	—	—	—	—	27	2.7	—	255
Anchovy, in oil, drained	5	20	42	5.78	0	1.94	—	.016	.073	.041	.176	3.98	—	46	.93	109	734
Bass	3 oz	85	82	15	0	1.98	—	.09	.03		3.25	1.9	—	—	.71	232	59
Bluefish	3 oz	85	105	17	0	3.6	338	.049	.068	.342	4.58	5.06	—	6	.41	316	51
Carp	3 oz	85	108	15	0	4.76	25	.008	.036	.162	1.3	1.34	1.4	35	1.05	283	42
Catfish	3 oz	85	99	15.5	0	3.62	—	.038	.09	—	.002	1.82	—	.34	.83	296	54
Clams	9 large	180	133	23	4.62	1.75	540	.18	.38	.14	89	3.17	—	83	25	564	100
Cod	3 oz	85	70	15	0	.57	34	.065	.055	.208	.772	1.75	.9	13	.32	351	46
Crab	3 oz	85	71	15.6	0	.51	20	.037	.037	.272	9.08	.934	1.8	39	.5	173	711
Eel	3 oz	85	156	15.7	0	9.9	2954	.128	.034	.057	2.55	2.98	1.3	17	.43	232	43
Flat fish, flounder and sole species	3 oz	85	78	16	0	1	28	.076	.065	.177	1.29	2.46	—	15	.3	307	69
Haddock	3 oz	85	74	16	0	.61	47	.03	.031	.255	1.02	3.23	—	28	.89	264	58
Halibut	3 oz	85	93	17.7	0	1.95	132	.051	.064	.292	1	11.97	—	40	.71	382	46
Herring	3 oz	85	134	15	0	7.68	80	.078	.198	.257	11.6	2.74	.6	49	.94	278	76
Kelp	1 T	14.2	—	—	5.53	.157	—	—	.046	—	—	.784	—	156	.014	753	429
Lobster	3 oz	85	77	16	.43	.76	—	.368	.041	—	.786	1.23	—	26	.54	236	272
Mackerel	3 oz	85	174	15.8	0	11.8	140	.15	.265	.339	7.4	7.72	.3	10	1.38	267	76
Oysters	6 medium	84	58	5.9	3.29	2.08	282	.128	.139	.042	16	1.1	—	38	5.63	192	94
Perch	3 oz	85	80	15.8	0	1.39	34	.08	.094	.2	.85	1.7	2.72	91	.78	232	64
Pike	3 oz	85	75	16.4	0	.58	60	.049	.054	.099	—	2.16	3.2	48	.47	220	33
Salmon	3 oz	85	121	16.9	0	5.39	34	.19	.32	.695	2.7	6.68	8.2	10	.68	417	37
Sardines, in oil, drained	2	24	50	5.9	0	2.75	54	.019	.054	.04	2.15	1.26	—	92	.7	95	121
Scallops	3 oz	85	75	14.3	2	.64	—	.01	.055	—	1.3	.978	—	21	.25	274	137
Shark	3 oz	85	111	17.8	0	3.83	198	.036	.053	—	1.27	2.5	—	29	.71	136	67
Shrimp	3 oz	85	90	17.3	.77	1.47	8.26	.024	.029	.088	.987	2.17	—	44	2.05	157	126
Smelt	3 oz	85	83	15	0	2.06	—	—	.102	—	2.92	1.23	—	51	.77	247	51
Snails	3 oz	85	117	20	6.6	.34	72	.022	.091	.291	7.7	.893	—	48	4.28	295	175
Snapper	3 oz	85	85	17.4	0	1.14	—	.039	.003	—	—	.241	—	27	.15	355	54
Swordfish	3 oz	85	103	16.8	0	3.41	101	.031	.081	.281	1.49	8.23	.9	4	.69	245	76
Trout	3 oz	85	126	17.7	0	5.62	49	.277	.261	1.43	6.6	7.6	.4	36	1.27	307	44
Tuna, in water	1 can	165	216	48.8	0	.83	130	.08	.19	.39	1.6	19	—	20	5.28	518	588
Whitefish	3 oz	85	114	16	0	4.98	2050	.128	.108	—	—	2.72	—	—	.31	269	43
SOUPS																	
Bean, black	1 c	247	116	5.64	19.8	1.51	506	.077	.054	.094	.02	.534	.8	45	2.16	273	1198
Bean, with frankfurters	1 c	250	187	9.99	22	6.98	869	.11	.065	.133	—	1.02	.9	86	2.34	477	1092
Beef bouillon	1 c	240	16	2.74	.1	.53	0	.005	.05	—	—	1.87	0	15	.41	130	782
Beef noodle	1 c	244	84	4.83	8.98	3.08	629	.068	.059	.037	.2	1.06	.3	15	1.1	99	952

	Measure	Mass	Energy	Protein	Carbohydrate	Fat	VITAMINS Vitamin A	Vitamin B$_1$	Vitamin B$_2$	Vitamin B$_6$	Vitamin B$_{12}$	Niacin	Vitamin C	MINERALS Calcium	Iron	Potassium	Sodium
		g	Cal	g	g	g	IU	mg	mg	mg	μg	mg	mg	mg	mg	mg	mg

SOUPS
(cont.)

	Measure	Mass	Energy	Protein	Carbohydrate	Fat	Vitamin A	Vitamin B$_1$	Vitamin B$_2$	Vitamin B$_6$	Vitamin B$_{12}$	Niacin	Vitamin C	Calcium	Iron	Potassium	Sodium
Chicken broth	1 c	244	39	4.93	.93	1.39	0	.01	.071	.024	.24	3.34	0	9	.51	210	776
Chicken, cream of	1 c	244	116	3.43	9.26	7.36	560	.029	.061	.017	—	.82	.2	34	.61	87	986
Chicken gumbo	1 c	244	56	2.64	8.37	1.43	136	.02	.05	.063	—	.664	4.9	24	.89	75	955
Chicken noodle	1 c	241	75	4.04	9.35	2.45	711	.053	.06	.027	—	1.38	.2	17	.78	55	1107
Chicken rice	1 c	241	60	3.53	7.15	1.91	660	.017	.024	.024	—	1.12	.1	17	.75	100	814
Clam chowder, New England	1 c	244	95	4.81	12.4	2.88	8	.02	.044	.083	8.01	.961	2	43	1.48	146	914
Minestrone	1 c	241	83	4.26	11.2	2.51	2337	.053	.043	.099	0	.942	1.1	34	.92	312	911
Mushroom, cream of	1 c	244	129	2.32	9.3	8.97	0	.046	.09	.015	.05	.725	1	46	.51	101	1031
Onion	1 c	241	57	3.75	8.18	1.74	0	.034	.024	.048	0	.6	1.2	26	.67	69	1053
Oyster stew	1 c	241	59	2.11	4.07	3.83	71	.022	.036	.012	2.19	.234	3.1	22	.98	49	980
Pea, split, with ham	1 c	253	189	10.3	28	4.4	444	.147	.076	.068	—	1.47	1.4	22	2.28	399	1008
Potato, cream of	1 c	244	73	1.74	11.4	2.36	288	.034	.037	.037	—	.539	0	20	.48	137	1000
Tomato	1 c	244	86	2.06	16.6	1.92	688	.088	.051	.112	0	1.41	66.5	13	1.76	263	872
Turkey noodle	1 c	244	69	3.9	8.63	1.99	292	.073	.063	.037	—	1.4	.2	12	.94	75	815
Vegetable, beef	1 c	244	79	5.58	10.1	1.9	1891	.037	.049	.076	.31	1.03	2.4	17	1.11	173	957
Vegetable, vegetarian	1 c	241	72	2.1	12	1.93	3005	.053	.046	.055	0	.916	1.4	21	1.08	209	823
VEGETABLES AND VEGETABLE JUICES Alfalfa, sprouts	1 c	33	10	1.32	1.25	.23	51	.025	.42	.011	0	.159	2.7	10	.32	26	2
Artichoke, globe	1 medium	128	65	3.4	15.3	.26	237	.1	.077	.143	0	.973	13.8	61	2.1	434	102
Asparagus	1 c	134	30	4.1	4.94	.3	1202	.15	.166	.2	0	1.5	44	28	.9	404	2
Black beans, dry	1 c	200	678	44.6	122	3	60	1.1	.4	—	0	4.4	—	270	15.8	2076	50
Black eye peas, cooked	1 c	165	178	13.4	29.9	1.3	580	.5	.18	.18	0	2.3	28	40	3.5	625	2
Beets	1 c	136	60	2	13.6	.2	28	.068	0.28	.06	0	.54	15	22	1.24	440	98
Beet greens	1 c	38	8	.7	1.5	.02	2308	.038	.08	.04	0	.152	—	46	1.2	208	76
Broccoli	1 c	88	24	2.6	4.6	.3	1356	.058	.1	.14	0	.56	82	42	.78	286	24
Brussels sprouts	1 c	88	38	3.3	7.88	.26	778	.12	.08	.19	0	.65	74	36	1.2	.342	22
Cabbage, common	1 c	70	16	.84	2.76	.12	88	.03	.02	.066	0	.2	33	32	.4	172	12
Cabbage, Chinese	1 c	70	9	1.05	1.53	.14	2100	.028	.049	—	0	.35	31.5	74	.56	176	45
Carrots	1 c	110	48	1	11	.2	30942	.1	.064	.16	0	1	10	30	.54	356	38
Carrot juice	1 c	227	96	2.47	22	—	24750	.13	.12	.534	0	1.35	20	8.3	1.5	767	105
Cauliflower	1 c	100	24	1.98	4.9	.18	16	.076	.058	.23	0	.634	71	28	.58	356	14
Celery	1 c	120	18	.8	4.36	.14	152	.036	.036	.036	0	.36	7.6	44	.58	340	106
Chard, Swiss	1 c	36	6	.64	1.34	.08	1188	.014	.032	—	0	.144	10.8	18	.64	136	76
Chives	1 T	3	1	.08	.11	.02	192	.003	.005	.005	0	.021	2.4	2	.05	8	0
Collards	1 c	186	35	2.9	7	.4	6194	.054	.119	.125	0	.696	43	218	1.16	275	52
Corn	1 c	154	132	4.96	29	1.8	432	.208	.09	.084	0	2.6	10.6	4	.8	416	23
Cucumber	1 c	104	14	.56	3	.14	46	.032	.02	.054	0	.321	4.8	14	.28	156	2
Eggplant	1 c	82	22	.9	5	.08	58	.074	.016	.078	0	.492	.14	30	.44	180	2
Endive	1 c	50	8	.62	1.68	.1	1026	.04	.038	.1	0	.2	3.2	26	.42	158	12
Garbanzos, dry	1 c	200	720	41	122	9.6	100	.62	.3	—	0	4	t	300	13.8	1594	52
Garlic	1 clove	3	4	.2	.9	.02	t	.01	t	—	0	t	t	1	t	16	1
Green beans	1 c	110	34	2	7.85	.013	735	.092	.116	.081	0	.827	17.9	41	1.14	230	6
Kale	1 c	67	33	2.21	6.7	.47	5963	.074	.087	.182	0	.67	80.4	90	1.14	299	29

VEGETABLES AND VEGETABLE JUICES (cont.)	Measure	Mass g	Energy Cal	Protein g	Carbohydrate g	Fat g	VITAMINS Vitamin A IU	Vitamin B$_1$ mg	Vitamin B$_2$ mg	Vitamin B$_6$ mg	Vitamin B$_{12}$ µg	Niacin mg	Vitamin C mg	MINERALS Calcium mg	Iron mg	Potassium mg	Sodium mg
Kidney beans, cooked	1 c	185	218	14.4	39.6	.9	10	.2	.11	—	0	1.3	—	70	4.4	629	6
Kohlrabi	1 c	140	38	2.38	8.68	.14	50	.07	.028	.21	0	.56	86.8	34	.56	490	28
Leeks	1	124	76	1.86	17.5	.37	118	.074	.037	.2	0	.496	14.9	73	2.6	223	25
Lentils, cooked	1 c	200	212	15.6	38.6	t	40	.14	.12	—	0	1.2	0	50	4.2	498	—
Lentil sprouts	1 c	77	81	6.9	17	.43	35	.176	.099	.146	0	.869	12.7	19	2.47	248	8
Lettuce, iceberg	1 c	75	10	.7	2.2	.12	250	.05	.05	.028	0	.148	5	15	.4	131	7
Lettuce, romaine	1 c	56	8	.9	1.3	.12	1456	.056	.056	—	0	.28	13.4	20	.62	162	4
Lima beans, cooked	1 c	170	208	11.6	40	.54	630	.238	.163	.328	0	1.77	17	54	4.2	969	29
Mung bean sprouts	1 c	104	32	3	6	.2	22	.088	.128	.092	0	.778	13.6	14	.94	154	6
Mushrooms	1 c	70	18	1.46	3	.3	0	.072	.3	.068	0	2.88	2.4	4	.86	260	2
Navy beans, cooked	1 c	190	224	14.8	40.3	1.1	0	.27	.13	—	0	1.3	0	95	5.1	790	13
Okra	1 c	100	38	2	7.6	.1	660	.2	.06	.2	0	1	21	82	.8	302	8
Onions, green	1 c	100	26	1.7	5.5	.14	5000	.07	.14	—	0	.2	45	60	1.88	256	4
Onions, mature	1 c	160	54	1.88	11.7	.42	0	.096	.016	.25	0	.16	13.4	40	.58	248	4
Parsley	1 c	60	26	2.2	5.1	.4	5100	.07	.16	.098	0	.7	103	122	3.7	436	27
Peas, green	1 c	146	118	7.9	21	.58	934	.387	.193	.247	0	3.05	58.4	36	2.14	357	7
Peas, split, cooked	1 c	200	230	16	41.6	.3	80	.3	.18	—	0	1.8	—	22	3.4	592	26
Peppers, sweet	1 c	100	24	.86	5.3	.46	530	.086	.05	.164	0	.54	128	6	1.2	196	4
Peppers, hot chili	1 c	75	30	1.5	7	.15	578	.068	.068	.21	0	.713	182	13	.9	255	5
Pickles, dill	1 large	100	11	.7	2.2	.4	100	t	.02	.007	0	t	6	26	1	200	1428
Pinto beans, cooked	1 c	190	663	43.5	121	2.3	0	1.6	.4	1	0	4.2	t	257	12.2	1870	19
Potato	1 c	150	114	3.2	25.7	.2	t	.15	.06	—	0	2.3	30	11	.9	611	5
Potato, baking, flesh & skin	1 large	202	220	4	32.8	.2	t	.15	.07	.7	0	2.7	31	14	1.1	782	6
Pumpkin	1 c	245	49	1.76	12	.17	2651	.076	.19	.139	0	1	11.5	37	1.4	564	3
Radish	10	45	7	.27	1.6	.24	3	.002	.02	.032	0	.135	10.3	9	.13	104	11
Sauerkraut	1 c	235	42	2.4	9.4	.33	120	.07	.09	.31	0	.5	33	85	1.2	329	1755
Soybeans, cooked	1 c	180	234	19.8	19.4	10.3	50	.38	.16	—	0	1.1	30.6	131	4.9	972	4
Soybean curd (tofu)	3.5 oz	100	72	7.8	2.4	4.2	0	.06	.03	—	0	—	—	100	5.2	—	354
Soybean milk	1 c	226	75	7.7	5	3.4	90	.18	.065	—	0	.5	—	47.5	1.8	—	—
Soybean sprouts	1 c	70	90	9	7.8	4.68	8	.238	.082	.124	0	.804	10.6	48	1.48	338	10
Spinach	1 c	55	14	1.8	2.4	.2	4460	.06	.11	.14	0	.3	28	51	1.7	259	39
Squash, summer	1 c	130	25	1.4	5.5	.28	530	.07	.12	.186	0	1.3	29	36	.5	263	1
Squash, winter	1 c	205	129	3.7	31.6	.4	8610	.1	.27	.18	0	1.4	27	57	1.6	945	2
Sweet potato	1	130	136	2	32	.38	26082	.086	.191	.334	0	.876	30	29	.76	265	17
Tomato	1	123	24	1.1	5.3	.26	1394	.074	.062	.059	0	.738	21.6	8	.59	254	10
Tomato juice	1 c	243	46	2.2	10.4	.2	1940	.12	.07	.366	0	1.9	39	17	2.2	552	486
Tomato paste	1 c	262	215	8.9	48.7	2	8650	.52	.31	.996	0	8.1	128	71	9.2	2237	100
Turnips	1 c	130	39	1.3	8.6	.13	t	.05	.09	.117	0	.8	47	51	.7	348	64
Turnip greens	1 c	55	15	.83	3	.17	4180	.039	.055	.145	0	.33	33	105	.61	163	22
Vegetable juice cocktail	1 c	242	41	2.2	8.7	.2	1690	.12	.07	.338	0	1.9	22	29	1.2	535	484
Water chestnuts	4 avg	25	20	.4	4.8	.1	0	.04	.05	—	0	.2	1	1	.2	125	5
Watercress	1 c	35	7	.8	1.1	.04	1720	.03	.06	.045	0	.3	28	53	.6	99	18
Yams	1 c	200	210	4.8	48.2	.4	t	.18	.8	.51	0	1.2	18	8	1.2	1508	17
Yellow wax beans	1 c	125	28	1.8	5.8	.3	290	.09	.11	.098	0	.6	16	63	.8	189	4

OBJECTIVES

SUPPLYING OUR WATER NEEDS

Upon completion of this unit, you will be able to:

1. List and use the units of the modernized metric system (SI) in measurements of length, volume, mass, and density. (A.1, B.1)

2. Discuss direct and indirect water uses and their importance for water conservation. (A.3, A.5, A.8, A.9)

3. Describe the function and operation of the hydrologic cycle and indicate the primary storage reservoirs of the Earth's water supply. (A.6, A.7)

4. Discuss some effects of water's unusual physical properties on plants and animals. (B.1)

5. Define the terms solution, solvent, and solute, and apply them in an example. (B.2)

6. Classify matter in terms of elements, compounds, and mixtures; distinguish among different types of mixtures (solutions, colloids, and suspensions) in a laboratory setting. (B.2–B.4)

7. Interpret the symbols and formulas in a balanced chemical equation in terms of atoms and molecules. (B.5)

8. Describe the three basic subatomic particles—proton, neutron, and electron (B.6). Identify the connection of molecular polarity to the solubility of a compound. (C.1)

9. Define the terms insoluble, unsaturated, saturated, and supersaturated, and calculate solution concentration as a percentage. (C.1, C.2)

10. Use solubility curves to describe the effect of temperature on solubility, and calculate percent saturation. (C.1, C.2, C.4, C.5)

11. Demonstrate the ability to organize and interpret environmental or other data in graphs or tables. (C.1, C.5, C.10)

12. Given the pH of a substance, classify it as acidic, basic, or neutral. (C.6)

13. Determine the formula and name of a simple ionic compound when provided with the anion's and cation's names and charges. (C.7)

14. Evaluate the risks of contaminants in our water supply, with particular attention to heavy-metal ions of lead, mercury, and cadmium. (C.9)

15. Compare and contrast natural and artificial water purification systems, and assess the risks and benefits of water softening and chlorination. (D.1, D.3–D.6)

CONSERVING CHEMICAL RESOURCES

Upon completion of this unit, you will be able to:

1. Compare and contrast science and technology. (Introduction)

2. State the law of conservation of matter, and apply the law by determining whether a given chemical equation is balanced. (A.2, A.3)

3. Describe the Spaceship Earth analogy, and apply it to the terms "throw away" and "using up." (A.5, A.6, A.7)

4. List common types and sources of municipal waste, and describe attempts to reuse and recycle waste. (A.5, A.6, C.4, C.5)

5. Define and give examples of renewable and nonrenewable resources. (A.5, A.7)

6. Distinguish between chemical and physical changes and/or properties when given specific examples of each. (B.1)

7. Classify selected elements as metals, nonmetals, or metalloids based on observations of their chemical and physical properties. (B.2, B.3)

8. Use the periodic table to (a) predict physical and chemical properties of an element; (b) write formulas for various compounds; (c) identify elements by their atomic masses and atomic numbers; and (d) locate periods and groups (families) of elements. (B.4, B.6)

9. Construct a workable periodic table and explain its organization, given chemical and physical properties of a set of elements. (B.5)

10. Compare the reactivities of selected elements, and explain the results in terms of the structure of their atoms. (B.7–B.9)

11. Discuss the development of new materials as substitutes for dwindling resources. (D.6)

12. Explain from a chemical viewpoint the problems and solutions involved in restoring the Statue of Liberty. (B.11)

13. List the three primary layers of our planet and some resources that are "mined" from each region. (C.1)

14. Write balanced chemical equations and relate them to the law of conservation of matter. (C.2)

15. Define the term mole, and calculate the molar mass of a compound when provided with its formula and the atomic masses of its elements. (C.3)

16. Outline the production of a metal from its ore (using copper as an example) and list four factors which determine the profitability of mining. (D.1)

17. Calculate the percent composition by mass of a specified element in a given compound. (D.2)

18. Define oxidation and reduction, and compare the three most common redox-reaction methods for separating metals from their ores. (D.4)

19. Use the concepts of the properties of materials to develop a currency for a new country. (E.1)

PETROLEUM: TO BURN? TO BUILD?

Upon completion of this unit, you will be able to:

1. Compare the usage of petroleum for "building" and "burning," and the benefits and burdens of each usage. (Introduction, A.1, A.2)

2. Identify regions of high petroleum usage and regions of petroleum reserves, and discuss the economic and political implications of petroleum supply and demand. (A.3)

3. Describe the chemical makeup of petroleum, its differences from other resources, and its refining. (Introduction, B.1, B.2)

4. Identify differences in density and viscosity among common petroleum products, and explain the relationship between the differences and the number of carbon atoms in their molecules. (B.3)

5. Describe the process of fractional distillation and list the five major fractions of petroleum distillation and typical products manufactured from each fraction. (B.1, B.2, B.4)

6. Describe the processes involved in ionic and covalent bonding. (B.6)

7. Name the first ten alkanes and draw structural and electron-dot formulas for each. (B.7)

8. State and explain the effect of carbon length and side groups on the boiling point of a hydrocarbon. (B.5, B.7, B.8)

9. Define the term isomer and draw structural formulas for at least three isomers of a given hydrocarbon. (B.8)

10. Trace the history of energy sources and consumption patterns in the United States, and account for major changes. (C.2)

11. Explain endothermic and exothermic reactions in terms of bond breaking and bond forming, and give examples of each type of reaction. (C.3)

12. Identify energy conversions in an automobile, and calculate savings resulting from increased automobile efficiency. (C.3)

13. Define the terms heat of combustion and specific heat, and calculate energies of various combustion reactions. (C.4–C.6)

14. Write balanced equations for the combustion of hydrocarbon fuels, including energy changes. (C.6)

15. Define the term octane number, state its relationship to grades of gasoline, and identify two ways of increasing octane number. (C.7)

16. Compare saturated and unsaturated hydrocarbons in terms of molecular models, formulas, structures, and physical and chemical properties. (D.1–D.3)

17. Identify the functional groups for common alcohols, ethers, carboxylic acids, and esters. (D.4, D.6)

18. Describe polymerization and give one example of addition and condensation reactions. (D.5)

19. Describe major sources of energy for the United States today and alternative sources of fuels and builder molecules for the future. (E.1, E.2)

UNDERSTANDING FOOD

Upon completion of this unit, you will be able to:

1. Compare the uses of food in terms of "building" and "burning." (Introduction, B.1)

2. Interpret food labels to analyze food components and the food pyramid to recognize the components of a healthy diet. (Introduction, A.1, A.2)

3. Distinguish malnutrition from undernutrition. (A.3)

4. Define calorie and joule, and calculate energy changes from calorimetry data. (B.1, B.2)

5. Correlate weight gain or loss with caloric intake and human activity. (B.3)

6. Identify key functional groups in fats, and write an equation for the formation of a typical fat. Distinguish between saturated and unsaturated fats, and relate the consumption of each to health. (B.4)

7. Compare and contrast mono-, di-, and polysaccharides in terms of structural formulas and properties; identify key functional groups in carbohydrates. (B.5)

8. Define and illustrate the concept of limiting reactant in biochemical examples and in calculations. (C.2)

9. Describe how functional groups in amino acids interact in protein formation. (C.3)

10. Describe five functions of proteins in the body. (C.3)

11. Discuss the concepts of essential amino acids, complete protein, and complementary protein with respect to a balanced diet. (C.3)

12. Distinguish water-soluble from fat-soluble vitamins (with specific examples of each) and discuss the implications of these differences in terms of dietary needs. (D.1)

13. Analyze the vitamin C content of foods by performing titrations. (D.2)

14. Identify minerals used in the body, and distinguish between macrominerals and trace minerals. (D.3)

15. Determine the iron content of foods by colorimetry. (D.4)

16. Discuss the relative risks and benefits of various types of food additives in terms of their purposes, and provide specific examples. (D.5)

17. Discuss the role of the Food and Drug Administration (FDA) and federal regulations in ensuring food safety. (D.5, D.6)

18. Interpret food labels in terms of Percent Daily Values. (E.1)

19. Compare and contrast two microwave dinners for their nutritional quality. (E.2)

NUCLEAR CHEMISTRY IN OUR WORLD

Upon completion of this unit, you will be able to:

1. Consider the risks and benefits of nuclear energy. (Introduction)

2. Identify the general regions and properties of the electromagnetic spectrum. Distinguish between ionizing and non-ionizing radiation and their biological effects. (A.2)

3. Describe the experiments of Roentgen, Becquerel, the Curies, and Rutherford, and explain how they led to modifications in the atomic model. (A.3, A.4, A.6)

4. Describe the properties and locations of the three major subatomic particles. (A.7)

5. Define the term isotope, and interpret isotope notation. (A.7, A.8)

6. Use molar masses and isotopic abundance data to calculate average mass and relative abundance of elements. (A.9)

7. Compare and contrast the general properties of alpha, beta, and gamma radiation, including penetrating power, and discuss safety considerations in terms of shielding abilities of cardboard, glass, and lead. (B.1, B.2)

8. Balance nuclear equations and use them to describe natural radioactive decay. (B.2)

9. Describe exposure to radiation and its effects in terms of rems and potential cellular damage, including radon in homes. (B.3–B.6)

10. Explain the concept of half-life and discuss the implications of half-life for natural radioactivity and nuclear waste disposal. (C.1, C.2)

11. Identify the beneficial uses of radioisotopes. (C.3)

12. Describe radiation detectors and their operating principles. (C.4, C.5)

13. Define nuclear transmutation and write a nuclear equation to illustrate the process. Describe the formation of synthetic elements by nuclear transmutation. (C.6)

14. Distinguish nuclear fission from nuclear fusion. (D.1, D.6)

15. Compare the energies produced by nuclear fission and by typical exothermic chemical reactions. (D.2)

16. Explain the energy effects of a chain reaction and compare a controlled and an uncontrolled reaction. (D.3, D.4)

17. Identify the main components of a nuclear power plant and describe their functions. (D.5)

18. Assess relative risks and benefits of various nuclear technologies (such as power generation, medical applications, and food irradiation). (E.1)

19. Discuss the problems and possible solutions associated with high-energy nuclear waste generation and long-term disposal. (E.2, F.1)

20. Compare and contrast the nuclear accidents at Three Mile Island and Chernobyl. (E.3)

LIVING IN A SEA OF AIR: CHEMISTRY AND THE ATMOSPHERE

Upon completion of this unit, you will be able to:

1. Describe common physical and chemical properties of air. (A.1, A.2)

2. Compare the chemical properties of nitrogen, oxygen, and carbon dioxide. (B.1)

3. Identify the major components of the troposphere and indicate their relative concentrations. (B.2)

4. Show how Avogadro's law and the concept of molar volume clarify the interpretation of chemical equations involving gases. (B.2)

5. Describe with words and mathematical equations the interrelationships among amount, temperature, volume, and pressure of a gas (Avogadro's, Charles' and Boyle's laws), and list one practical application of each law. (B.2, B.4–B.7)

6. Define and apply in appropriate situations the terms molar volume, standard temperature and pressure (STP), Kelvin temperature scale, and absolute zero. (B.2, B.5–B.7)

7. Sketch or graph the relationship between altitude and air pressure. (B.3)

8. Discuss air pressure and explain how to measure it. (B.4)

9. Account for the gas laws and ideal gas behavior in terms of the kinetic molecular theory of gases. (B.7)

10. Compare the various components of solar radiation and their energies. (C.1)

11. Describe how reflection, absorption and reradiation of solar radiation account for the Earth's energy balance. (C.2)

12. Explain how differing heat capacities and reflectivities of various land covers and water can influence local climates. (C.3)

13. Describe the greenhouse effect, its natural incidence and causes, and the significance of industrial contributions. (C.2, C.4, C.5)

14. Use graphical extrapolation to predict future CO_2 concentrations, and outline assumptions and problems associated with such predictions. (C.5)

15. Compare the production of CO_2 from combustion with that from respiration. (C.4, C.6)

16. Describe the function of the stratospheric ozone layer and how human activities can affect it. Identify CFC replacements. (C.8)

17. List the major categories of air pollutants and discuss the relative contributions of various human and natural factors to each category. (D.1–D.3)

18. Describe major general strategies for controlling pollution, and specific strategies for particulates. (D.4, D.5)

19. Describe chemical reactions and geographic and meteorological factors that contribute to photochemical smog. (D.6)

20. Interpret graphs and tables related to automotive-induced air pollution. (D.7, D.8)

21. Explain the role of activation energy in a chemical reaction, and give an example of how a catalyst affects it. (D.8)

22. Describe the role of catalytic converters in reducing automotive emissions of unburned hydrocarbons, carbon monoxide, and nitrogen oxides. (D.8)

23. Describe sources and consequences of acid rain. (D.9, D.10)

24. Define the terms acid and base, give examples of each, describe their formation with balanced ionic equations, and relate hydrogen ion concentration to the pH scale. (D.11)

25. Describe the factors affecting indoor air quality and four types of indoor air pollutants. (E.1, E.2)

PERSONAL CHEMISTRY AND CHOICES

Upon completion of this unit, you will be able to:

1. Describe the components of risk assessment and apply them to smoking. (A.1)

2. Identify the major chemical components of cigarette smoke. (A.2, A.3)

3. Provide examples of correlation, and determine the causal relationship between the members of a given pair of events. (A.4)

4. Define epidemiology, and describe some benefits and limitations of epidemiological studies. (A.5)

5. Define homeostasis and give examples of how it is related to maintaining good health. (B.1)

6. Describe the major elements of the human body and their function in maintaining health. (B.2)

7. Explain how enzymes work and list several factors that may alter their effectiveness. (B.3–B.5)

8. Describe cellular energy production and storage, including the role of ATP. (B.6)

9. Define and give examples of acids and bases, and use net ionic equations to describe a neutralization reaction. (C.1–C.5)

10. Describe the components of a buffer and explain how it prevents acidosis and alkalosis. (C.3–C.5)

11. Apply the concept of "like dissolves like" to skin cleansing and the function of soap. (D.1, D.2)

12. Sketch the parts of human skin and describe their functions. (D.2)

13. Describe the effect of sunlight on skin and the effectiveness of commercial sunscreens in relation to sunscreen protection factor. (D.3, D.4)

14. Describe hair structure, the types of bonding in hair protein, and the effects of various hair treatment chemicals on hair. (D.6–D.8)

15. Distinguish between drugs and toxins, and describe circumstances where a substance's usual effect on homeostasis may be reversed. (E.1, E.6)

16. Use the concept of receptors to account for drug specificity and for the action of narcotic analgesics. (E.1)

17. Contrast the benefits and burdens associated with aspirin use and those associated with the use of alcohol. (E.2, E.3)

18. Outline the effects of illicit drugs on the human body. (E.4)

19. Discuss the role of antigen–antibody complexes in protecting the body against infectious organisms. (E.5)

20. Use the concept of synergism to explain the hazards of combining drugs and medicines. (E.6)

21. Relate what is risk and the various components of risk assessment. (F.1–F.3)

THE CHEMICAL INDUSTRY: PROMISE AND CHALLENGE

Upon completion of this unit, you will be able to:

1. List the functions of the chemical industry and the general categories of industrial products. (A.1)

2. Outline the types of products produced by the chemical industry. (A.1)

3. Evaluate the potentially positive and negative impacts of a chemical industry on a community. (A.2, D.2, E.1)

4. Contrast responsibilities of the public, the federal government, and of industry in preserving the quality of life in a community. (A.2, D.2)

5. Compare natural and synthetic products, providing examples of each. (B.2)

6. Analyze a fertilizer sample for its major components, and describe their importance (particularly nitrogen compounds) in agriculture. (Introduction, B.1, B.2)

7. Use colorimetry to quantify phosphate content in fertilizer samples. (B.3)

8. Apply oxidation–reduction concepts to nitrogen fixation in the Haber–Bosch process. (B.4)

9. Use electronegativity values to determine oxidation states. (B.4)

10. Describe factors that must be controlled in the equilibrium synthesis of ammonia. (B.5)

11. Trace the history and development of explosives, including the contributions of Alfred Nobel. (B.6)

12. Develop and evaluate voltaic cells using the activity series of common metals. (C.1, C.2)

13. Use the concept of half-reactions to describe commercial electrochemical cells, including their charging and discharging reactions. (C.2)

14. Demonstrate the technique of electroplating. (C.3)

15. Describe the industrial applications of electrolysis for brine decomposition and for aluminum production. (C.4)

16. Identify key considerations involved in the development of a new chemical process or product. (D.1)

17. Outline the major divisions and departments of the two industrial chemical companies in this unit, and explain their interrelationships. (D.3)

GLOSSARY

A

absolute zero
the temperature at which the movement of molecules from place to place ceases; the coldest temperature possible $(-273.16\ °C)$

acid
molecular substance or other chemical that releases $H^+(aq)$ ions in aqueous solution

acidosis
harmful condition in which blood pH stays below 7.35

activation energy
minimum energy required for successful collision of reactant particles in a chemical reaction

active site
in biochemistry, the site on an enzyme where the substrate molecule is made ready for reaction

activity series
ranking of elements in order of chemical reactivity

addition polymer
polymer (such as polyethene) formed by addition reactions at double bonds

addition reaction
a reaction at the double (or triple) bond in an organic molecule that results in adding or bonding atoms to each atom of the double (or triple) bond; one type of polymerization

adsorption
the process of attracting and holding something on the surface (of charcoal, for example)

aeration
mixing of air (particularly, oxygen gas) into a liquid, as in water flowing over a dam

aerobic bacteria
oxygen-consuming bacteria

alcohol
organic compound whose molecules contain one or more OH groups

alkalosis
harmful condition in which blood pH stays above 7.4

alkane
hydrocarbon having a general formula C_nH_{2n+2} whose molecules contain only single covalent bonds

alkene
hydrocarbon whose molecules contain a double covalent bond

alkyne
hydrocarbon whose molecules contain a triple covalent bond

alloy
solid solution consisting of atoms of different metals

alpha keratin
key structural protein unit of hair

alpha particle (ray)
helium nucleus emitted during radioactive decay

amino acid
organic compound whose molecules contain an amino ($—NH_2$) and a carboxyl ($—COOH$) group; proteins are polymers of amino acids

amylase
enzyme in saliva that catalyzes breakdown of starch to glucose

anaerobic bacteria
bacteria that do not require oxygen to live

anaerobic glycolysis
cellular process for quick release of energy from glucose by non-oxygen-consuming reactions; lactic acid is produced

anion
ion possessing a negative charge

anode
electrode at which oxidation occurs in an electrochemical cell

antibody
complementary protein created by the body to inactivate specific foreign protein molecules (antigens)

antigen
foreign protein that triggers body's defense mechanisms to produce antibodies

aquifer
porous rock structure that holds water beneath the Earth's surface

aromatic compound
compound such as benzene whose molecules are cyclic and can be represented as having alternating double and single bonds between carbon atoms

atmosphere
all the air surrounding the Earth

atmosphere (atm)
a unit of pressure, represented by a column of mercury 760 mm high; approximately 15 pounds per square inch or 100 kPa

atomic mass
mass of an atom

atomic number
number of protons in an atom; distinguishes atoms of different elements

atoms
smallest particles possessing the properties of an element

Avogadro's law
equal volumes of gases at the same temperature and pressure contain the same number of molecules

B

background radiation
radiation from naturally radioactive sources in the environment

base
chemical that yields OH^-(aq) ions in aqueous solution

beta particle (ray)
electron emitted during radioactive decay

biodegradable
able to be broken down into simpler substances by organisms in the environment

biomass
matter in plant materials

biomolecules
large molecules found only in living systems

biopolymers
polymers made by organisms

biosphere
a combination of portions of the Earth's waters, land, and atmosphere that supports living things

boiling point
the temperature and pressure at which a substance changes from the liquid to the gaseous state

Boyle's law
at constant temperature, the product of the pressure and volume of a given gas sample is a constant

branched-chain alkane
alkane that consists of molecules in which one or more carbon atoms are bonded to three or four other carbon atoms; for example,

$$CH_3-\underset{\underset{\displaystyle CH_3}{|}}{CH}-\underset{\underset{\displaystyle CH_3}{|}}{CH}-CH_3$$

buffer solution
solution that resists changes in pH; contains a weak acid and a salt of that acid, or a weak base and its salt

C

calorie
a unit of heat energy

Calorie (Cal)
an energy unit used to express food energy;
1 Cal = 1000 cal or 1 kcal

calorimetry
technique for determining heat of reaction or other thermal properties, and for finding caloric value of foods

carbohydrate
energy-rich compound composed of carbon, hydrogen, and oxygen; examples are starch and sugar

carbon chain
carbon atoms linked to one another, forming a stringlike sequence in a molecule

carboxylic acids
organic compounds that contain the —COOH group

carcinogen
substance that causes cancer

catalyst
substance that speeds up a chemical reaction but is itself unchanged

catalytic converter
reaction chamber in auto exhaust system designed to reduce harmful emissions

cathode
electrode at which reduction occurs in an electrochemical cell

cathode ray
beam of electrons emitted from cathode when electricity passes through evacuated tube

cation
ion possessing a positive charge

cellular respiration
oxidation of glucose or other energy-rich substances in living cells to produce CO_2, H_2O, and energy

cellulose
polysaccharide composed of chains of glucose molecules; makes up fibrous and woody parts of plants

Celsius degree (°C)
a degree on the Celsius temperature scale, 1.8 times larger than a Fahrenheit degree

ceramics
materials made by heating or "firing" clay or components of certain rocks; includes bricks, glass, and porcelain

chain reaction
in nuclear fission, reaction that produces enough neutrons to allow the reaction to continue

Charles' law
at constant pressure, the volume of a given gas sample is directly proportional to the kelvin temperature

chemical bond
force that holds atoms or ions together in chemical compounds

chemical change
change in matter resulting in a change in the identity of one or more substances

chemical compound
substance composed of two or more elements that cannot be separated by physical means

chemical equation
combination of chemical formulas that represents what occurs in a chemical reaction, such as
$2\ H_2(g) + O_2(g) \rightarrow 2\ H_2O(g)$

chemical equilibrium
condition when forward and reverse reactions occur at same rate, and concentrations of all reactants and products remain unchanged

chemical formula
combination of symbols that represents elements in a substance, with subscripts showing the number of atoms of each element; for example, the formula for ammonia is NH_3

chemical property
property of a substance related to a chemical change undergone by the substance

chemical reaction
change in matter in which one or more chemicals are transformed into new or different chemicals

chemical symbols
one- or two-letter abbreviations for the names of chemical elements

chlorination
adding chlorine to a water supply to kill harmful organisms

chlorofluorocarbons (CFCs)
carbon compounds of two to four carbon atoms with varying numbers of chlorine, fluorine, and hydrogen atoms; implicated in stratospheric ozone depletion

cis-trans **isomerism**
molecules with the same molecular formula, but containing the same atoms or groups of atoms arranged either on the same side of the plane of the C=C double bond (*cis*) or on opposite sides of the plane of the bond (*trans*)

coefficient
number preceding a formula in a chemical equation; specifies the relative number of units participating in the reaction

coenzyme
molecule or ion that assists an enzyme in performing its function

colloid
mixture containing macro-size particles that are small enough to remain suspended

colorimetric method
method for determining concentration of a solution by observing color intensity

combustion
burning

complementary proteins
two or more proteins that, in combination only, include all essential amino acids

complete protein
a single protein containing adequate amounts of all the essential amino acids

compound
substance composed of two or more elements that cannot be separated by physical means

concentration
quantity of solute dissolved in a specific quantity of solvent or solution

condensation
conversion of a substance from a gaseous to the liquid or solid state

condensation polymer
polymer formed by condensation reactions; for example, polyester

condensation reaction
chemical combination of two molecules, accompanied by loss of water or another small molecule, such as,

$$CH_3OH + HO \underset{O}{\overset{\|}{C}} CH_3 \longrightarrow CH_3O \underset{O}{\overset{\|}{C}} CH_3 + H_2O$$

condensed formula
formula such as $CH_3CH_2CH_3$; in contrast to structural formula

conductor
material that allows electricity to flow through it

conservation of matter, law of
matter is neither created nor destroyed in chemical reactions

control
in an experiment, a setup duplicating all conditions except the variable being tested

correlated
happening together; scientists often identify and seek explanations for correlated events

corrosion
deterioration or "eating away" of a material

covalent bond
a force that holds two atoms tightly to each other, found when the two atoms share one or more electron pairs

cracking
process in which hydrocarbon molecules from petroleum are converted to smaller molecules

critical mass
mass of fissionable material needed to sustain a nuclear chain reaction

crude oil
petroleum as it is pumped from underground

cryogenics
studies of the chemistry and physics of materials and systems at very low temperatures

current
flow of electrons

cuticle
tough outer layer in hair

cycloalkane
saturated hydrocarbon whose molecules contain carbon atoms joined in a ring

D

data
objective pieces of information, often the information gathered in experiments

denaturation
alteration of protein shape and function by disruption of folding and coiling in molecules

density
mass per unit volume of a given material

dermis
inner layer of the skin

diatomic
molecule made up of two atoms

dipeptide
compound made from two amino acids

disaccharide
compound made from two simple sugars; for instance, maltose (made from two glucose units)

distillate
condensed product of distillation

distillation
method of separating substances using differences in their boiling points

double covalent bond
bond in which four electrons are shared by two bonded atoms

dynamic equilibrium
in a reversible reaction, the state of product formation occurring at the same rate as product decomposition

E

electrical conductivity
the ability to conduct an electric current

electrical potential
potential for moving or pumping electric charge in an electrical circuit or by an electrochemical cell; measured in volts

electrochemical cell
device for carrying out electrolysis or producing electricity from a chemical reaction

electrodes
two strips of metal or other conductors serving as contacts between the solution or molten salt and the external circuit in an electrochemical cell; reaction occurs at each electrode

electrolysis
to break apart or decompose a compound by electricity

electromagnetic radiation
radiation moving at the speed of light, ranging from low-energy radio waves to high-energy cosmic and gamma rays; includes visible light

electromagnetic spectrum
energy with wave properties ranging from low-energy radio waves to high-energy gamma waves

electrometallurgy
use of electrical energy to process metals or their ores

electron
negatively charged particle present in all atoms

electron dot formula
formula for a substance in which dots representing the outer electrons in each atom show the sharing of electron pairs between atoms

electronegativity
tendency of bonded atoms to attract electrons in molecules

electroplating
deposition of a thin layer of metal on a surface by electrolysis

elements
fundamental chemical substances from which all other substances are made

endorphins
natural painkillers produced in the brain

endothermic
a process requiring energy

endpoint
point during a titration at which the reaction is complete

enkephalins
natural painkillers produced in the brain

enzyme
catalyst for a biochemical reaction

epidermis
outer layer of the skin

essential amino acid
one of eight amino acids that the human body cannot synthesize; must be included in the diet

esters
organic compounds containing the —COOR group, where R represents any stable arrangement of carbon and hydrogen atoms

ether
organic compound containing the functional group —O—

evaporation
conversion of a substance from the liquid to the gaseous state

exothermic
an energy-releasing process

extrapolation
estimate of a value beyond the known range (the continuation of a curve on a graph past the measured points, for example)

F

family (periodic table)
vertical column of elements in the periodic table; also called a group; members of a family share similar properties

fat
lipid resulting from reaction of glycerol and fatty acids; storage form for food energy in animals

fatty acid
organic compound whose molecules consist of a long hydrocarbon chain and a —COOH group; combined with glycerol in fats

fibrous protein
protein whose molecules form ropelike or sheetlike structures; found in hair, muscles, skin

filtrate
liquid collected after filtration

filtration
separation of solid particles from a liquid by passing the mixture through a material that retains the solid particles

fluorescence
emission of visible light from a material following its exposure to ultraviolet radiation

force
the cause of a body's motion or weight, brought about by its mass and by gravity

formula unit
group of atoms or ions represented by chemical formula of a compound; simplest unit of an ionic compound

fossil fuel
petroleum, natural gas, or coal

fraction (petroleum)
mixture of petroleum-derived substances of similar boiling points and other properties

fractional distillation
a technique by which a mixture is separated into its components by boiling and condensing the component with the lowest boiling point, leaving the other components behind

freezing point
the temperature at which a substance changes from the liquid to the solid state

frequency
number of vibrations or cycles per unit of time

functional group
atom or group of atoms that imparts characteristic properties to an organic compound; —Cl, —OH, or —COOH, for example

G

gamma ray
high-energy electromagnetic radiation emitted during radioactive decay

gaseous state
state of matter having no fixed volume or shape

Geiger counter (radiation counter)
device that produces an electrical signal in the presence of ionizing radiation

globular protein
protein whose molecules assume ball shapes and are water soluble because of polar and ionic groups on surface; may function as hormone, enzyme, or carrier protein

glycogen
polymer made of repeating glucose units; synthesized in liver and muscles as reserve source of glucose

gram (g)
SI unit of mass commonly used in chemistry (kilogram is SI base unit for mass)

greenhouse effect
retention of energy at or near the Earth's surface as carbon dioxide and other atmospheric gases capture escaping radiation and return it to the Earth's surface; result is surface warming

greenhouse gases
atmospheric gases that allow sunlight to reach the Earth, but prevent heat from leaving it

groundwater
water that collects underground

group (periodic table)
See family

H

Haber–Bosch process
industrial process for catalyzed synthesis of ammonia from nitrogen and hydrogen

half-cell
metal (or other electrode material) and its surrounding solution of ions in a voltaic cell

half-life
time needed for decay of one-half the atoms in a sample of radioactive material

half-reaction
half of oxidation–reduction reaction in which electrons are either lost or gained; for example, the process that occurs in one half-cell of a voltaic cell

hard water
water containing relatively high concentrations of calcium (Ca^{2+}), magnesium (Mg^{2+}), or iron(III) (Fe^{3+}) ions

heat capacity
quantity of heat required to raise the temperature of a given sample of matter by 1 °C when in a given state

heat of combustion
quantity of thermal energy released when a specific amount of a substance burns

heat of fusion
quantity of heat required to convert a specific amount of a solid to a liquid at its melting point

heavy metals
metals of high atomic mass, generally from fifth or sixth row of the periodic table

heterogeneous
not uniform throughout, as in a heterogeneous mixture

homeostasis
maintenance of balance in all body systems

homogeneous
uniform throughout, as in a homogeneous mixture

hormone
biomolecule that serves as a specific messenger to stimulate biochemical activity at specific sites in the body

hydrocarbons
molecular compounds composed solely of carbon and hydrogen

hydrogen bonding
attraction between molecules, or between parts of the same molecule, involving hydrogen atoms and strongly electron-attracting atoms such as nitrogen or oxygen

hydrologic cycle
circulation of water between the Earth's atmosphere and crust

hydrometallurgy
water-based methods for processing metals or their ores

hydronium ion
H_3O^+, a hydrogen ion bonded to a water molecule

I

ideal gas
gas that behaves as predicted by kinetic molecular theory

infrared radiation
electromagnetic radiation of slightly lower energy than visible light; raises temperature of objects that absorb it

intensity (radiation)
measure of quantity of radiation per unit time

intermediate (chemical)
product of chemical industry used to synthesize consumer products or other chemicals; sulfuric acid is an intermediate in the manufacture of certain detergents

intermolecular forces
forces holding molecules together

interpolation
inserting a value between the known values in a series (such as reading a part of a curve between two measured points)

ion
an atom or group of atoms that has become electrically charged by gaining or losing electrons

ion exchange
process used to purify water in which hard water ions exchange for soft water ions of an ion exchange resin

ionic bond
attraction between oppositely charged ions in an ionic compound

ionic compound
substance composed of positive and negative ions

ionizing radiation
electromagnetic radiation or high-speed particles possessing enough energy to ionize atoms and molecules; emitted during radioactive decay

irradiation
treatment (of food, for instance) with radiation

isomer
compound having the same molecular formula, but a different structural formula, than another compound

isotopes
atoms of the same element having different numbers of neutrons

K

kinetic energy
energy associated with the motion of an object

kinetic molecular theory of gases
theory that accounts for properties of gases based on kinetic energy and constant random motion of molecules

L

law of conservation of energy
energy can change form, but is never created or lost in chemical reactions

law of conservation of matter
matter cannot be created or destroyed in chemical reactions

length
linear distance; the SI base unit of length is the meter (m)

limiting reactant
starting substance used up first as a chemical reaction occurs

lipid
fat or other member of a class of biomolecules not soluble in water

liquid state
state of matter with fixed volume but no fixed shape

liter (L)
unit of volume; equal to 1 dm^3, 1000 cm^3, or 1000 mL

M

macromineral
essential mineral present in amount of 5 g or more in adult human body

malleable
material's ability to be flattened without shattering

malnourishment
receiving inadequate amounts of essential nutrients such as protein, vitamins, minerals

mass
amount of matter in something

mass number
sum of the number of protons and neutrons in an atom of a given isotope

melanin
body pigment responsible for dark skin and dark hair

mesosphere
region of atmosphere outside stratosphere

metal
element having certain properties: shiny, ductile, conductive, malleable

metalloid
element with properties intermediate between those of metals and nonmetals

meter (m)
SI base unit of length

microfibril
bundle of coiled protein chains; a component of hair, for instance

milliliter (mL)
unit of volume; equal to 1 cm^3

millimeters of mercury (mmHg)
pressure unit: 1 atm = 760 mmHg = 101 kPa

minerals
essential micronutrient elements such as calcium, sodium, iodine, and iron

minimal erythemal dose (MED)
number of hours of sunlight at a given time and place needed to turn skin barely pink

mixture
combination of substances in which each substance retains its separate identity

molar concentration
concentration of a solution expressed in moles of solute per liter of solution

molar heat of combustion
thermal energy released by burning one mole of a substance

molar mass
mass (usually in grams) of one mole of a substance

molar volume
volume occupied by one mole of a substance; at STP molar volume of a gas is 22.4 L

molarity
concentration of a solution expressed as moles of solute per liter of solution; also called molar concentration

mole (mol)
amount of substance or chemical species equal to 6.02×10^{23} units, where the units may be atoms, molecules, formula units, electrons, or other specified entities; chemist's "counting" unit

molecular structure
arrangement and bonding of atoms in a molecule

molecule
smallest unit of a substance retaining the properties of the substance; composed of two or more atoms joined by covalent bonds

monomer
compound whose molecules react to form a polymer

monosaccharide
simple sugar, such as glucose

N

narcotic analgesic
drug that relieves intense pain

negative oxidation state
negative number assigned to atom in a compound when that atom has greater control of its electrons than it has as a free element

net ionic equation
equation showing only those chemicals that participate in a reaction involving ions in aqueous solution

neuron
nerve cell

neutralization
reaction of an acid with a base in which the characteristic properties of both are destroyed

neutral solution
neither acidic nor basic

neutron
neutral particle present in nuclei of most atoms

newton (N)
a unit of force in the metric system; roughly equal to the force exerted on your hand by a 100-g bar of soap

nitrogen fixation
conversion of nitrogen gas (N_2) to nitrogen compounds usable by plants

nonconductor
material that does not allow electricity to flow through it

non-ionizing radiation
electromagnetic radiation possessing insufficient energy to ionize atoms or molecules; for example, visible light

nonmetal
element having certain properties: nonlustrous, brittle, and nonconductive

nonpolar
having no electrical asymmetry or polarity, as in a nonpolar molecule

nonpolar interaction
weak attraction between nonpolar chemical groups

nonrenewable resource
resource that will not be replenished by natural processes during the time frame of human experience

nuclear fission
splitting of one atom into two smaller atoms; undergone by uranium-235 when bombarded with neutrons

nuclear fusion
combining of two atomic nuclei to form a single more massive nucleus

nuclear radiation
particles and energy emitted from radioactive atoms

nucleus, atomic
dense central region in an atom; contains all protons and neutrons

nutrients
components of food needed in the diet

O

octane number
rating indicating combustion quality of gasoline

ore
rock or mineral from which it is profitable to recover a metal or other useful substance

organic compound
compound composed mainly of carbon and hydrogen atoms; a hydrocarbon or a compound derived from a hydrocarbon

oxidation
any process in which electrons are lost or the extent of electron control decreases

oxidation–reduction (redox) reaction
reaction in which oxidation and reduction occur

oxygenated fuels
oxygen-containing fuel additives, such as methanol, that increase octane rating and reduce pollutants

P

patina
surface film or coating, such as the stable green coating on copper exposed to the atmosphere

pepsin
enzyme that aids in digesting protein

peptide
chain of amino acids; part of a protein

peptide linkage
—CONH— linkage formed by reaction of the —NH$_2$ group of one amino acid and the —COOH group of another amino acid; linkage between amino acid residues in proteins

periodic law
when elements are arranged in order of increasing atomic number, elements with similar properties occur at regular intervals

periodic table
table in which elements, arranged in order of increasing atomic number, are placed so that those with similar properties are near each other

periods (periodic table)
horizontal rows of elements in the periodic table

petrochemical
substance produced from petroleum or natural gas

petroleum
liquid fossil fuel composed mainly of hydrocarbons, but also containing compounds of nitrogen, sulfur, and oxygen, along with small amounts of metal-containing compounds

pH
number representing acidity or alkalinity of an aqueous solution; at 25 °C, solution with pH 7 is neutral, pH < 7 is acidic, pH > 7 is basic

photochemical smog
smog produced when sunlight interacts with nitrogen oxides and hydrocarbons in atmosphere

photon
packet of energy present in electromagnetic radiation

photosynthesis
process by which green plants make sugars from carbon dioxide and water in the presence of sunlight

physical change
change in matter in which the identity of the substance involved is not changed, such as the melting of ice

physical property
property that can be observed or measured without changing the identity of a sample of matter; for example, color, boiling point

polar
having electrical poles, or regions of positive and negative charge, as in a polar molecule

polyatomic ion
ion containing two or more atoms, such as SO_4^{2-}

polymer
substance whose large molecules are composed of many identical repeating units

polysaccharide
polymer made from simple sugar molecules; starch, for example

polyunsaturated
organic molecules with several double bonds per molecule

positive oxidation state
positive number assigned to atom in a compound when that atom has less control of its electrons than it has as a free element

positron
particle with mass of electron but possessing a positive charge

precipitate
insoluble solid substance that has separated from a solution

pressure
force applied to one unit of surface area

primary air pollutant
pollutant in the form originally emitted to the atmosphere

product
substance formed in a chemical reaction

protease
enzyme that aids digestion of proteins

proteins
polymers made from amino acids; important compounds in body such as hair, nails, muscle, enzymes, hormones

proton
positively charged particle present in nuclei of all atoms

pyrometallurgy
use of thermal energy (heat) to process metals or their ores

R

radioactive decay
emission of alpha, beta, or gamma rays by unstable isotopes

radioactivity
spontaneous decay of unstable atomic nuclei accompanied by emission of ionizing radiation

radioisotope
radioactive isotope

reactant
starting substance in a chemical reaction

reactants
substances that are the starting materials of a chemical reaction

receptors
proteins in membranes of key body cells, shaped to receive the molecule of a hormone, drug, or other activator and, having done so, to activate chemical processes within the cell

recycling
reprocessing materials in manufactured items so they can be reused as raw materials for manufacturing new items

reducing agent
substance that causes another substance to be chemically reduced, that is, to gain electrons

reduction
any process in which electrons are gained or the extent of electron control increases

reference solution
solution of known composition used as a comparison

reflectivity
surface's property of returning radiation that strikes it

rem
unit indicating power of ionizing radiation to cause damage to human tissue (Roentgen equivalent man)

renewable resource
resource that is replenished by natural processes in the time frame of human experience

reversible reaction
chemical reaction in which reverse reaction can occur simultaneously with forward reaction

rickets
disease caused by lack of vitamin D; occurs in absence of exposure to sunlight, which helps body to produce vitamin D

S

salicylates
family of painkillers that includes aspirin (acetylsalicylic acid)

saturated
organic compound that has only single bonds

saturated fat
fat whose molecules contain no carbon–carbon double bonds

saturated hydrocarbon
hydrocarbon consisting of molecules in which each carbon atom is bonded to four other atoms

saturated solution
solution in which the solvent has dissolved as much solute as it can stably retain at a given temperature

science
a group of disciplines that gather, organize, and analyze knowledge about natural phenomena and natural objects

scintillation counter
sensitive radiation-measuring device; produces flashes of light in the presence of ionizing radiation

sewage treatment plant
installation built for post-use cleaning of municipal water

shell
an energy level surrounding the nucleus of an atom

single covalent bond
bond in which two electrons are shared by the two bonded atoms

solid
state of matter having a fixed volume and fixed shape

solubility
quantity of a substance that will dissolve in a given quantity of solvent to form a saturated solution

solute
dissolved substance in a solution, usually the component present in the smaller amount

solution
homogeneous mixture of two or more substances

solution concentration
quantity of solute dissolved in a specific quantity of solvent or solution

solvent
dissolving agent in a solution, usually the component present in the largest amount

specific heat
quantity of heat needed to raise the temperature of 1 g of a material by 1 °C

spectator ions
ions that are present but do not participate in a reaction in solution

spectrum
range of radiation waves, from low to high energy

STP
standard temperature and pressure: 0 °C and 1 atm

starch
polysaccharide made by plants to store glucose

state
form—gas, liquid, or solid—in which matter is found

straight-chain alkane
alkane consisting of molecules in which each carbon atom is linked to no more than two other carbon atoms, such as $CH_3CH_2CH_2CH_3$

stratosphere
region of atmosphere outside troposphere

strong force
force of attraction between particles in atomic nucleus

structural formula
chemical formula showing the arrangement of atoms and covalent bonds in a molecule

subscript
character printed below a line of type; in H_2O, for example, the subscript 2 indicates the number of H atoms

substrate
reactant molecule or ion in an enzyme-catalyzed biochemical reaction

sunscreen protection factor (SPF)
rating system that estimates the number of hours of out-door protection a commercial sunscreen may give

supersaturated solution
solution containing a higher concentration of solute than a saturated solution at the given temperature

superscript
character printed above a line of type; in Cl^-, the super-script $-$ indicates the charge of the chloride ion

surface water
water on the surface of the ground

suspension
mixture containing such large, dispersed particles that it appears cloudy; muddy water, for instance

symbol
one- or two-letter expression that represents an element; the symbol Na represents sodium

synergistic interaction
combination of interactions that produces a total effect greater than the sum of the individual interactions; for example, combined effect of air pollutants

synthetic
not natural; for example, a substance created industrially from petroleum

T

technology
application of science to create useful goods and services

temperature
degree of hotness or coldness of a substance, as measured on a thermometer

tetrahedron
a regular triangular pyramid; the four bonds of each carbon atom in molecules of alkanes point to the corners of a tetrahedron

thermosphere
outermost region of the Earth's atmosphere

titration
laboratory technique used to determine the concentration of a solution, or the amount of a substance in a sample

toxin
substance harmful to the body

trace mineral
essential mineral present in quantities of less than 5 g in adult human body

tracer, radioactive
radioactive isotope used to follow movement of material; used in medicine to detect abnormal functioning in body, for example

transmutation
conversion of one element to another; unknown before discovery of radioactivity

transuranium element
element having an atomic number higher than that of uranium

triglyceride
an ester whose molecules were formed by combination of glycerol with three fatty acid molecules; a fat

tripeptide
compound made from three amino acids

troposphere
region of atmosphere from the Earth's surface to 10 km outside it

turbidity
cloudiness

Tyndall effect
caused by reflection of light from suspended particles in a colloid

U

ultraviolet radiation
electromagnetic radiation of higher energy than visible light; can cause tissue damage

undernourishment
receiving less food than needed to supply bodily energy needs

unsaturated
organic compound with at least one double bond per molecule

unsaturated fat
fat whose molecules contain carbon–carbon double bonds

unsaturated hydrocarbon
hydrocarbon molecules containing double or triple bonds; for example, alkenes, alkynes

unsaturated solution
solution containing a lower concentration of solute than a saturated solution at the given temperature

V

viscosity
measure of a fluid's resistance to flow

visible radiation
electromagnetic radiation detectable by human eye

vitamins
biomolecules needed in small amounts for body function; must be provided in food or as food supplement

vitrification
formation of a glasslike substance

voltaic cell
electrochemical cell in which a spontaneous chemical reaction is used to produce electricity

W

water softening
removal from water of ions that cause its hardness (*see* hard water)

X

X-rays
high-energy electromagnetic radiation; normally unable to penetrate bone or lead, but can penetrate less-dense materials

Z

zero oxidation state
neither oxidized nor reduced; the oxidation state of an uncombined element

INDEX

Page numbers in **boldface** indicate pages with definitions.

Charles' law, **398**
as fluids, **375**
greenhouse, **226, 408,** 410–412
hazardous, 97, 221, 574
kinetic molecular theory of, 399–402
molecular attraction in, 63, 401–402
noble, 113, 177–178, 327
pressure and, 386–387, 393–395, 541, 546
solubility, 43–45, 49–50
temperature–volume relationships, 396–399, 402
in troposphere, 385–386, 434
Gasohol, **206,** 215
Gasoline, 23, 64, 204
additives, 206–207, 213, 217
combustion, 94, 127, 412
consumption, 88, 156–157
energy conversion, 195–197
fractional distillation, 169–170, 174–175, 203
octane ratings, 205–206
Gay-Lussac, Joseph, 397
Geiger, Hans, 305–306
Geiger–Mueller counter, 314–315
Generally Recognized as Safe (GRAS) list, 285
Genes, 536
Geothermal energy, 191, 226
Ghioso, Albert, 341
Glaciers, 15, 385
Glass, 99, 133–134, 571
uses, 317–318, 531, 534
Global warming, 226, 369, 412, 414, 417
Glucose, 251–253, 467, 472
aqueous pH, 443
in carbon cycle, 48, 378, 411
formulas, 252–253, 485
D-glucose as benzene substitute, 574
oxidation, 259–260, 470–471, 478
Glycerol, 247, 487
Glyceryl tripalmitate, 247, 487
Glycine, 264
Glycogen, 252–253
Gold, 139, 141, 344, 564
reactivity, 117, 555
Graphite, 102, 119, 353
as raw material, 146
rods as electrodes, 144, 557–558, 563

GRAS (Generally Recognized as Safe) list, 285
Greenhouse effect, **408,** 411
Greenhouse gases, **226, 408**
human activity and, 411–412, 414
sunlight and, 408, 410
Groundwater, **14,** 14–15
Groups, chemical, **113**
bonding between, 499–500
functional, **214,** 264, 469

H

Haagen-Smit, Arie J., 434
Haber, Fritz, 541
Haber–Bosch process, 541–542, 546–547
Hahn, Otto, 346
Hair, human, 496–497, 499–501
styling, **501**
Half-cell (electrochemistry), **554,** 557
Half-life (atomic decay), **331**–333
selected isotopes, 331, 364, 370
Hall, Charles Martin, 566
Hall–Héroult process, 566–567
Hallucinogens, **509**
HCFCs. *See* Hydrochlorofluoro-carbons (HCFCs)
HDLs. *See* High-density lipoproteins (HDLs)
Health, human, 459, 494, 519
air quality and, 451
fluoridation and, 75
limiting reactants and, 259, 261, 266, 276
nutrients for, 11, 237, 261, 271
sewage treatment and, 464
water quality and, 2–4, 11–12, 15, 34
Heat, specific, **199**–200
Heat capacity, **409**–410
Heat of combustion, **198**–202, 206, 350
Heavy metal ions, 76
toxicity, 59–62, 328, 429
Heavy metals, 34, 59
Helium, 113, 123, 385, 406
electron shells, 177–178
Hemoglobin, 263, 458
in blood buffer system, 479
ionic bonding in, 500
Heptane, 181, 184
in gasoline, 204–205
properties, 176, 201

Heroin, 507
Heterogeneity, 25
Hex- (root), 181
Hexane, 181, 184, 204, 212
HFCs. *See* Hydrofluorocarbons (HFCs)
High-density lipoproteins (HDLs), 250
Histidine, 266
Homeostasis, **464**–465, 478–482
Homogeneity, 25, 29
Hormones, 247, 263, 275
adrenaline, 504
testosterone, 510
Héroult, Paul-Louis, 566
Humidity, 385, 445
Hydrazine, 544
Hydrocarbons, **156,** 204
in crude oil, 169–170, 174
halogenated, 38, 77, 408, 418–422, 446
heat of combustion, 198, 202
isomerization, **207**
nonmethane (NMHC), 426
saturated, **179**–185, 209
substituted, 209–210, 408, 418–421
VOCs, 428, 435, 445, 447, 450
Hydrochloric acid, 497–499
alkalosis and, **481**
formation and dissociation, 475–477
in stomach fluid, 55, 443, 481, 510
uses, 55, 279, 531–532
Hydrochlorofluorocarbons (HCFCs), **421**
Hydrofluorocarbons (HFCs), **421**
Hydrogen, 23, 111, 465, 542
atoms, 28–29, 32, 54–55, 178, 541–542
bonding, 180, **483**–484, 499, 501, 504
dehydrogenases and, 469
isotopes, 331, 337, 353
molecules, 30, 32, 125, 178, 542
in troposphere, 385–386
Hydrogen-2. *See* Deuterium
Hydrogen-3. *See* Tritium
Hydrogen ions, 441–442
acids and, 273, 381, 475–477
pH, 442
Hydrogen peroxide, 31, 125
decomposition, 468
as reactant, 381, 383

VISUAL CREDITS

PHOTOGRAPHS

WATER

Opener: page 1: SuperStock; **page 2:** © U.S. Fish and Wildlife Service, photo by W. French; **page 5:** © Kendall/Hunt Publishing Co., Greg Nauman Photographer; **page 6, fig. 1:** © Kendall/Hunt Publishing Co., Greg Nauman Photographer; **page 8, fig. 2:** © Kendall/Hunt Publishing Co., Greg Nauman Photographer; **page 22:** © Tom Bean, The Stock Market; **page 23:** © State of Maine, Dept. of Transportation; **page 24, fig. 10:** © Oscar Palmquist/ Lightwave; **page 25, fig. 11:** Dr. Peter Cooke, Electron Microscope Unit, Eastern Regional Research Center, Agricultural Research Service, USDA; **page 26:** © Kip Peticolas, Fundamental Photographs; **page 28:** © Kendall/Hunt Publishing Co., Greg Nauman Photographer; **page 29 top:** © Kendall/Hunt Publishing Co., Greg Nauman Photographer; **page 32, fig. 16:** © Kendall/Hunt Publishing Co., Greg Nauman Photographer; **page 33, fig. 17:** © Kendall/Hunt Publishing Co., Greg Nauman Photographer; **page 37:** © American Chemical Society, Collette Mosley Photographer; **page 41:** © Tom Bean, The Stock Market; **page 48:** © Dean Hulse/Rainbow; **page 56:** Molecular Simulations, Inc.; **page 57:** © Dr. Jeremy Burgess/Photo Researchers; **page 61:** © Linda Moore/Rainbow; **page 65:** © Kendall/Hunt Publishing Co., Greg Nauman Photographer; **page 66:** © U.S. EPA; **page 68:** © Oscar Palmquist/Lightwave; **page 70, fig. 27:** © Kendall/Hunt Publishing Co., Greg Nauman Photo; **page 73:** Courtesy of Diedrich Technologies; **page 75:** © Metropolitan Water District of Southern California; **page 76:** © City of Jacksonville, Florida; **page 77:** © PPG Industries, Inc.; **page 80:** © Bob Daemmrich Photography; **page 83:** © Grant Heilman Photography; **page 85:** © Roy Orsch, The Stock Market.

RESOURCES

Opener: page 86: NASA; **page 88 left:** © Dan McCoy/Rainbow; **right:** © Oscar Palmquist/Lightwave; **page 89 top:** © Alain Nogues/Sygma; **bottom:** © Kelly R. Foster, The Stock Market; **page 91:** © Kendall/Hunt Publishing Co., Greg Nauman Photographer; **page 93, fig. 2 and fig. 3:** © Kendall/Hunt Publishing Co., Greg Nauman Photographer; **page 98:** © Jose Fuste Raga, The Stock Market; **page 100:** © Susie Leavines; **page 102:** © The World Bank/IFC/MIGA; **page 104:** City of New York, Dept. of Sanitation; **page 105 top:** © Les Christman, Visuals Unlimited; **bottom left:** © Tom Kelly; **bottom right:** © Tom Kelly; **page 107:** © A. Copeland, Visuals Unlimited; **page 109:** © 1996, Jeffrey O'Connor; **page 116:** © Linda Moore/Rainbow; **page 121:** © George Disario, The Stock Market; **page 125:** © Inland Steel Company; **page 128:** © Kendall/ Hunt Publishing Co., Greg Nauman Photographer; **page 131:** © Comstock INC/ Michael Thompson; **page 132:** © Steve Elmore/Tom Stack & Associates; **page 137 top:** Joel E. Arem; **bottom:** © Copper Development Association, Inc.; **page 143:** Courtesy of the Aluminum Association; **page 144:** © 1996, Jeffrey O'Connor; **page 145:** © Sprint/United Administration Services Group, Visual Communication Center; **page 146 potter's wheel:** © Bill Stanton/Rainbow; **electric kiln:** © S.E. Bryrne/Lightwave; **pottery bowls:** © Dan McCoy/ Rainbow; **fired bricks:** © D. Cunningham, Visuals Unlimited; **plasma spray:** © David Parker, Science Photo Library, Photo Researchers, Inc.; **ceramic turbine blades:** © Dick Luria Photography, Inc./Photo Researchers, Inc.; **electronic components:** © Adam Hart, Science Photo Library, Photo Researchers, Inc.;

making turbine blades: © Will & Deni McIntyre, Photo Researchers, Inc.; **micrograph of graphite:** Polycarbon, Inc.; **kaolin:** ECC International

PETROLEUM

Opener: page 154: GALA/SuperStock; **page 156:** © State of California Energy Commission; **page 157:** © Aldo Matrocola/Lightwave; **page 160:** © American Chemical Society, Matt Haas Photographer; **page 161:** © 96 Tom & DeAnn McCarthy, The Stock Market; **page 164:** American Petroleum Institute; **page 166:** © ARCO Library; **page 170:** © Oscar Palmquist/Lightwave; **page 172:** © Kendall/Hunt Publishing Co., Greg Nauman Photographer; **page 180, fig. 11:** © Kendall/Hunt Publishing Co., Greg Nauman Photographer; **fig. 12:** © Kendall/Hunt Publishing Co., Greg Nauman Photographer; **page 185:** Texas Mid-Continent Oil & Gas, Courtesy of American Petroleum Institute; **page 186:** © Kendall/Hunt Publishing Co., Greg Nauman Photographer; **page 187:** Courtesy of Drake Museum, American Petroleum Institute; **page 188:** Courtesy of Exxon Corporation, American Petroleum Institute; **page 192:** © Grant Heilman Photography; **page 199, fig. 19:** © Kendall/Hunt Publishing Co., Greg Nauman Photographer; **page 202:** Courtesy of U.S. Air Force; **page 205:** © Wes Thompson, The Stock Market; **page 211:** © Kendall/Hunt Publishing Co., Greg Nauman Photographer; **page 212:** © Kendall/Hunt Publishing Co., Greg Nauman Photographer; **page 213 top:** © Kendall/Hunt Publishing Co., Greg Nauman Photographer; **center:** © Kendall/Hunt Publishing Co., Greg Nauman Photographer; **bottom:** © Kendall/Hunt Publishing Co., Greg Nauman Photographer; **page 215:** Allied Signal, Inc.; **page 218:** © The Dow Chemical Company; **page 220:** © Brent Jones; **page 221:** © Kendall/Hunt Publishing Co., Greg Nauman Photographer; **page 223:** © Dan McCoy/Rainbow; **page 225 top:** © U.S. Department of Interior, Bureau of Reclamation, Photo by Tom Fredmann; **lower left:** © National Mining Association; **lower right:** © National Mining Association; **page 229:** © National Mining Association

FOOD

Opener: page 232: © 94 Jim Foster, The Stock Market; **page 234:** Courtesy of United Nations; **page: 236:** © Produce Marketing Association; **page 240:** 1996, Jeffrey O'Connor; **page 242:** © Greg Davis, The Stock Market; **page 245:** © Joel Dexter/Lightwave; **page 246:** Bruce Alexander, PC&F; **page 251:** Bruce Alexander, PC&F; **page 256:** © Don W. Fawcett, Visuals Unlimited; **page 261:** Bruce Alexander, PC&F; **page 262:** Dale Farnham, Dept. of Agronomy, Extension Service, Iowa State University; **page 263 right:** © Charles Cupton, The Stock Market; **page 263 upper left:** Bruce Alexander, PC&F; **page 268:** Bruce Alexander, PC&F; **page 273:** © 1996, Jeffrey O'Connor; **page 274:** © American Chemical Society, Matt Haas Photographer; **page 276:** Morton Salt package used by permission of Morton International Inc., Morton Salt; **page 278:** © Kendall/Hunt Publishing Co., Greg Nauman Photographer; **page 281:** © American Chemical Society, Matt Haas Photographer; **page 284:** © National Down Syndrome Society; **page 286:** Bruce Alexander, PC&F; **page 289:** © American Chemical Society, Matt Haas Photographer; **page 290:** Bruce Alexander, PC&F

NUCLEAR

Opener: page 294: © Chris Priest/Science Photo Library, Photo Researchers, Inc.; **page 296:** © Steven Peters, Tony Stone Images; **page 298:** © American Chemical Society; **page 301 top:** Courtesy of Dr. Sidney L. Horowitz/School of Dentistry and Oral Surgery/Columbia University; **bottom:** American Cancer Society, Photographed by James Morehead, Source: Dr. Cooney; **page 304:** Courtesy of

Rutherford Museum, McGill University, Montreal, Canada; **page 310:**
© Kendall/Hunt Publishing Co., Greg Nauman Photographer; **page 314:** © Will
McIntyre, Photo Researchers, Inc.; **page 315:** Oxford Instruments Inc., Nuclear
Measurement Group; **page 316, fig. 8 and fig. 9:** © 1996, Jeffrey O'Connor;
page 318: © David Parker/Science Photo Library, Photo Researchers, Inc.; **page
332:** © Diego Goldberg/Sygma; **page 336:** SIU Biomedical Communications/
Photo Researchers, Inc.; **page 337:** GE Medical Systems; **page 340:** © Fermilab,
Visuals Unlimited; **page 341:** Courtesy of Lawrence Berkeley Laboratory; **page
342:** © Ernie Carpenter, Chemical & Engineering News, American Chemical
Society; **page 346 left:** University of Chicago; **right:** AIP Niels Bohr Library;
page 348: U.S. Department of Energy; **page 349, fig. 21,** Bruce Alexander,
PC&F; **fig 22:** U.S. Department of Energy; **page 352 top:** © Visuals Unlimited;
fig. 26: U.S. Department of Energy; **page 354, fig. 27:** © Dan McCoy/Rainbow;
page 357: © Jon Jacobson; **page 360:** Courtesy of Nordion International, Inc.;
page 363: U.S. Ecology; **page 364:** Sandia National Laboratories; **page 366:**
U.S. Nuclear Regulatory Commission; **page 367:** © Sygma; **page 371:** Courtesy
of Argonne National Laboratory

AIR

Opener: page 372: National Severe Storms Laboratory, National Oceanic and
Atmospheric Administration; **page 374:** Courtesy of Diedrich Technologies, Inc.;
page 377: © Kendall/Hunt Publishing Co., Greg Nauman Photographer; **page
379:** © Bill Banaszewski, Visuals Unlimited; **page 380:** Courtesy of Praxair;
page 381 left: Courtesy of Praxair; **right:** Courtesy of Praxair; **page 383, fig. 2:**
© Kendall/Hunt Publishing Co., Greg Nauman Photographer; **page 385:** NASA;
page 388: © Joan Baron, The Stock Market; **page 392, fig. 4:** Bruce Alexander,
PC&F; **page 393:** Courtesy of Princo Instruments, Inc.; **page 395:** National
Severe Storms Laboratory, National Oceanic and Atmospheric Administration;
page 396, fig. 7: © Kendall/Hunt Publishing Co., Greg Nauman Photographer;
bottom: © Kendall/Hunt Publishing Co., Greg Nauman Photographer; **page 401:**
© Robert Mort, Tony Stone Images; **page 405:** Courtesy of United Nations; **page
412:** Courtesy William Albert Allard/ National Geographic Society; **page 416:**
© 1996, Jeffrey O'Connor; **page 418:** Courtesy Robinair Division, SPX
Corporation; **page 419:** NASA; **page 424:** National Snow and Ice Center,
Science Source/Photo Researchers; **page 425:** © Alan Parcairn/Grant Heilman
Photography; **page 427:** NASA; **page 437:** © Thomas Braise, The Stock Market;
page 439: © Field Museum of Natural History; **page 444:** Courtesy of Argonne
National Laboratory; **page 448:** Culver Collection/SuperStock

CHOICES

Opener: page 452: © Leverett Bradley/FPG International Corp.; **page 454:**
Bruce Alexander, PC&F; **page 456:** © Oscar Palmquist/Lightwave; **page 459:**
© Terry Qing/FPG International Corp.; **page 460:** © Henley & Savage, The
Stock Market; **page 470:** © Joe Sohm, The Stock Market; **page 474:** © Patti &
Milt Putnam, The Stock Market; **page 480:** © Kendall/Hunt Publishing Co.,
Greg Nauman Photographer; **page 481:** © Guy Gillette/Photo Researchers, Inc.;
page 483: © Joe Barabam, The Stock Market; **page 486:** © Oscar
Palmquist/Lightwave; **page 492:** © Paul Barton, The Stock Market; **page 494:**
© Kendall/Hunt Publishing Co., Greg Nauman Photographer; **page 496:**
Courtesy of The Gillette Company; **page 497, fig. 15:** © Kendall/Hunt
Publishing Co., Greg Nauman Photographer; **page 501:** Zukas/Courtesy Modern
Salon Magazine; **page 503 top:** Partnership for a Drug-Free America; **bottom:**
Courtesy of Odyssey House; **page 504:** U.S. Dept. of Justice/Drug Enforcement
Administration

INDUSTRY

Opener: page 520: © Gary Arruda Photography; **page 524, top left:** © Genetech, Visuals Unlimited; **bottom:** Courtesy of Lonza Incorporated; **page 525, top center:** © Roger Tully, Tony Stone Images; **top right:** © Roger Tully, Tony Stone Images; **top left:** Bruce Alexander, PC&F; **bottom right:** © 94 Kunio Owaki, The Stock Market; **bottom left:** Bruce Alexander, PC&F; **page 527:** Chemical Manufacturers Association; **page 530:** Bruce Alexander, PC&F; **page 533, sodium:** © Andrew McClenagham, Science Source/Photo Researchers; **lithium:** © Rich Treptow, Science Source/ Photo Researchers; **strontium:** © Andrew McClenagham, Science Source/Photo Researchers; **calcium:** © Jerry Mason, Science Source/Photo Researchers; **page 535:** National Severe Storms Laboratory; **page 536:** ConAgra Fertilizer Co.; **page 539 top:** © 1996 Jeffrey O'Connor; **bottom:** © Kendall/Hunt Publishing Company, Greg Nauman Photographer; **page 541 top:** UPI/Corbis-Bettmann; **bottom:** UPI/Corbis-Bettmann; **page 542:** © Kendall/Hunt Publishing Company, Greg Nauman Photographer; **page 545:** © Randall Hyman; **page 547:** © Donald Johnston, Tony Stone Images; **page 548:** Ira Wyman/Sygma; **page 549:** © The Nobel Foundation; **page 556, fig. 6a and fig. 6b:** © 1996, Jeffrey O'Connor; **page 558:** © Kendall/Hunt Publishing Co., Greg Nauman Photographer; **page 560 bottom:** Bruce Alexander, PC&F; **middle left:** Courtesy of General Motors Advanced Technology Vehicles; **page 563, fig. 10 and fig. 11:** © 1996, Jeffrey O'Connor; **page 566:** AP Photo/Shayna Brennan; **page 567, top:** © Palmer/ Kane, Inc. 1986, The Stock Market; **bottom:** © Bruce Foster, Tony Stone Images; **page 568, middle left:** © Jack Fields/Photo Researchers, Inc.; **bottom:** © Ron Sherman/ Reynolds Metals; **page 570:** Courtesy of Stauffer Chemical Company

LINE ART

AIR

Page 420, fig. 16: Used with permission of Dr. James G. Anderson, Harvard University

Concept	ChemCom Unit							
	Wat.	Res.	Petro.	Food	Nuc.	Air	Choices	Ind.
Metric (SI) measurement	I	U	E	U	U	U	U	U
Scale and order of magnitude	I	U	U	U	U	U	U	U
Physical and chemical properties	I	E	E	U	E	E	E	E
Solids, liquids, and gases	I	U	E		U	E	U	U
Solutions and solubility	I	E	U	U	U	U	E	U
Elements and compounds	I	E	E	E	E	U	U	U
Nomenclature	I	E	E	E	E	U	U	U
Formula and equation writing	I	E	E	E	U	U	U	U
Atomic structure	I	E	E		E			
Chemical bonding	I	U	E	E		U	E	U
Shape of molecules	I		E	U			E	U
Ionization	I	U	E		E	E	E	E
Periodicity		I/E/U						
Mole concept		I	E	U	E	E	U	U
Stoichiometry		I	E	E		U	U	U
Energy relationships		I	E	E	E	E	E	E
Acids, bases, and pH	I			E		E	E	U
Oxidation–reduction		I		U		U	U	E
Reaction rate/kinetics				I	E	U	E	U
Gas laws						I/E/U		
Equilibrium								I/U
Chemical analysis	I	E	E	E		U	U	U
Synthesis			I			U		E
Biochemistry				I	U		E	
Industrial chemistry	I	E	E	E	E	E	E	E
Organic chemistry			I	E			E	
Nuclear chemistry					I/E/U			

CODE: I = Introduced   E = Elaborated   U = Used